PROCEEDINGS

OF THE FIFTH INTERNATIONAL CONGRESS

OF

MATHEMATICIANS

CAMBRIDGE UNIVERSITY PRESS
London: FETTER LANE, E.C.
C. F. CLAY, Manager

Edinburgh: 100, PRINCES STREET
Berlin: A. ASHER AND CO.
Leipzig: F. A. BROCKHAUS
New York: G. P. PUTNAM'S SONS
Bombay and Calcutta: MACMILLAN AND CO., Ltd.

PROCEEDINGS

OF THE FIFTH INTERNATIONAL CONGRESS

OF

MATHEMATICIANS

(Cambridge, 22—28 August 1912)

EDITED BY THE GENERAL SECRETARIES OF THE CONGRESS

E. W. HOBSON

SADLEIRIAN PROFESSOR OF PURE MATHEMATICS IN THE UNIVERSITY OF CAMBRIDGE

AND

A. E. H. LOVE

SEDLEIAN PROFESSOR OF NATURAL PHILOSOPHY IN THE UNIVERSITY OF OXFORD

VOL. I.

PART I REPORT OF THE CONGRESS

PART II LECTURES

COMMUNICATIONS (SECTION I)

Cambridge:

at the University Press

1913

Cambridge:
PRINTED BY JOHN CLAY, M.A.
AT THE UNIVERSITY PRESS

PART I

REPORT OF THE CONGRESS

THE ORGANIZING COMMITTEE

Sir G. H. Darwin, *President*
Sir J. Larmor, *Treasurer*
Prof. E. W. Hobson } *General Secretaries*
Prof. A. E. H. Love }
Dr H. F. Baker
Mr A. Berry

INTERNATIONAL COMMITTEE

Appell, P.
Baker, H. F.
Bendixson, I. O.
Bjerknes, V.
Blaserna, P.
Bôcher, M.
Burkhardt, H.
Castelnuovo, G.
Darboux, G.
Darwin, G. H.
Dini, U.
Enriques, F.
Fejér, L.
Fields, J. C.
Forsyth, A. R.
Galdeano, Z. G.
Galitzin, Prince B.
Geiser, C. F.
Hensel, K.
Hilbert, D.
Hill, M. J. M.
Hobson, E. W.
Kikuchi (Baron)
Klein, F.
König, J.
Lamb, H.
Larmor, J.
Levi-Civita, T.
Liapounov, A. M.
Lindelöf, E.
Lindemann, F.
Lorentz, H. A.
Love, A. E. H.
Macdonald, H. M.
Mittag-Leffler, G.
Moore, E. H.
Moulton, F. R.
Painlevé, P.
Peirce, B. Osgood
Picard, E.
Pierpont, J.
Poussin, C. de la Vallée
Rayleigh (Lord)
Schmidt, E.
Segre, C.
Stéphanos, C.
Teixeira, F. G.
Volterra, V.
Weber, H.
Webster, A. G.
Whittaker, E. T.
Wirtinger, W.
Zeuthen, H. G.

INTRODUCERS FOR THE VARIOUS SECTIONS

Sections I and II.

Baker, Dr H. F., St John's College, Cambridge.

Bennett, G. T., Emmanuel College, Cambridge.

Berry, A., King's College, Cambridge.

Bromwich, Dr T. J. I'A., St John's College, Cambridge.

Elliott, Prof. E. B., 4, Bardwell Road, Oxford.

Grace, J. H., Peterhouse, Cambridge.

Hardy, G. H., Trinity College, Cambridge.

Richmond, H. W., King's College, Cambridge.

Section III.

Jeans, J. H., Trinity College, Cambridge.

Lamb, Prof. H., 6, Wilbraham Road, Fallowfield, Manchester.

Love, Prof. A. E. H., 34, St Margaret's Road, Oxford.

Hopkinson, Prof. B., Adams Road, Cambridge. (*For Engineering Mathematics.*)

Bowley, Prof. A. L., Northcourt Avenue, Reading, and Edgeworth, Prof. F. Y., All Souls College, Oxford. (*For Statistical, Economic, and Actuarial Mathematics.*)

Stratton, F. J. M., Gonville and Caius College, Cambridge. (*For Mathematical Astronomy.*)

Section IV.

Russell, Hon. B. A. W., Trinity College, Cambridge. (*For philosophical questions.*)

Rouse Ball, W. W., Trinity College, Cambridge. (*For historical questions.*)

Godfrey, C., Royal Naval College, Osborne, Isle of Wight. (*For didactical questions.*)

OFFICERS OF THE CONGRESS

Honorary President:

Lord Rayleigh.

President:

Sir G. H. Darwin.

Vice-Presidents:

W. von Dyck.
L. Fejér.
R. Fujisawa.
J. Hadamard.
J. L. W. V. Jensen.
P. A. MacMahon.
G. Mittag-Leffler.
E. H. Moore.
F. Rudio.
P. H. Schoute.
M. S. Smoluchowski.
V. A. Steklov.
V. Volterra.

Secretaries:

E. W. Hobson.
A. E. H. Love.

DELEGATES

Austria. *Government.* Prof. E. Czuber.

Jednota česleých mathematiku v Prace. Prof. B. Bydžovský.

Brazil. *Government of the United States of Brazil.* Prof. E. de B. R. Gabaglia.

Bulgaria. *University of Sofia.* Prof. A. V. Sŏurek.

France. *L'école polytechnique.* Prof. J. Hadamard.

L'institut des actuaires français. M. A. Quiquet.

German Empire. *Deutsche Mathematiker Vereinigung.* Prof. W. v. Dyck and Prof. A. Gutzmer.

Kgl. Bayer. Akademie der Wissenschaften. Prof. W. v. Dyck, Prof. A. Pringsheim, and Prof. H. Burkhardt.

Great Britain. *Board of Education.* Mr W. C. Fletcher.

Greece. *L'Université nationale et l'Université Kapodistrienne.* Prof. N. Hatzidakis, Prof. G. Rémoundos, and Prof. C. Stéphanos.

Hungary. *University of Kolozsvár (Hungary).* Prof. L. Klug.

Italy. *Circolo matematico di Palermo.* Members of the Consiglio Direttivo attending the Congress.

Japan. Prof. R. Fujisawa.

Servia. Prof. M. Petrovitch.

LIST OF MEMBERS

The Cambridge address of visiting Members is in italics.

Abbott, P., 5, West View, Highgate Hill, London, N. 19[a], *Market Hill.*
Abbott, Mrs.
Abraham, Professor Max., R. Ist. tecnico sup., Milan. *St John's College.*
Ackermann-Teubner, Dr B. G. A., Poststrasse 3, Leipzig. c/o *Professor Sir G. H. Darwin, Newnham Grange.*
Adams, C. W., Haileybury College, Hertford. *Trinity College.*
Airey, J. R., Claremont Road, Forest Gate, E. *St John's College.*
Albeggiani, Prof. M. L., Salita Banditore 4, Palermo.
Alfsen, Professor M., Lycée Municipal, Christiania. *Trinity College.*
Allardice, Professor R. E., Stanford University, California, U.S.A. *King's College.*
Allcock, W. B., Emmanuel College, Cambridge.
Allen, E. S., Harvard University, Cambridge, U.S.A. *Gonville and Caius College.*
Alvarez y Ude, Professor J. G., Cerdán, 2, 3°, Zaragoza.
Amaldi, Professor Ugo, Via Caselle 6, Modena.
Amieux, Professor Mm[lle] Anna, 4, rue de la Sorbonne, Paris. *Blue Boar Hotel.*
Amieux, Mme V. M.
Anderegg, Prof. F., Oberlin, Ohio, U.S.A. *Christ's College.*
Andreoli, Professor G., Via dei Mille 50, Naples. *Emmanuel College.*

Arany, Professor Daniel, VII. Kerepesi-út 38 II. 25, Budapest. 8, *Benet Street.*
Arany, Mme.
Arber, Dr E. A. N., 52, Huntingdon Road, Cambridge.
Arber, Mrs A., D.Sc.
Archibald, Professor R. C., 9, Charles Field St, Providence (R.I.), U.S.A. *King's College.*
Archilla, Professor Pedro, Madrid.
Archilla y Salido, Professor F., Universidad, Madrid.
Armstrong, Prof. G. N., Ohio Wesleyan University, Delaware, U.S.A. *Emmanuel College.*
Ashford, C. E., M.V.O., Royal Naval College, Dartmouth.
Atwater, Miss S. M., c/o Dr A. B. Chace, Brown University, Providence, U.S.A. *Newnham College.*

Baker, Dr H. F., F.R.S., St John's College, Cambridge.
Balak Ram, (I.C.S.) Thána, India. *St John's College.*
Ball, W. W. Rouse, Elmside, Grange Road, Cambridge.
Barnes, The Rev. Dr E. W., F.R.S., Trinity College, Cambridge.
Barrell, Professor F. R., 1, The Paragon, Clifton, Bristol. *Pembroke College.*
Bateman, Professor H., Bryn Mawr College (Penn.), U.S.A. 23, *Park Parade.*
Bateman, Mrs.
Beckh-Widmanstetter, Dr H. A. von, Schulgasse 1, III Stock, Vienna. *Gonville and Caius College.*
Beke, Professor Dr E., Universität, Budapest. *Christ's College.*
Beliankin, Professor J., Technical Institüt, Charkow. 18, *Sidney Street.*
Bell, W. G., Huntly, Herschel Road, Cambridge.
Bellom, Professor M., 6 rue Daubigny, Paris.
Bendixson, Professor I. O., Drottninggatan 87, Stockholm. *Bull Hotel.*
Bennett, G. T., Emmanuel College, Cambridge.
Bernstein, M. Serge, Aptekarsky 22, Charkow. *St John's College.*
Berry, A., Meadowside, Grantchester Meadows, Cambridge.
Berry, Mrs.
Bertram, F. G. L., Board of Education, Whitehall. c/o *J. M. Edmonds*, 24, *Halifax Road.*
Bianchedi, Romulo, Calle Arenales 1980, Buenos Aires.
Bioche, Professor C., Rue notre-dame-des-champs 56, Paris. *Emmanuel College.*
Birtwistle, G., 10, Newnham Terrace, Cambridge.
Birtwistle, Mrs.
Blacker, Miss W. E., 20, Victoria Square, Clifton, Bristol. *Newnham College.*
Blaess, Professor Dr A. V., Heinrichsstrasse 140, Darmstadt. *Gonville and Caius College.*
Blaschke, Dr W., Karlsplatz 19, Greifswald. *King's College.*
Bliss, Professor G. A., University, Chicago. *Bull Hotel.*
Bliss, Mrs.
Block, Professor Dr H. G., Observatorium, Lund. *St John's College.*
Blumberg, Dr H., 268, Stockton Street, Brooklyn, New York, U.S.A. *Trinity College.*
Blumenthal, Professor O., Rutscherstrasse 48, Aachen. *Trinity College.*
Boberil, Vicomte R. du, rue d'Antibes 114, Cannes.
Bobylev, Professor D. K., Wassiliewsky Ostrov, Line 15 Haus 44, St Petersburg.

Bôcher, Professor Maxime, Harvard University, Cambridge (Mass.). *Emmanuel College.*
Bogyó, Professor S., Munhacsy 22, Budapest. 8, *Benet Street.*
Bogyó, Mme.
Bolza, Hans, Würtzburg. *Gonville and Caius College.*
Bolza, Professor O., Reichsgrafenstrasse 10, Freiburg i. B. *University Arms Hotel.*
Bolza, Frau.
Bompiani, Dr E., Piazza S. Bernardo 106, Rome. *Emmanuel College.*
Bonifacio, Professor J. A., Academia Polytech., Porto. *Castel Hotel.*
Bonney, Rev. Professor T. G., Sc.D., F.R.S., 9, Scroope Terrace, Cambridge.
Bonolis, Professor A., Corso Amedeo 121, Livourne. *Blue Boar Hotel.*
Bonolis, Signora.
Boon, F. C., The College, Dulwich, S.E. 6, *Benet Street.*
Boon, Mrs.
Bordiga, Professor G., R. Università, Padova.
Bordmane, Professor E. P., Cevallos 269, Buenos Aires.
Borel, Professor É., 45, rue d'Ulm, Paris. 24, *S. Andrew's Street.*
Borel, Mme.
Borger, R. L., 1105, W. Oregon Street, Urbana, Ill., U.S.A. *Trinity College.*
Boulad, Farid, Egyptian State Railways, Cairo. *St John's College.*
Bourlet, Professor C., 56, rue Raynouard, Paris. *Emmanuel College.*
Boutroux, Professor P., The University, Poitiers. *Kenmare House*, 74, *Trumpington Stree*
Bowley, Professor A. L., Northcourt Avenue, Reading. *Christ's College.*
Brauch, Herr K., Lindenstrasse 13, Ludwigsburg. 22, *Maid's Causeway.*
Brewster, G. W., Oundle School, Northants. *King's College.*
Brocard, Professor H., 75, rue des Ducs-de-Bar, Bar-le-Duc.
Brodén, Professor T., Universitet, Lund. *Castel Hotel.*
Brodetsky, S., Trinity College, Cambridge.
Brodie, R. R., 6, St Andrew Square, Edinburgh.
Broggi, Professor Ugo, 53, Maipu, Buenos Aires.
Bromwich, Dr T. J. I'A., F.R.S., 1, Selwyn Gardens, Cambridge.
Bromwich, Mrs.
Brouwer, Professor L. E. J., University, Amsterdam. *Christ's College.*
Brown, Professor E. W., F.R.S., Yale University, New Haven, Conn. *Christ's Colleg*
Brückner, Professor Dr Max, Georgstrasse 30, Bautzen, Sachsen. *King's College.*
Brunt, D., Trinity College, Cambridge.
Bryan, Professor G. H., F.R.S., Plas Gwyn, Bangor, N. Wales. *Peterhouse.*
Budmani, Professor P., Castelferretti, Ancona. *Blue Boar Hotel.*
Budmani, Mm^lle^.
Burkhardt, Professor Dr H., Prinzenstrasse 13, München. *Christ's College.*
Burnside, Professor W., F.R.S., The Croft, Bromley Road, Catford, S.E.
Burnside, Professor W. S., Trinity College, Dublin.
Butt, E. H., Lower School of John Lyon, Harrow-on-the-Hill. *Trinity College.*
Byck, Mm^lle^ Marie, Rue Bolchaya Wasilkowskaya, N. 28, logm. 8, Kieff. *Bull Hote*
Byck, M. A.
Byck, Mm^lle^ Lydie.
Bydžovský, Professor Dr B., Česká Universita, Prag. *St John's College.*

Cahen, Professor E., Rue Cortambert 46, Paris.
Cameron, J. F., End House, Lady Margaret Road, Cambridge.
Cameron, Mrs.
Cameron, Miss J. F., Newnham College, Cambridge.
Campo d'Isola, Marquis de, c/o Prof. G. Mittag-Leffler, Högskola, Stockholm. *University Arms Hotel.*
Cardinaal, Professor Dr J., Vude Delft 47, Delft. *Trinity College.*
Cardinaal Ledeboer, Mme. *Newnham College.*
Carey, Professor F. S., University, Liverpool. c/o *W. W. Rouse Ball, Elmside, Grange Road.*
Carson, G. E. St L., Knox House, Tonbridge. 18, *Portugal Street.*
Carson, Mrs.
Carus, Dr Paul, La Salle, Illinois, U.S.A. *Trinity College.*
Carus, Edward H. *Trinity College.*
Carvallo, M., L'école polytechnique, Rue Descartes, Paris.
Caspar, Dr M., Gartenstr. 37, Ravensburg, Germany. 4, *Clarendon Street.*
Castellarnau, Don J. M., Madrid.
Castelnuovo, Professor G., Via Boncompagni 16, Rome. *Peterhouse.*
Cavallini, Sig. B., Via Porta Po. 36, Ferrara. *St John's College.*
Cave-Browne-Cave, Miss F. E., Girton College, Cambridge.
Chace, Dr A. B. (Chancellor), Brown University, Providence, U.S.A. *Christ's College.*
Chace, Mrs. *Newnham College.*
Chadwick, P. M., West Brae, Enfield, Middlesex. c/o *Rev. Prof. C. Anderson Scott, The Knott, S. Margaret's Road.*
Chapman, H. W., 24, Pollard Road, Whetstone, N. *Gonville and Caius College.*
Chappell, E. M., St Leonard's School, St Andrews, Fife. 9, *Fitzwilliam Street.*
Cher, Mme Marie, Pension Mme Jeanne, 123, Bd. Montparnasse, Paris. *Newnham College.*
Clariana y Ricart, Professor, Calle Ariban 133, piso 2, 1[a], Barcelona.
Clevers, Principal E., Athénée royal, Ghent. *Peterhouse.*
Coddington, Dr Emily, 24 West 58th Street, New York.
Coddington, Mrs Jefferson.
Collier, Miss A. B., Newnham College, Cambridge.
Comrie, P., F.R.S.E., 19, Craighouse Terrace, Edinburgh. *King's College.*
Comstock, Prof. D. F., Institute of Technology, Boston (Mass.), U.S.A. 4, *Bateman Street.*
Conant, Professor L. L., Worcester Polytechnic Institute, Worcester, U.S.A. *University Arms Hotel.*
Conant, Mrs.
Coolidge, Professor J. L., Harvard University, Cambridge (Mass.). *Emmanuel College.*
Cortés, Professor M. F., Madrid. *King's College.*
Cosserat, M. F., 36, Avenue de l'observatoire, Paris.
Cowley, Dr Elizabeth B., Vassar College, Poughkeepsie, New York, U.S.A. *Newnham College.*
Cramer, Direktor H., Sasbacherstrasse, 20, Achern, Baden. 15, *King's Parade.*
Cramer, Mme.
Crathorne, Dr A. R., 1113, South Fourth Street, Champaign, U.S.A. *Emmanuel College.*
Crathorne, Miss Anne. *Newnham College.*
Cunningham, Lt.-Col. A. J. C., 20, Essex Villas, Kensington, W. *King's College.*

Cunningham, E., Wayside, Huntingdon Road.
Cunningham, Mrs.
Curjel, H. W., "La Mexicana," Apartado 651, Mexico. *Trinity College.*
Czuber, Professor E., IV Karlsplatz 13, Vienna. *King's College.*

Dakin, A., 16, Florence Park, Westbury Park, Bristol. *Gonville and Caius College.*
Darboux, Professor G., La Sorbonne, Paris.
Darwin, C. G., The University, Manchester. *Trinity College.*
Darwin, Dr Francis, F.R.S., 10, Madingley Road.
Darwin, Professor Sir George H., K.C.B., F.R.S., The Grange, Newnham.
Darwin, Horace, F.R.S., J.P., The Orchard, Huntingdon Road, Cambridge.
Davis, H. J., Grammar School, Oswestry, Salop. *St John's College.*
Dean, Miss Dorothy, Girton College, Cambridge.
Degenhart, Dr H., Zieblandstrasse 39 I/1, München. *Christ's College.*
de Gramont de Lisparre, M. A., 62, rue de Ponthieu, Paris. *Trinity College.*
De la Vallée Poussin, Professor C., 149, rue de la station, Louvain.
Del Re, Professor A., R. Università, Naples.
De Lury, Professor A. T., The University, Toronto. *King's College.*
De Montessus de Ballore, Vicomte R., 15, Boulevard Bigo-Danel, Lille.
Denizot, Professor Dr A., Hochschule, Lwów (Lemberg). *King's College.*
de Zylinski, Dr E., Hainholzweg 14, Göttingen. *Trinity College.*
Dickstein, Professor S., Marszalkowskastrasse 117, Warsaw. *Bull Hotel.*
Dickstein, Mme.
Dickstein, Mmlle Julia.
Dingler, Dr H., Grünewaldstrasse 15, Aschaffenburg, Germany. *Christ's College.*
Dinkhauser, Professor Dr T., Innsbruck. 6, *Clarendon Street.*
Dintzl, Professor Dr E., Gersthoferstrasse 126, Vienna. *King's College.*
Dixon, Professor A. C., F.R.S., Hurstwood, 52, Malone Park, Belfast. c/o *Dr T. J. I'A. Bromwich,* 1, *Selwyn Gardens.*
Dixon, A. L., Merton College, Oxford. *Trinity College.*
Dmitrieff, M. Alexius, V. Gymnasium, Kieff. 62, *Jesus Lane.*
Dodds, J. M., Peterhouse, Cambridge.
Dolinski, Professor M., Felix-Mottlstrasse 19, Vienna. *Kenmare House, Trumpington Street.*
Dolinski, Frau.
Domsch, Professor Dr Paul, Reichstrasse 34, Chemnitz. *King's College.*
Dougall, Dr J., Kippen, Stirlingshire.
Dow, Miss Jeannie, 22, Belgrave Terrace, Newcastle-on-Tyne. *Newnham College.*
Drach, Professor J., Université, Toulouse. *Peterhouse.*
Duclout, Professor Jorge, Guvadvia 329, Escritorio 55, Buenos Aires.
Dufton, A., 28, Springfield Mount, Leeds. *Trinity College.*
Dyck, Professor Dr W. von, Hildegardstrasse 5, München. *Christ's College.*

Echegaray, Professor J., Universidad, Madrid.
Edgeworth, Professor F. Y., All Souls College, Oxford. *Trinity College.*
Eisenhart, Professor L. P., Princeton University, U.S.A. 72, *Jesus Lane.*
Ekman, Professor V. W., Bytaregatan, 22, Lund. *St John's College.*

Elliott, C., North Street, Oundle, Northants. 15, *Portugal Place.*
Elliott, Professor E. B., F.R.S., 4, Bardwell Road, Oxford. 20, *Trumpington Street.*
Elliott, Mrs.
Elliott, Miss D. M.
Ellis, H. D., 12, Gloucester Terrace, Hyde Park, W. 17, *Silver Street.*
Ellis, Miss F. M.
Enriques, Professor F., R. Università, Bologna. *Peterhouse.*
Esson, Professor W., F.R.S., 13, Bradmore Road, Oxford. c/o *W. W. Rouse Ball, Elmside, Grange Road.*
Evans, Professor G. C., Rice Institute, Houston, Texas. *Gonville and Caius College.*
Evans, W. D., 125, Millbrook Road, Southampton. *King's College.*
Ewald, Dr P. P., Friedländer Weg 7, Göttingen. *Gonville and Caius College.*

Fano, Professor G., Corso Vittorio Emanuele 105, Turin.
Fehr, Professor H., Rue du Florissant 110, Geneva. *Christ's College.*
Fejér, Professor Dr L., Falk Miksa-utca 15, Budapest. *Trinity College.*
Fekete, Dr M., 1, Fehérvári-út $\frac{30}{32}$ III. 3, Budapest. *King's College.*
Feÿtes, M. J. J., 64, Franklin Street, The Hague, Holland. *Christ's College.*
Fiedler, Professor Dr E., Englischviertelstrasse 57, Zürich. *Christ's College.*
Field, Miss Kathleen, Adelaide Square, Bedford. *Girton College.*
Fields, Professor J. C., University, Toronto. *Christ's College.*
Fine, Professor H. B., Princeton, New Jersey, U.S.A. *Christ's College.*
Finsterbusch, Professor J., Mühlpfortstrasse 74^{II}, Zwickau. *King's College.*
Fitting, Professor Dr F., Ringstrasse 17, München-Gladbach. *Trinity College.*
Fletcher, W. C., Board of Education, Whitehall, S.W. *St John's College.*
Florez, Director Juan, Madrid.
Föppl, Dr L., Bergstrasse 17, Göttingen. *Gonville and Caius College.*
Forsyth, Dr A. R., F.R.S., Hotel Rembrandt, London.
Fort, Dr Tomlinson, Mount Airy, Georgia, U.S.A. *Gonville and Caius College.*
Franklin, J., R.N. College, Osborne, I.W. *St John's College.*
Fréchet, Professor M., Université, Poitiers. *Gonville and Caius College.*
Freeborn, E. W., The Cottage, Harrow-on-the-Hill. *Emmanuel College.*
Frizell, Professor A. B., McPherson College, McPherson (Kans.), U.S.A. 6, *Green Street.*
Fujisawa, Professor R., University, Tokyo. *Bull Hotel.*

Gabaglia, Professor E. de B. R., L'école polytechinigin, Rio de Janeiro. *Bull Hotel.*
Galdeano y Yanguas, Professor Z. Garcia, Cervantes 5, Zaragoza. *Christ's College.*
Galitzin, Prince B., Académie Impériale des Sciences, St Petersburg. *University Arms Hotel.*
Gallop, E. G., Gonville and Caius College, Cambridge.
Gama, Professor V., Observatoire, Tacuyaba, Mexico.
Garstang, T. J., Dunhill House, Petersfield, Hants. *Trinity College.*
Gaztelu, Professor Luis (Marquis de Echandia), Madrid.
Echandia, Marquise de.
Genese, Professor R. W., Isallt, Aberystwyth. *St John's College.*
Gérardin, M. A., 32 quai Claude le Lorrain, Nancy. *Bull Hotel.*

Gerrans, H. T., Worcester College, Oxford. c/o *Professor E. W. Hobson, F.R.S., The Gables, Mount Pleasant.*
Gerrans, Mrs.
Gianfranceschi, Rev. Professor G., Panisperna 89, Rome. *Blue Boar Hotel.*
Gibb, David, 15, South Lauder Road, Edinburgh. 10, *King's Parade.*
Gibb, Mrs.
Gibbins, N. M., Victoria College, Jersey.
Gibson, Professor G. A., The University, Glasgow. *Emmanuel College.*
Giorgi, Professor G., Corso Vittorio Emanuele 39, Rome.
Godfrey, C., M.V.O., Park Gate, East Cowes, I.W. *Trinity College.*
Godfrey, Mrs. c/o *J. Eaden, Little Shelford.*
Goldziher, Professor Dr C., Holls utca 4, Budapest. c/o *Mrs Gibson and Mrs Lewis, Castle-brae, Chesterton Lane.*
Goodwill, G., The Plain, Wetherall, Carlisle. 13, *St John's Road.*
Grace, J. H., F.R.S., Peterhouse, Cambridge.
Graf, Professor Dr J. H., Wylerstrasse 10, Bern. *Christ's College.*
Grant, Miss Kathleen, Clough Hall, Newnham College, Cambridge.
Grassmann, Professor H., Frankfurterstrasse 53^{I}, Giessen. 10, *Jesus Lane.*
Grassmann, Frau.
Graustein, W. C., 19, Arlington Street, Cambridge, Mass., U.S.A. *Trinity College.*
Gray, Dr J. G., F.R.S.E., 11, University, Glasgow.
Gray, Miss.
Greaves, J., Christ's College, Cambridge.
Greene, Miss E., 7, Glenluce Road, Blackheath, S.E. *Newnham College.*
Greenhill, Sir A. G., F.R.S., 1, Staple Inn, London, W.C. *St John's College.*
Greenstreet, W. J., The Woodlands, Burghfield Common, nr Mortimer, Berks. *St John's College.*
Griffith, I. O., 49, Banbury Road, Oxford. *Lion Hotel.*
Griffith, Mrs.
Griffith, Miss M. J., 45, Hereford Road, Acton, W. *Newnham College.*
Grossmann, I., Göttingen. *St John's College.*
Grossmann, Professor Dr M., Herrenbergstrasse 1, Zürich. *Bull Hotel.*
Grossmann, Frau.
Guccia, Professor Dr G. B., 30, Via Ruggiero Settimo, Palermo.
Guichard, Professor L., Collège de Barbezieux, Charente. *Trinity College.*
Guldberg, Professor A., Sörkedalsveien, Christiania. 74, *Trumpington Street.*
Guldberg, Mme.
Günther, Professor N. M., Petersburg Side, Great Perspective 13, St Petersburg.
Günther, Mme.
Gutzmer, Professor Dr A., Wettinerstrasse 17, Halle (Saale). *University Arms Hotel.*
Gutzmer, Frau.
Gutzmer, Frl. Irene.
Gwatkin, Miss E. R., 8, Scroope Terrace.

Hadamard, Professor J., Collège de France, Paris. *King's College.*
Hagen, Rev. Father J. G., S.J., Specola Vaticana, Rome. c/o *Baron A. von Hügel, Croft Cottage, Barton Road.*

Hall, Professor E. H., Harvard University, Cambridge (Mass.). 3, *Bridge Street.*
Hall, Mrs.
Hall, Miss Constance.
Halphen, Professor C., 8 bis. Chaussée de la Muette, Paris.
Hammond, Miss Marjorie, The Firs, Sunte Avenue, Haywards Heath. *Girton College.*
Hancock, Miss E. M., 28, Bergholt Crescent, Stamford Hill, N. *Newnham College.*
Harding, P. J., 13, Old Oak Road, Acton, W. *Sidney Sussex College.*
Hardy, G. H., F.R.S., Trinity College, Cambridge.
Harris, A., Board of Education, Whitehall. *Clare College.*
Harvey, Miss Ella, Rock Bank, Milnsow, Rochdale. 12, *Guest Road.*
Harvey, Miss Ethel.
Haselfoot, C. E., 7, Northmoor Road, Oxford.
Haskell, Professor M. W., California University, Berkeley, U.S.A. *Bull Hotel.*
Haskell, Mrs.
Hassé, H. R., The University, Manchester. *St John's College.*
Hatzidakis, Professor N., Xenokratrystrasse 29^{II}, Athens. 83, *Regent Street.*
Havelock, Dr T. H., Rockliffe, Gosforth, Newcastle-on-Tyne. *St John's College.*
Hensel, Professor Kurt, Universität, Marburg. c/o *J. F. Cameron, End House, Lady Margaret Road.*
Hensel, Mme.
Hermann, A., Rue de la Sorbonne 6, Paris.
Heywood, Dr H. B., Old Summer House, Bromley, Kent. *Gonville and Caius College.*
Hicks, Professor W. M., F.R.S., Leamhurst, Ivy Park Road, Sheffield.
Hill, Professor M. J. M., Sc.D., F.R.S., 18, Ferncroft Avenue, Hampstead, N.W. *Peterhouse.*
Hilton, H., 73, Platt's Lane, Hampstead, N.W. *Peterhouse.*
Hinks, A. R., The Observatory, Cambridge.
Hobson, Professor E. W., Sc.D., F.R.S., The Gables, Mount Pleasant, Cambridge.
Hobson, Mrs.
Hodson, Dr F., Bablake School, Coventry. *Trinity College.*
Holden, A., Sunnyhill, Fulwood, Sheffield. 6, *Benet Street.*
Holden, Mrs.
Holden, Miss Annie.
Holmgren, Professor E. A., Arkosund, Norrköping, Sweden. *Trinity College.*
Hopkinson, Professor B., F.R.S., Adams Road, Cambridge.
Horsburgh, E. M., F.R.S.E., A.M.I.C.E., The University, Edinburgh. *Peterhouse.*
Hostinský, Dr B., Kr. Vinohrady, Moravska 40, Prague. *St John's College.*
Huard, Professor A., Lycée Henri IV, Paris. *King's College.*
Hubrecht, J. B., The Hill House, Coton.
Hubrecht, Mme M.
Hudson, Professor W. H. H., 34, Birdhurst Road, Croydon. *St John's College.*
Hudson, Miss. *Newnham College.*
Huntington, Professor E. V., Harvard University, Cambridge (Mass.). 15, *Holland Street.*
Huntington, Mrs.
Hunton, Professor S. W., Mount Allison University, New Brunswick, Canada. c/o *A. Berry, Meadowside, Grantchester Meadows, Cambridge.*
Hutchinson, A., Aysthorpe, Newton Road, Cambridge.

Inglis, C. E., King's College, Cambridge.
Ionescu, Professor I., Str. Călusei, 23, Bucharest.
Itelson, Herr G., Kurfürstenstr. 40, Berlin. *Trinity College.*

Jack, J., Angel Court, Fortrose.
Jackson, C. S., 25, Nightingale Place, Woolwich. *Trinity College.*
Jackson, Mrs. c/o *Rev. Professor H. M. Gwatkin*, 8, *Scroope Terrace.*
Jackson, Miss E. M. H.
Jackson, Miss I. V. H.
Jackson, Miss M. M., 34, Park Road, Sittingbourne. *Newnham College.*
Janisch, Professor E., Deutsche Technische Hochschule, Prag. *King's College.*
Janiszewski, Dr Z., Marjensstad 6, Warsaw. 14, *Earl Street.*
Jensen, J. L. W. V., Frederiksberg Allé 68, Copenhagen. *Trinity College.*
Joffe, S. A., 55, Cedar Street, New York, U.S.A.
Johnston, W. J., University College, Aberystwyth. *Trinity College.*
Jolliffe, A. E., Corpus Christi College, Oxford. *Peterhouse.*
Jones, Miss C. M. J., 15, The Mount, York. *Newnham College.*
Jones, Miss E. E. Constance, Mistress of Girton College, Cambridge.
Jourdain, P. E. B., The Lodge, Girton, Cambridge.
Juel, Professor Dr C., Østerbrogade 74[4], Copenhagen. 19 A, *Market Hill.*
Juel, Mme.
Junius, P., Bochum, Westfalen. *Trinity College.*
Justschenko, M. Paul Tchità, Siberia, Russia. *Bull Hotel.*
Justschenko, M. A.

Kapteyn, Professor Dr W., R. Universiteit, Utrecht. *Christ's College.*
Kármán, Professor F. von, Bimbó-út. 10, Budapest. *Gonville and Caius College.*
Kármán, Dr T. von, Göttingen. *Gonville and Caius College.*
Kasner, Professor E., Columbia University, New York, U.S.A. *Gonville and Caius College.*
Kellogg, Professor O. D., University of Missouri, Columbia and Rolla, U.S.A. *Trinity College.*
Kellogg, Mrs. *Newnham College.*
Keyser, Professor C. J., Columbia University, New York, U.S.A. 3, *Bridge Street.*
Keyser, Mrs.
Kirkman, J. P., Clanyrafon, Bedford. *Emmanuel College.*
Klug, Professor Dr Lipót, Universität, Kolozsvár, Hungary. *St John's College.*
Knopp, Dr Konrad, Richardstrasse 1, Berlin-Lichterfelde. 37, *Market Hill.*
Knott, Professor C. G., 42, Upper Gray Street, Edinburgh. *Emmanuel College.*
Koch, Professor H. von, Djursholm, Stockholm. *Christ's College.*
Koch, Mme von. *Newnham College.*
König, Dr Dénes, Vámházkörút. 5, Budapest. *Trinity College.*
Kousnetzoff, U., Sucharevskaia-Sadovaie 4, 5, Moscow. *Trinity College.*
Krazer, Professor Dr A., Westendstrasse 57, Karlsruhe.
Krazer, Frl. Marianne.
Krygowski, Professor Dr Z., Tech. Hochschule, Lemberg. *King's College.*
Kubota, Professor T., Imperial University, Tohoku, Japan. *St John's College.*

Kuhn, Professor Dr H. W., Ohio State University, Columbus, U.S.A. *Trinity College.*
Kundt, Frl. Frieda, 50, Ansbacherstr., Berlin, W. *Old Hall, Newnham College, Cambridge.*
Kürschák, Professor J., Föh. Albrecht-út. 17, Budapest. 36, *Trinity Street.*
Kürschák, Mme.

Lágos Campos, Professor B. Casilla 173, Talca, Chili.
Laisant, Mons C.-A., Rue du Conseil 5, Asnières (Seine). *Christ's College.*
Lamb, Professor Horace, F.R.S., 6, Wilbraham Road, Fallowfield, Manchester. *Trinity College.*
Lampe, Professor Dr O. E., Fasanenstrasse 67, Berlin, W.
Lancelin, M., Rue Boissonnade 3, Paris.
Lanchester, F. W., 53, Hagley Road, Edgbaston, Birmingham. *University Arms Hotel.*
Landau, Professor E., Herzbergerschaussee 48, Göttingen. *University Arms Hotel.*
Landau, Frau.
Landis, Professor W. W., Dickinson College, Carlisle, Pa., U.S.A. *Bull Hotel.*
Langevin, Professor Paul, 11 rue de la Pitié, Paris. 4, *Park Parade.*
Larmor, Professor Sir Joseph, M.P., Sec. R.S., St John's College, Cambridge.
Law, Miss R. M., The Grange, Irthington, Carlisle. 9, *Fitzwilliam Street.*
Lawson, G., Waid House, Anstruther, N.B. *King's College.*
Leahy, Professor A. H., 92, Ashdell Road, Sheffield. *Pembroke College.*
Leathem, J. G., St John's College, Cambridge.
Lebeuf, Professor A., Observatoire, Besançon. 7, *Downing Place.*
Lebeuf, Mme.
Leonard, Miss Agatha, 2, Cloverdale Lawn, Cheltenham. *Newnham College.*
Lepeschkin, W., Bern. *Christ's College.*
Leuschner, Professor Dr A. O., 1816 Scenic Avenue, Berkeley, U.S.A.
Levi-Civita, Professor T., R. Università, Padova. *St John's College.*
Levy, Professor Albert, 86 rue de Rennes, Paris.
Lewis, T. C., West Home, West Road, Cambridge.
Lewis, Mrs.
Lhermitte, Professor J. L., 39 rue F. Bourquelot, Provins, France. 44, *Victoria Park.*
Liapounov, Professor Alexandre M., K. Akademie der Wissenschaften, St Petersburg.
Liapounov, Mme.
Lietzmann, Dr W., Obere Sehlhofstrasse 39, Barmen. *Gonville and Caius College.*
Lindelöf, Professor E. Sandvikskajen 15, Helsingfors. *Trinity College.*
Lischka, Frl. M., c/o Prof. M. Dolinski. *Newnham College.*
Littlewood, J. E., Trinity College, Cambridge.
Liveing, Dr G. D., F.R.S., The Pightle, Newnham, Cambridge.
Lorente, Don J. M., Madrid.
Love, Professor A. E. H., F.R.S., 34, St Margaret's Road, Oxford. *St John's College.*
Love, Miss Blanche E. 20, *Trumpington Street.*
Luiggi, Professor C. L., 81, Via Sardegna, Rome. *Peterhouse.*
Lusin, Professor N. N., Universitet, Moscow.
Lusin, Mme.

Macaulay, Dr F. S., 19, Dewhurst Road, Brook Green, London, W. *University Arms Hotel.*

Macaulay, W. H., King's College, Cambridge.
M'Bride, J. A., Allan Glen's School, Glasgow. *Peterhouse.*
Macdonald, Professor H. M., F.R.S., The University, Aberdeen. c/o *G. Birtwistle, 10, Newnham Terrace.*
Macfarlane, Professor Dr A., 317, Victoria Avenue, Chatham (Ontario), Canada. *Christ's College.*
MacGregor, Professor J. G., F.R.S., The University, Edinburgh.
McLaren, S. B., The University, Birmingham.
McLaren, Miss.
McLaren, Miss.
McLennan, Professor J. C., The University, Toronto. *Christ's College.*
MacMahon, Major P. A., F.R.S., 27, Evelyn Mansions, Carlisle Place, London. c/o *Professor H. F. Newall, F.R.S., Madingley Rise, Cambridge.*
MacRobert, T. M., Cong. Manse, Dreghorn, N.B. *Trinity College.*
Maggi, Professor G. A., R. Università, Pisa.
Mangoldt, Professor Dr H. von, Hermannshöfer Weg 8, Danzig-Langführ. *Christ's College.*
Mansion, Professor P., Quai de Dominicains, Ghent.
Marsden, H. K., Eton College, Windsor. *Trinity College.*
Martin, Dr A., 918, N. Street, N.W., Washington, U.S.A.
Masson, Mm^{lle} Dr Renée, 23, Place du Marché, Carouge, Geneva. *Newnham College.*
Mataix, Professor Carlos, Madrid.
Mayall, R. H. D., The Birches, Cavendish Avenue, Cambridge.
Mayall, Mrs.
Meidell, M. Birger, Norske Liv, Drammensveien 21, Christiania. *Trinity College.*
Meiklejohn, J. B., 10, Rosewood Terrace, Dundee. 8, *Little St Mary's Lane.*
Meiklejohn, Mrs.
Meissner, Professor Dr E., Freudenbergstrasse 69, Zürich. *Trinity College.*
Mercer, J., Clewer, Gt Shelford, Cambridge.
Mercer, Mrs.
Mercer, J. W., R.N. College, Dartmouth. 5, *St Edward's Passage.*
Meyer, Mm^{lle} Helène, c/o M. H. Tripier, Rue Alphonse-de-Neuville 17, Paris.
Meyer, Miss M. T., Girton College, Cambridge.
Michalski, M. S., 73, Wspotna, Warsaw.
Michel, Professor Dr P., Borgo degli Albizi 16, Firenze.
Miller, Professor Dayton C., Case School of Applied Science, Cleveland, U.S.A. *St John's College.*
Mingot Shelly, Don J., Madrid.
Mittag-Leffler, Professor G., Högskola, Stockholm. *University Arms Hotel.*
Mittag-Leffler, Mme.
Mollison, W. L., 5, Cranmer Road, Cambridge.
Montel, M. Paul, Boulevard de Vaugirard 57, Paris. 4, *Park Parade.*
Moore, Professor C. L. E., Institute of Technology, Boston (Mass.). *Christ's College.*
Moore, C. N., 1123, East 3rd Street, Cincinnati, U.S.A. *Trinity College.*
Moore, Professor E. H., The University, Chicago, U.S.A. *Christ's College.*
Mordell, L. J., St John's College. 4, *Clarendon Street.*

Moreno, Don A., Sociedad matemática Española, San Bernardo 51, Madrid. *Kenmare House, Trumpington Street.*
Morley, Professor Frank, Johns Hopkins University, Baltimore, U.S.A. *King's College.*
Moulton, Professor F. R., The University, Chicago, U.S.A. 14, *New Square.*
Moulton, Mrs.
Muirhead, Dr R. F., 64, Great George Street, Hillhead, Glasgow. *Emmanuel College.*
Mulder, Herr P., Slotlaan 62, Zeist. *Trinity College.*
Müller, Professor Dr R., Wittmannstrasse 38, Darmstadt. *Gonville and Caius College.*
Mumelter, Professor Dr M., Bozen Sparkasse, Austria.
Munro, A., Queens' College, Cambridge.
Myller, Professor A., Jasi. 5, *Pembroke Street.*
Myller, Madame.

Naïto, Professor J., Lycée supérieur, Tokyo. *Trinity College.*
Naraniengar, Professor M. T., Mallesvaram, Bangalore, S. India.
Narayan, Professor L., Benares, India.
Neville, E. H., Trinity College, Cambridge.
Newall, Professor H. F., F.R.S., Madingley Rise, Cambridge.
Newbold, W., F.R.A.S., 7, Broadwater Down, Tunbridge Wells. 17, *Market Hill.*
Nicholson, Dr J. W., Trinity College, Cambridge.
Niven, Sir W. D., K.C.B., F.R.S., Eastburn, St John's Road, Sidcup.
Nunn, Dr T. P., Training College, Southampton Row, London.

Obeso, Rev. Professor Dr T. M., S.J., Apartado, No. 1, Bilbao, Spain. *Leahoe, Hertford.*
Obraszoff, M. Z., 41, Wass Ostroff, 10th line, St Petersburg.
Octavio de Toledo, Professor Luis, Universidad, Madrid. *King's College.*
Ohm, D. McKay, 54, Hawkshead Street, Southport. *St John's College.*
Orr, Professor W. McF., F.R.S., Innisfail, Bushy Park Road, Terenure, Dublin.
Otamendi, Professor A. D., Moreno 1230, Buenos Aires.

Padoa, Professor A., R. Instituto, Genova. 28, *Richmond Road.*
Padoa, Signora Elisabetta.
Padoa, Sig. Baldo.
Palmer, G. W., Christ's Hospital, West Horsham.
Parsons, Miss Rachel. c/o *Sir G. H. Darwin, F.R.S., The Grange, Newnham.*
Peace, J. B., Emmanuel College, Cambridge.
Peano, Professor G., Via Barbaroux 4, Torino. *Trinity College.*
Peddie, Professor W., Rosemount, Broughty Ferry, N.B. *St John's College.*
Peel, T., Magdalene College, Cambridge.
Peñalver, Professor P., Universidad, Seville. *King's College.*
Peirce, Professor B. Osgood, Harvard University, Cambridge (Mass.). *Emmanuel College.*
Peirce, Mrs Osgood. *Newnham College.*
Peirce, Miss Jessie L. Osgood. *Newnham College.*
Peirce, Miss Emily Osgood. *Newnham College.*
Pérès, M. J., 34, rue de Bretagne, Caen (Calvados). *Emmanuel College.*
Pérez, Professor J. S., Madrid.

Pérez del Pulgar, Rev. Professor J. A., S.J., Madrid.
Perrin, Professor Jean, 106, boulevard Kellermann, Paris. 18, *Hertford Street.*
Perrin, Madame.
Perron, Professor O., Steinlachstrasse 5, Tübingen. *Bull Hotel.*
Perron, Frau.
Petrovitch, Professor M., 26, Kossantch-Venac, Belgrade. *Blue Boar Hotel.*
Phragmén, Professor E., The University, Stockholm. *Bull Hotel.*
Piaggio, H. T. H., University College, Nottingham. *St John's College.*
Pick, Miss M., Royal Holloway College, Englefield Green, Surrey. *Newnham College.*
Pickford, A. G., Hulme Grammar School, Oldham. *St John's College.*
Pidduck, F. B., Queen's College, Oxford. *Emmanuel College.*
Pidoll, Herr Max F. von, Reitmorstrasse 12, München. *St John's College.*
Pincherle, Professor S., Via del Mille 9, Bologna.
Polakov, Professor Alexis, 1, Uschakowsky pereulok 22, Ostogenka, Moscow. *Gonville and Caius College.*
Prebble, Miss Edith, 196, Adelaide Road, Brockley, S.E. *Newnham College.*
Prestwich, A. R., The Friary, Richmond, Yorks. 14, *King's Parade.*
Prestwich, Mrs.
Price, E. A., Henleaze, East Cowes, I.W. *Trinity College.*
Pringsheim, Professor A., Arcisstrasse 12, München. 17, *Market Hill.*
Pringsheim, Frau.
Pringsheim, Dr Peter. *Trinity College.*
Proudlove, R., 32, Hutt Street, Spring Bank, Hull. *St John's College.*
Przeborsky, Professor A., L'université, Kharkoff. *Peterhouse.*
Punier, Mm^{lle}. c/o *Miss C. M. Richardson,* 9, *Fitzwilliam Street.*
Punnett, Miss M., Training College, Southampton Row, London, W.C. 15, *Earl Street.*

Quiquet, M. A., Boulevard St Germain 92, Paris. *King's College.*

Rabinovich, G., Gretcheskaja 25, Odessa. *Peterhouse.*
Rados, Professor Dr G., Ferencz-Körut 38, Budapest.
Radou, Dr J., Huschkagasse 22, Vienna. *Christ's College.*
Ramsey, A. S., Howfield, Buckingham Road, Cambridge.
Ramsey, Professor Rolla R., Indiana University, Bloomington, U.S.A. 1, *Warkworth Terrace.*
Ramsey, Mrs.
Ranson, F. M., R.N. College, Osborne, I.W. *St John's College.*
Rátz, Professor L., Városligeti fasor 19–23, Budapest. *Bull Hotel.*
Rawlinson, J. F. P., K.C., M.P., 5, Crown Office Row, Temple, London, E.C. *University Arms Hotel.*
Rayleigh, Rt. Hon. Lord, O.M., F.R.S., Terling Place, Witham, Essex. c/o *The Master, St John's College.*
Rayleigh, Lady.
Reese, Professor Hubert M., University of Missouri, Columbia, U.S.A. *Trinity College.*
Reilly, Miss Marion, Bryn Mawr College (Pen.), U.S.A.
Reina, Professor V., R. Università, Rome.

Rémoundos, Professor G., 54, rue Spyridion-Tricoupis, Athens. 100, *Regent Street.*
Rémoundos, Madame.
Richard, Professor J. A., 100, rue de Fonds, Chateauroux, Indre.
Richardson, Miss C. M., Lindern House, Bexhill-on-Sea. 9, *Fitzwilliam Street.*
Richardson, Professor R. G. D., Brown University, Providence, Rhode Island, U.S.A. *Trinity College.*
Richardson, Mrs. *Newnham College.*
Richmond, H. W., F.R.S., King's College, Cambridge.
Riebesell, Dr P., Klosterallee 100[II], Hamburg. *King's College.*
Riesz, Professor F., University, Kolozsvár, Hungary. *Trinity College.*
Riesz, Professor Dr Marcel, Université, Stockholm. *Trinity College.*
Rietz, Dr H. L., 708, S. Goodwin Avenue, Urbana, Ill., U.S.A. *Trinity College.*
Robb, Dr A. A., 57, Sidney Street, Cambridge.
Robson, A., Marlborough College, Wilts. 3, *Park Street.*
Rogers, Professor L. J., 6, Hollin Lane, Leeds. *King's College.*
Rose, Madame. c/o *J. B. Hubrecht, The Hill House, Coton, Cambridge.*
Rosenblatt, Dr A., Basteistrasse 19, Krakau. *Emmanuel College.*
Rottenburg, H., 13, Madingley Road, Cambridge.
Rousseau, Professor G. T., 37, rue J. Bte. Baudin, Dijon. 44, *Victoria Park.*
Routh, G. R. R., Board of Education, Whitehall. *Newnham Cottage, Queens' Road.*
Rudio, Professor F., Dolderstrasse 111, Zürich. *Christ's College.*
Rueda, Professor C. Jiménez, Universidad, Madrid.
Runge, Professor C., Wilhelm-Weberstrasse 21, Göttingen. c/o *Prof. H. F. Newall, F.R.S., Madingley Rise, Cambridge.*
Runge, Frau.
Runge, Frl. Iris.
Russell, Hon. B. A. W., Trinity College, Cambridge.
Russyan, Professor César, L'université, Kharkoff. 1, *Silver Street.*

Sadek, M., Union Society, University College, Gower Street, W.C. *Kenmare House, 74, Trumpington Street.*
Safford, Professor Dr F. H., University of Pennsylvania, Philadelphia, U.S.A. 1, *Silver Street.*
Safford, Mrs.
Saltykow, Professor Dr N. N., L'université, Kharkoff. *University Arms Hotel.*
Saltykow, Mme.
Sampson, Professor R. A., F.R.S., Royal Observatory, Edinburgh. *St John's College.*
Savage, T., St Patrick's College, Armagh. 15, *Norwich Street.*
Savitch, Professor S. E., Nizolaewsnäia 35, St Petersburg. *Gonville and Caius College.*
Schilling, Professor Dr F., Technische Hochschule, Danzig. 43, *New Square.*
Schilling, Frau.
Schlesinger, Professor L., Bismarckstrasse 40, Giessen. *Emmanuel College.*
Schoute, Professor P. H., Universiteit, Groningen. *Trinity College.*
Schur, Professor Dr I., Prinzregentenstrasse 6, Wilmersdorf, Berlin. *The Lion Hotel.*
Schur, Frau.
Scott, R. F. (The Vice-Chancellor), The Lodge, St John's College, Cambridge.
Segre, Professor C., Corso Vittorio Emanuele 85, Turin.

Seibert, Professor H., Lehenerstrasse 1, Salzburg. *Gonville and Caius College.*
Severi, Professor F., Fuori Porta Pontecorvo, Padova. *St John's College.*
Sharpe, J. W., Woodroffe, Portarlington Road, Bournemouth.
Shaw, H. Knox, The Observatory, Helwan, Egypt. *Sidney Sussex College.*
Shaw, T. Knox, Sidney Sussex College, Cambridge.
Sheppard, Dr W. F., Braybrooke, Worcester Road, Sutton, Surrey. *Trinity College.*
Sheppard, Mrs.
Siddons, A. W., Garlands, Harrow-on-the-Hill. *Christ's College.*
Silberstein, Professor L., R. Università, Rome. *Emmanuel College.*
Sinzov, Professor D. M., L'université, Kharkoff. *Gonville and Caius College.*
Smith, C., Master of Sidney Sussex College, Cambridge.
Smith, Professor David E., Teachers' College, Columbia University, New York, U.S.A. *University Arms Hotel.*
Smith, Mrs.
Smith, Miss E. M., Ladies' College, Cheltenham. *Newnham College.*
Smoluchowski, Professor Dr Maryan S., Dlugosza 8, Lemberg. *Christ's College.*
Snell, Mrs, 29, Beech Hill Road, Sheffield. 26, *Petty Cury.*
Snell, Miss Dora.
Snyder, Professor V., 217, University Avenue, Ithaca, N.Y., U.S.A. *Trinity College.*
Snyder, Mrs. *Newnham College.*
Soler, Professor E., R. Università, Padova.
Somigliana, Professor C., R. Università, Torino. *Emmanuel College.*
Sommer, Professor Dr J., Johannistal 2, Langführ-Danzig. *Trinity College.*
Sommerville, D. M. Y., University, St Andrews, N.B. *Trinity College.*
Sòs, Dr E., Eötvös-u. 1. F., Budapest.
Sǒurek, Professor A. V., Ul. Sĭpka 20, Sofia, Bulgaria. *Trinity College.*
Speiser, Professor A., Stephansplen 7, Strassburg. *Emmanuel College.*
Spiers, A. H., Gresham's School, Holt.
Stäckel, Professor Dr P., Stefanienstrasse 7, Karlsruhe i. B. *University Arms Hotel.*
Stäckel, Frau Elisabeth.
Stäckel, Frl. Hildegard.
Stankiewitsch, Professor Jean, Universitet, Moscow. *Trinity College.*
Stecker, Professor H. F., Pennsylvania State College, U.S.A. 5, *Melbourne Place.*
Steffensen, Dr J. F., A. F. Ibsensvej 26, Hellerup, Denmark. *Gonville and Caius College.*
Steffensen, Frau. *Newnham College.*
Steggall, Professor J. E. A., Woodend, Perth Road, Dundee. c/o *Dr J. G. Frazer, St Keynes, Grange Road.*
Steinmetz, Dr C. P., Wendell Avenue, Schenectady, U.S.A.
Steklov, Professor V. A., Rue Zwerinskäia 6/8, log. 5, St Petersburg. *University Arms Hotel.*
Steklov, Mme.
Stéphanos, Professor C., 20, rue Solon, Athens. *The Lion Hotel.*
Stephen, Miss H. B., Newnham College, Cambridge.
Sterneck, Professor Dr Robert von, Merangasse 35, Graz. *St John's College.*
Stevensen, Mrs. c/o *J. M. Dodds, Peterhouse.*
Stirling, Rt. Hon. Sir J., F.R.S., Finchcocks, Goudhurst. c/o *Prof. Sir G. H. Darwin, K.C.B., F.R.S., The Grange, Newnham.*

Stoney, Miss E. A., 20, Reynolds Close, Hampstead Way, N.W. *Newnham College.*
Stott, W., 155, Withens Lane, Liscard, Cheshire. *Trinity College.*
Stradling, W., R.N. College, Osborne, I.W. *St John's College.*
Strassburg University Library.
Stratton, F. J. M., Gonville and Caius College, Cambridge.
Study, Professor E., Argelanderstrasse 126, Bonn. c/o *A. Berry, Meadowside, Grantchester Meadows.*
Stuyvaert, M., 44, rue des Chanoines, Ghent. *Peterhouse.*
Suppantschitsch, Professor R., Rosengasse 4, Vienna. 7, *Manor Street.*
Swift, Miss M. C., 62, Buston Terrace, Jesmond, Newcastle-on-Tyne.
Szász, Dr Ottó, Jeréz-körńt 30, Budapest. *King's College.*

Tarroni, Rev. Professor E. J., S.J., Madrid. *Leahoe, Hertford.*
Taylor, G. I., Trinity College, Cambridge.
Teixeira, Professor F. G., Academia Polytech., Porto. *Castel Hotel.*
 Teixeira, Mm^{lle}.
Terracini, Professor Dr A., Corso Siccardi 16, Turin. *The University Arms Hotel.*
 Terracini, Mme.
 Terracini, Dr B.
Terradas é Illa, Professor, Córcega 331, Ent., Barcelona. *King's College.*
Thaer, Professor Dr A., Hamburg 36. *Christ's College.*
Third, Dr J. A., Spier House, Beith, Ayrshire. *Blue Boar Hotel.*
 Third, Mrs.
Thompson, C. H., Queen's College, Oxford.
Thomson, Professor Sir J. J., O.M., F.R.S., Holmleigh, West Road, Cambridge.
Thurber, Dr C. H., 9, St Martin's Street, Leicester Square, London, W.C. *University Arms Hotel.*
 Thurber, Miss G. R.
Tietze, Professor Dr H., Parkstrasse 32, Brünn. *Ingleside, Christ's Piece.*
 Tietze, Mme.
Timoschenko, Professor S., Electrotechnical Institute, St Petersburg. *St John's College.*
Toja, Sig. Guido, Fondiaria Incendio, Vita Infortuni, Firenze.
Torner de la Fuente, Professor Jorge, Ingeniero de Montes, Escorial, Spain.
 Torner, Mme de.
Torroja y Caballé, Professor E., Universidad, Madrid.
Tötössy, Professor B. von, Népsrinházu 22, Budapest. *Christ's College.*
Traynard, Professor C. E., 8, rue Charles Nodier, Besançon. *Trinity College.*
Trimble, C. J. A., Thornton House, Christ's Hospital, West Horsham. *Gonville and Caius College.*
Tripier, M. H., rue Alphonse-de-Neuville 17, Paris. *The Bull Hotel.*
 Tripier, Mme Marianne.
Turner, Professor H. H., F.R.S., University Observatory, Oxford. c/o *Mrs Routh, Newnham Cottage.*
 Turner, Mrs.
Tzitzeica, Professor G., Strada Scaune 33, Bucharest. *The Lion Hotel.*
 Tzitzeica, Mme.

Vacca, Dr G., Via del Boschetto 40, Rome.
Valentin, Dr G., Burggrafenstrasse 6, Berlin, W. *St John's College.*
Valentiner, Dr H., Sit Knūdsvy 31/2S, Copenhagen.
Van der Heyden, A. F., 3, St John's Terrace, Middlesbrough.
Van der Kamp, Dr H., Middelburg, Holland. *Christ's College.*
Van Dorsten, Professor Dr R. H., Voorschoterlaan 74, Rotterdam. *Trinity College.*
Varićak, Professor Dr V., Franz Josephspl. 6, Agram, Austria-Hungary. *Gonville and Caius College.*
Varley, Miss A. E., Homerton College, Cambridge.
Vasiljev, Professor A. V., Imp. Universität, St Petersburg.
Vegas y Puebla Collado, Professor M., Madrid.
Venn, Dr J., F.R.S., Vicarsbrook, Chaucer Road, Cambridge.
Veronese, Professor G., R. Università, Padova.
Volterra, Professor V., Via in Lucina 17, Rome. *Peterhouse.*
Vuibert, M. P., Boulevard St Germain 63, Paris. *Gonville and Caius College.*

Wahlin, G. E., Obergasse, Schonau, Amt, Heidelberg. 4, *Fitzwilliam Street.*
Wahlin, Mme.
Wallis, A. J., Corpus Christi College, Cambridge.
Watson, F. B., 55, Ash Grove, Headingley, Leeds.
Watson, Miss L.
Watson, G. N., Trinity College, Cambridge.
Webster, Professor A. G., Clark University, Worcester (Mass.), U.S.A. *Emmanuel College.*
Webster, Miss. *Newnham College.*
Weitzenböck, Dr K., Kaiserstrasse 7, Bonn. *Trinity College.*
Welmin, M. Anatole. c/o Professor J. Beliankin.
Western, Dr A. E., 35, Essex Street, Strand, London, W.C.
White, Sir W. H., K.C.B., F.R.S., Cedarcroft, Putney Heath, S.W. c/o *The Master, Christ's College.*
Whitehead, Dr A. N., F.R.S., 17, Carlyle Square, Chelsea, S.W. *G (New Court) Trinity College.*
Whittaker, Professor E. T., Sc.D., F.R.S., The University, Edinburgh. c/o *W. W. Rouse Ball, Elmside, Grange Road.*
Wieleitner, Professor Dr H., Luisenstrasse 31, Pirmasens.
Wilkinson, Rev. Canon M. M. U., Reepham Rectory, Norwich. 11, *King's Parade.*
Wilkinson, Miss Geraldine M.
Wilkinson, Miss Maud F. U.
Willaert, Rev. Professor Dr F., S.J., 22, Boulevard St Michel, Brussels. *Emmanuel College.*
Wilson, Professor A. H., Haverford College (Penn.), U.S.A. 5, *Market Street.*
Wilson, Mrs.
Wilson, Professor D. T., Case School, Cleveland, U.S.A. *Emmanuel College.*
Wilson, Professor E. B., Institute of Technology, Boston (Mass.), U.S.A.
Wilson, T., Rivers Lodge, Harpenden, St Albans. *University Arms Hotel.*
Wiman, Professor A., K. Universitet, Upsala.
Winter, M. Maximilien, 29 Avenue Kléber, Paris.

Wolkow, Professor A., Starokonüshenny per 15_4, Moscow.

Wood, P. Worsley, Emmanuel College.

Wood, Professor Ruth G., 249, Crescent Street, Northampton (Mass.), U.S.A. *Newnham College.*

Woollcombe, Dr R. L., 14, Waterloo Road, Dublin.

Young, Professor J. W. A., The University, Chicago, U.S.A. *Bull Hotel.*
Young, Mrs.

Yule, G. U., 28, Great Ormond Street, London, W.C.

Zaboudski, Professor N., Wilensky 3, St Petersburg.

Zaremba, Professor S., Czersvony Pradnik, près Cracovie, Austria. *Christ's College.*

Zermelo, Professor Ernst, Schönberggasse 9, Zürich. *Trinity College.*

Zervos, Professor P., Rue Acharnon 41, Athens. *Emmanuel College.*

Zimmerman, C. D. A., Middelburg, Holland. *Christ's College.*

Ziwet, Professor A., University of Michigan, Ann Arbor, U.S.A. *Gonville and Caius College.*

Zöllich, Dr H., Spandauer Berg 6, Berlin-Westend.

LADIES' COMMITTEE

Mrs Montagu Butler.
Miss Collier.
Lady Darwin (*President*).
Mrs Hobson.
Miss Constance Jones.
Mrs Scott.
Lady Thomson.
Mr Hinks (*Secretary*).

DISTRIBUTION OF MEMBERS ACCORDING TO NATIONALITY

	Members	Members of family	Total
Argentine	5	—	5
Austria	20	3	23
Belgium	5	—	5
Brazil	1	—	1
Bulgaria	1	—	1
Canada	5	—	5
Chili	1	—	1
Denmark	4	2	6
Egypt	2	—	2
France	39	6	45
Germany	53	17	70
Greece	4	1	5
Holland	9	1	10
Hungary	16	3	19
India	3	—	3
Italy	35	6	41
Japan	3	—	3
Mexico	2	—	2
Norway	3	1	4
Portugal	2	1	3
Roumania	4	1	5
Russia	30	10	40
Servia	1	—	1
Spain	25	2	27
Sweden	12	2	14
Switzerland	8	2	10
United Kingdom	221	49	270
United States of America	60	27	87
	574	134	708

REGULATIONS FOR THE FIFTH INTERNATIONAL CONGRESS OF MATHEMATICIANS

I. At the first General Meeting the Chair shall be taken by the President of the Organizing Committee. The Meeting shall proceed to elect the following Officers of the Congress:

A President,
Vice-Presidents,
General Secretaries.

II. The President of the Congress or one of the Vice-Presidents shall preside at each of the succeeding General Meetings.

III. One of the Introducers nominated by the Organizing Committee shall preside at the first Meeting of each Section. At such Meeting the Section shall appoint a Secretary and one or more Assistant Secretaries. The Secretaries shall remain in office for the whole time of the Congress. At each Meeting the members present shall elect the President for the next Meeting.

IV. The Organizing Committee shall settle the order in which the communications to each Section shall be read. This order may however be modified by a vote of the Section concerned.

V. The reading of a Communication shall not occupy more than twenty minutes. During a discussion a speaker shall not be allowed more than ten minutes, nor shall he speak more than once on the same subject without special permission from the President of the Section.

VI. The speakers are requested to furnish the Secretary of the Section with a brief résumé of their remarks immediately after the conclusion of the discussion. The Sectional Secretary, at the end of each Meeting, shall draw up and send to the General Secretary the titles of the papers read for publication in the Journal of the following day. A complete report containing the abstracts of the Communications and of the subsequent discussions shall be drawn up by the Secretary of the Section before the end of the Congress.

VII. The Lectures and Communications read at the Congress shall be collected in the Volume of Proceedings. Authors should deliver the texts of their Lectures and Communications to the General Secretary of the Congress not later than the end of the Congress. Those Lectures or Communications which are written in French, German, or Italian should be type-written (except formulae).

CINQUIÈME CONGRÈS INTERNATIONAL DES MATHÉMATICIENS

RÈGLEMENT DU CONGRÈS

I. La première séance génerale sera présidée par le Président du Comité d'organisation. A cette séance on procédera à la constitution du Bureau définitif, qui comprendra :

Un Président,
Des Vice-Présidents,
Des Secrétaires généraux.

II. Les séances générales successives seront présidées par le Président du Congrès ou par un des Vice-Présidents.

III. La première réunion de chaque Section sera présidée par un des Introducteurs désigné par le Comité d'organisation. A cette séance, la Section nommera un Secrétaire et un ou plusieurs Secrétaires adjoints. Les Secrétaires resteront en charge pendant toute la durée du Congrès. A chaque séance, les membres présents éliront le Président de la séance suivante.

IV. Le Comité d'organisation établira l'ordre des lectures de chaque Section. Cet ordre pourra, toutefois, être modifié par le vote des Sections respectives.

V. L'exposé d'un Mémoire ne pourra pas dépasser la durée de vingt minutes. Pendant la discussion, un orateur ne pourra pas parler plus de dix minutes, ni prendre la parole plus d'une fois sur le même sujet, sans autorisation spéciale du Président.

VI. Les orateurs sont priés de transmettre au Secrétaire de la Section, aussitôt après l'exposé, un résumé succinct des sujets traités. Le Secrétaire de la Section, après la clôture de la séance, rédigera et remettra au Secrétaire général un extrait du procès-verbal, qui sera imprimé dans le Bulletin du jour suivant. Le procès-verbal complet, contenant le résumé des exposés et des discussions, devra être rédigé par le Secrétaire de la Section avant la fin du Congrès.

VII. Les Conférences et les Communications faites au Congrès seront réunies dans le volume des comptes rendus. MM. les auteurs sont priés de bien vouloir remettre le texte de leurs Communications au Secrétaire Général du Congrès, au plus tard à la fin du Congrès. Les Conférences et les Communications en français, allemand, ou italien, devront être écrites (sauf les formules) à la machine à écrire.

V INTERNATIONALER MATHEMATIKER-KONGRESS

REGLEMENT FÜR DEN KONGRESS

I. Die erste Generalversammlung wird von dem Präsidenten des Organisations-Komitees geleitet werden. In derselben wird die Zusammensetzung des Leitungs-Komitees festgesetzt werden, welches bestehen wird aus:

Einem Präsidenten,
Den Vice-Präsidenten,
Den Generalsekretären.

II. Die folgenden Generalversammlungen werden von dem Präsidenten des Kongresses oder von einem der Vice-Präsidenten geleitet werden.

III. In der ersten Sitzung von jeder Sektion wird ein vom Organisationskomitee ernannter Einführender den Vorsitz führen. In dieser Sitzung wird die Section einen Sekretär und einen oder mehrere Hülfssekretäre ernennen. Die ernannten Sekretäre werden während der ganzen Dauer des Kongresses im Amt bleiben. An jeder Sitzung werden die anwesenden Mitglieder den Präsidenten der folgenden Sitzung erwählen.

IV. Das Organisations-Komitee wird die Reihenfolge der Vorträge in jeder Sektion festsetzen; dieselbe kann aber durch die Abstimmung der respektiven Sectionen geändert werden.

V. Der Vortrag eines Referates darf nicht länger als zwanzig Minuten dauern. Während der Discussion darf ein Redner nicht länger als zehn Minuten sprechen. Er darf auch ohne besondere Erlaubniss des Präsidenten nicht mehr als einmal über denselben Gegenstand reden.

VI. Die Redner werden gebeten, dem Sectionssekretär, sogleich nach Schluss des Vortrages, einen kurzen Auszug ihrer Ausführungen zu überreichen. Nach Schluss der Sitzung soll der Sekretär der Section einen Auszug des Sitzungsprotokolles abfassen und ihn dem Generalsekretär überreichen, damit dieser Auszug im Journal des folgenden Tages erscheint. Das vollständige Protokoll, mit dem Auszug der gehaltenen Vorträge und der Diskussionen, soll vom Sekretär vor dem Schluss des Kongresses abgefasst werden.

VII. Die Vorträge und Mitteilungen an den Kongress werden im Band der Verhandlungen zusammengefasst. Die Verfasser werden gebeten das Manuskript ihrer Vorträge spätestens bis Ende des Kongresses dem Generalsekretär zu übergeben. Die Vorträge und Mitteilungen in französischer, deutscher, oder italienischer Sprache sollten (mit Ausnahme der Formeln) mit der Schreibmaschine geschrieben sein.

V CONGRESSO INTERNAZIONALE DEI MATEMATICI

REGOLAMENTO DEL CONGRESSO

I. La prima seduta plenaria sarà presieduta dal Presidente del Comitato organizzatore. In essa si procederà alla costituzione del Seggio, il quale comprenderà:

Un Presidente,
Alcuni Vice-Presidenti,
Alcuni Segretari generali.

II. Le sedute plenarie successive saranno presiedute dal Presidente del Congresso o da uno dei Vice-Presidenti.

III. La prima adunanza di ciascuna Sezione sarà presieduta da uno degli Introduttori designato dal Comitato organizzatore. In questa seduta la Sezione nominerà un Segretario ed uno o più Segretari aggiunti. I Segretari resteranno in carica per tutta la durata del Congresso. All' ogni adunanza i membri presenti eleggeranno il Presidente della seduta successiva.

IV. Il Comitato organizzatore stabilirà l' ordine delle letture di ciascuna Sezione; questo però potrà esser modificato dal voto delle Sezioni rispettive.

V. La lettura di una comunicazione non potrà durare più di venti minuti. Durante la discussione gli oratori non potranno tener la parola più di dieci minuti, nè potranno prenderla più di una volta sullo stesso argomento senza speciale permesso del Presidente.

VI. Gli oratori sono pregati di trasmettere al Segretario della Sezione, appena compiuta la lettura, un breve sunto degli argomenti trattati. Il Segretario della Sezione, terminata la seduta, redigerà e consegnerà al Segretario generale un estratto del processo verbale destinato alla stampa nel Bollettino del giorno successivo. Il processo verbale completo, continente il sunto delle letture fatte e delle discussioni avvenute, dovrà esser redatto dal Segretario della Sezione avanti la fine del Congresso.

VII. Le Conferenze e le Comunicazioni lette al Congresso saranno raccolte nel Volume degli Atti. Si pregano gli Autori di voler consegnare il testo delle loro letture al Segretario generale del Congresso, non più tardi della fine del Congresso. Per gli scritti in lingue francese, tedesco, o italiano, è richiesto l' uso della machina da scrivere (tranne che per le formole).

PROCEEDINGS OF THE CONGRESS

Wednesday, August 21

At 9.30 p.m. the Members of the Congress were received by Sir G. H. Darwin, President of the Cambridge Philosophical Society, and were presented to Mr R. F. Scott, Vice-Chancellor of the University, at a conversazione held in the Combination Room and Hall of St John's College.

Thursday, August 22

The opening meeting of the Congress was held at 10.0 a.m.

Sir G. H. Darwin, President of the Cambridge Philosophical Society, spoke as follows:

Four years ago at our Conference at Rome the Cambridge Philosophical Society did itself the honour of inviting the International Congress of Mathematicians to hold its next meeting at Cambridge. And now I, as President of the Society, have the pleasure of making you welcome here. I shall leave it to the Vice-Chancellor, who will speak after me, to express the feeling of the University as a whole on this occasion, and I shall confine myself to my proper duty as the representative of our Scientific Society.

The Science of Mathematics is now so wide and is already so much specialised that it may be doubted whether there exists to-day any man fully competent to understand mathematical research in all its many diverse branches. I, at least, feel how profoundly ill-equipped I am to represent our Society as regards all that vast field of knowledge which we classify as pure mathematics. I must tell you frankly that when I gaze on some of the papers written by men in this room I feel myself much in the same position as if they were written in Sanskrit.

But if there is any place in the world in which so one-sided a President of the body which has the honour to bid you welcome is not wholly out of place it is perhaps Cambridge. It is true that there have been in the past at Cambridge great pure mathematicians such as Cayley and Sylvester, but we surely may claim without undue boasting that our University has played a conspicuous part in the advance of applied mathematics. Newton was a glory to all mankind, yet we Cambridge men are proud that fate ordained that he should have been Lucasian Professor here. But as regards the part played by Cambridge I refer rather to the men of the last hundred years, such as Airy, Adams, Maxwell, Stokes, Kelvin, and other lesser lights, who have marked out the lines of research in applied mathematics as studied in this University. Then too there are others such as our Chancellor, Lord Rayleigh, who are happily still with us.

Up to a few weeks ago there was one man who alone of all mathematicians might have occupied the place which I hold without misgivings as to his fitness; I mean Henri Poincaré. It was at Rome just four years ago that the first dark shadow fell on us of that illness which has now terminated so fatally. You all remember the dismay which fell on us when the word passed from man to man "Poincaré is ill." We had hoped that we might again have heard from his mouth some such luminous address as that which he gave at Rome; but it was not to be, and the loss of France in his death affects the whole world.

It was in 1900 that, as president of the Royal Astronomical Society, I had the privilege of handing to Poincaré the medal of the Society, and I then attempted to give an appreciation of his work on the theory of the tides, on figures of equilibrium of rotating fluid and on the problem of the three bodies. Again in the preface to the third volume of my collected papers I ventured to describe him as my patron Saint as regards the papers contained in that volume. It brings vividly home to me how great a man he was when I reflect that to one incompetent to appreciate fully one half of his work yet he appears as a star of the first magnitude.

It affords an interesting study to attempt to analyze the difference in the textures of the minds of pure and applied mathematicians. I think that I shall not be doing wrong to the reputation of the psychologists of half a century ago when I say that they thought that when they had successfully analyzed the way in which their own minds work they had solved the problem before them. But it was Sir Francis Galton who showed that such a view is erroneous. He pointed out that for many men visual images form the most potent apparatus of thought, but that for others this is not the case. Such visual images are often quaint and illogical, being probably often founded on infantile impressions, but they form the wheels of the clockwork of many minds. The pure geometrician must be a man who is endowed with great powers of visualisation, and this view is confirmed by my recollection of the difficulty of attaining to clear conceptions of the geometry of space until practice in the art of visualisation had enabled one to picture clearly the relationship of lines and surfaces to one another. The pure analyst probably relies far less on visual images, or at least his pictures are not of a geometrical character. I suspect that the mathematician will drift naturally to one branch or another of our science according to the texture of his mind and the nature of the mechanism by which he works.

I wish Galton, who died but recently, could have been here to collect from the great mathematicians now assembled an introspective account of the way in which their minds work. One would like to know whether students of the theory of groups picture to themselves little groups of dots; or are they sheep grazing in a field? Do those who work at the theory of numbers associate colour, or good or bad characters with the lower ordinal numbers, and what are the shapes of the curves in which the successive numbers are arranged? What I have just said will appear pure nonsense to some in this room, others will be recalling what they see, and perhaps some will now for the first time be conscious of their own visual images.

The minds of pure and applied mathematicians probably also tend to differ from one another in the sense of aesthetic beauty. Poincaré has well remarked in his *Science et Méthode* (p. 57):

"On peut s'étonner de voir invoquer la sensibilité à propos de démonstrations mathématiques qui, semble-t-il, ne peuvent intéresser que l'intelligence. Ce serait oublier le sentiment de la beauté mathématique, de l'harmonie des nombres et des formes, de l'élégance géometrique. C'est un vrai sentiment esthétique que tous les vrais mathématiciens connaissent. Et c'est bien là de la sensibilité."

And again he writes:

"Les combinaisons utiles, ce sont précisément les plus belles, je veux dire celles qui peuvent le mieux charmer cette sensibilité spéciale que tous les mathématiciens connaissent, mais que les profanes ignorent au point qu'ils sont souvent tentés d'en sourire."

Of course there is every gradation from one class of mind to the other, and in some the aesthetic sense is dominant and in others subordinate.

In this connection I would remark on the extraordinary psychological interest of Poincaré's account, in the chapter from which I have already quoted, of the manner in which he proceeded in attacking a mathematical problem. He describes the unconscious working of the mind, so that his conclusions appeared to his conscious self as revelations from another world. I suspect that we have all been aware of something of the same sort, and like Poincaré have also found that the revelations were not always to be trusted.

Both the pure and the applied mathematician are in search of truth, but the former seeks truth in itself and the latter truths about the universe in which we live. To some men abstract truth has the greater charm, to others the interest in our universe is dominant. In both fields there is room for indefinite advance; but while in pure mathematics every new discovery is a gain, in applied mathematics it is not always easy to find the direction in which progress can be made, because the selection of the conditions essential to the problem presents a preliminary task, and afterwards there arise the purely mathematical difficulties. Thus it appears to me at least, that it is easier to find a field for advantageous research in pure than in applied mathematics. Of course if we regard an investigation in applied mathematics as an exercise in analysis, the correct selection of the essential conditions is immaterial; but if the choice has been wrong the results lose almost all their interest. I may illustrate what I mean by reference to Lord Kelvin's celebrated investigation as to the cooling of the earth. He was not and could not be aware of the radio-activity of the materials of which the earth is formed, and I think it is now generally acknowledged that the conclusions which he deduced as to the age of the earth cannot be maintained; yet the mathematical investigation remains intact.

The appropriate formulation of the problem to be solved is one of the greatest difficulties which beset the applied mathematician, and when he has attained to a true insight but too often there remains the fact that his problem is beyond the reach of mathematical solution. To the layman the problem of the three bodies seems so simple that he is surprised to learn that it cannot be solved completely, and yet we know what prodigies of mathematical skill have been bestowed on it. My own work on the subject cannot be said to involve any such skill at all, unless indeed you describe as skill the procedure of a housebreaker who blows in a safe-door with dynamite instead of picking the lock. It is thus by brute force that this tantalising problem has been compelled to give up some few of its secrets, and great as has been the labour involved I think it has been worth while. Perhaps this work too has done something to encourage others such as Störmer[1] to similar tasks as in the computation of the orbits of electrons in the neighbourhood of the earth, thus affording an explanation of some of the phenomena of the aurora borealis. To put at their lowest the claims of this clumsy method, which may almost excite the derision of the pure mathematician, it has served to throw light on the celebrated generalisations of Hill and Poincaré.

I appeal then for mercy to the applied mathematician and would ask you to consider in a kindly spirit the difficulties under which he labours. If our methods are often wanting in elegance and do but little to satisfy that aesthetic sense of which I spoke before, yet they are honest attempts to unravel the secrets of the universe in which we live.

We are met here to consider mathematical science in all its branches. Specialisation has become a necessity of modern work and the intercourse which will take place between us in the course of this week will serve to promote some measure of comprehension of the work which is being carried on in other fields than our own. The papers and lectures which you will hear will serve towards this end, but perhaps the personal conversations outside the regular meetings may prove even more useful.

Mr R. F. Scott, Vice-Chancellor of the University of Cambridge, spoke as follows:

GENTLEMEN, It is my privilege to-day on behalf of the University of Cambridge and its Colleges to offer to Members of the Congress a hearty welcome from the resident body.

Sir George Darwin has dwelt on the more serious aspects of the meeting and work of the Congress, may I express the hope that it will also have its lighter and more personal side? That we shall all have the privilege and pleasure of making the personal acquaintance of many well known to us both by name and by fame, and that those of our visitors who are not familiar with the College life of Oxford and Cambridge will learn something of a feature so distinctive of the two ancient English Universities. If the Congress comes at a time when it is not possible to see the great body of our students either at work or at play, the choice of date at least renders it possible that many of our visitors may enjoy for a time that Collegiate life which has so many attractions.

I see that one of the Sections of the Congress deals with historical and

[1] *Videnskabs Selskab*, Christiania, 1904.

didactical questions. Those members of the Congress who are interested in these subjects will have an opportunity of learning on the spot something of our methods in Cambridge, and of the history of our chief Mathematical Examination, the Mathematical Tripos, and of its influence on the study and progress of Mathematics both in Cambridge and Great Britain.

The researches of Dr Venn seem to point to the fact that until it was altered at a very recent date the Mathematical Tripos represented something like the oldest example in Europe of a competitive Examination with an order of merit. Those who are interested in such matters of history will find much to interest them in Mr Rouse Ball's *History of the Study of Mathematics at Cambridge.*

The subject is to me and I hope to others an interesting one. There can be no doubt that the Examination and the preparation for it has had a profound influence on Mathematical studies at Cambridge.

Many Cambridge mathematicians, as the names given by Sir George Darwin testify, studied mathematics for its own sake and with the view of extending the boundaries of knowledge. Many others, probably the great majority, studied mathematics with their eyes fixed upon the Mathematical Tripos, with the view in the first place of being examined and afterwards of acting as examiner in it.

The tendency at Cambridge has been to give great minuteness to the study of any particular branch of mathematics. To stimulate the invention of what we call "Problems," examples of more general theories. If I may borrow a simile from the study of Literature the tendency was to produce critics and editors rather than authors or men of letters, followers rather than investigators. The effect must I think be obvious to any one who compares Cambridge Text Books and Treatises with those of the Continental Schools of Mathematics. I may illustrate what I mean by referring to the *Mathematical Problems* of the late Mr Joseph Wolstenholme, a form of work I believe without a parallel in the mathematical literature of other nations. The fashion is fading away, but while you are in Cambridge I commend it to your notice.

Professor E. W. Hobson, Senior Secretary of the Organizing Committee, stated that the number of persons who had joined the Congress up to 10.0 p.m. on Wednesday, August 21st, was 670, the number of representatives of different countries being as follows: Argentine 4, Austria 19, Belgium 4, Bulgaria 1, Canada 4, Chili 1, Denmark 5, Egypt 2, France 42, Germany 70, Great Britain 250, Greece 5, Holland 9, Hungary 19, India 3, Italy 38, Japan 3, Mexico 1, Norway 4, Portugal 3, Roumania 5, Russia 38, Servia 1, Spain 25, Sweden 13, Switzerland 9, United States 82. He also called the attention of the Members of the Congress to the exhibition of books, models and machines (chiefly calculating machines) arranged in two rooms of the Cavendish Laboratory.

The first general meeting of the Congress was held at 2.30 p.m.

On the motion of Prof. Mittag-Leffler, seconded by Professor Enriques, Sir G. H. Darwin was elected President of the Congress.

On the motion of the President it was agreed that Lord Rayleigh be made Honorary President of the Congress (Président d'honneur).

On the motion of the President, Vice-Presidents of the Congress were elected as follows:—W. von Dyck, L. Fejér, R. Fujisawa, J. Hadamard, J. L. W. V. Jensen, P. A. MacMahon, G. Mittag-Leffler, E. H. Moore, F. Rudio, P. H. Schoute, M. S. Smoluchowski, V. A. Steklov, V. Volterra.

On the motion of the President, General Secretaries of the Congress were elected as follows:—E. W. Hobson, A. E. H. Love.

Sir G. Greenhill made the following statement in regard to the work of the International Commission on the teaching of mathematics:

The statement I have to make, Sir, to the Congress, is given in the formal words following:

1. The International Commission on the *Teaching of Mathematics* was appointed at the Rome Congress, on the recommendation of the Members of Section IV.

2. The several countries, in one way or another, have recognised officially the work, and have contributed financial support.

3. About 150 reports have been published, and about 50 more will appear later.

4. The Commission will report in certain Sessions of Section IV.

5. The Commission hopes to be continued in power, in order that the work now in progress may be brought to completion. A Resolution to this effect will be offered at the final Meeting of the Congress.

At 3.30 p.m. Prof. F. Enriques delivered his lecture "Il significato della critica dei principii nello sviluppo delle matematiche."

At 5.0 p.m. Prof. E. W. Brown delivered his lecture "Periodicities in the Solar System."

At 9 p.m. Section IV met.

Friday, August 23

The various Sections met at 9.30 a.m.

At 3.30 p.m. Prof. E. Landau delivered his lecture "Gelöste und ungelöste Probleme aus der Theorie der Primzahlverteilung und der Riemannschen Zetafunktion."

At 5 p.m. Prince B. Galitzin delivered his lecture "The principles of instrumental seismology."

At 9 p.m. the Members of the Congress were received at a conversazione in the Fitzwilliam Museum by the Chancellor of the University, The Rt. Hon. Lord Rayleigh, O.M.

Saturday, August 24

The various Sections met at 9.30 a.m.

At 3.30 p.m. Prof. É. Borel delivered his lecture "Définition et domaine d'existence des fonctions monogènes uniformes."

At 5 p.m. Sir W. H. White delivered his lecture "The place of mathematics in engineering practice."

Sunday, August 25

At 3 p.m. the Members of the Congress were received at an afternoon party in the Garden of Christ's College by the President of the Congress, Sir G. H. Darwin.

At 9 p.m. the Members of the Congress were invited to be present at an Organ Recital in the Chapel of King's College.

Monday, August 26

The various Sections met at 9.30 a.m.

A meeting of Section IV (*b*) was held at 3 p.m.

A meeting of Section I was held at 3.30 p.m.

In the afternoon Members of the Congress made an excursion to Ely and visited the Cathedral. Other Members visited the works of the Cambridge Scientific Instrument Company and the University Observatory.

At 9 p.m. the Members of the Congress were entertained in the Hall and Cloisters of Trinity College by the Master and Fellows of the College.

Tuesday, August 27

The various Sections met at 9.30 a.m.

At 3.30 p.m. Professor Bôcher delivered his lecture "Boundary problems in one dimension."

At 5 p.m. Sir J. Larmor delivered his lecture "The dynamics of radiation."

In the afternoon a number of Members of the Congress proceeded to the Mill Road Cemetery for the purpose of depositing a wreath upon the grave of the late Professor A. Cayley. An address was delivered by Professor S. Dickstein. Other Members visited the works of the Cambridge Scientific Instrument Company.

At 9 p.m. the final meeting of the Congress was held.

The President read a telegram from Prof. V. Volterra regretting that family reasons prevented his attending the Congress, and proceeded to speak as follows:

LADIES AND GENTLEMEN,

We meet to-night for the final Conference of the present Congress. The majority of those present in the room will have been aware that a procession was formed to-day to lay a wreath on the tomb of Cayley. This has touched the hearts of our University. It had, I believe, been suggested that a more permanent wreath should have been deposited; but no such things can be obtained at short notice, and the final arrangement adopted is that a silver wreath shall be made and presented to the University, whose authorities will I am sure gratefully accept it and deposit it in some appropriate place, where it will remain as a permanent memorial of the recognition accorded by the mathematicians of all nations to our great investigator. The subscribers have entrusted the carrying out of this to our Organizing Committee in Cambridge.

I will now explain the order in which we think that it will be convenient for us to carry out our business of to-night. At the last meeting at Rome various resolutions were adopted, and I shall draw your attention to all of them which may give rise to any further discussion to-night, and I shall then successively call on speakers who may have resolutions to propose. In doing this the order of subjects will be followed as I find them in the *procès-verbal* of the Roman Congress. After these matters are decided, opportunity will of course be afforded to any of our members who may have new subjects on which they have proposals to make.

The first resolution at Rome concerned the work of the International Commission on the Teaching of Mathematics, and a resolution will be proposed as to this.

The second resolution was one as to the unification of vectorial notations. I learn from M. Hadamard that exchange of views has taken place on this subject during the last four years, but that it has not been found possible to arrive at any definite conclusions. No resolution will be proposed on the present occasion, but it is hoped that by the time the next Congress takes place something may have been achieved and that the matter will be brought forward again.

It was proposed at Rome that a constitution should be formed for an International Association of Mathematicians. I have not heard that any proposal will be made to-night and I do not hesitate to express my own opinion that our existing arrangements for periodical Congresses meet the requirements of the case better than would a permanent organisation of the kind suggested.

There has been a resolution as to the improvement and unification of the methods of pure and applied mathematics. This subject seems to be sufficiently taken cognisance of as part of the work of the Commission on Teaching, and I cannot think that any further action on our part is needed.

Next there followed a resolution as to the publication of the works of Euler, and a resolution as to this will be proposed to-night.

In this connection I would remark that a complete edition of the works of the immortal Herschel is in course of publication by the Royal Society.

An important matter has to be determined to-night, namely the choice of the place and of the time of the next Congress, and I shall call on Professor Mittag-Leffler to speak to this subject, and of course others may also speak if they desire.

Opportunity will then be afforded to any others who may have proposals to bring forward. When these matters of business are decided I shall say a few words as to the Congress which is now terminating.

The following resolution was moved by Mr C. Godfrey, seconded by Prof. W. v. Dyck and carried *nem. con.*:

> That the Congress expresses its appreciation of the support given to its Commission on the Teaching of Mathematics by various governments, institutions, and individuals; that the Central Committee composed of F. Klein (Göttingen), Sir G. Greenhill (London) and H. Fehr (Geneva) be continued in power and that, at its request, David Eugene Smith of New York be added to its number; that the Delegates be requested to continue their good offices in securing the cooperation of their respective governments, and in carrying on the work; and that the Commission be requested to make such further report at the Sixth International Congress, and to hold such conferences in the meantime, as the circumstances warrant.

A French translation of the resolution was read to the meeting by M. Bioche.

In seconding the resolution Prof. v. Dyck spoke as follows:

I wish to second the motion, but in doing so you will allow me to insist with a few words upon the prominent work done by the International Committee on Teaching of Mathematics during these last four years. None of us who were present in Rome could even imagine what an immense labour was to be undertaken when Dr D. E. Smith proposed a comparative investigation on mathematical teaching.

Now, by the activity of the splendid organisation of the Central Committee, under the guidance of Klein, Greenhill, and Fehr, with the worthy help of D. E. Smith, every country in nearly every part of the world has contributed in its own department to the *Reports for Cambridge*—so that there were about 150 different volumes with about 300 articles brought before the Congress—papers which were not all read but were aptly spoken about by the collaborators.

So we will congratulate the Committee upon the work already done, and we have to express our most hearty thanks both to the Central and Local Committees and the collaborators.

But furthermore we have to congratulate ourselves that this Committee *will remain still in charge* and will continue and finish the work.

For us, the outsiders, the series of reports is like a series of a very large number of coefficients to be calculated. And the problem arises now to find the principles under which they may be grouped and compared with each other according to their individuality and their quality. Whom could we better entrust with that problem than *this acting committee*, which has been at work all this time, and to whom we are even now so deeply obliged?

The following resolution was moved by Prof. E. Gutzmer and carried:

> In accordance with a wish that has been repeatedly expressed by successive International Congresses of Mathematicians, and in particular, in accordance with the resolution adopted at Rome, concerning the publication of the collected works of Leonhard Euler, the fifth International Congress of Mathematicians, assembled at Cambridge, expresses its warmest thanks

to the Schweizerische Naturforschende Gesellschaft for their efforts in inaugurating the great work, and for the magnificent style in which the five volumes already published have been completed. The Congress expresses the hope that the scientific world will continue to exhibit that sustained interest in the undertaking which it has hitherto shewn.

Im Anschluss an die Verhandlungen der früheren Internationalen Mathematiker Kongresse, insbesondere an den Beschluss des 4. Kongresses in Rom, betreffend die Herausgabe der sämtlichen Werke Leonhard Eulers bringt der 5. Internationale Kongress zu Cambridge der Schweizerischen Naturforschenden Gesellschaft seinen wärmsten Dank für die tatkräftige Inangriffnahme des grossen Unternehmens zum Ausdruck und verbindet damit zugleich seine hohe Anerkennung für die monumentale Ausgestaltung, die sie dem Werke in den bereits vorliegenden fünf Bänden hat angedeihen lassen. Der Kongress spricht die Erwartung aus, dass der Euler-Ausgabe auch fernerhin die Unterstützung nicht fehlen werde, die ihn bisher schon in so dankenswerter Weise von der ganzen wissenschaftlichen Welt, insbesondere von den grossen Akademien, zu teil geworden ist.

Prof. G. Mittag-Leffler presented an invitation to the Congress to hold its next meeting at Stockholm in 1916. The following is the text of the invitation:

Au nom des membres de la 1ière classe de l'Académie royale des sciences de Suède, au nom de la rédaction suédoise du journal *Acta Mathématica* ainsi que de tous les géomètres suédois, j'ai l'honneur d'inviter le congrès international des mathématiciens à se réunir à Stockholm en l'année 1916.

Notre august souverain le roi Gustave m'a gracieusement confié la charge d'exprimer au congrès qu'il lui souhaiterait avec plaisir la bienvenue dans sa capitale et qu'il serait prêt à le prendre sous son haut patronage pendant son séjour à Stockholm.

Nous autres nous nous estimerions très heureux si le congrès voulait accepter notre invitation, et nous ferons tout ce qui est en notre pouvoir pour rendre le séjour des membres dans notre pays aussi agréable et aussi instructif que possible.

Prof. E. Beke presented an invitation to the Congress to hold its meeting of 1920 at Budapest. The following is the text of the invitation:

Au nom des mathématiciens hongrois, j'ai l'honneur d'inviter le septième Congrès International à venir siéger en 1920 dans la capitale de la Hongrie—Budapest.

Tout en sachant que c'est le Congrès de Stockholm qui aura à décider de notre invitation, nous la présentons déjà ici, conformément à un excellent usage adopté par les Congrès précédents.

Je suis autorisé à vous annoncer que les institutions scientifiques compétentes ainsi que le gouvernement royal hongrois nous donneront leur concours effectif et tout leur appui.

La patrie des Bolyai et sa belle capitale seront fières de pouvoir offrir leur hospitalité aux savants représentants des sciences mathématiques, qui voudront nous honorer de leur présence.

On the motion of the President the invitation to Stockholm was accepted *nem. con.*

The President stated that the Congress noted with gratitude the invitation to Budapest, but the decision as to the place of the next meeting after that of 1916 would properly be made at Stockholm.

Prof. C. Stéphanos expressed the hope that the Congress would meet in Athens in 1920 or 1924.

It was resolved that the following telegram be sent to Lord Rayleigh:

Le cinquième congrès international des mathématiciens en terminant ses travaux adresse à l'illustre Chancelier de l'Université de Cambridge, au grand créateur dans le domaine des sciences mathématiques et des sciences physiques, l'expression respectueuse de ses hommages et de son admiration.

The President then spoke as follows:

We have come to the end of a busy week, and I have the impression that the papers and lectures which you have heard have been worthy of the occasion. I trust too that you will look back on the meeting as a week of varied interests. The weather has been such that in a more superstitious age we should surely have concluded that heaven did not approve of our efforts; but fortunately to-day we regard it rather as a matter for the consideration of Section III (*a*) to decide why it is that solar radiation acting on a layer of compressible fluid on the planet should have selected England as the seat of its most unkindly efforts in the way of precipitation. Notwithstanding this I cannot think that I have wholly misinterpreted the looks and the words of those of whom I have seen so much during these latter days, when I express the conviction that you have enjoyed yourselves. There is much of the middle ages in our old Colleges at Cambridge, and it is only at Oxford that you can find any parallel to what you have seen here. Many of you will have the opportunity to-morrow under the guidance of Professor Love of seeing the wonderful beauties of Oxford, and I express the hope that the weather may be such as to make us Cambridge men jealous of the good fortune of Oxford.

I feel assured that all of you must realise how long and arduous are the preparations for such a Congress as this. I believe that the arrangements made for your reception have been on the whole satisfactory, and I wish to tell you how much you owe in this respect to Professor Hobson. For months past he has been endeavouring to do all that was in his power to render this meeting both efficient and agreeable. During the last few weeks he has been joined by Professor Love from Oxford, and they have both been busy from morning to night at countless matters which needed decision. You are perhaps aware that our Parliament in its wisdom has decided that coal-miners shall not be allowed to work for more than eight hours a day. There has been no eight hours bill for the Secretaries of this Congress, and if I were to specify a time for the work of Hobson and Love I should put it at sixteen hours a day. As President of the Organizing Committee and subsequently of the Congress I wish to express my warm thanks to them for all that they have done. Before closing the meeting I shall ask them to say a few words, and Professor Hobson will take this opportunity of telling you something as to the final numbers of those attending.

Sir Joseph Larmor has undertaken the financial side of our work. His work although less arduous than that of the Secretaries has been not less responsible.

Each department of the social arrangements has been in the charge of some one man, and I want to thank them all for what they have done. Mr Hinks kindly served me as my special aide-de-camp, and has also been of inestimable service to Lady Darwin and the Committee of Ladies in entertaining the ladies who are present here. May I be pardoned if I say that I think the reception by the Chancellor and Lady Rayleigh at the Fitzwilliam Museum was a brilliant one, and I think you should know that every detail was carried out at the suggestion and under the care of Mr Hinks. His work was not facilitated by the fact that a number of things had to be changed at the last minute on account of the bad weather, but I doubt whether any traces of the changes made will have struck you.

Then I desire also to express our warm thanks to Mr A. W. Smith who was in charge of the reception room which has proved so convenient an institution. He had, as Assistant Secretary, countless other matters to which to attend and he has carried out all these with the highest success.

Finally I am sure that I may take on myself as your President to express to the Authorities of the University our gratitude for the use of these rooms for the meeting, and to the Committee (called by us the Syndicate) of the Fitzwilliam Museum, responsible for many valuable collections, for the loan to the Chancellor of the Museum for our reception. We also desire to acknowledge the pleasure we had in the beautiful reception given to us by the Master and Fellows of Trinity College, to the Master in person for the interesting lecture which he gave to the ladies, and to Colonel Harding and Sir C. Waldstein for their kindness in receiving the ladies at their country houses.

Prof. Hobson stated the number of Members of the Congress as follows:

Number of Members of the Congress	708
Number of effective Members	574

He also thanked the Members of the Congress of all nations for their courtesy in their correspondence with the General Secretaries during the time of preparation for the Congress.

Prof. G. Mittag-Leffler then spoke as follows:

MESDAMES ET MESSIEURS,

Les étrangers qui ont pris part à ce cinquième congrès international des mathématiciens qui vient d'être clos m'ont chargé d'être auprès de nos collègues et hôtes anglais l'interprète de leur vive et chaleureuse reconnaissance pour l'accueil charmant que nous avons reçu. C'est avec un plaisir particulier que nous avons séjourné en cette ville remplie de grands souvenirs scientifiques, berceau de cette illustre université où les anciennes coutumes et la pensée moderne ont pu s'unir comme nulle part ailleurs. Grâce à l'excellente direction du comité d'organisation les forces scientifiques variées et puissantes qui ont été réunies ici ont pu se faire valoir de la manière la plus féconde. Nous tous qui avons pris part aux travaux du congrès en avons reçu des suggestions nouvelles et abondantes pour nos propres travaux à venir.

Pour les mathématiques les congrès sont d'une importance peut-être plus grande que pour les autres sciences. La mathématique, la science du nombre, la science des sciences même lorsque elle prend en aide la représentation géométrique, ou lorsqu'elle tache de s'adapter à l'expérience extérieure, ne traite pourtant au fond que d'abstractions pures. Voilà ce qui rend l'étude des idées communiquées seulement par l'imprimé bien plus laborieuse que dans les autres sciences dont l'objet est plutôt concret. Il s'en suit que l'étude de la littérature dans les mathématiques, plus encore peut-être que dans ces autres sciences, trouve un complément précieux dans l'échange verbal des idées. Je crois donc interpréter nos vœux unanimes en exprimant l'espoir que le cinquième congrès international des mathématiciens ne soit qu'un terme dans une série jamais interrompue de congrès pareils, renouvelés touts les quatre ans.

Je crois de même exprimer la pensée commune en affirmant que la manière si admirable dont le comité d'organisation du congrès de Cambridge a su préparer nos réunions nous sera à l'avenir un modèle à suivre. C'est à notre illustre président Sir George Darwin, à nos sécrétaires infatigables et doués de ce sens pratique qu'on reconnait chez les anglais les professeurs Hobson et Love, au trésorier Sir J. Larmor et aux autres membres du comité que nous devons ce résultat. Nous les remercions de plein cœur en leur affirmant que notre séjour ici sera pour chacun de nous un souvenir inoubliable.

Wednesday, August 28

Members of the Congress took part in an excursion to Oxford, and a reception at Hatfield House, on the invitation of Lord Salisbury.

PROCEEDINGS OF THE SECTIONS

Thursday, August 22

Section IV

The Section met at 9 p.m. when the following lecture was delivered and illustrated by lantern slides:

P. J. HARDING: The history and evolution of arithmetic division.

Friday, August 23

Section I. (*Arithmetic, Algebra, Analysis.*)

The Section met at 9.30 a.m. Professor E. B. Elliott was in the Chair. Dr T. J. I'A. Bromwich was elected Secretary, and Prof. I. Bendixson and Prof. J. C. Fields were elected Assistant Secretaries. Prof. E. Landau was elected Chairman for Saturday, August 24.

The Chairman addressed the Section as follows:

LADIES AND GENTLEMEN,

It is a great honour to be asked to preside to-day over this Section I of the Fifth International Congress of Mathematicians. My first pleasant duty is to address a few words of welcome to the Section. The many distinguished Analysts and Arithmeticians who are now honouring their fellow-workers in Cambridge by their presence will I trust carry back to their own countries pleasant memories of their stay in this famous university town, and a strong sense of the vitality of Mathematical investigation in its English home. On the other side they will I am sure leave here lasting memories of good work done in promoting combined effort for the advancement of Mathematics. There was a time not long ago when British Mathematicians may have been thought too self-centred. If the judgment were ever correct, it is so no longer. We are alive to what is being done elsewhere, and now aim at cooperation. Our Academical methods are being modified. The severity of examination competition has been relaxed. We shall not give it up entirely: for we have to think not only of providing proficient mathematicians, but also of using mathematical training for the development of men, for the cultivation of exactness of thought, and the power of grasping situations and dealing with problems that arise in life. But the furtherance of mathematical thought is no longer secondary. Perhaps in times past we were too much occupied with exercises of skill and ingenuity, too much, some of us, engrossed with the struggle for absolute perfection in the mastery of limited curricula, too serenely contented with the older analysis. But our slowness in assimilating new ideas by adequate study of the writings of the great masters of the newer analysis is over. Here as elsewhere a younger mathematician now realises that, after grounding himself in common knowledge, he must choose his department of higher study, must acquaint himself by prolonged effort with original authorities, and then produce for himself. A Congress like this will, I feel sure, greatly assist those who are inculcating the sound doctrine.

There is one special reason why I value the opportunity which has been given me of saying these few words of welcome. Like yourselves, ladies and gentlemen from abroad, I am a guest at Cambridge. Not being a Cambridge man, I can let myself say that, while Cambridge has had the honour of inviting you and has the pleasure of entertaining you, the welcome is extended to you by all the mathematicians of the United Kingdom. Our other universities, and my own of Oxford in particular, have Mathematical Faculties of which they are proud, and they do not sink their individualities in their consciousness of the greatness of that of Cambridge. But we owe many times more to Cambridge in the domain of mathematics than Cambridge owes to us. We come here for inspiration, and not infrequently for men. The small band of Oxonians here now will go back grateful to their Cambridge hosts, and yours, for one more benefit conferred in the opportunity of meeting you.

I will delay you no longer, ladies and gentlemen, except to refer in one word to the unspeakable loss sustained in the death of Henri Poincaré, whom in this Section we think of as the prince of analysts, and whom we had hoped to see here to-day. A higher power has ordered otherwise.

Papers were read as follows:

DRACH, J.: Sur l'intégration logique des équations différentielles.

HARDY, G. H. and LITTLEWOOD, J. E.: Some problems of diophantine approximation.

Prof. E. Landau spoke on this paper.

MOORE, E. H.: On the fundamental functional operation of a general theory of linear integral equations.

Prof. J. Hadamard spoke on this paper.

BERNSTEIN, S.: Sur les recherches récentes relatives à la meilleure approximation des fonctions continues par des polynômes.

MACFARLANE, A.: On vector-analysis as generalized algebra.

JOURDAIN, P. E. B.: The values that certain analytic functions can take.

In the absence of the author this paper was presented by Mr G. H. Hardy.

KÜRSCHÁK, J.: Limesbildung und allgemeine Körpertheorie.

Prof. J. Hadamard spoke on this paper.

In the absence of the author, Prof. L. Silberstein, his paper "Some applications of Quaternions" was taken as read.

Section II. (*Geometry.*)

This Section met at 9.30 a.m. Dr H. F. Baker was in the Chair. Mr A. L. Dixon was elected Secretary, and Dr W. Blaschke and Dr E. Bompiani were elected Assistant Secretaries. Prof. F. Severi was elected Chairman for Saturday, August 24.

The Chairman addressed the Section as follows:

LADIES AND GENTLEMEN,

We are met here this morning to begin the work of the Section of Geometry. I believe it has been the custom that the President of the first meeting of a Section should be named by the Organizing Committee, and I have been asked to undertake the duty. I beg you to permit me to make a few introductory remarks.

It will be the duty of the meeting, before we break up, to choose a President for the meeting of to-morrow of this Section of Geometry. It will also be necessary to elect a secretary for the Section of Geometry, and one or more assistant secretaries; the secretaries will hold their offices until the end of the Congress. Before I sit down I will suggest to you names for your consideration; and my remarks are to some extent directed to giving reasons for the name I intend to suggest for President to-morrow.

We in England have known geometers. Here Cayley lived and worked; and all of you know the name of Cayley, as you know the name of Salmon, who lived in Dublin, but was in close relation with Cayley. To-day we have Sir Robert Ball, who has written a large book on the theory of linear complexes. I desire to express our regret that he is prevented by illness from being present. We have also many younger geometers, working in various directions. I am sure that these all join with me in saying how much honoured we in this country feel by the presence at the Congress of so many distinguished geometers from other lands. It is in order that we may express our gratification at their presence that I wish to say some words before the papers are communicated.

Of the recent progress of geometry in several directions you will hear from those who will present papers in this Section. There is, however, one side of geometry with which my own studies in the theory of algebraic functions have brought me into contact—the Theory of Algebraic Curves and Surfaces. I should like to say to you that I think extraordinary advances have recently been made in this regard, and to mention in a few words some of the more striking results. The history of the matter seems to me extremely interesting and, as has often happened before, the success has been achieved by the union of two streams of thought, which, though having a common origin, had for some time flowed apart.

It is a commonplace that the general theory of Higher Plane Curves, as we now understand it, would be impossible without the notion of the genus of a curve. The investigation by Abel of the number of independent integrals in terms of which his integral sums can be expressed may thus be held to be of paramount importance for the general theory. This was further emphasized by Riemann's consideration of the notion of birational transformation as a fundamental principle.

After this two streams of thought were to be seen.

First Clebsch remarked on the existence of an invariant for surfaces, analogous to the genus of a plane curve. This number he defined by a double integral; it was to be unaltered by birational transformation of the surface. Clebsch's idea was carried on and developed by Noether. But also Brill and Noether elaborated in a geometrical form the results for plane curves which had been obtained with transcendental considerations by Abel and Riemann. Then the geometers of Italy took up Noether's work with very remarkable genius, and carried it to a high pitch of perfection and clearness as a geometrical theory. In connexion therewith there arose the important fact, which does not occur in Noether's papers, that it is necessary to consider a surface as possessing two genera; and the names of Cayley and Zeuthen should be referred to at this stage.

But at this time another stream was running in France. Picard was developing the theory of Riemann integrals—single integrals, not double integrals—upon a surface. How long and laborious was the task may be judged from the fact that the publication of Picard's book occupied ten years—and may even then have seemed to many to be an artificial and unproductive imitation of the theory of algebraic integrals for a curve. In the light of subsequent events, Picard's book appears likely to remain a permanent landmark in the history of geometry.

For now the two streams, the purely geometrical in Italy, the transcendental in France, have united. The results appear to me at least to be of the greatest importance.

Will you allow me to refer to some of the individual results—though with the time at my disposal I must give them roughly and without defining the technical terms?

Castelnuovo has shewn that the deficiency of the characteristic series of a linear system of curves upon a surface cannot exceed the difference of the two genera of the surface. Enriques has completed this result by shewing that for an algebraic system of curves the characteristic series is complete. Upon this result, and upon Picard's theory of integrals of the second kind, Severi has constructed a proof that the number of Picard integrals of the first kind upon a surface is equal to the difference of the genera. The names of Humbert and of Castelnuovo also arise here. Picard's theory of integrals of the third kind has given rise in Severi's hands to the expression of any curve lying on a surface linearly in terms of a finite number of fundamental curves. Enriques shewed that the system of curves cut upon a plane by adjoint surfaces of order $n-3$, when n is the order of the fundamental surface, if not complete, has a deficiency not exceeding the difference of the genera of the surface. Severi has given a geometrical proof that this deficiency is equal to the difference of the genera, a result previously deduced by Picard, with transcendental considerations, from the assumption of the number of Picard integrals of the first kind. The whole theory originally arose, as has been said, with Clebsch's remark of a numerical invariant of birational transformation; conversely it is a matter of the profoundest geometrical interest to state in terms of invariants the sufficient conditions for the birational transformation of one surface into another. I might make reference to the Zeuthen-Segre invariant, which has been extended by Castelnuovo to the case of an algebraic system of curves. I will allow myself to mention, out of a vast number of results, one striking theorem. Enriques and Castelnuovo have shewn that a surface which possesses a system of curves for which what may be called the canonical number, $2\pi - 2 - n$, where π is the genus of a curve and n the number of intersections of two curves of the system, is negative, can be transformed birationally to a ruled surface. There is one other result I will refer to. On the analogy of the case of plane curves, and of surfaces in three dimensions, it appears very natural to conclude that if a rational relation, connecting, say, $m+1$ variables, can be resolved by substituting for the variables rational functions of m others, then these m others can be so chosen as to be rational functions of the $m+1$ original variables. Enriques has recently given a case, with $m=3$, for which this is not so.

These results, here stated so roughly, are, you see, of a very remarkable kind. They mean, I believe, that the theory of surfaces is beginning a vast new development. I have referred to them to emphasize the welcome which we in England wish to express to our distinguished foreign guests, whose presence here will, we believe, stimulate English geometry to a new activity. In particular we are very glad to have Professor Severi present with us in this room this morning. I will venture to propose to you that he be asked to act as President of this Section of Geometry to-morrow, Saturday.

I will also propose that for Secretaries you appoint Mr A. L. Dixon of Oxford, Herr Blaschke of Vienna, and Signor Bompiani of Rome.

[These proposals were adopted.]

Papers were read as follows:

BROUWER, L. E. J.: Sur la notion de "classe" de transformations d'une multiplicité.

MORLEY, F.: On the extension of a theorem of W. STAHL.

EISENHART, L. P.: Certain continuous deformations of surfaces applicable to the quadrics.

BOMPIANI, E.: Recent progress in projective differential geometry.

NEVILLE, E. H.: The general theory of moving axes.

BRÜCKNER, M.: Ueber Raumteilung durch 6 Ebenen und die Sechsflache.

STÉPHANOS, C.: Sur l'équivalent analytique du problème des principes de la géométrie.

A paper by Dr A. Martin "On rational right-angled triangles" was taken as read.

Section III (a). (*Mechanics, Physical Mathematics, Astronomy.*)

The subsection met at 9.30 a.m. Prof. H. Lamb was in the Chair. Mr F. J. M. Stratton was elected Secretary, and Prof. G. Andreoli and Dr L. Föppl were elected Assistant Secretaries. Prince B. Galitzin was elected Chairman for Saturday, August 24.

The Chairman addressed the subsection as follows:

You will not expect that I should say more than a few words before we enter on the business of the meeting. But as I have the privilege of being, like most of those present, a visitor and a guest, I may be allowed, or rather you will wish me, to give expression to a thought which is doubtless present to the minds of those who have followed the migrations of the Congress from place to place. The University of Cambridge has a long and glorious record in connection with our special subject of Mechanics. I need not repeat the words which fell from our President yesterday; but it must be a matter of peculiar interest and satisfaction to the members of our Section that they are at length assembled in a place so intimately associated with the names of illustrious leaders in our science.

One other point I would ask leave to touch upon. In spite of the process of subdivision which has been carried out, the field covered by our Section is still a very wide one. It has been said that there are two distinct classes of applied mathematicians; viz. those whose interest lies mainly in the purely mathematical aspect of the problems suggested by experience, and those to whom on the other hand analysis is only a means to an end, the interpretation and coordination of the phenomena of the world. May I suggest that there is at least one other and an intermediate class, of which the Cambridge school has furnished many examples, who find a kind of aesthetic interest in the reciprocal play of theory and experience, who delight to see the results of analysis verified in the flash of ripples over a pool, as well as in the stately evolutions of the planetary bodies, and who find a satisfaction, again, in the continual improvement and refinement of the analytical methods which physical problems have suggested and evoked? All these classes are represented in force here to-day; and we trust that by mutual intercourse, and by the discussions in this Section, this Congress may contribute something to the advancement of that Science of Mechanics, in its widest sense, which we all have at heart.

Papers were read as follows:

TURNER, H. H.: On double lines in periodograms.

Professors R. A. Sampson and Sir Joseph Larmor took part in the discussion.

MOULTON, F. R.: Relations among families of periodic orbits in the restricted problem of three bodies.

Professors T. Levi-Civita, Sir George Darwin and E. W. Brown took part in the discussion.

FÖPPL, L.: Stabile Anordnungen von Elektronen im Atom.

Professors Abraham, v. Kármán and Lamb took part in the discussion.

SMOLUCHOWSKI, M. S.: On the practical applicability of Stokes's law of resistance and the modifications of it required in certain cases.

Professors Lamb, Sampson and Webster and Mr Cunningham took part in the discussion.

LOVE, A. E. H.: The application of the method of W. Ritz to the theory of the tides.

Professors Turner, Sampson and Lamb took part in the discussion.

In the absence of Prof. A. O. Leuschner his paper on "The Laplacian orbit methods" was taken as read.

Section III (b). (*Economics, Actuarial Science, Statistics.*)

The subsection met at 9.30 a.m. Prof. F. Y. Edgeworth was in the Chair. Prof. A. L. Bowley was elected Secretary. Dr W. F. Sheppard was elected Chairman for Saturday, August 24.

The Chairman addressed the subsection as follows:

The first duty of the Chairman is, on behalf of English members of the Congress, to welcome visitors from foreign countries. There is a particular propriety in the expression of such a welcome by our subsection. For we are particularly benefited by the presence of visitors from distant countries. The advantage which is obtained by exchange of ideas with original minds educated in different ways is at a maximum when the subjects dealt with, like most of ours, are somewhat dialectical and speculative. Speaking of visitors, I cannot forget that many of the English members of the Congress, like myself, are visitors to Cambridge. Visitors of all nationalities will be unanimous in expressing their appreciation of Cambridge hospitality. Towards our subsection this hospitality has a peculiar delicacy. It is not merely that we are admitted to the privileges of Hall and Common-room. We have received an invitation even more grateful to some of us. There are some classes of us who have hitherto, so to speak, "sat below the salt" at the feast of reason. Economic Science is now in the position of that humble but deserving one who was invited to "go up higher." It is a proud day for Mathematical Economics, the day on which it enters the Congress of Mathematicians *pari passu* with other mathematical subjects. There is a propriety in this recognition being made at Cambridge, in which University Dr Marshall has shown that mathematical reasoning in Economics may be not only brilliant but fruitful. The cultivators of our second branch, Mathematical Statistics, have also reason to be satisfied with their reception. Cambridge has indeed lately shown her appreciation of their work in a very solid fashion by establishing a lectureship in Statistics. I trust that the designated lecturer will take part in our debates.

As for the Actuarial branch of our subsection the Science founded by Halley, like the sun, does not require recognition. Still even in the midst of rigid actuarial formulae there is an element of the calculus of probabilities. Thus even with respect to this department what may be described as the more human side of mathematics is now recognised. The Calculus of Probabilities has been described by purists of old as the opprobrium of mathematics. It will be the part of this subsection, with the valuable cooperation of foreign statisticians, to dispel this prejudice.

Papers were read as follows:

LEHFELDT, R. A.: Equilibrium and disturbance in the distribution of wealth.
The Chairman spoke on the subject.

AMOROSO, L.: I caratteri matematici della scienza economica.
The paper was read by the Chairman, and Professors G. C. Evans and Bowley and the Chairman took part in the discussion.

SHEPPARD, W. F.: Reduction of errors by means of negligible differences.
Dr Sòs and Dr H. L. Rietz took part in the discussion.

Section IV (a). (*Philosophy and History.*)

The subsection met at 9.30 a.m. The Hon. B. A. W. Russell was in the Chair. Prof. E. V. Huntington and Prof. M. Fréchet were elected Secretaries. Prof. A. Gutzmer was elected Chairman for Saturday, August 24.

The Chairman addressed the subsection as follows:

LADIES AND GENTLEMEN,

In opening the meetings of this Section, I desire to say one word of welcome to the distinguished visitors whom we are glad to see amongst us. The philosophy of mathematics has made extraordinarily rapid advances in recent times, and I am happy to see that many of those to whom these advances owe most are taking part in our meetings. Some unavoidable absences are to be deplored; among these, the illustrious name of Georg Cantor will occur to all. I had hoped, but in vain, that we might have been honoured by the presence of Frege, who, after many years of indomitable perseverance, is now beginning to receive the recognition which is his due. In common with other sections, we cannot but feel how great a loss we have sustained by the death of Henri Poincaré, whose comprehensive knowledge, trenchant wit, and almost miraculous lucidity gave to his writings on mathematical philosophy certain great qualities hardly to be found elsewhere. The work of the pioneers has been great, not only through its actual achievement, but through the promise of an exact method and a security of progress of which, I am convinced, the papers and discussions which we are to hear will afford renewed evidence.

I will now no longer stand between you and the proper business of the meeting.

Papers were read as follows:

ITELSON, G.: Bemerkungen über das Wesen der Mathematik.
Prof. S. Dickstein and the Chairman took part in the discussion of this paper.

In the absence of Prof. G. Vacca his paper "Sul valore della ideografia nella espressione del pensiero; differenze caratteristiche tra ideografia e linguaggio ordinario" was taken as read.

ZERMELO, E.: Ueber axiomatische und genetische Methoden bei der Grundlegung mathematischer Disciplinen.

BLUMBERG, H.: A set of postulates for arithmetic and algebra.

Prof. A. Padoa, Dr A. N. Whitehead and Dr A. A. Robb took part in the discussion of this paper.

HUNTINGTON, E. V.: A set of postulates for abstract geometry, expressed in terms of the simple relation of inclusion.

Prof. G. Peano, Prof. M. Fréchet, Prof. A. Padoa, the Chairman and Dr A. N. Whitehead took part in the discussion of this paper.

PADOA, A.: La valeur et les rôles du principe d'induction mathématique.

The Chairman and Dr G. Itelson took part in the discussion of this paper.

Section IV (b). (*Didactics.*)

The subsection met at 9.30 a.m. Mr C. Godfrey was in the Chair. Prof. G. A. Gibson was elected Secretary, and Messrs J. Franklin and E. A. Price were elected Assistant Secretaries. Prof. A. Gutzmer was elected Chairman for the joint meeting of Sections IV (*a*) and IV (*b*), and Prof. E. Czuber for the subsequent meeting.

The Chairman addressed the subsection as follows:

GENTLEMEN,

This is the opening meeting of Subsection IV (*b*), the subsection engaged in the discussion of didactical questions. The subsection will hold five meetings. Three of these meetings will be taken up by proceedings arising out of the activities of the International Commission on Mathematical Teaching—namely the meetings of this morning, of Monday afternoon, and of Tuesday morning. Two meetings remain, those of Saturday morning and Monday morning. On Saturday morning this subsection will join with the subsection for philosophical questions to discuss the very important question—how far it is expedient to introduce into school teaching the consideration of the fundamentals of mathematics.

After the words of welcome spoken by Sir George Darwin yesterday, no further words of mine should be needed to make our visitors from abroad feel that they are 'at home' among us. But it is fitting that I should avail myself of this occasion to offer to our visitors a very special welcome on behalf of the Mathematical teachers of this country. We Mathematical teachers welcome you, first because we are glad to have you with us and because we are glad to have the opportunity of making new friendships. We welcome you for another reason—because there is much that we can learn from you in the exercise of our craft. M. Bourlet has expressed the opinion that it is futile to transplant the teaching methods of one country into another, and to expect that these methods will always flourish in a new environment. I agree with his remarks; but I repeat that we have much to learn from you, and I assure you that many of us propose so to learn.

It is a matter of deep regret to all of us that our natural leader, Professor Klein, is unable to be present at this Congress. I will not anticipate the resolution of regret that Sir George Darwin will submit to you. For myself, I have done my best to acquaint myself with Professor Klein's views on Mathematical teaching, with which I am strongly in sympathy. If I may try to characterize in mathematical language the leading *motif* of the movement of which Professor Klein is the leader, it is this—

that mathematical teaching is a function of two variables: the one variable is the subject-matter of mathematics, the other variable is the boy or girl to whom the teaching is addressed; the neglect of this second variable is at the root of most of the errors that Professor Klein combats.

I learn from a letter addressed to Sir George Darwin that there is one matter which interests Professor Klein greatly and that he would have desired to call the attention of the subsection to it. It is the publication of the Encyclopaedic work *Die Kultur der Gegenwart* which is in course of compilation under his direction. This work will consist of a series of volumes in which every branch of culture is explained by experts in non-technical language, so that the articles will be within the reach of the reader of general education. This undertaking does not, it is true, appertain to education in the narrower sense of the word, but it does not seem too great an extension of the word to regard it as belonging to our special division. Dr Klein remarks in his letter that it was a matter of much difficulty to determine how so specialised a subject as mathematics could be made a suitable one for memoirs of the general character described, but he is glad to say that a good beginning has been made by Dr Zeuthen of Copenhagen in an article on the Mathematics of Classical Times and of the Middle Ages. Those who are interested in this will be able to see copies of the article in the Exhibition.

The meeting will now be asked to receive the report of the International Commission, and I hope that I shall be allowed to delegate my duties as Chairman to Professor D. E. Smith, to whose initiative the creation of the International Commission is due.

At the suggestion of the President of the Congress a telegram was despatched to Prof. F. Klein to express on behalf of the International Commission on the Teaching of Mathematics their regret at his absence and their best wishes for his recovery.

Prof. D. E. Smith was called to the Chair.

Prof. H. Fehr presented the printed report on the work of the International Commission.

The Reports from the participating countries were presented with a few explanatory remarks by the delegates mentioned below.

Germany	Prof. A. Gutzmer (Halle)
Austria	Prof. E. Czuber (Vienna)
Belgium	Principal E. Clevers (Ghent)
Denmark	Prof. H. Fehr
Spain	Prof. Toledo (Madrid)
United States	Prof. J. W. A. Young (Chicago)
France	Prof. C. Bourlet (Paris)
Greece	Prof. H. Fehr
Holland	Prof. J. Cardinaal (Delft)
Hungary	Prof. E. Beke (Budapest)

British Isles	Mr C. S. Jackson (Woolwich)
Italy	Prof. G. Castelnuovo (Rome)
Japan	Prof. R. Fujisawa (Tokyo)
Norway	Prof. M. Alfsen (Christiania)
Portugal	Prof. F. J. Teixeira (Oporto)
Roumania	Prof. G. Tzitzeica (Bucharest)
Russia	Prof. H. Fehr
Sweden	Prof. H. Fehr
Switzerland	Prof. H. Fehr (Geneva).

Also the following associated Countries:

Brazil	Prof. E. de B. R. Gabaglia (Rio de Janeiro)
Servia	Prof. M. Petrovitch (Belgrade).

These Reports are printed in the Publications of the Central Committee of the International Committee on Mathematical Instruction. 2nd Series, fasc. I. November 1912.

Saturday, August 24

Section I. (*Arithmetic, Algebra, Analysis.*)

Prof. E. Landau in the Chair:—Dr M. Riesz was elected an Additional Assistant Secretary. Prof. É. Borel was elected Chairman for Monday, August 26.

Papers were read as follows:

BATEMAN, H.: Some equations of mixed differences occurring in the theory of probability and the related expansions in series of Bessel's functions.

PETROVITCH, M.: Fonctions implicites oscillantes.

HADAMARD, J.: Sur la série de STIRLING.

SCHLESINGER, L.: Ueber eine Aufgabe von HERMITE aus der Theorie der Modulfunktionen.

FIELDS, J. C.: Direct derivation of the Complementary Theorem from elementary properties of rational functions.

FRIZELL, A. B.: Axioms of ordinal magnitudes.

PADOA, A.: Une question de maximum ou de minimum.

STERNECK, R. VON: Neue empirische Daten über die zahlentheoretische Funktion $\sigma(n)$.

The Chairman spoke on this paper.

Section II. (*Geometry.*)

Prof. F. Severi in the Chair:—Prof. F. Morley was elected Chairman for Monday, August 26.

Papers were read as follows:

ESSON, W.: The characters of plane curves.

DRACH, J.: Résumé de recherches géométriques.

GROSSMANN, M.: Die Zentralprojection in der absoluten Geometrie.

SCHOUTE, P. H.: On the characteristic numbers of the polytopes $e_1 e_2 \dots e_{n-2} e_{n-1} S(n+1)$ and $e_1 e_2 \dots e_{n-2} e_{n-1} M_n$ of space S_n.

KASNER, E.: Conformal geometry.

TZITZEICA, G.: Sur les surfaces isothermiques.

Section III (a). (*Mechanics, Physical Mathematics, Astronomy.*)

Prince Galitzin in the Chair:—Prof. T. Levi-Civita was elected Chairman for the meeting on Monday, August 26.

Papers were read as follows:

BENNETT, G. T.: The balancing of the four-crank engine.

Prof. F. Morley and Sir W. H. White took part in the discussion.

KÁRMÁN, TH. VON: Luftwiderstand und Wirbelbewegung.

Professors H. Lamb, C. Runge and M. S. Smoluchowski took part in the discussion.

BROMWICH, T. J. I'A.: Some theorems relating to the resistance of compound conductors.

Prof. H. M. Macdonald took part in the discussion.

EWALD, P. P.: Dispersion and double-refraction of electrons in rectangular grouping (crystals).

Dr T. H. Havelock and Prof. W. Peddie took part in the discussion.

MILLER, D. C.: The graphical recording of sound waves; effect of free periods of the recording apparatus.

Prof. A. G. Webster took part in the discussion.

TERRADAS, E.: On the motion of a chain.

Section III (b). (*Economics, Actuarial Science, Statistics.*)

Dr W. F. Sheppard in the Chair:—Dr J. F. Steffensen was elected Chairman for Monday, August 26.

Papers were read as follows:

QUIQUET, A.: Sur une méthode d'interpolation exposée par Henri Poincaré, et sur une application possible aux fonctions de survie d'ordre *n*.

Questions were asked by Dr Sòs and Dr Goldziher.

STEFFENSEN, J. F.: On the fitting of MAKEHAM'S curve to mortality observations.

Dr Sheppard, Dr Goldziher, Prof. Edgeworth, Prof. Bowley and Dr Sòs took part in the discussion.

In the absence of Mr J. H. Peek, his paper "Application of the Calculus of Probabilities in calculating the amount of securities &c. in practice of the Dutch State Insurance Office" was taken as read.

In the absence of Mr R. R. Brodie, his paper "Curves of certain functions involving compound interest and mortality" was summarized by Prof. Edgeworth.

Sections IV (a) **and IV** (b). JOINT MEETING. (*Philosophy, History, Didactics.*)

Prof. A. Gutzmer in the Chair.

Papers were read as follows:

WHITEHEAD, A. N.: The principles of mathematics in relation to elementary teaching.

SUPPANTSCHITSCH, R.: Le raisonnement logique dans l'enseignement mathématique secondaire et universitaire.

Prof. C. Bourlet and Prof. A. Padoa took part in the discussion of this paper.

Section IV (a). (*Philosophy and History.*)

Prof. A. Gutzmer in the Chair:—Prof. A. Padoa was elected Chairman for Monday, August 26.

In the absence of Prof. E. Burali-Forti, his paper "Sur les lois générales de l'algorithme des symboles de fonction et d'opération" was taken as read.

In the absence of Mr P. E. B. Jourdain, his paper "Isoid relations and theories of irrational number" was taken as read.

Papers were read as follows:

PEANO, G.: Proposizioni esistenziale.

Prof. E. Zermelo and Dr G. Itelson took part in the discussion of this paper.

ZERMELO, E.: Ueber eine Anwendung der Mengenlehre auf die Theorie des Schachspiels.

MUIRHEAD, R. F.: Superposition as a basis for geometry; its logic, and its relation to the doctrine of continuous quantity.

Prof. E. V. Huntington spoke on the subject of the paper.

Section IV (b). (*Didactics.*)

Professor E. Czuber in the Chair:—Professors C. Bourlet and J. W. A. Young were elected Chairmen for the meeting to be held in the morning of Monday, August 26.

Papers were read as follows:

HILL, M. J. M.: The teaching of the theory of proportion.

Mr C. Godfrey, Mr G. St L. Carson, Mr T. J. Garstang and Mr W. G. Bell took part in the discussion of this paper.

HATZIDAKIS, N.: Systematische Recreationsmathematik in den mittleren Schulen.

Prof. C. Bourlet spoke on the subject of the paper.

Monday, August 26

Section I. (*Arithmetic, Algebra, Analysis.*)

Prof. É. Borel in the Chair:—The Section agreed to adjourn at noon and to meet again at 3.30 p.m. Prof. E. H. Moore was elected Chairman for the afternoon meeting.

Papers were read as follows:

ELLIOTT, E. B.: Some uses in the theory of forms of the fundamental partial fraction identity.

KOCH, H. VON: On regular and irregular solutions of some infinite systems of linear equations.

Prof. Borel asked a question on this paper.

WHITTAKER, E. T.: On the functions associated with the elliptic cylinder in harmonic analysis.

SALTYKOW, N.: Sur l'intégration des équations partielles.

RÉMOUNDOS, G.: Sur les singularités des équations différentielles.

Afternoon meeting.

Prof. E. H. Moore in the Chair:—Prof. H. von Koch was elected Chairman for Tuesday, August 27.

Papers were read as follows:

HILL, M. J. M.: The continuation of the hypergeometric series.

CUNNINGHAM, A.: On MERSENNE'S numbers.

Prof. A. Gérardin spoke on the subject.

The papers by Prof. J. Drach and Prof. S. Bernstein which were read on Friday, August 23, were further discussed.

Section II. (*Geometry.*)

Prof. F. Morley in the Chair:—Prof. J. Drach was elected Chairman for Tuesday, August 27.

Papers were read as follows:

SOMMERVILLE, D. M. Y.: The pedal line of the triangle in non-Euclidean geometry.

Professors Coolidge and Schoute took part in the discussion.

HOSTINSKÝ, B.: Sur les Hessiennes successives d'une courbe du troisième degré.

FINSTERBUSCH, J.: Geometrische Maxima und Minima mit Anwendung auf die Optik.

Prof. Schoute took part in the discussion.

HUDSON, Miss H. P.: On binodes and nodal curves.

Mr Berry spoke on this paper.

STUDY, E.: The conformal representation of convex domains.

Section III (a). (*Mechanics, Physical Mathematics, Astronomy.*)

Prof. T. Levi-Civita in the Chair:—Prof. P. Stäckel was elected Chairman for Tuesday, August 27.

Papers were read as follows:

ABRAHAM, M.: Das Gravitationsfeld.

Prof. L. Silberstein took part in the discussion.

M^C^LAREN, S. B.: Aether, matter and gravity.

Professors Abraham and Webster took part in the discussion.

SOMIGLIANA, C.: Sopra un criterio di classificazione dei massimi e dei minimi delle funzioni di più variabili.

ESSON, W.: On a law of connection between two phenomena which influence each other.

In order to allow more time for discussion of other papers, the paper of Prof. L. Silberstein "Self-contained electromagnetic vibrations of a sphere as a possible model of the atomic store of latent energy" was, at the request of the author, taken as read.

The Section adjourned to the Cavendish Laboratory where the following paper was read:

THOMSON, SIR J. J.: Multiply-charged atoms.

Section III (b). (*Economics, Actuarial Science, Statistics.*)

Dr J. F. Steffensen in the Chair.

Papers were read as follows:

SHEPPARD, W. F.: The calculation of moments of an abrupt frequency distribution.

Prof. A. L. Bowley and Dr H. L. Rietz took part in the discussion.

EDGEWORTH, F. Y.: A method of representing statistics by analytic geometry.

Mr Stott and Dr Sheppard took part in the discussion.

ARANY, D.: Contribution to Laplace's theory of the generating function.

Dr Steffensen took part in the discussion.

GÉRARDIN, A.: Statistique des vingt séries parues du Répertoire Bibliographique des Sciences Mathématiques.

Section IV (a). (*Philosophy and History.*)

Prof. A. Padoa in the Chair:—Prof. F. Rudio was elected Chairman for Tuesday, August 27.

In the absence of Le Vicomte R. du Boberil, his paper entitled "Réflexions sur la loi de l'attraction" was briefly presented by Mr W. W. Rouse Ball.

In the absence of Prof. G. Loria, his paper "Intorno ai metodi usati dagli antichi greci per estrarre le radici quadrate" was taken as read.

In the absence of Mr P. E. B. Jourdain, his paper "Fourier's influence on the Conceptions of mathematics" was taken as read.

Papers were read as follows:

DYCK, W. VON. Ueber den Mechaniker und Ingenieur GEORG VON REICHENBACH.

ITELSON, G.: THOMAS SOLLY of Cambridge als Logistiker.

Mr W. W. Rouse Ball presented communications from Prof. H. G. Zeuthen and Sir Thomas L. Heath, announcing that an edition of the collected works of Paul Tannery is in preparation.

The discussion of Prof. G. Peano's paper read on Saturday, August 24, was continued by Prof. E. H. Moore, with a reply by Prof. Peano.

The discussion of Prof. A. Padoa's paper read on Friday, August 23, was continued by Hon. B. A. W. Russell, with a reply by Prof. Padoa.

Section IV (b). (*Didactics.*)

Prof. C. Bourlet in the Chair.

The following paper was read:

GÉRARDIN, A.: Sur quelques nouvelles machines algébriques.

Lt.-Col. Cunningham spoke on the subject of the paper.

In the absence of Prof. B. Lágos Campos, his paper "De l'importance de la Cosmographie et de son enseignement aux écoles secondaires" was taken as read.

In the absence of M. H. le Chatelier, his paper "L'enseignement des mathématiques à l'usage des ingénieurs" was taken as read.

Prof. J. W. A. Young in the Chair.

The following papers were read:

CARSON, G. ST L.: The place of deduction in elementary Mechanics.

Miss Punnett read Dr Nunn's paper on "The Calculus as a subject of School instruction."

The following speakers took part in the discussion on the paper:—Mr P. J. Harding, Mr G. St L. Carson, Dr P. Riebesell, Prof. Gibson, Prof. C. Bioche.

Afternoon session.

Sir J. J. Thomson in the Chair:—Prof. R. Fujisawa and Mr C. Godfrey were elected Chairmen for Tuesday, August 27.

The Report of Sub-Commission B of the International Commission on the Teaching of Mathematics was presented by Prof. C. Runge, Subject: The mathematical education of the physicist in the university.

The following speakers took part in the discussion on the report:—Prof. P. Stäckel, Prof. C. Bourlet, Prof. F. Enriques, Sir G. Greenhill, Prof. A. G. Webster, Prof. É. Borel, Sir J. Larmor, Prof. C. Bioche, Prof. A. E. H. Love, Prof. E. W. Hobson, Prof. G. A. Gibson, and Sir J. J. Thomson. Prof. Runge replied on the discussion.

Tuesday, August 27

Section I. (*Arithmetic, Algebra, Analysis.*)

Prof. H. von Koch in the Chair.

Papers were read as follows:

EVANS, G. C.: Some general types of functional equations.

BECKH-WIDMANSTETTER, H. A. VON: Eine neue Randwertaufgabe für das logarithmische Potential.

PEDDIE, W.: A mechanism for the solution of an equation of the *n*th degree.

WILKINSON, M. M. U.: Elliptic and allied functions; suggestions for reform in notation and didactical method.

ZERVOS, P.: Sur les équations aux dérivées partielles du premier ordre à quatre variables.

RABINOVITCH, G.: Eindeutigkeit der Zerlegung in Primzahlfaktoren in quadratischen Zahlkörpern.

In the absence of their authors the following papers were taken as read:

VOLTERRA, V.: Sopra equazioni di tipo integrale.

PEEK, J. H.: On an elementary method of deducing the characteristics of the partial differential equation of the second order.

MARTIN, A.: On powers of numbers whose sum is the same power of some number.

Section II. (*Geometry.*)

Prof. J. Drach in the Chair.

Papers were read as follows:

JANISZEWSKI, Z.: Ueber die Begriffe "Linie" und "Fläche."

KÖNIG, D.: Zur analysis situs der Doppelmannigfaltigkeiten und der projektiven Räume.

SINZOV, D.: On the theory of connexes.

HATZIDAKIS, N.: On pairs of Frenetian trihedra.

WEITZENBÖCK, R.: Ueber das sechs-Ebenenproblem im R_4.

Sections III (a) and III (b). JOINT MEETING. (*Astronomy and Statistics.*)

Prof. P. Stäckel in the Chair.

The following paper was read:

SAMPSON, R. A.: On the Law of Distribution of Errors.

Section III (a). (*Mechanics, Physical Mathematics, Astronomy.*)

Prof. P. Stäckel in the Chair.

The following papers were read:

LAMB, H.: On wave-trains due to a single impulse.

HAGEN, J. G.: How the Atwood machine proves the rotation of the earth even quantitatively.

BLASCHKE, W.: Reziproke Kräftepläne zu den Spannungen in einer biegsamen Haut.

BOULAD, F.: Extension de la notion des valeurs critiques aux équations à quatre variables d'ordre nomographique supérieur (Nomographie).

BRODETSKY, S.: Integrals in Dynamics and the problem of three bodies.

DENIZOT, A.: Contribution à la théorie de la chute des corps, en ayant égard à la rotation de la Terre.

BLUMENTHAL, O.: Ueber asymptotische Integration von Differentialgleichungen mit Anwendung auf die Berechnung von Spannungen in Kugelschalen.

In the absence of the author the following paper was taken as read:

DOUGALL, J.: The method of transitory and permanent modes of equilibrium in the theory of thin elastic bodies.

Section IV (a). (*Philosophy and History.*)

Prof. F. Rudio in the Chair.

The following papers were read:

RUDIO, F.: Mitteilungen über die Eulerausgabe.

HARDING, P. J.: The geometry of Thales.

GÉRARDIN, A.: Note historique sur la théorie des nombres.

ENESTRÖM, G.: Resolution relating to the publication of G. Valentin's general Bibliography of mathematics.

JOURDAIN, P. E. B.: The ideas of the "fonctions analytiques" in LAGRANGE'S early work.

VACCA, G.: On some points in the history of the infinitesimal calculus; relations between English and Italian mathematicians.

Section IV (b). (*Didactics.*)

Prof. R. Fujisawa and Mr C. Godfrey in the Chair.

The following papers were read:

GOLDZIHER, C.: Remarks on a bibliography on the teaching of mathematics.

SMITH, D. E.: Report of Sub-Commission A of the International Commission on the Teaching of Mathematics:—Intuition and experiment in mathematical teaching in secondary schools.

GODFREY, C.: Report on Methods of intuition and experiment in English secondary schools.

International recommendations as to future work of the Commission on the teaching of Mathematics were discussed.

PART II

LECTURES AND COMMUNICATIONS

IL SIGNIFICATO DELLA CRITICA DEI PRINCIPII NELLO SVILUPPO DELLE MATEMATICHE

DI FEDERIGO ENRIQUES.

SOMMARIO.—I: Introduzione.—II: Il continuo e i procedimenti infinitesimali nell'antichità.—III: La fondazione del calcolo infinitesimale.—IV: La critica dei concetti infinitesimali e i nuovi sviluppi sul calcolo delle variazioni.—V: Le funzioni arbitrarie e la moderna elaborazione del concetto del continuo.—VI: Lo sviluppo intensivo delle Matematiche: le equazioni e i numeri immaginarii.—VII: La teoria delle funzioni algebriche secondo Riemann e la critica dei principii della Geometria.—VIII: Nuovi sviluppi dell'algebra.—IX: Conclusioni; pragmatismo e naturalismo matematico.—X: Le Matematiche come istrumento e come modello della scienza.

I. *Introduzione.*

La critica dei principii è all'ordine del giorno fra i matematici contemporanei. L'analisi approfondita dei concetti di limite e di funzione, le ricerche che hanno come punto di partenza la teoria delle parallele e la geometria non-euclidea, quelle più recenti che si riattaccano alla fondazione della geometria proiettiva e all'*Analysis situs*, gli sviluppi sulle varietà a più dimensioni, sulle trasformazioni e sui loro gruppi; finalmente la teoria degl'insiemi e le speculazioni sull'infinito e l'infinitesimo attuale, cui si connettono le geometrie non-archimedee, hanno sollevato tanti problemi che toccano le profonde radici dell'edifizio matematico e attraggono, per diversi motivi, gli spiriti filosofici.

Nell'ambito di una scienza eminentemente conservatrice, che, da duemila anni, offre lo spettacolo di una continuità ininterrotta di costruzioni progredienti senza demolizione, le critiche innovatrici, di colore rivoluzionario, svegliano forse un interesse emotivo più forte che in qualsiasi altro campo dello scibile, ove le crisi si succedono visibilmente in modo periodico. A questo interesse emotivo si deve non soltanto la resistenza che le nuove idee incontrano presso il pubblico non preparato a comprenderle, ma più ancora la seduzione che esse esercitanò su tanti spiriti, pronti a passare, per naturale reazione psicologica, dalla meraviglia e dallo sbigottimento alla fede e all'entusiasmo, per il mondo nuovo che si dischiude ai loro occhi.

Di qui il fenomeno singolare, a cui abbiamo più volte assistito: la propagazione delle idee critiche attraverso piccole cerchie di lavoratori e d'interpreti, che, sviluppandone fino agli estremi le conseguenze logiche, compiono intorno a sè un vero apostolato, illusi forse che la nuova verità ad essi scoperta debba segnare un radicale rivolgimento del pensiero matematico, e instaurare una nuova êra nella sua storia.

Dobbiamo ringraziare la moltiplicità delle Chiese se la propaganda che si svolge fervidamente intorno a noi non ci toglie il senso della relatività e ci permette di conservare qualche fede anche nella vecchia matematica.

Ora le discussioni più vive suscitate dai nuovi campi di indagine e soprattutto i nuovi atteggiamenti dello spirito critico, pongono naturalmente un problema d'ordine filosofico e storico: quale sia il valore proprio della critica dei principii e quale posto le spetti nell'ordine dei progressi della nostra scienza. Tutte le questioni particolari di valutazione, per riguardo a diversi indirizzi di analisi e di ricerca, sembrano dominate da quel problema generale che, sia pure in diversi modi, ogni lavoratore, riflettendo sul proprio lavoro, è indotto a porre a sè stesso.

II. *Il continuo e i procedimenti infinitesimali nell'antichità.*

La storia ci offre a questo riguardo un primo insegnamento istruttivo: la critica dei principii non è affatto un fenomeno nuovo che caratterizzi la produzione matematica dei tempi nostri; all'opposto essa è parte essenziale dell'elaborazione dei concetti che in ogni tempo prepara o accompagna il progresso della scienza e la sua più estesa applicazione.

La perfezione universalmente ammirata della opera d'Euclide si rivela appunto allo storico come il frutto maturo di una lunga critica, che si svolge durante il periodo costruttivo della geometria razionale da Pitagora ad Eudosso. Tanta finezza e profondità d'idee si dispiega in quel movimento critico, che talune vedute non poterono essere comprese se non in tempi recentissimi, quando gli sviluppi della nostra stessa critica ci condussero a superare veramente, anche in questa direzione, il pensiero greco.

Allora in particolare ha cominciato a palesarsi nella sua propria luce il significato dei metodi e dei principii mercè cui i Greci stessi riuscirono a vincere i paradossi che sembra incontrare naturalmente chi riflette sull'infinito; giacchè le difficoltà che a questo riguardo travagliarono lungamente i matematici e i filosofi dell'antichità, sono le medesime che ebbe a sperimentare il Rinascimento nel periodo costruttivo dell' analisi infinitesimale, ed anche dopo la costituzione sua fino alla critica più recente.

La fondazione di una teoria della misura per opera della scuola pitagorica sollevò per la prima volta la questione del continuo geometrico. I pitagorici ponevano a fondamento di quella teoria un elemento indivisibile dello spazio, il punto dotato di estensione finita; intanto il rapporto incommensurabile della diagonale al lato del quadrato suscitava sui loro passi una insuperabile contraddizione.

Tuttavia P. Tannery ha mostrato che soltanto la critica degli Eleati riuscì a vincere definitivamente l'erroneo concetto dei pitagorici. Come generalmente avviene per riguardo a certe costruzioni astratte, il paradossale aspetto negativo degli argomenti di Zenone (Achille e la tartaruga!) dovette colpire l'immaginazione del gran pubblico, e questa impressione è passata nella corrente della tradizione letteraria ov' è tutt'ora dominante. Ma il valore positivo di codesta critica è di avere schiuso la via ad una esatta veduta del continuo e ad una teoria delle grandezze incommensurabili.

La quale viene fondata da Eudosso di Cnido mercè l'introduzione del postulato, comunemente citato col nome di Archimede, che serve di base alla trattazione generale delle proporzioni, esposta nel V libro d'Euclide.

La critica di Eudosso permetteva in pari tempo di dare una base rigorosa ai procedimenti infinitesimali impiegati dagli antichi per la misura delle aree e dei volumi. Infatti sul suo postulato, Eudosso stesso fondava il processo d'esaustione, e se ne serviva per dimostrare i resultati sui volumi della piramide e del cono, già trovati da Democrito.

Il pubblico matematico, assetato di rigore, plaudì all'opera di Eudosso, e la testimonianza di Archimede (nell'opera ritrovata da Heiberg) ci fa noto che, appunto in omaggio al rigore, non era lecito citare altro autore di tali dottrine fuori di colui che era riuscito a rimuovere ogni obbiezione stabilendo il resultato con logica impeccabile. Tanto varrebbe, nota argutamente lo Zeuthen, attribuire la scoperta del calcolo infinitesimale a Cauchy, che dette l'ultima risposta ai dubbi sollevati dall'uso degl'infinitesimi!

Il metodo d'esaustione fu dunque il termine consapevole a cui si arrestò lo sviluppo dei procedimenti infinitesimali presso i Greci, ma i concetti che ad esso soggiaciono e che erano volontariamente banditi per l'esigenza del rigore, vengono alla luce ad ogni passo nell'opera d'Archimede. E la lettera che questi scrisse ad Eratostene ci rivela come appunto i metodi dell'analisi infinitesimale, la riduzione del continuo ad una somma d'un numero finito di termini, gli servissero di guida nella scoperta, mentre l'esposizione dei risultati, condotta col processo d'esaustione, gli permetteva di soddisfare alle esigenze del pubblico scientifico.

III. *La fondazione del calcolo infinitesimale.*

Le idee d'Archimede sono riprese e approfondite nel Rinascimento da Galileo e da Keplero, ai quali si riattacca la prima sistemazione organica di esse che è la geometria degli indivisibili di Bonaventura Cavalieri.

Il grande geometra italiano pone il fecondo principio chè le superficie e i volumi si possono riguardare come somme di un numero infinito d'elementi indivisibili, che sono rispettivamente linee o superficie, e ne trae conseguenze assai generali ed importanti. Attaccato da Guldino nel 1640, mostra che il suo metodo si riduce all'esaustione degli antichi; esso non è che una *finzione,* utile per la rapida soluzione dei problemi, e non involge alcuna *ipotesi* contraria al concetto tradizionale del continuo.

Frattanto i metodi infinitesimali vengono alla luce secondo diversi aspetti; la difficoltà fondamentale di coglierne il vero significato filosofico determina appunto le divergenze di questi conati, che vanno di pari passo cogli acquisti positivi.

Torricelli e Roberval ottengono la tangente colla composizione dei movimenti; e questi decompono le superficie e i solidi in una moltiplicità indefinita di rettangoli o di prismi decrescenti secondo una certa legge.

I nuovi procedimenti investigati da Cavalieri, Fermat, Descartes, Roberval, ricevono un più alto sviluppo nell'*Aritmetica degli infiniti* di Wallis, da cui segue il primo esempio di rettificazione d'una curva, e poi per opera di Mercator che determina mediante una serie l'area compresa fra l'iperbole e i suoi asintoti; infine Barrow, il maestro di Newton che si riattacca specialmente a Galileo e a Torricelli, pone in piena luce il carattere inverso delle operazioni relative alla determinazione dell'area e della tangente d'una curva.

Occorreva ancora scoprire che quest'ultima è un'operazione *diretta* che può essere semplicemente compiuta. Si giunge così alla costituzione organica dell'analisi infinitesimale moderna, cioè al metodo delle flussioni e fluenti di Newton e al calcolo differenziale e integrale di Leibniz.

Una grande elaborazione concettuale, che profonda le sue radici nel più antico pensiero dei Greci, presiedette dunque all'acquisto che rimarrà come titolo d'onore dello spirito umano; il concetto dell'infinito e dell'infinitesimo potenziale che, liberato più tardi da ogni oscurità, diverrà la solida base del calcolo, rappresenta per così dire la sintesi dell'ipotesi pitagorica, ripresa come finzione da Cavalieri, e della critica negativa di Zenone convertita in un procedimento rigoroso di dimostrazione mercè il postulato di Eudosso. La sintesi diverrà logicamente perfetta quando Cauchy sarà riuscito a conciliare le divergenze di vedute separanti ancora per lungo tempo i newtoniani e i leibniziani, come diremo più avanti. Frattanto per apprezzare in tutta la sua vastità il lavoro d'analisi compiuto, occorre tener presente l'elaborazione dei principii della Meccanica che vi si accompagna. La stessa idea fondamentale che costituisce il passaggio dal finito all' infinito e dal discreto al continuo determina il disegno generale della scienza moderna, cioè il principio di un determinismo universale che scompone i processi naturali in una serie continua di *cause elementari*, e ritrova così nella forma delle equazioni differenziali gl'invarianti che costituiscono l'oggetto di una rappresentazione razionale della realtà. La metafisica razionalistica delle scuole di Descartes e di Leibniz appare, da questo punto di vista, come un ramo grandioso della stessa critica dei principii ond'è uscito il calcolo infinitesimale.

IV. *La critica dei concetti infinitesimali e i nuovi sviluppi sul calcolo delle variazioni.*

Ho detto che lungamente, dopo Newton e Leibniz, proseguì la critica tendente a dare una base logica all'analisi infinitesimale, la cui fecondità si mostrava ogni giorno più meravigliosa.

Il metodo di Newton, che introduce le flussioni come velocità, toccò primo ad un assetto rigoroso mercè la critica di Maclaurin e di D'Alembert; i quali, eliminando il concetto dinamico per svolgere puramente i principii analitici, riconobbero il suo fondamento logico nella teoria dei limiti. La fondazione newtoniana riesce così all'ordinario calcolo delle derivate. Tuttavia la rapidità consentita dall'uso dell'infinitesimo, a cui si conformano le notazioni leibniziane più generalmente adottate, faceva ancora desiderare una giustificazione piena della ipotesi fondamentale che s'incontra per questa via, cioè, del principio di Leibniz, che "si possono trascurare gl'infinitesimi di fronte alle quantità finite e gl'infinitesimi d'ordine superiore di fronte a quelle d'ordine inferiore."

La difficoltà di comprendere logicamente questo principio, che sembra rompere collo spirito d'esattezza delle Matematiche, travagliava ancora Lagrange, per cui impulso l'Accademia di Berlino nel 1784 bandiva un concorso volto ad ottenere una sistemazione rigorosa dell'analisi infinitesimale, e premiava poi la memoria di Lhuilier tendente in sostanza ad eliminare la caratteristica feconda del calcolo di Leibniz.

C'era da temere che gli scrupoli logici ancora una volta, come nel mondo antico, prendessero il disopra sulla fecondità dei metodi. Ma questi avevano ormai una base troppo larga nella coscienza matura dei progressi acquisiti. Nacque tosto un movimento di reazione che produsse le *Réflexions sur la métaphysique du calcul infinitésimal* di S. Carnot; qui è contenuta l'idea che il principio di Leibniz trovi la sua giustificazione, come regola practica, nella considerazione dell'arbitrarietà dell'infinitesimo; e più tardi Cauchy dimostrava che codesta considerazione, a prescindere da altre riflessioni meno chiare di Carnot, basta da sola a legittimare il Calcolo di Leibniz, e a stabilirne l'identità colla teoria dei limiti.

Se ora vuolsi convenientemente apprezzare il valore della nuova critica che riesce a fondare logicamente le basi del calcolo infinitesimale, occorre riflettere all'intima connessione che lega codesta critica ad altri positivi sviluppi nella mente dei nominati matematici, ricordando che, appunto dal tentativo di definire rigorosamente i principii del calcolo, Lagrange fu condotto all'introduzione delle funzioni analitiche che, per opera di Cauchy, Riemann, Weierstrass, ricevettero organica trattazione.

Ancora è da menzionare il legame della critica dei principii con un'altra grande scoperta di Lagrange, il calcolo delle variazioni, che è appunto un'estensione dei concetti infinitesimali allo studio delle funzioni dipendenti da altre funzioni e dei loro massimi e minimi.

In una serie di conferenze, tenute recentemente a Parigi, Vito Volterra* ha illustrato i progressi della teoria di queste funzioni di linee, mostrando come l'evoluzione delle idee fondamentali del calcolo infinitesimale si prosegua qui nella fiorente costruzione della dottrina delle equazioni integrali e integro-differenziali; onde questa mirabile estensione di concetti si palesa come un frutto maturo di quella stessa critica che riconoscemmo come fermento attivo dei progressi delle Matematiche durante due millennii di storia.

V. *Le funzioni arbitrarie e la moderna elaborazione del concetto del continuo.*

Ad allargare il campo della critica nel secolo decimonono interviene l'estensione del concetto di funzione, guadagnata per una parte attraverso il problema delle corde vibranti e gli studi di D'Alembert e di Fourier, d'altra parte mercè la considerazione qualitativa degli algoritmi d'integrazione.

La prima via suggerisce il concetto di Dirichlet della funzione arbitraria, la seconda, con Abel e Jacobi, porta alla introduzione effettiva di funzioni più generali nell'Analisi.

Appunto la considerazione di queste, ed in genere la veduta più larga degli algoritmi analitici come operazioni, porge un interesse positivo alle ricerche critiche concernenti la convergenza delle serie, la continuità e la derivabilità delle funzioni, ecc.; cioè a quelle speculazioni sui principii della teoria delle funzioni arbitrarie che furono spinte innanzi da Riemann, Weierstrass, Dini, ecc.

* Cfr. *Revue du Mois*, 10 mars 1912.

Il vedere tra i fondatori di questa critica quegli stessi matematici che hanno constituito l'organismo della teoria delle funzioni analitiche, ci fa comprendere il nesso profondo fra due campi di studio che taluno ama talora contrapporre come due indirizzi delle Matematiche.

In realtà se le funzioni analitiche sono nate nella mente di Lagrange per rispondere ai dubbi intorno ai fondamenti del calcolo infinitesimale, il loro progresso appare sempre legato a preoccupazioni critiche della medesima specie; basti rammentare che il più bel resultato della teoria delle funzioni analitiche è la determinazione di esse per mezzo dei loro punti singolari nel piano, e che a base dei teoremi d'esistenza che le concernono si trova il principio di Riemann-Dirichlet.

Ora per approfondire in generale le questioni d'esistenza in rapporto al concetto esteso delle funzioni, delle serie, ecc., occorre una nuova analisi del continuo che conduce ad un complemento essenziale della critica antica. Alludo al postulato della continuità e al nuovo assetto della dottrina dei numeri irrazionali considerati nella loro integrità. Questa dottrina si confonde infatti per Weierstrass colla teoria generale della convergenza delle serie, per Cantor e per Dedekind colla determinazione delle condizioni d'esistenza dei limiti. L'intimo rapporto ond'essa è legata al concetto dell'arbitrarietà delle funzioni si palesa negli sviluppi critici di Cantor sugli insiemi e sulla loro potenza.

Per questi sviluppi e per le speculazioni più recenti (di Veronese, Hilbert, ecc.) sul considetto continuo non-archimedeo, il pensiero moderno sembra avere svolto fino alle sue estreme conseguenze l'analisi iniziata quasi duemilacinquecent'anni or sono coll'ipotesi pitagorica. Certo il concetto del continuo rimane per noi quello stesso che Eudosso ed Archimede ponevano a base delle loro costruzioni, ma arricchito di un nuovo principio esistenziale che è in rapporto all'estensione dei concetti delle Matematiche; e il significato di questo principio viene messo in tutta la sua luce grazie ai suddetti sviluppi non-archimedei, divenuti parte integrante dell'esplorazione critica del continuo.

VI. *Lo sviluppo intensivo delle Matematiche: le equazioni e i numeri immaginarii.*

Le considerazioni precedenti mirano sopratutto al progresso *estensivo* delle Matematiche, mostrando in ordine ad esso l'ufficio della critica dei principii. Le idee suggerite primitivamente da un'intuizione ristretta si affinano coll'analisi delle condizioni di validità e diventano atte a fecondare un campo di problemi sempre più vasto. A questa estensione che è un aspetto del progresso scientifico presiede appunto il pensiero critico, inteso come strumento di sapere positivo.

Ma lo sviluppo delle Matematiche non avviene soltanto nel senso estensivo, bensì anche secondo una direzione che può dirsi *intensiva.*

Allargare la posizione dei problemi riuscendo a sottomettere all'analisi un campo ognor più vasto di rapporti reali, non dispensa dall'approfondire i problemi antichi, proseguendone una risoluzione effettiva con mezzi determinati.

Così la considerazione più generale dei numeri irrazionali lascia posto ad una teoria dei numeri razionali od interi o di particolari specie d'irrazionali; e mentre le equazioni algebriche trovano la loro naturale estensione nelle equazioni differenziali, e queste nelle equazioni a derivate parziali, nelle equazioni integrali ed integro-differenziali, ciascuna di tali classi di problemi dà luogo ad uno sviluppo intensivo proprio.

La veduta estesa porge a questo stesso sviluppo un criterio fondamentale, cioè il principio di *relatività*, per cui la risoluzione cercata viene messa in rapporto con dati mezzi, e si converte in una classificazione gerarchica dei varii tipi di problemi secondo un ordine di difficoltà crescente.

Ora la Matematica, considerata da tale punto di vista, tocca il suo punto culminante nello sviluppo dell'Algebra, intesa—in modo ampio—come teoria generale dei problemi *qualitativi* che sorgono in rapporto al gruppo delle operazioni razionali (equazioni e funzioni algebriche, funzioni ellittiche e abeliane, equazioni algebrico-differenziali, ecc.), o come primo ramo di una teoria qualitativa delle funzioni.

L'ufficio della critica dei principii che abbiamo riconosciuto nella estensione dei concetti e dei problemi, si discopre non meno essenziale per riguardo a questo indirizzo delle Matematiche, dove—a primo aspetto—potrebbe apparire meno evidente.

Riportiamoci col pensiero alle origini della teoria delle equazioni algebriche. Le equazioni di 2° grado trovansi risolute in veste geometrica nel libro II d'Euclide e la loro risoluzione si connette alla scoperta degli incommensurabili, cui si è già innanzi accennato. Attraverso gli Arabi quella dottrina assunse la forma propriamente algebrica, che mette in evidenza il problema generale delle equazioni di grado superiore.

E da questa rappresentazione sorgono i numeri negativi (già incontrati dall'Indiano Bhâskara nel 1114); i quali saranno ripresi dai matematici dei secoli XV e XVI, Pacioli, Cardano, Stiefel; e, dopo Harriot e Descartes, adottati come numeri ordinali o ascisse d'una retta.

L'uso dei simboli non era ancora familiare ai matematici italiani del secolo decimosesto, i quali—volgendosi alla trattazione delle equazioni cubiche—vedono in esse il soggiacente problema geometrico. Scipione dal Ferro e Niccolò Tartaglia scoprono le regole per la risoluzione delle equazioni stesse, distinte allora in tre classi*; e quelle regole sono riprese e svolte da Girolamo Cardano e Raffaele Bombelli.

Orbene il caso irriducibile delle equazioni di terzo grado apre la via alla considerazione dell'immaginario, cioè al problema critico del valore e del significato che si può conferire alla radice quadrata d'un numero negativo ed al suo uso nei calcoli. Il progresso ulteriore dell'Algebra esige che piena luce sia fatta su questo delicato concetto; la profonda elaborazione che primo ne ha data il Bombelli resta pressochè incompresa fino a Leibniz e a Wallis, e—ripigliata da questi matematici—riceve uno sviluppo pieno coll'interpretazione trigonometrica del De Moivre; tuttavia la critica si affatica ancora a cercare un significato concreto dei numeri complessi, e vi riesce colla nota rappresentazione geometrica di Wessel, Argand, Gauss.

Allora soltanto, sulla base della critica compiuta, si asside la dimostrazione del teorema fondamentale che un'equazione di grado n ha n radici; teorema ricercato dai

* $x^3+px=9$, $x^3=px+9$, $x^3+9=px$. Cfr. D. Gigli, *Dei numeri complessi...*, in F. Enriques, *Questioni riguardanti le Matematiche elementari*, Zanichelli, Bologna, 1912.

matematici del secolo decimottavo e segnatamente dal D'Alembert (1746), rigorosamente stabilito da Gauss nel 1789.

Se ora si riflette al posto che l'immaginario riceve nella teoria delle funzioni, si è condotti a comprendere in tutta la sua larghezza il valore della critica iniziata colle speculazioni del Bombelli, e suscitata da un problema determinato come quello delle equazioni cubiche.

VII. *La teoria delle funzioni algebriche secondo Riemann e la critica dei principii della Geometria.*

Procediamo a considerare la storia degli sviluppi dell'Algebra, e riscontriamo che ogni passo innanzi si lega ugualmente con una critica toccante i concetti fondamentali della scienza matematica.

Mentre le equazioni del 4° grado si riconducono a quelle del 3°, lo studio delle equazioni di grado superiore al 4° conduce, con Ruffini et Abel, alla dimostrazione della impossibilità di risolvere per radicali l'equazione del 5° grado e quindi alla dottrina generale della risolubilità algebrica secondo Galois. Ebbene questa dottrina che—estesa poi in vari sensi—ha fecondato tutti i rami delle Matematiche, e con Lie è riuscita a porgere la base di una classificazione razionale delle equazioni differenziali, è finalmente una critica di taluni elementari concetti: ordine, operazione o corrispondenza, gruppo di operazioni.

Il posto di questi concetti per riguardo ai principii delle Matematiche, ed in ispecie della Geometria, appare chiaramente nell'opera che—per diversi riguardi—può riconoscersi come centrale nello sviluppo delle Matematiche del secolo decimonono: dico l'opera di Bernardo Riemann.

La straordinaria attività creatrice di questo pensatore s'illumina di una più viva luce a chi investighi il legame profondo fra le ricerche onde uscì la dottrina generale delle funzioni algebriche e dei loro integrali, la sua critica dei concetti del Calcolo e quella—di carattere più largamente filosofico—che tocca i principii della Geometria e pone le basi di essa nell'*Analysis situs.* Appunto il rapporto fra le proprietà invarianti per trasformazioni birazionali delle funzioni algebriche e la connessione delle corrispondenti superficie riemanniane costituisce la scoperta dominante in quel campo di studii.

Inoltre si deve alla sintesi di Riemann, che le pure speculazioni dei geometri non-euclidei vengano riattaccate all'organismo della realtà matematica, formante oggetto della storia (forme differenziali quadratiche). Quindi innanzi il rapporto fra la critica dei principii della Geometria e lo sviluppo delle dottrine matematiche apparirà più chiaramente sotto molteplici aspetti. Il nodo di questo rapporto sta nella Geometria proiettiva che da Poncelet a Möbius, a Steiner, a Staudt, si svolge, non solo come dottrina delle proiezioni e come metodo di riduzione, ma anche come critica dei concetti e dei rapporti spaziali, riuscendo ad una trattazione qualitativa di questi indipendente dalle nozioni metriche. Il valore di tale sviluppo per riguardo ai problemi filosofici concernenti i principii risulta chiaro dall'opera di Beltrami, Schläfli, Cayley, Klein, ecc. Quanto alla sua importanza in ordine al progresso costruttivo delle Matematiche, basti additare la nuova forma data ai problemi concernenti le funzioni algebriche, e i resultati che ne conseguono.

Infatti i continuatori di questo indirizzo riemanniano (e segnatamente Clebsch e Noether) hanno rinnovato la dottrina mercè una considerazione più astratta della Geometria proiettiva delle curve e delle superficie algebriche, dalla quale è scaturita finalmente una posizione più generale degli stessi problemi dell'Algebra. Così per es. un sistema d'equazioni soddisfatto da un numero finito di soluzioni, viene ad essere riguardato come avente un grado invariante per un cambiamento continuo dei parametri, in forza della convenzione geometrica che estende lo spazio coi punti impropri, e porta quindi ad annoverare fra le soluzioni effettive anche le soluzioni asintotiche ed in particolare anche ad eliminare taluni casi d'incompatibilità.

Ora è essenziale avvertire che non soltanto la Geometria proiettiva come edifizio costruito, ma appunto le indagini sulla sua fondazione debbono essere ritenute come elemento essenziale della nuova teoria delle funzioni algebriche. Infatti uno dei concetti metodologici principali appare qui la considerazione astratta della Geometria proiettiva come sistema ipotetico-deduttivo caratterizzato dai postulati, cioè il fecondo principio che, generalizzando la dualità scoperta da Gergonne, permette di considerare certi sistemi di enti o di funzioni come diverse interpretazioni di quella Geometria.

Trattisi p. es. di ricercare i gruppi finiti di sostituzioni lineari sopra una variabile. Sotto l'aspetto geometrico si ha dunque a fare coi gruppi di omografie sulla retta. Ma la totalità ∞^3 delle omografie sulla retta forma un sistema lineare che può essere ritenuto in senso astratto come uno spazio proiettivo ordinario; basta perciò chiamare "punti" le omografie stesse e "rette" i "fasci d'omografie." La rappresentazione che così risulta, studiata da Stephanos, mette in evidenza una quadrica immagine delle omografie degeneri e un punto che corrisponde all'identità. Il nostro problema si riduce a determinare i gruppi d'omografie che lasciano ferma una quadrica e un punto non appartenente ad essa. Con una trasformazione lineare immaginaria la quadrica si muterà in una sfera e il punto nel suo centro. I gruppi cercati corrisponderanno ai gruppi di rotazioni della sfera, cioè (com'è noto) ai gruppi dei poliedri regolari!

Ora il principio generale della Geometria proiettiva astratta assume tutta la sua estensione grazie al concetto degli spazi a più dimensioni, e così diventa possibile di trattare come "spazi," cioè di tradurre nei termini della Geometria proiettiva generale, le serie g_n^r di gruppi di punti sopra una curva, i sistemi lineari di curve sopra una superficie, ecc.

VIII. *I nuovi sviluppi dell'Algebra moderna.*

Ho accennato all'Algebra in veste geometrica, che contempla le equazioni e i sistemi d'equazioni a più incognite, di fronte alle trasformazioni birazionali; e credo di non cedere ad una predilezione individuale, affermando che essa si trova oggi al posto più avanzato sulla linea del progresso intensivo delle Matematiche, come continuatrice legittima della grande tradizione dei problemi algebrici, di cui ho porto innanzi una rapida veduta. Tanto più che quasi tutti i rami della Matematica qualitativa, dalle funzioni abeliane alle funzioni automorfe e alle equazioni algebrico-differenziali (secondo gli sviluppi di Poincaré e Painlevé) vi si riattaccano intimamente.

Si consideri dunque un teorema pertinente a quella teoria, p. es. il teorema che l'annullamento del genere porge la condizione per la risoluzione d'un'equazione

$$f(xy) = 0,$$

mediante funzioni razionali d'un parametro:

$$x = \phi(t), \quad y = \psi(t)$$

(Clebsch), e che la risoluzione effettiva si ottiene con operazioni razionali e tutt'al più coll'estrazione di una radice quadrata a partire dai coefficienti di f (Noether). Un siffatto enunciato porge una risposta pienamente determinata a una questione del pari determinata; nulla sembra più remoto dal campo proprio della critica dei principii; eppure la storia di quell'acquisto suppone—come abbiamo accennato—una lunga elaborazione di concetti: dal numero irrazionale all'immaginario, dalle permutazioni alle superficie più volte connesse, dalla Geometria proiettiva al principio di dualità generalizzato quale scaturisce dalla contemplazione logica dell'edifizio geometrico!

Se ora taluno stimasse che l'elaborazione critica dei concetti s'incontri in quelle dottrine algebriche soltanto per fondare una base, su cui la costruzione proseguirà poi senz'altro rapporto colla critica stessa, ch'ei rifletta su altri ulteriori sviluppi; e così sarà tratto a riconoscere come la teoria invariantiva della superficie esiga un'analisi delicata, tendente allo stesso scopo per cui s'introducono nella Geometria proiettiva i punti impropri, cioè a rimuovere casi eccezionali d'invarianza (convenzioni sui buchi delle curve o sui punti-base dei sistemi lineari, sulla riducibilità o irriducibilità, ecc.); e d'altra parte vedrà che le stesse idee dominanti l'estensione del campo dei numeri, cioè l'introduzione dei numeri fratti, negativi o immaginarii, trovano una feconda applicazione nel concetto più largo delle funzioni ϕ di grado $n-4$, *aggiunte* ad un'equazione di grado n

$$f(xyz) = 0,$$

le quali sono invarianti di f rispetto alle trasformazioni birazionali (Clebsch-Noether). Infatti l'estensione della teoria degl'invarianti di Clebsch-Noether si è fatta appunto nel senso di considerare per così dire le ϕ *virtuali* che—nel caso del genere $p=0$—porgono talora funzioni ϕ^n effettivamente esistenti, le quali (almeno per $n = 2, 3, 4, 6$) costituiscono dei *nuovi invarianti* di f. E mi sia lecito ricordare che mercè questi invarianti si è potuta assegnare in forma semplice e determinata la condizione per la risoluzione razionale di $f(xyz) = 0$, la condizione per la trasformazione di $f(xyz) = 0$ in un'equazione fra due variabili $F(xy) = 0$, la condizione perchè $f(xyz) = 0$ possegga un gruppo continuo di trasformazioni birazionali in se stessa, ecc.

Infine lo studio delle trascendenti connesse con un campo algebrico a due dimensioni, ha messo Picard e Poincaré in faccia alle difficoltà concernenti la connessione delle varietà a quattro dimensioni; e deve ritenersi che talune difficoltà non ancora superate verranno sciolte il giorno che la critica dei principii della Geometria avrà approfondito il più alto problema che ancora appaia insoluto nel suo campo, dando una base geometrica pura all'edifizio dell'*Analysis situs*.

IX. *Conclusioni: pragmatismo e naturalismo matematico.*

La tesi enunciata in principio mi sembra ormai sufficientemente dimostrata: la critica dei principii fa parte integrante della storia degli sviluppi delle Matematiche, così dal punto di vista estensivo come dal punto di vista intensivo; essa è il processo di elaborazione e di definizione dei concetti che tende ad estendere i dati dell'intuizione a campi sempre più vasti e così ad allargare la posizione dei problemi e a preparare strumenti più penetranti per recare risposta determinata a più profonde questioni.

Ora questa veduta storica suppone in qualche modo una legge di sviluppo delle Matematiche, rispetto a cui assegna—per così dire—un fine naturale alla critica dei principii. Ed intanto il progresso di questa critica stessa sembra all'opposto far scaturire l'illimitata *arbitrarietà della costruzione matematica.*

Già vedemmo che le funzioni, assunte altra volta come dato di una realtà naturale (la potenza, la radice, l'esponenziale, il logaritmo, il seno, ecc.), cedono il posto alle funzioni generali nel senso di Dirichlet, che sono corrispondenze arbitrarie. Le proprietà fondamentali dei numeri non appaiono più l'espressione di assiomi necessarii, ma—sopratutto per l'analisi di Cantor e di Peano—divengono condizioni arbitrarie con cui si definiscono certi insiemi ordinati; p. es. il principio d'induzione matematica perde il suo valore di canone logico per rappresentare solo una condizione costruttiva della serie ben ordinata di oggetti cui corrispondono i numeri interi, tantochè la negazione del principio dà luogo—in quella serie—all'esistenza di punti-limiti cui corrispondono numeri ordinali transfiniti.

La Geometria, che aveva visto allargare il campo delle sue possibilità colla trattazione delle ipotesi non euclidee e cogli spazî a più dimensioni, diventa ormai suscettibile di un'estensione illimitata, sicchè non vi è più gruppo di oggetti dotato di proprietà qualsiasi che non possa rivendicare il nome di "spazio."

La scuola logica non ha mancato di lumeggiare il significato delle rivoluzione compiuta. Si sono detronizzati gli assiomi; rotto l'incantesimo della loro investitura per diritto divino, cioè il loro fondamento in una evidenza o necessità naturale dello spirito umano, essi sono divenuti dei semplici postulati, non più principi o membri d'un'aristocrazia gentilizia, ma funzionari elettivi di una repubblica democratica, che possono essere revocati o sostituiti per motivi di economia o di semplice rinnovamento.

Un Aristofane potrebbe anche trovare che l'arbitrio illimitato di scelta rischia di convertire questa democrazia in una vera demagogia; che le funzioni *disoneste* prendono troppo spesso il posto delle semplici ma oneste funzioni soddisfacenti ai teoremi del Calcolo infinitesimale, che talune costruzioni di Geometrie bizzarre (giustificate dapprima come mezzo per investigare certi rapporti di subordinazione) affermano la libertà dell'idea ispiratrice al modo stesso che le forme di governo succedentisi nel Principato di Monaco, auspice Rabagas.

Eppure anche le esagerazioni un po'barocche a cui conduce l'odierna critica dei principii sono servite a diffondere una giusta idea del valore della Logica e per contrapposto lasciano indovinare il valore che altri elementi non logici assumono nella conoscenza matematica. L'importanza della veduta della Logica così messa in luce risulta già dal fatto che quell'indirizzo critico ha suscitato un vasto movimento filosofico, propagatosi—ai nostri giorni—sotto il nome di *pragmatismo.* Infatti di padre di quel pragmatismo filosofico che è riuscito finalmente ad una reazione antiscientifica, è proprio il pragmatismo dei logici matematici che, armati della critica dei principii, rivendicano il carattere di definizioni dei postulati e ne desumono l'arbitrarietà della costruzione matematica, contro una concezione che potrebbe chiamarsi *naturalistica,* secondo la quale gli enti delle Matematiche esistono fuori di noi, al pari delle specie viventi delle scienze naturali, come oggetto di scoperta e di osservazione.

Ebbene se il pragmatismo logico-matematico riesce a combattere vittoriosamente il naturalismo ed il realismo ingenuo che vi soggiace, quel pragmatismo a sua volta viene vinto dalla storia. La storia degli sviluppi delle Matematiche ci ha mostrato appunto il lavoro della critica logica in una elaborazione secolare di concetti.

Pertanto alle conseguenze che si vorrebbero trarre dalla veduta dei postulati come definizioni implicite, la storia oppone che le definizioni stesse degli enti matematici non sono arbitrarie poichè appariscono come frutto di un lungo processo d'acquisto e di uno sforzo assiduo rivelante alcuni motivi generali della ricerca.

Vi è una tradizione di problemi e vi è un ordine che presiede ai progressi estensivi ed intensivi della scienza; perciò solo vi è una materia propria delle Matematiche, che le definizioni mirano a rispecchiare; onde l'arbitrio del definitore non sembra diverso da quello dell'architetto che dispone le pietre di un edifizio elevantesi secondo un armonico disegno.

Opera d'Architettura è infatti la scienza matematica; non realtà che si offra allo sguardo di un osservatore esterno come qualcosa di dato, ma processo che si fa dallo spirito umano, e pur rivela la realtà stessa dello spirito creatore.

Così dunque l'atto di volontà che il matematico rivendica ognora più libero nella posizione dei problemi, o nella definizione dei concetti o nell'assunzione delle ipotesi, non può mai significare arbitrio, ma solo facoltà di avvicinarsi da più lati, per approssimazioni successive, a non so che ideale implicito nel pensiero umano, cioè ad un ordine e ad un'armonia che ne riflette le intime leggi.

Se questa è la conclusione che emerge da una veduta storica della scienza e della critica, il pragmatismo logico matematico, lungi dall'aprire un'èra di costruzioni fantastiche moltiplicantisi all'infinito quasi per giuoco o per bizzarria, avrà dato alla ricerca una coscienza più elevata dei suoi scopi; e d'altra parte, purificando la Logica, avrà dimostrato l'insufficienza di essa e la necessità di approfondire gli altri elementi psicologici che conferiscono significato e valore alla costruzione matematica.

X. *Le Matematiche come istrumento e come modello della scienza.*

Alla nostra veduta idealistica, che sembra nascere da una considerazione esclusiva delle Matematiche pure, altri potrebbe contrapporre una veduta apparentemente più larga in cui le Matematiche stesse sieno ritenute, non più come un oggetto di per sè stante, ma come strumento della scienza naturale. Senonchè questo concetto (che a taluno può essere suggerito, per reazione, dalle esagerazioni del pragmatismo logico) porterebbe ad impoverire singolarmente il campo dell'attività matematica. Esso ci ricondurrebbe alla veduta di Fourier che rimproverava Abel e Jacobi di studiare le equazioni e funzioni algebriche anzichè volgersi di preferenza al movimento del calore; rimprovero cui bene rispose lo Jacobi che scopo della scienza è unicamente l'onore dello spirito umano e che a questo titolo una questione di numeri non vale meno che una questione relativa all'ordine dell'universo.

Se è necessario portare argomenti in appoggio alla veduta di Jacobi, basta—io credo—riflettere come appunto le difficoltà della teoria dei numeri abbiano attratto in ogni tempo gl'intelletti più alti, compresi i fondatori del sistema del mondo e della Meccanica.

Ma la ristrettezza dell'anzidetta considerazione delle Matematiche si rivela meglio sul suo proprio terreno, mercè una veduta approfondita del posto che alle Matematiche spetta nell'ordine dello scibile.

Se la larghezza delle applicazioni del calcolo ha potuto avvalorare l'idea che le Matematiche sieno soltanto uno *strumento* della cognizione fisica, in ogni tempo i più alti pensatori riconobbero in esse un *modello* della scienza.

Questo concetto porge una più giusta veduta del rapporto che intercede fra il progresso dello spirito matematico e il progresso scientifico in generale. Bene avverte il Poincaré che appunto lo spirito matematico, indipendente dalla potenza degli algoritmi, è attivo nelle feconda intuizione delle analogie di Faraday, che guida Maxwell alle sue memorabili scoperte. E nello stesso senso può dirsi p. es. che la Termodinamica è tutta intera un'opera matematica, sebbene si svolga in gran parte indipendentemente, non dai concetti, ma dai resultati del calcolo.

L'ispirazione matematica si rivela del pari più largamente in altri rami di scienza, che pure non toccano propriamente ad una fase di sviluppo matematico.

Per tal modo si può dire che il progresso tenda a realizzare l'ideale scientifico di Platone, di Descartes e di Leibniz, che pone *le Matematiche modello della scienza.*

Questa veduta, allargante il valore delle Matematiche nell'ordine universale dello scibile, restituisce anche tutto il suo valore al libero sviluppo della teoria pura, rivendicato dal secolo decimonono.

Ora al lume della concezione idealistica delle Matematiche, quella veduta acquista un significato nuovo rispetto alla posizione *realistica* dei nominati filosofi.

Quando Platone costruiva il mondo delle Idee ad immagine della classificazione delle forme geometriche; quando Galileo, Descartes, Leibniz, foggiavano una nuova realtà dinamica, i cui invarianti sono rapporti di successione, cioè leggi naturali; questi filosofi proiettavano nel mondo esterno il processo interiore del loro spirito, ed in quel mondo credevano di riconoscere le cause elementari, come dato semplice della realtà stessa.

Oggi la critica gnoseologica, connessa all'investigazione dei principii di cui si è innanzi discorso, ci avverte che il modello matematico della scienza ha un significato diverso; non si tratta di scoprire la profonda struttura metafisica del reale, ma di riconoscere le forme dell'attività spirituale che atteggia la realtà sensibile nella costruzione scientifica, secondo le intime leggi dello spirito umano.

Così le Matematiche, che per Platone, Descartes, Leibniz offrivano il fondamento di una filosofia della natura, elevantesi ad una grandiosa metafisica razionalistica, oggi, mercè il possente risveglio della critica contemporanea, suscitano una nuova filosofia dello spirito, cioè una gnoseologia che deve rivelare il pensiero a se stesso indagando le profonde armonie psicologiche ond'esso si atteggia nella continuità della storia.

E da questo lato la critica dei principii promette di recare nuovi resultati importanti; dopo avere illuminato il carattere proprio della Logica, essa riuscirà ad approfondire lo studio degli elementi intuitivi di diverso ordine che conferiscono alle Matematiche il loro inesauribile valore.

PERIODICITIES IN THE SOLAR SYSTEM

BY ERNEST W. BROWN.

The treatment of problems in celestial mechanics has undergone a variety of phases since the history of the subject opened with the discoveries of Isaac Newton. He and his immediate successors regarded them almost entirely from one point of view. They desired to know the motions and positions of the members of the solar system at any time in the past or future. The numerical treatment of the problems of gravitation was thus the first and chief consideration.

All attempts to find expressions for the coordinates in terms of a finite number of functions whose numerical values had been tabulated, ended in failure. We now know that, in general, no such expressions exist. It is necessary to make use of infinite series. In the majority of cases these series consist solely of harmonic terms —the only form, since it was taken as an axiom deduced from observations in the past, that all the motions of the bodies must be recurrent. Exceptions to this rule were allowed in the cases of those motions whose periods were so long that a few terms of their development in powers of the time gave results sufficiently accurate for the purposes required. These series consist, therefore, of development by harmonic terms for the shorter periods and by powers of the time for the longer ones.

The methods used were constructed mainly with a view of obtaining the harmonic or partly harmonic series to the required degree of accuracy. The best method was that which demanded the least calculation, algebraic and numerical. This was soon found to differ with different bodies. For the motions of the planets certain methods came to be employed; for those of the satellites, and in particular for that of the moon, other methods. All of them depended to a greater or less extent on knowledge, obtained in various ways, of what should be the final form of the expressions.

Since infinite series are employed, the question of their convergence arises. By the astronomer who is mainly concerned with the numerical applications this question is usually left aside. For him the series are either possible or impossible according as they give correct numerical values of the functions for certain values of the variables or not. In this he is guided by observation and past experience. We now know that the convergence of a series is not a necessary condition for obtaining a good approximation to the function it replaces. To demand that a series shall be convergent is to limit unnecessarily the range of its application.

We thus come to a second stage in the development of the subject, namely, the examination from the logical point of view of the series and expressions used by the astronomer. In order to make such an examination it does not follow that one must use his methods. What is chiefly needed is the examination of the expressions finally obtained rather than that of the methods by which they are obtained. The older methods are frequently not well adapted for such an examination; hence the necessity for devising new ones. Are these new methods of no interest whatever to the student who has in view only the numerical solution of the special problems presented by the solar system? As far as historic time is concerned they have perhaps but little bearing from the astronomical point of view. They have not so far assisted materially in gaining better methods of calculation; but if one tries to penetrate into the remote past and future, it is seen that the numerical methods fail to give any certain information, and the failure arises mainly because there is no indication as to whether the expressions obtained are possible for great values of the time as well as for small values. In other words, we do not know whether the expressions lie between finite limits for all values of the argument. Connected with this is the question of stability. If we could prove that our series were convergent the stability would be proved, but the fact that they are divergent tells us nothing about the stability. The divergence may be a mathematical defect only.

A third line of investigation is the discovery of possible modes of motion under the law of gravitation. It is somewhat remarkable that while most of the investigations in this direction have been made independently of the applications, these applications have arisen about the same time. The numerical knowledge of stellar systems from the new applications of the spectroscope, certain photo-metric determinations and accurate direct measurements, is furnishing material for this new branch of stellar mechanics. One very remarkable case, that of Laplace's three particles, has found its application in the Trojan group of asteroids in the solar system within the last decade.

A fourth branch of the subject is the investigation of the properties of the differential equations which occur in the problems of celestial mechanics. While many portions of this branch have direct applications to physical problems, the point of view is rather that of the pure mathematician. It may be regarded as the theory of a certain class of differential equations.

The methods used in these four branches are in most cases devised each for a particular purpose, and in approaching the study of any problem one naturally chooses the one best adapted for the purpose. This is more particularly the case with a specific problem in which a fixed degree of accuracy is required. Devices which give formulas admitting of a simple theoretical treatment are frequently not well adapted for the purposes of computation; and the converse is true. In a problem which must require much computation, it is advisable to use any useful resources provided by theory; formulae for numerical calculation which will also permit of the consideration of theoretical questions will generally involve much more computation than formulae devised for numerical calculation only. One may even be justified in using processes, the logic of which is doubtful, if the object in view is the discovery of hidden phenomena. The justification of such processes may be regarded as another problem for which different devices should be employed.

The methods used for the numerical solution of the problems presented by the solar system illustrate what has just been said. For comparison with observation the numerical results deduced from the Newtonian law are satisfactory. But the time-range of the observations is limited and the theoretical examination of the methods makes it doubtful whether we may deduce phenomena in the remote past or the remote future from expressions which serve for a limited period. In this examination the periodicity of the expressions is important. Nearly all the difficulties, both numerical and theoretical, originate in the attempt to use expressions the arguments of which depend on two or more periods. In order to set forth these difficulties it is necessary to go somewhat more into detail.

It has been mentioned that the usual mode of expression of the coordinates is by infinite series of harmonic terms. Each of these terms will have a different period. The periods of the various terms may be related or not related to one another, but practically all the arguments are linear combinations, with integral coefficients, of a finite number of angles. Hence the periods connected with the motion of any one body are reduced to the consideration of a finite number of them, this number varying from one upwards according to the number of bodies that are under consideration.

Astronomers have also been accustomed to recognize several different kinds of oscillations in the motions of the celestial bodies:—those of shor period, of long period, librations, and the somewhat misleading secular changes. These are in reality relative and not absolute terms, and there is no sharp line of division between them. In the applications to the motion of any one body, we have one principal period, that of the revolution of the body about some centre, and this is usually taken as the basis of comparison. If the motion be expressed as a sum of harmonic terms, we have a number of periods which are nearly the same or are less than this principal period. These are called short. Those periods which are much longer than the principal period are classified as long period terms. Some of them are, however, so long that it is more convenient in the calculations to expand the corresponding sines or cosines in powers of the time. This is allowed in a comparison of the results with observation when but few terms of the expansion are needed; they are then called secular. These terms are frequently mingled with terms which are really secular, that is, with terms which are not properly expressible by harmonic series; but as we have no certain knowledge of any real secular terms occurring in the motions of the bodies of the solar system, and arising solely from gravitational action, these secular terms constitute a form of mathematical expression available only for a limited time. Finally, we have a class of terms denoted as librational. Properly speaking these are not different from the harmonic terms. Their presence depends mainly on the mode of representation of the motion. In the ordinary mode the argument of a harmonic term varies directly with the time, and it makes complete angular revolutions. In certain exceptional cases it is more convenient to use arguments which oscillate about a mean value. The terms are then called librations. In reality the motion can in general still be expressed by harmonic terms, but there is a discontinuity in passing from the harmonic form to the librational form: a discontinuity, not only in the mathematical representation, but also, from certain points of view, in the motion itself.

Long period terms in the motion of any body arise from several sources. With given forces depending only on the distances there will be a number of so-called natural periods in the motion. If one of these is long or if two of them are nearly equal or are nearly in the ratio of two integers so that their combination will produce a long period term, the integrations will generally produce a small divisor which will raise the magnitude of the corresponding coefficients. It is usual, however, to consider the motions of all the bodies but one as known. In this case the periods of the known motions are to be considered as forced; but the effect under consideration is practically the same whether the oscillation periods are forced or free.

If one of the forced and one of the natural periods are so nearly equal that the divisor becomes very large, the numerical processes used may cease to be convergent. In this case the representation has to be changed. It is assumed that they are exactly equal on the average and that oscillations about this average value take place. The oscillations become librational.

Let us return to the general problem in which the motions of all the bodies are supposed to be unknown. The periods then depend on the masses and the initial conditions: all of these may have any values whatever and may vary continuously. If we have a function expanded into a sum of harmonic terms containing all multiples of, say, two periods, we may always find one term whose period may be made as long as we please by choosing a convergent for the ratio of the two periods which are measured to a certain degree of accuracy only. If before integration the coefficient is not so related to this ratio that it vanishes when the period is infinite, we have either terms which are not harmonic or else there is a libration. In any case an arbitrarily small change in the masses or the initial conditions will alter the representation. In other words, the convergence, if it exists, is not uniform with respect to the masses and the initial conditions.

This is the case with the usual mathematical representation, but it does not follow that there is a discontinuity or instability in the actual motion. For example, in certain cases where we proceed by continued approximation it is found that in the earlier steps, the radius vector of the body is but little affected by a certain long period force, while the longitude may be greatly changed after the expiration of a long interval. Such changes are not unstable from the physical point of view, but the position of a celestial body based on a given degree of accuracy at the start becomes less accurate with the increase of the time. If purely periodic convergent expressions could be obtained, the accuracy of the positions would only depend on the accuracy with which the measurement of the initial conditions could be made. Since the actual theories of the motions of the planets and satellites possess these defects, it is necessary to examine how far they represent the motions of the bodies.

Short period terms simply produce rapid oscillations. A long-period term has a similar effect provided the interval over which our observations range is not short compared with the period of the term. If this interval is short the term is usually expanded in powers of the time, and the expansion may give rise to apparently secular motions. If the term is present in the longitude of a body, the first two powers of the time are absorbed in the arbitrary constants determined by observation. If we can neglect the second and higher powers of the time this is practically equivalent to neglecting the term altogether. In the case of the radius vector we can only neglect the term provided the portions dependent on the first and higher powers of the time are insensible. The advantage of this mode of treatment lies in the fact that in general the small divisors which give rise to the large coefficients in a long period term disappear when expansion is made in powers of the time. The extent to which calculations should be carried depends therefore, first, on the degree of accuracy required at any given interval from the epoch of the observations, and secondly, on the interval of time for which the expressions are to be used. Hence we can always neglect terms which have coefficients smaller than a certain limit and also terms whose periods are longer than a certain interval, and these limits can be determined beforehand.

The problem of the motion of the moon regarded from these points of view is sufficiently simple. If we consider the attractions of the sun and earth alone there are four principal periods: the month and the year, which are obtained directly from observation, and two other periods (deduced from theory) which are known as the period of revolution of the apse—about nine years—and that of the revolution of the node—about nineteen years. All other periods in the motion arise from linear integral combinations of the motions of the angles which have these four periods. They are approximately in the ratio:

$$1 : 13\tfrac{1}{3} : 118 : 210.$$

The month may be regarded as a natural period. The year in the motion of the moon is a forced period and the other two may be regarded as natural periods belonging to the system of three bodies*.

In the integration of the equations it is necessary to consider the cases where large coefficients arise from the presence of small divisors. These small divisors occur mainly through the forced period being long and through two of the three natural periods being also long. As we have to proceed to numerous approximations, we have a variety of combinations of the four periods that are possible. Fortunately the two principal periods, the month and the year, are far from being nearly equal and at least thirteen multiples of the mean solar motion are required to produce one multiple of the mean lunar motion. In this problem high multiples of any motion are accompanied by correspondingly small coefficients before integration. The only case arising from near commensurability of two periods which is likely to produce a large coefficient is that formed from a combination of the two latter periods. The motion of the perigee plus twice the motion of the node gives a term whose period is 180 years or nearly 2400 months. This term, however, has such a small coefficient

* In this statement the motion of the solar apse is neglected.

before integration that it is insensible to observation with even the most modern refinements.

The moon is also affected by the attractions of the planets and each planet introduces a new angle depending upon its average period. With the five planets which produce observable effects there are altogether nine angles whose combinations have to be considered. The difficulties chiefly arise from the fact that the periods of these planets are all of the same order of magnitude and that the series which express their attractions converge quite slowly. Nevertheless the examination of the possible combinations is much less troublesome than one might expect. By the use of certain theoretical properties of the motion, it is not difficult to carry the examination sufficiently far to make certain that no combinations giving long periods will sensibly effect the motion beyond those known. What has been said concerning an assumed degree of accuracy at a given interval of time from the epoch, will make it clear that there will be a certain stage at which we can definitely say that no further combinations can effect the motion to an observable degree within the interval of time for which observations exist. It is on this basis that the statement may be made with considerable confidence that no well-recognized gravitational cause can explain any observed deviation of the moon from its computed theoretical orbit.

The motions of the planets are treated in the same manner. In considering the actions of two planets and the sun only, we have two principal periods which are in general of the same order of magnitude, namely, their periods of revolution round the sun. These are measured by a few years or by a not very small fraction of a year, while the periods of revolution of their perihelia and nodes are measured by tens or hundreds of thousands of years. For observational purposes it is sufficient to express the latter in powers of the time, the short period deviations being retained in the harmonic form.

Let us express the ratio of the two principal periods by a series of convergents to their mean motions n, n': namely

$$\frac{n}{n'} = \frac{p_1}{q_1}, \qquad \frac{p_2}{q_2}, \qquad \frac{p_3}{q_3} \cdots.$$

There will then be a series of terms with the motions

$$nq_1 - n'p_1, \qquad nq_2 - n'p_2 \ldots;$$

such terms will be of long period in comparison with the two principal periods but the earlier convergents will in general give periods short in comparison with those of the perihelia and nodes. For large values of p, q, the coefficients are small before integration so that we can proceed to find all the possible terms which can sensibly affect the motion within a given interval of time. The matter is not quite so simple as it is in the motion of the moon with the attractions of the planets neglected. The series of harmonic terms before integration have coefficients which converge quite slowly and we have practically to deal with a quintuply infinite series expanded in powers of five parameters. The rates of convergence along powers of these parameters differ considerably with different bodies. Along powers of one parameter,—the ratio of the mass of the disturbing planet to that of the sun,—the convergence is very rapid: along powers of another, the ratio of the distances from the sun, it is correspondingly slow; the known expansions of elliptic motion furnish the other parameters. As a result, we have a large number of terms with coefficients of observable magnitude and consequently a large number of combinations of angles to be considered.

In the majority of cases the ratio of $qn - pn'$ to n or n' is not so small for small values of p and q that we get a very large term of long period. The principal exception to this occurs in the theory of Jupiter and Saturn, the periods of which are nearly twelve and thirty years respectively; the values $p = 5$ and $q = 2$ produce an inequality of period 880 years with coefficients in longitude amounting to nearly half a degree in the motion of Jupiter and to four-fifths of a degree in the motion of Saturn. While a case of this kind is troublesome the search for other long periods is much more simple; for, whenever a close early convergent exists, the next convergent requires such large values of p and q that it becomes a simple matter to examine the few possible cases.

These considerations indicate that while the theoretical solution of these problems leaves much to be desired, numerical expressions, deduced from the law of gravitation and the observed initial conditions, can be obtained which shall satisfy all observational needs. The existing theories for the planets and the moon appear to fulfil the demands of astronomers in this respect. Thus any well-marked differences between the theories and the observations may with good reason be put down, not to defects in the theories, but to forces, gravitational or otherwise, which are either unknown or are known only in a speculative sense. I refer for example, to the ring of small bodies inside the orbit of Mercury, assumed by Newcomb and Seeliger to explain the outstanding motion of the perihelion of Mercury, or to the doubtful hypothesis of a resisting medium, sometimes assumed to account for certain apparently secular changes. The problems presented by the deviations of the moon from its theoretical orbit will be mentioned below.

The gravitational theories for the seven to eight hundred known asteroids are far more difficult. Most of these small bodies lie between the orbits of Mars and Jupiter and are subject to large perturbations by the latter planet; Saturn may under certain circumstances also produce considerable changes in the orbits. The great majority of the periods of revolution round the sun are between two and four times that of Jupiter. The ratios of the mean distances to that of Jupiter lie between $\frac{3}{4}$ and $\frac{2}{5}$, so that the series in powers of this ratio converge slowly. Moreover many of the asteroids have large eccentricities and inclinations, thus requiring expansions in powers of parameters whose mean values may approach $\frac{1}{2}$.

Many of the bodies have periods whose ratios to that of Jupiter approach quite closely to commensurability. The most important of the fractions thus arising are $\frac{4}{1}, \frac{7}{2}, \frac{3}{1}, \frac{8}{3}, \frac{5}{2}, \frac{7}{3}, \frac{9}{4}, \frac{2}{1}, \frac{9}{5}$. In these cases we obtain large perturbations of long period. These periods are generally not long enough to permit of expansions in powers of the time but are sufficiently long to give rise to large coefficients.

The calculation of tables for so many bodies under these circumstances, even for a quite rough degree of accuracy, would be an enormous task and such tables exist only for a limited number of the bodies. The usual method is to take the elliptic elements which satisfy the motion of the asteroid for a given date and then to find by mechanical quadratures the changes for succeeding dates as they may be required. This method, while satisfactory for observational purposes, is of little value from the theoretical point of view. It enables the observer to pick up and identify the asteroid from time to time and thus to accumulate a store of observations. At present it appears that the material thus accumulated has not a value which is commensurate with the labour of obtaining it. Nevertheless it would seem that our generation has the obligation of providing for the possible needs of the future even if those needs may be, to a certain extent, doubtful, since we ourselves are profiting greatly by the observations of those of bygone generations. One may, in certain directions, see uses to which these observations may be put. Individually they furnish many cases of the three and four-body problem not presented elsewhere in the solar system. Collectively, they form a stream for statistical investigations, and in this direction the rings of Saturn may assist in a consideration of any hypotheses which may be brought forward. For both purposes observations extending over a long period are needed. In a word these bodies constitute the laboratory of the gravitational astronomer, the very variety of the opportunities for experiment constituting, perhaps, their chief value.

To return to the theory. As long as the period of revolution of the asteroid and that of Jupiter have not a ratio which is very near a proper fraction whose terms are small, the work is not very different from that necessary for the planets. A near approach to commensurability between these periods produces in time large disturbances. Suppose we follow a set of asteroids whose periods gradually approach nearer and nearer to one of the principal commensurable ratios. The principal long period inequalities become larger and larger but as long as we confine our work to a limited interval of time the disturbance does not tend to become infinite. It appears to reach a maximum value. At this point the ratio becomes exact and an oscillation about the exact ratio commences. As we follow the series further, these libratory oscillations diminish in extent and appear, so far as we know, to reach a case of motion in which the librations are small. It is to be remarked however that the observations of the asteroids do not show any certain case of libration; the well-known gaps in the distribution are precisely at those places where librations should exist.

In the mathematical investigation of these critical cases it is found that, in the first approximation, the radius vector lies between limits which are small and are never crossed. The disturbance of the longitude may vary to any extent. In the limiting case between libration and non-libration it would seem that the longitude becomes practically indeterminate but at this place the radius vector takes its mean value. The case is analogous to that of a rod which can just make complete revolutions in a vertical circle. At the highest point it comes theoretically to rest but as there are other forces continually acting to produce small disturbances the actual effect will be an apparent absence of regular periodicity. Whether the first approximation on which these results depend gives a sufficiently close idea of the motion of a librating asteroid is not certain but it appears probable that it should do so. The theory of the question is difficult.

If the motions are commensurable the period of the asteroid and the period of Jupiter must be both multiples of one period. It therefore appears as if one of our periodicities had disappeared. The period of the asteroid, however, is replaced by that of the libration. The analogy of the libration to that of an oscillating pendulum, while not complete, is sufficient to show the nature of the motion. One point is of importance. When the ratio is near that of the fraction $\frac{2}{1}$ the libration can never be indefinitely small and consequently this second oscillation must always exist. Now we know that when the arc through which a pendulum oscillates is not very small the period of oscillation depends on the length of the arc. The same is true in libration. The amplitude of the oscillation is an arbitrary constant and if we neglect its square the period is dependent only on the forces. If, however, there are forces which cause the amplitude to increase, the period will increase also. The period of a libration is very sensitive to any change in the forces.

Let us now introduce the action of Saturn. It has already been mentioned that the periods of Jupiter and Saturn are nearly in the ratio of 5 to 2. Thus, every asteroid which has a period nearly commensurable with that of Jupiter will also have one nearly commensurable with that of Saturn. In every case of libration or approximate commensurability, there will be a long period term due to the action of Saturn. If the ratio of the periods of Jupiter and Saturn had been exact the problem would perhaps be more simple. The theoretical difficulties arise owing to the slight difference from a commensurable ratio. The period due to this difference is about 880 years. It seems certain that this must largely increase the extent of a libration.

There is another feature. A large libration has a long period and if this period becomes commensurable with 880 years we may get a secondary inequality of much longer period and this may be so large as to produce a secondary libration. That this is likely to happen may be gathered from the fact that the period of the principal libration can so easily vary when it is not very small. This secondary libration may attain very large values. The question arises as to whether two such independent librations can exist. Even if this secondary libration does not exist we shall get a long period inequality with a very large coefficient and this may tend to make the primary libration unstable or impossible. It is a well-known fact that there are no asteroids yet discovered which librate about this particular ratio. All attempts hitherto made seem to show that there is no instability when the action of the planet Jupiter alone is considered. It is possible that the action of the planet Saturn may be responsible partly for the gap in the distribution. Another way of putting this question consists in the statement that commensurability between the periods of Jupiter and the asteroid limits the range of cases of stable motion. The action of Saturn seems to cause a further limitation.

How do these principles apply to other cases of libration in the solar system? Three of the satellites of Jupiter have exact integral relations between their periods of revolution around the planet. It might seem then that the theory is not verified in the case of the satellites of Jupiter. The matter is, however, different. The masses of these three bodies are of similar order of magnitude and their influence on one another is not very different. Thus a near approach to commensurability can be effective in causing librations in the motions of each of them. In the case of Jupiter,

Saturn and an asteroid moving round the Sun, the mass of the asteroid is so minute compared with those of Jupiter and Saturn that it probably cannot compel those two planets to librate and thus possibly to render its own orbit stable. On the other hand, if we consider two of the satellites of Saturn, Titan and Hyperion, we find there a libration. The fourth body may be taken to be one of the particles constituting the ring. The mass of the latter must be supposed to be very small compared with those of the two satellites. Here then we have a case of a four-body problem in which there might be libration periods between three of them. Observations of Saturn's ring show dark spaces at the places where librational phenomena occur, and we conclude from this fact that there are few or no particles in the ring which can perform this libration. The question, if analogous to the case of the three satellites of Jupiter, would differ only in the magnitude of the actions. Is there a complete absence of such particles in the dark spaces or is the range of stable orbits very small so that these portions of the ring have few particles and hence that their aggregate reflecting power is too minute for our limited powers of observation? It seems certain that the action of Hyperion alone is sufficient to produce a difference in the illumination round a libration region by means of the large perturbations which it must produce in the individual particles of the ring, but it does not seem quite sufficient on gravitational grounds to explain completely the observed differences of illumination exhibited by the divisions of the ring.

The consideration of the stability of systems of three or more bodies is involved. Although it is possible and indeed probable that we have a wide range of stable cases of motion of the three-body problem it would seem as if the addition of a fourth body must limit the range of stability. This refers not only to the so-called secular stability but stability within periods which are comparable with that of our observational data. The whole question needs a full examination from the gravitational point of view. Until this has been given it is inadvisable to introduce hypotheses which may after all not be necessary. It has happened not infrequently, that the law of gravitation has been modified or some new force assumed in order to explain an observed anomaly in the motion of a body and that finally gravitation was found to be a sufficient cause. We can be justified in thinking that such may prove to be the case in the statistical problems of the asteroids and of the ring of Saturn.

Certain of the anomalies in the motion of the moon are still unexplained. Briefly stated they belong to three classes. First, the unexplained secular changes in the mean period and possibly in the periods of the apse and node. With reference to these it may be noted that, under certain circumstances, a gravitational cause producing a secular change in the mean period will also produce secular changes in the periods of the apse and node of the same order of magnitude. Hence, if we find from observation changes in the latter a clue may be furnished as to the cause of the changes in the former. For the mean motion alone, tidal friction is possibly a sufficient explanation of the difference between observation and gravitational theory.

Secondly, two inequalities in the longitude, one with a period of some 300 years and a coefficient of about fifteen seconds of arc; another with a period of some 60 years and a coefficient of about three seconds of arc; a number of inequalities of shorter period and with smaller coefficients. With reference to these latter it is quite possible that many of them may disappear when the observations are compared with more accurate tables; their irregularity indicates this and a number of small terms to be included may account for them. With reference to the two larger observed terms, if one looks at them apart from the theory they do not seem to be very great; but it is to be remembered that in the last investigation into the motion of the moon an attempt was made to take into account every known gravitational force and to find every coefficient greater than $\frac{1}{100}$ of a second of arc attached to terms with periods of less than 4000 years. The great unknown periodic term is therefore more than a thousand times the magnitude of the greatest term not included in the theory. The work was moreover carried, in the great majority of cases, so far as to include coefficients greater than one-thousandth of a second of arc in longitude. Further, the largest inequality produced by planetary action is of about the same size and period as the large empirical term and several hundred due to this cause, most of them very small, are included in the theory. No indication of a new large term appeared, nor does it seem likely that omissions or errors of this magnitude have been made by all those who have investigated the subject.

In searching for causes, it is first necessary to notice that the 60-year term is probably more important than the 300-year term since it will require a larger force. The period of the force is probably short. It is natural to look for a minute force and this will be much smaller if it has a period of nearly a month than if it be of long period. Information on this point may be obtainable from an analysis of differences between theory and observation of the short period terms in the moon's motion. If, for example, the force has a period of a few years, the short period terms in the moon's motion are likely to be of the same order of magnitude as the long period terms. Hence the importance of including in the lunar tables all short period terms which may separately or collectively influence the motion. This carries with it the necessity of continuing the daily observations of the moon with the greatest possible accuracy. These short period terms constitute the third class of unexplained differences between theory and observation; they will probably disappear when the terms of the second class can be properly included in the theory.

There are certain tests which should always be made in discussing the effect of any assumed force. In nearly all cases the mean value of an assumed force is not zero—its effect is usually sensible when the effect of the variable portion is so. The mean value causes additions to the mean motions of the perigee and node. Now nearly within the limits of the errors of observation, these theoretical and observed mean motions agree. Hence any assumed force which produces motions of the perigee and node much greater than the errors of observation must be rejected.

The effect of the assumed force on the motion of the earth round the sun must also be computed. Unless the large coefficients in the moon's motion are due to near commensurability of the period of the force with one of the natural lunar periods, the effect on the earth's motion will be as great or greater than that on the moon's motion. These facts indicate that the large empirical terms are either due to forces with periods having near commensurability with one of the lunar periods or forces which have their origin within the earth-moon system.

While the effect of approximate or exact commensurability between two periods has far reaching effects from the point of view of the theory, the practical advantages to be gained from it are also great. Dr Cowell's success in his analysis of the observations of the moon has largely arisen from the use which he has made of the numerical relations between the lunar periods. An equally extensive use of the same principle is being made in the construction of new tables of the moon's motion. We adopt in fact, an expression for a number of harmonic terms which is periodic after many revolutions, and correct it only at long intervals. Another view of the process is the separation of the short period changes from those of long period, a separation which is the more effective whenever we use a convergent to the ratio of the two periods whose terms are small compared with those of the succeeding convergent. This idea, already found to be of advantage in purely numerical problems, is also applicable to problems which are partly theoretical. One of the disadvantages of the osculating ellipse is the presence of periods of different orders of magnitude in the perturbations. Expressions for the coordinates in which the short period terms are separated from those of long period have many advantages in the numerical tabulation of the orbit, and these advantages are increased if the separation be made at the outset, namely, in the expressions assumed for the solution of the differential equations of motion. It is not supposed that we may get convergent series in this way but that series may be found which can be used over longer intervals of time than those, for example, which are ordinarily employed in the calculation of special perturbations.

The President has spoken of the achievements of Henri Poincaré but I cannot conclude without laying my personal tribute to his memory. Although he contributed greatly to the advance of almost every department of mathematics, I shall not, perhaps, be accused of undue prejudice if I say that the impression he has left on Celestial Mechanics is the greatest of all. His insight, his wide knowledge and the fertility of his resources were brought together in a massed attack on the problems of the subject. "Les Méthodes Nouvelles de la Mécanique Céleste" must remain a classic for many succeeding generations. We deplore his departure while we hope to build on the foundations laid down by so sure a hand.

GELÖSTE UND UNGELÖSTE PROBLEME AUS DER THEORIE DER PRIMZAHLVERTEILUNG UND DER RIEMANNSCHEN ZETAFUNKTION

VON EDMUND LANDAU.

Das Organisationskomitee dieses Kongresses war so freundlich, mich vor einigen Monaten aufzufordern, in einer allgemeinen Sitzung einen Vortrag über die Entwicklung eines Kapitels der Mathematik zu halten, dessen Auswahl mir vorbehalten blieb. Ich glaubte, im Sinne dieser ehrenvollen Aufforderung zu handeln, wenn ich das Gebiet wählte, dem die Mehrzahl, ungefähr zwei Drittel, meiner bisherigen eigenen Publikationen angehört, und in dem ich daher neben dem Bericht über die Leistungen anderer auch einige eigene Gedanken entwickeln kann: die Lehre von der Verteilung der Primzahlen und die damit in engem Zusammenhang stehende Theorie der Riemannschen Zetafunktion. Der Vortrag wird sich jedoch nicht auf diese Probleme allein beschränken, sondern gleichzeitig zu verwandten Fragen der analytischen Zahlentheorie und der speziellen Funktionentheorie Bezug nehmen.

Ich weiss, dass die Kenntnis der Zahlentheorie wenig verbreitet unter den Mathematikern ist, und dass speziell die Schwierigkeit der Methoden der analytischen Zahlentheorie nicht viele Fachgenossen angelockt hat, sich mit den schönen Ergebnissen dieser Disziplin vertraut zu machen. Ich werde daher in diesem Vortrag keine Kenntnisse aus diesem Gebiet voraussetzen und so sprechen, als ob ich vor einer Korona stände, welche von diesen Dingen noch nichts weiss. Natürlich erwartete ich von vornherein unter Ihnen manchen Meister gerade dieses Gebietes, von dem ich früher nur zu lernen hatte.

Was eine Primzahl ist, weiss jeder; es sind die Zahlen 2, 3, 5, 7 u. s. w., welche genau zwei Teiler besitzen, nämlich die Zahl 1 und sich selbst. Aus ihnen lassen sich alle Zahlen > 1 durch Multiplikation zusammensetzen, sogar eindeutig. Das ist der Fundamentalsatz der Zahlentheorie und auch jedem aus dem Elementarunterricht bekannt. Daraus allein folgt noch nicht, dass es unendlich viele Primzahlen gibt; denn schon die eine Primzahl 2 erzeugt durch Multiplikation mit sich unendlich viele Zahlen, und es wäre daher nicht ausgeschlossen, dass endlich viele Primzahlen genügen, um alle Zahlen zu erzeugen. Aber bereits Euklid hat vor mehr als zwei Jahrtausenden bewiesen, dass es unendlich viele Primzahlen gibt. Wenn also p_n die $n$te Primzahl bezeichnet, so hat dies für jedes positive ganze n eine Bedeutung, und p_n wächst natürlich mit n ins Unendliche. Es wäre vom Standpunkte der heutigen Zahlentheorie aus ein unbescheidenes Verlangen, p_n durch n mit Hilfe eines geschlossenen Ausdrucks darstellen zu wollen, der etwa nur aus denjenigen Funktionen zusammengesetzt ist, mit denen die Mathematik üblicherweise operiert, und der noch dazu einfache Bauart hätte. Vielmehr ging die Fragestellung dahin, p_n für grosse n näherungsweise durch eine der einfachen Funktionen von n darzustellen. Näherungsweise in dem präzisen Sinn, dass der Quotient für $n=\infty$ den Limes 1 hat. Diese Fragestellung ist gleichwertig mit der folgenden. x sei eine positive Grösse, ganz oder nicht; $\pi(x)$ bezeichne die Anzahl der Primzahlen $\leqq x$. Nach Euklid wächst $\pi(x)$ mit x ins Unendliche. Das Problem lautet, $\pi(x)$ mit einer einfachen Funktion so in Beziehung zu setzen, dass der Quotient den Limes 1 hat, d. h. dass er wirklich einen Limes besitzt und dieser gleich 1 ist. Gauss, Legendre und Dirichlet vermuteten, dass

$$\pi(x) \sim \frac{x}{\log x}$$

ist. Das Zeichen $\sim$ (sprich: asymptotisch gleich) bedeutet eben, dass der Quotient für $x=\infty$ gegen 1 strebt. Diese höchst bemerkenswerte Vermutung ist völlig identisch mit der anderen, die sich an meine ursprüngliche Fragestellung anlehnt:

$$p_n \sim n \log n.$$

Es wäre also
$$\lim_{n=\infty} \frac{p_n}{n \log n} = 1,$$

folglich
$$\lim_{n=\infty} \frac{p_{n+1}}{p_n} = 1.$$

Letzteres ist natürlich nur ein Korollar und sagt weniger aus; ist es doch z. B. auch für die Menge der Zahlen n^2 statt der p_n erfüllt. Ich bemerke aber gleich, dass mir für dies Korollar kein Beweis bekannt ist, der nicht zugleich auch den schärferen Satz liefert. Gauss, Legendre und Dirichlet gelang es nicht, den Satz über $\pi(x)$ zu beweisen, den ich den Primzahlsatz nennen will. Und doch hat Gauss ihn schon als ungefähr fünfzehnjähriger Knabe vermutet, wie er in einem Briefe berichtet, den er als 72-jähriger Mann über diese Dinge an Encke geschrieben hat.

Was aber Dirichlet betrifft, so liegt eine seiner berühmtesten Leistungen im Primzahlgebiet. Er hat im Jahre 1837 für die Primzahlen einer beliebigen arithmetischen Progression unter Überwindung grosser Schwierigkeiten das geleistet, was für die Primzahlen überhaupt schon durch Euklid bekannt war. Es seien k und l positive ganze Zahlen, aber ohne gemeinsamen Teiler; man betrachte alle Zahlen $ky+l$, wo y die Werte 0, 1, 2, 3, ... durchläuft; also z. B. ($k=100$, $l=19$) die Zahlen 19, 119, 219, 319, ... Dirichlet hat bewiesen, dass in jeder solchen arithmetischen Reihe unendlich viele Primzahlen vorkommen. Die Hauptschwierigkeit bei seinem Beweise bestand darin zu zeigen, dass gewisse unendliche Reihen, deren Konvergenz trivial ist, eine von 0 verschiedene Summe besitzen. Diese Klippe überwand er auf genialem Umwege durch Heranziehung der Theorie der Klassenzahl quadratischer Formen. Heute kann man dies Nichtverschwinden allerdings nach Herrn Mertens direkt auf wenigen Zeilen beweisen; aber sonst ist Dirichlets Beweis in keinem wesentlichen Punkte vereinfacht worden.

Schon Legendre hatte den Dirichletschen Satz vermutet und zugleich einen weitergehenden, den auch Dirichlet nennt, ohne ihn beweisen zu können. Es seien zwei solche arithmetische Reihen mit derselben Differenz, aber mit verschiedenen Anfangsgliedern gegeben, nämlich z. B. die oben genannte und 77, 177, 277, 377, ... $\pi_1(x)$ sei die Anzahl der Primzahlen der ersten Progression bis x, wird also nach dem Dirichletschen Satz mit x unendlich; $\pi_2(x)$ sei das entsprechende für die zweite. Dann vermuten Legendre und Dirichlet, dass

$$\pi_1(x) \sim \pi_2(x)$$

sei.

Ich kehre nun zunächst zum allgemeinen Primzahlproblem zurück. Ein englischer Gelehrter, Hargreave, hat zuerst eine heuristische Plausibelmachung des Primzahlsatzes publiziert, die er selbst nicht etwa als Beweis angesehen wissen wollte. Der berühmte russische Mathematiker Tschebyschef hat bald danach bewiesen, dass der Quotient

$$\pi(x) : \frac{x}{\log x}$$

von einem gewissen x an grösser ist als eine positive Konstante und kleiner als eine gewisse endliche Konstante. Man vermutete also, dass er den Limes 1 hat, und Tschebyschef hat bewiesen, dass sein $\limsup\limits_{x=\infty}$ endlich, ausserdem, dass er $\geqq 1$ ist, und er hat bewiesen, dass der $\liminf\limits_{x=\infty} > 0$, und ausserdem, dass er $\leqq 1$ ist. Bei Tschebyschefs Untersuchungen spielen neben $\pi(x)$ noch zwei andere Funktionen $\vartheta(x)$ und $\psi(x)$ eine Rolle, die folgendermassen erklärt sind: $\vartheta(x)$ ist die Summe

$$\vartheta(x) = \sum_{p \leqq x} \log p$$

der natürlichen Logarithmen aller Primzahlen bis x; $\psi(x)$ ist die Summe

$$\psi(x) = \sum_{p^m \leqq x} \log p,$$

erstreckt über alle Primzahlpotenzen bis x, wo also für jede 1te, 2te, ... Primzahlpotenz der Logarithmus ihrer Basis in die Summe aufgenommen wird. Für die Quotienten $\frac{\vartheta(x)}{x}$ und $\frac{\psi(x)}{x}$ bewies Tschebyschef genau dasselbe, was ich oben über den Quotienten $\pi(x) : \frac{x}{\log x}$ gesagt habe; er bewies es sogar zuerst für diese handlicheren Ausdrücke; daraus folgt aber jenes ohne Schwierigkeit. Überhaupt ist der damals vermutete Primzahlsatz ein unmittelbares Korollar jeder der beiden Vermutungen

$$\vartheta(x) \sim x$$

und

$$\psi(x) \sim x,$$

wie man leicht einsehen kann, und wie übrigens in einer Abhandlung Ihres berühmten Landsmanns Sylvester vom Jahre 1891 besonders hervorgehoben worden ist. Sie können aus dieser Andeutung entnehmen, dass auch im Jahre 1891 das Problem noch nicht gelöst war; und doch bin ich in meiner historischen Auseinandersetzung erst bei Tschebyschef, d. h. in der Mitte des 19ten Jahrhunderts.

Ich gehe jetzt zu meinem grossen Landsmann und Göttinger Vorgänger Riemann über, dem wir im Primzahlgebiete eine kurze, keines der darin enthaltenen Hauptprobleme lösende und doch bahnbrechende Arbeit aus dem Jahre 1859 verdanken. Wir würden alle nichts im Primzahlgebiete erreicht haben, wenn uns Riemann nicht den Weg gewiesen hätte. Übrigens stellte sich Riemann ein anderes Ziel als den Beweis jener asymptotischen Relation, die ich Primzahlsatz nannte. Es handelt sich bei ihm um einen gewissen expliziten Ausdruck für eine mit $\pi(x)$ eng verwandte Funktion; nämlich eine unendliche Reihe von Integralen, bei der in jedem Glied eine nicht reelle Nullstelle einer von Riemann neu in die Analysis eingeführten Funktion vorkommt, der sog. Zetafunktion. Weder bewies Riemann, dass diese Nullstellen existieren; noch—ihre Existenz vorausgesetzt—dass seine unendliche Reihe konvergiert; noch—die Richtigkeit seiner Formel vorausgesetzt—dass daraus der Primzahlsatz folgt; ein anderes Problem, welches in seiner Arbeit gestellt ist, ist bis heute ungelöst. Und doch hat Riemann den späteren Forschern durch die Einführung der Zetafunktion und durch das, was er in seiner Abhandlung bewiesen und nicht bewiesen hat, das Werkzeug in die Hand gegeben, mit dem später der Primzahlsatz und vieles andere dem mathematischen Wissen hinzuerobert wurde. Daher ist es sehr wichtig, etwas genauer bei der Riemannschen Abhandlung zu verweilen. Riemann betrachtet die unendliche Reihe

$$\zeta(s) = \sum_{n=1}^{\infty} \frac{1}{n^s},$$

wo s eine komplexe Variable ist, deren Abszisse ich stets σ, deren Ordinate ich stets t nennen werde: $s = \sigma + ti$. Es bedeutet x^s dabei $e^{s \log x}$, wo der reelle Wert des Logarithmus gemeint ist. Es ist leicht zu zeigen, dass die unendliche Reihe in der Halbebene $\sigma > 1$ konvergiert, sogar absolut konvergiert und dort eine reguläre analytische Funktion von s darstellt. Das ist uns heute nach dem Weierstrassschen Doppelreihensatz trivial; Riemann begründete es direkt. Es entsteht nun die Frage, ob $\zeta(s)$ über die Gerade $\sigma = 1$ fortgesetzt werden kann. Riemann bewies, dass die Funktion $\zeta(s)$ in der ganzen Ebene bis auf den Punkt $s = 1$ regulär ist, und dass $s = 1$ Pol erster Ordnung mit dem Residuum 1 ist. $\zeta(s) - \frac{1}{s-1}$ ist also eine ganze transzendente Funktion. Riemann bewies ferner die Funktionalgleichung

$$\zeta(1-s) = \frac{2}{(2\pi)^s} \cos \frac{s\pi}{2}\, \Gamma(s)\, \zeta(s),$$

welche uns lehrt, dass wir die Funktion in der ganzen Ebene beherrschen, wenn wir sie in der Halbebene $\sigma \geqq \frac{1}{2}$ gut genug studiert haben. Für $\sigma > 1$ besteht die leicht beweisbare Identität

$$\zeta(s) = \prod_p \frac{1}{1 - \frac{1}{p^s}},$$

wo p alle Primzahlen in beliebiger Reihenfolge durchläuft. Sie sagt nichts anderes aus als die Tatsache der eindeutigen Zerlegbarkeit der zusammengesetzten Zahlen in Primfaktoren; wenn nämlich der auf p bezügliche Faktor in die (sogar für $\sigma > 0$ konvergente) geometrische Reihe

$$\frac{1}{1 - \frac{1}{p^s}} = 1 + \frac{1}{p^s} + \frac{1}{p^{2s}} + \frac{1}{p^{3s}} + \ldots$$

entwickelt wird und alle diese Reihen multipliziert werden, so kommt für $\sigma > 1$ (wie formal klar und auch leicht zu rechtfertigen ist) genau die Reihe

$$\sum_{n=1}^{\infty} \frac{1}{n^s}$$

heraus, da ja jedes n eindeutig $2^a 3^b 5^c \dots$ ist, wo alle Exponenten ganze Zahlen $\geqq 0$ sind. Diese Produktdarstellung, welche natürlich für Riemann die Veranlassung war, $\zeta(s)$ als Hilfsmittel in die Primzahltheorie einzuführen, lehrt, dass $\zeta(s)$ rechts von $\sigma = 1$ keine Nullstelle besitzt. Die Funktionalgleichung transformiert diese Halbebene in $\sigma < 0$, und Riemann konnte leicht aus ihr ablesen, dass $\zeta(s)$ zwar in den Punkten $-2, -4, -6, \dots, -2q, \dots$ je eine Nullstelle erster Ordnung besitzt (ich will diese die trivialen Nullstellen nennen), dass aber sonst in der Halbebene $\sigma < 0$ keine Nullstelle gelegen ist. Jede sonst noch etwa vorhandene Nullstelle gehört daher dem Streifen $0 \leqq \sigma \leqq 1$ an. Dass auf der reellen Strecke 0 bis 1 keine Wurzel liegt, ist leicht einzusehen. Auf Grund der Tatsache, dass die Funktion für reelle s reell ist, in Verbindung mit der Riemannschen Funktionalgleichung ist ersichtlich, dass alle etwa im Streifen vorhandenen Nullstellen symmetrisch zur reellen Achse und symmetrisch zur Geraden $\sigma = \frac{1}{2}$ liegen. Riemann sprach nun, ohne irgend eine derselben beweisen zu können, folgende 6 Vermutungen aus; sie sind nicht unabhängig, und ich wähle diese Formulierung, weil die nachfolgenden historischen Erörterungen dann verständlicher sind.

(I) Es gibt unendlich viele Nullstellen von $\zeta(s)$ im Streifen $0 \leqq \sigma \leqq 1$.

(II) Wenn für $T > 0$ unter $N(T)$ die Anzahl der Nullstellen des Rechtecks

$$0 \leqq \sigma \leqq 1, \quad 0 \leqq t \leqq T$$

verstanden wird, die natürlich endlich ist und nach (I) mit T unendlich wird, so ist

$$N(T) = \frac{1}{2\pi} T \log T - \frac{1 + \log(2\pi)}{2\pi} T + O(\log T);$$

unter $O(g(T))$ verstehe ich immer eine Funktion, deren Quotient durch $g(T)$ für alle hinreichend grossen T absolut genommen unterhalb einer endlichen Schranke liegt.

(III) [was übrigens aus II reichlich folgt] Wenn ρ alle nicht trivialen Wurzeln von $\zeta(s)$ durchläuft, so ist

$$\sum_{\rho} \frac{1}{|\rho|^2}$$

konvergent.

Hieraus folgt in heutiger Bezeichnungsweise, dass die Weierstrasssche Produktdarstellung der ganzen Funktion $(s-1)\zeta(s)$ die Gestalt

$$(s-1)\zeta(s) = e^{K(s)} \prod_{\mathfrak{w}} \left(1 - \frac{s}{\mathfrak{w}}\right) e^{\frac{s}{\mathfrak{w}}}$$

hat, wo $\mathfrak{w}$ alle Wurzeln von $\zeta(s)$ durchläuft und $K(s)$ eine ganze transzendente Funktion ist.

Riemann vermutete weiter:

(IV) $K(s)$ ist eine lineare Funktion von s; in heutiger Bezeichnungsweise: die ganze Funktion hat endliches Geschlecht und zwar das Geschlecht 1.

(V) Die Wurzeln von $\zeta(s)$ im Streifen $0 \leqq \sigma \leqq 1$ haben alle den reellen Teil $\frac{1}{2}$.

(VI) Es besteht eine gewisse Identität für $\pi(x)$, die ich oben schon erwähnt habe, aber hier nicht aufschreiben will, zumal sie nur das Ende eines Seitenweges darstellt und für die anderen Fortschritte der Primzahltheorie nicht von Bedeutung ist.

Die Riemannschen Anregungen lagen 34 Jahre brach. Erst im Jahre 1893 gelang es Herrn Hadamard, nachdem er zu diesem Zwecke uns als Abschluss früherer Ansätze von Poincaré und Laguerre die Theorie der ganzen Funktionen endlichen Geschlechtes geschenkt hatte, die Riemannschen Vermutungen (I), (III) und (IV) zu beweisen, also die Existenz jener geheimnisvollen Nullstellen, die Konvergenz der Summe der absoluten Werte ihrer reziproken Quadrate, und dass die Funktion $(s-1)\zeta(s)$ das Geschlecht 1 hat. Herr Hadamard betont besonders, dass er die Riemannsche Vermutung (II) nicht beweisen konnte, nicht einmal die Existenz des Limes

$$\lim_{T=\infty} \frac{N(T)}{T \log T}.$$

Des weiteren hat sich Herr von Mangoldt zwei grosse Verdienste um die Primzahltheorie erworben. Erstens bewies er im Jahre 1895, von den Hadamardschen Resultaten ausgehend, durch Hinzufügung einer langen Reihe weiterer Schlüsse die Riemannsche Primzahlformel (VI), und zweitens bewies Herr von Mangoldt im Jahre 1905 die Riemannsche Vermutung (II) über $N(T)$. Für beide von Mangoldtschen Resultate habe ich übrigens später viel kürzere Beweise angegeben.

Von den 6 Riemannschen Vermutungen blieb also allein (V) offen; und diese Frage, ob wirklich die nicht trivialen Nullstellen $\beta+\gamma i$ von $\zeta(s)$ alle den reellen Teil $\frac{1}{2}$ haben, ist bis heute ungelöst. Bewiesen wurde nur 1896 durch die Herren Hadamard und de la Vallée Poussin, dass $\beta<1$, also wegen der Funktionalgleichung $0<\beta<1$ ist, und 1899 durch Herrn de la Vallée Poussin, dass bei passender Wahl einer absoluten positiven Konstanten c stets

$$\beta < 1 - \frac{1}{c \log|\gamma|}$$

ist, also

$$\frac{1}{c \log|\gamma|} < \beta < 1 - \frac{1}{c \log|\gamma|}$$

ist. Von dem Streifen $0 \leqq \sigma \leqq 1$ ist also auf beiden Seiten ein bestimmtes Flächenstück, das oben und unten immer dünner wird, aber doch unendlichen Flächeninhalt hat, herausgeschnitten, so dass dort $\zeta(s) \neq 0$ ist. Für die ersten Nullstellen hat allerdings Herr de la Vallée Poussin 1899 bewiesen, dass sie auf der Geraden $\sigma=\frac{1}{2}$ liegen. In besonders geschickter Weise hat kürzlich, 1912, Herr Backlund diesen Beweis für die ersten 58 Nullstellen, nämlich alle zwischen den Ordinaten -100 und 100 gelegenen, geführt; diese Wurzeln ergeben sich ausserdem als einfache Wurzeln. Mehr weiss man nicht über die Nullstellen der Zetafunktion.

Nun zurück zur Primzahltheorie! Auf die Hadamardschen Resultate gestützt haben unabhängig und gleichzeitig im Jahre 1896 Herr Hadamard und Herr de la Vallée Poussin den Primzahlsatz bewiesen, auf ganz verschiedenen Wegen abgesehen von der gemeinsamen Grundlage. Einen dritten, auch hierauf basierenden Beweis gab Herr von Koch 1901. Ich habe später, im Jahre 1903, einen vierten Beweis angegeben, welcher nicht nur viel kürzer ist, sondern von jener heute klassisch gewordenen Hadamardschen Theorie der ganzen Funktionen keinen Gebrauch macht. Dieser letztere Umstand war von grösster Bedeutung für die Theorie der Primideale eines algebraischen Zahlkörpers, welche die Primzahltheorie als Spezialfall enthält. Hier weiss man von der zugehörigen verallgemeinerten Zetafunktion bis heute nicht, ob sie in der ganzen Ebene existiert, und meine Methode war daher die erste und bis heute einzige, welche zum Beweise des sogenannten Primidealsatzes führt, den ich 1903 entdeckt habe, mit dem Wortlaut: In jedem algebraischen Körper gibt es asymptotisch gleich viele Primideale, deren Norm $\leqq x$ ist, indem eben diese Anzahl für jeden Körper $\sim \frac{x}{\log x}$ ist; doch will ich in diesem Vortrag nur von Primzahlen reden und nicht weiter auf Körper und Ideale abschweifen. Ich erwähnte vorhin Herrn de la Vallée Poussin als einen der beiden Forscher, denen unabhängig die Lösung des klassischen Gauss-Legendre-Dirichletschen Problems geglückt ist; der Primzahlsatz lässt sich auch

$$\pi(x) \sim Li(x)$$

(sprich: Integrallogarithmus von x) schreiben, indem bekanntlich

$$\frac{x}{\log x} \sim \int_2^x \frac{du}{\log u}$$

und dies Integral plus einer additiven Konstanten $= Li(x)$ ist. Das folgende Resultat, welches den Primzahlsatz enthält, hat Herr de la Vallée Poussin allein entdeckt, nämlich den Satz

$$\pi(x) = Li(x) + O\left(\frac{x}{\log^q x}\right),$$

wo q eine beliebig grosse Konstante ist. Er bewies sogar

$$\pi(x) = Li(x) + O\left(xe^{-\alpha\sqrt{\log x}}\right),$$

wo α eine bestimmte positive Konstante ist.

Nun kehre ich zur arithmetischen Reihe zurück. Die Herren Hadamard und de la Vallée Poussin bewiesen 1896 unabhängig für die Anzahl $\pi_1(x)$ der Primzahlen in der Progression $ky + l$

$$\pi_1(x) \sim \frac{1}{\phi(k)} \frac{x}{\log x},$$

woraus durch Division die Richtigkeit der Legendre-Dirichletschen Vermutung folgt, dass die Anzahlen für zwei Progressionen mit der Differenz k asymptotisch gleich sind. Herr de la Vallée Poussin konnte, ohne es besonders anzuführen, sogar

$$\pi_1(x) = \frac{1}{\phi(k)} Li(x) + O\left(xe^{-\alpha\sqrt{\log x}}\right)$$

beweisen, wo α eine nur von k und l abhängige oder, was auf dasselbe hinauskommt, nur von k abhängige Konstante bezeichnet; ich bewies dies später kürzer und sogar mit absolut konstantem α.

Nun sei noch von einem anderen analogen Problem die Rede, das ich bisher nicht gestreift habe. Es sei eine quadratische Form $au^2 + buv + cv^2$ gegeben und dabei a, b, c teilerfremd, ferner $a > 0$ im Falle $b^2 - 4ac < 0$. Es sind dabei u, v ganzzahlige Variable. Dirichlet hat einen Beweis dafür skizziert, dass die quadratische Form unendlich viele Primzahlen darstellt, ohne den Hauptpunkt, das Nichtverschwinden gewisser Reihen, genauer auszuführen. Das tat erst Herr Weber

1882; aus einem 1909 herausgegebenen Manuskripte E. Scherings ist ersichtlich, dass auch dieser einen vollständigen Beweis besessen hat. Herr de la Vallée Poussin bewies nun 1897 durch eine lange und scharfsinnige Schlusskette sogar das Analogon zum Primzahlsatz, und meine allgemeinen Untersuchungen über Primideale in sogenannten Idealklassen lieferten 1907 durch passende Spezialisierung den de la Vallée Poussinschen Satz und sogar einen schärferen, nämlich als Anzahl der durch die Form darstellbaren Primzahlen $\leqq x$

$$\frac{1}{h} Li(x) + O\left(x e^{-\sqrt[\gamma]{\log x}}\right),$$

wo γ eine positive Konstante ist und h die Klassenzahl oder (für sog. zweiseitige Klassen) ihr Doppeltes bezeichnet.

Nun habe ich bisher in diesem Vortrag abwechselnd von zwei ganz getrennten Forschungsobjekten gesprochen, einerseits von den Primzahlen, andererseits von der Riemannschen Zetafunktion; allgemeiner ausgedrückt: einerseits von gewissen zahlentheoretischen Funktionen, andererseits von gewissen analytischen Funktionen. Welches ist die Brücke? Wieso hat speziell das Studium der Zetafunktion zum Beweise des Primzahlsatzes geführt?

Es sei a_n eine beliebige zahlentheoretische Funktion; dann kann ich formal ohne Rücksicht auf Konvergenz die unendliche Reihe

$$\sum_{n=1}^{\infty} \frac{a_n}{n^s}$$

aufschreiben. Es besteht nun ein gewisser Zusammenhang zwischen den Eigenschaften dieser Reihe als analytischer Funktion von s und dem Verhalten der Summe

$$\sum_{n=1}^{[x]} a_n$$

für grosse x; $[x]$ bezeichnet die grösste ganze Zahl $\leqq x$. Ehe ich diesen Zusammenhang andeute, will ich an einem Beispiel zeigen, welche Zahlenmenge a_n für das Primzahlproblem ausschlaggebend ist. Ich sagte schon, dass alles auf das Studium der Funktion $\psi(x)$, d. h. der Funktion

$$\psi(x) - [x] = \sum_{p^m \leqq x} \log p - \sum_{n=1}^{[x]} 1,$$

ankommt; da habe ich also zu setzen:

$$a_n = \begin{cases} \log p - 1 & \text{für } n = p^m \ (m \geqq 1), \\ -1 & \text{sonst.} \end{cases}$$

Die zugehörige Funktion ist nun aber, wie leicht aus der Produktdarstellung von $\zeta(s)$ folgt, für $\sigma > 1$

$$\sum_{n=1}^{\infty} \frac{a_n}{n^s} = \sum_{p,\, m} \frac{\log p}{p^{ms}} - \sum_{n=1}^{\infty} \frac{1}{n^s} = -\frac{\zeta'(s)}{\zeta(s)} - \zeta(s).$$

Nun denken wir uns a_n wieder allgemein. Über das Konvergenzgebiet einer solchen sog. Dirichletschen Reihe

$$\sum_{n=1}^{\infty} \frac{a_n}{n^s}$$

hat Herr Jensen im Jahre 1884 den fundamentalen Satz entdeckt, dass es eine Halbebene

$$\sigma > \gamma$$

ist, ganz analog, wie das Konvergenzgebiet einer Potenzreihe

$$\sum_{n=1}^{\infty} a_n x^n$$

ein Kreis ist; nämlich Konvergenz rechts von $\sigma=\gamma$, Divergenz links von $\sigma=\gamma$, wobei zwei extreme Fälle möglich sind: Konvergenz überall, d. h. $\gamma=-\infty$, und Konvergenz nirgends, d. h. $\gamma=+\infty$. Potenzreihen und Dirichletsche Reihen sind beides Spezialfälle einer allgemeineren Reihenkategorie

$$\sum_{n=1}^{\infty} a_n e^{-\lambda_n s},$$

wo

$$\lambda_1 < \lambda_2 < \ldots, \quad \lim_{n=\infty} \lambda_n = \infty$$

ist, und bei der Herr Jensen die Existenz einer Konvergenzhalbebene gleichfalls bewies; nämlich die Dirichletsche Reihe der Spezialfall $\lambda_n = \log n$, die Potenzreihe der Spezialfall $\lambda_n = n$, $e^{-s}=x$ als Variable angesehen; ich will aber hier nicht von diesem allgemeinen Typus λ_n sprechen. In der Konvergenzhalbebene stellt, wie Herr Cahen 1894 als leichte Anwendung der Sätze über gleichmässige Konvergenz zeigen konnte, die Dirichletsche Reihe eine reguläre analytische Funktion dar; Herr Cahen konstatierte auch, analog zur bekannten Cauchyschen Darstellung des Konvergenzradius einer Potenzreihe, dass hier im Falle $\gamma \geqq 0$ die sog. Konvergenzabszisse γ die untere Grenze aller c ist, für welche die Relation

$$\sum_{n=1}^{[x]} a_n = O(x^c)$$

richtig ist. Die Konvergenzabszisse der Dirichletschen Reihe gibt uns also Aufschluss über das Anwachsen der summatorischen Funktion

$$\sum_{n=1}^{[x]} a_n$$

in Bezug auf Potenzen von x als Vergleich. Zum Beweise des Primzahlsatzes mussten allerdings feinere Vergleichsskalen hinzugenommen werden. Denn die blosse Tatsache, dass die Dirichletsche Reihe für

$$-\frac{\zeta'(s)}{\zeta(s)} - \zeta(s)$$

ihre Konvergenzabszisse $\leqq 1$ hat (und mehr weiss man bis heute nicht über diese Zahl!) besagt nur, dass $\psi(x) - x = O(x^c)$ für jedes $c > 1$ ist, was trivial ist; man will aber

$$\lim_{x=\infty} \frac{\psi(x)-x}{x} = 0$$

beweisen.

Da ich einmal von der Analogie der Dirichletschen Reihen mit den Potenzreihen gesprochen habe, so möchte ich nicht unterlassen, auch zweier Unterschiede Erwähnung zu tun.

Erstens: Aus der Konvergenz einer Potenzreihe in einem Punkte folgt bekanntlich ihre absolute Konvergenz in jedem Punkte, welcher näher am Mittelpunkt liegt. Die Potenzreihe konvergiert also absolut im Innern ihres Konvergenzkreises. Bei Dirichletschen Reihen ist dies nicht der Fall; sondern es folgen—allgemein gesprochen—von links nach rechts drei Gebiete auf einander: Eine Halbebene der Divergenz, ein Streifen bedingter Konvergenz, dessen Dicke übrigens höchstens 1 ist, und eine Halbebene absoluter Konvergenz. Beispiel:

$$\sum_{n=1}^{\infty} \frac{(-1)^{n+1}}{n^s} = \frac{1}{1^s} - \frac{1}{2^s} + \frac{1}{3^s} - \frac{1}{4^s} + \ldots$$

divergiert für $\sigma < 0$, konvergiert bedingt für $0 < \sigma < 1$, absolut für $\sigma > 1$. Übrigens ist diese Funktion $= (1 - 2^{1-s}) \zeta(s)$.

Zweiter Unterschied: Auf dem Konvergenzkreis einer Potenzreihe muss mindestens eine singuläre Stelle der Funktion liegen. Bei Dirichletschen Reihen braucht dies nicht einmal in beliebiger Nähe der Konvergenzgeraden der Fall zu sein. Das soeben genannte Beispiel stellt sogar eine ganze Funktion dar.

Um nun von meinem Beweise des Primzahlsatzes einen skizzenhaften Begriff zu geben, so will ich nur folgendes sagen: Wenn eine Dirichletsche Reihe

$$\sum_{n=1}^{\infty} \frac{a_n}{n^s} = f(s)$$

für $\sigma > 1$ absolut konvergiert, so ist es ganz leicht, für $x \geqq 1$ bei jedem $b > 1$ und Integration über die unendliche Gerade $\sigma = b$ die Identität

$$\sum_{n=1}^{[x]} a_n \log \frac{x}{n} = \frac{1}{2\pi i} \int_{b-\infty i}^{b+\infty i} \frac{x^s}{s^2} f(s)\, ds$$

nachzuweisen, welche einen genauen Ausdruck für eine mit der zu untersuchenden Funktion

$$\sum_{n=1}^{[x]} a_n = A(x)$$

eng zusammenhängende summatorische Funktion liefert. Die linke Seite jener Identität ist nämlich $\int_1^x \frac{A(u)}{u} du$. Unter dem Integral rechts kommt x als Parameter vor; wenn es gelingt, den Integrationsweg auf Grund des Cauchyschen Satzes durch einen links von der Geraden $\sigma = 1$ verlaufenden Integrationsweg zu ersetzen, so ist der Integrand in jedem festen Punkt des neuen Weges $o(x)$, d. h. so beschaffen, dass der Quotient durch x für $x = \infty$ gegen Null strebt; unter Umständen, die eben beim Primzahlproblem und Primidealproblem glücklicherweise eintreten, kann man aber zeigen, dass das ganze Integral $o(x)$ ist; so erhalte ich z. B., wenn ich den Ansatz auf die oben genannte Funktion

$$f(s) = -\frac{\zeta'(s)}{\zeta(s)} - \zeta(s)$$

anwende, auf Grund gewisser Hilfssätze von mir über die Zetafunktion

$$\int_1^x \frac{A(u)}{u} du = o(x),$$

woraus man leicht durch elementare Schlüsse zu

$$A(x) = o(x),$$

d. h. hier

$$\psi(x) - [x] = o(x),$$

$$\psi(x) \sim x$$

übergehen kann und somit den Primzahlsatz erhält. Man kann übrigens allgemein auch $A(x)$ statt $\int_1^x \frac{A(u)}{u}\,du$ mit $f(s)$ in Verbindung bringen, durch die für nicht ganze $x>1$ giltige Identität

$$A(x)=\frac{1}{2\pi i}\int_{b-\infty i}^{b+\infty i}\frac{x^s}{s}f(s)\,ds.$$

Wegen der nur bedingten Konvergenz des Integrals ist es schwieriger, diese Identität den asymptotischen Schlüssen zu Grunde zu legen. Aber in diesem Jahre 1912 ist es mir gelungen, auch auf diesem Weg zum Ziel zu gelangen.

Ich kehre zurück zu der vorher angedeuteten Beziehung zwischen der Grössenordnung der Summe und der Konvergenz der zugehörigen Dirichletschen Reihe. Ich erinnere nochmals daran, dass das Konvergenzgebiet der Reihe, welches für das Studium der Summe ausschlaggebend ist, nicht durch die obere Grenze der Abszissen der singulären Punkte bestimmt ist, wenn es auch natürlich nicht weiter reichen kann. Es fragt sich also, welche Bedingungen zur Regularität hinzukommen müssen, damit man mit Sicherheit schliessen kann: eine Dirichletsche Reihe, deren Konvergenz sagen wir für $\sigma>1$ bekannt ist, konvergiert sogar sagen wir für $\sigma>\tau$, wo τ eine bestimmte Zahl <1 ist.

In dieser Richtung habe ich die erste Entdeckung gemacht. Es war schon bekannt, dass, wenn σ_0 grösser ist als die Konvergenzabszisse einer Dirichletschen Reihe $f(s)$, in der Halbebene $\sigma>\sigma_0$ bei positiv oder negativ ins Unendliche rückendem t gleichmässig

$$f(\sigma+ti)=O(|t|)$$

ist. Ich habe nun den Satz bewiesen: "Es sei $a_n=O(n^\epsilon)$ für jedes $\epsilon>0$, also die Reihe

$$\sum_{n=1}^{\infty}\frac{a_n}{n^s}=f(s)$$

für $\sigma>1$ absolut konvergent. Die analytische Funktion, welche durch $f(s)$ definiert ist, sei für $\sigma\geqq\eta$ regulär, wo η eine bestimmte Zahl des Intervalls $0<\eta<1$ ist, und in der Halbebene $\sigma\geqq\eta$ sei gleichmässig

$$f(s)=O(|t|^a),$$

wo a eine Konstante ist. Dann ist die Dirichletsche Reihe über $\sigma=1$ hinaus konvergent." Herr Schnee ging dann weiter und bewies: Wenn $0\leqq a<1$ ist, so ist die Reihe sicher für $\sigma>\frac{\eta+a}{1+a}$ konvergent; übrigens habe ich hier später die Einschränkung $a<1$ ohne Modifikation der Behauptung fortgebracht. Der Schneesche Satz enthält speziell: Falls für $\sigma\geqq\eta$

$$f(s)=O(|t|^a)$$

bei jedem noch so kleinen positiven a richtig ist, so ist die Reihe für $\sigma>\eta$ konvergent. Wendet man die Schneesche Beweismethode auf die (in $s=1$ einen Pol besitzende) Funktion

$$-\frac{\zeta'(s)}{\zeta(s)}$$

an, so erhält man auf Grund gewisser Eigenschaften der Zetafunktion einen Beweis des Satzes, den zuerst Herr von Koch 1901 auf anderem Wege bewiesen hatte: Unter der Annahme der Richtigkeit der Riemannschen Vermutung (V) ist

$$\psi(x) = x + O(x^{\frac{1}{2}+\epsilon})$$

mit jedem $\epsilon > 0$, und

$$\pi(x) = Li(x) + O(x^{\frac{1}{2}+\epsilon})$$

mit jedem $\epsilon > 0$. *Mutatis mutandis,* falls die obere Grenze der reellen Teile der Nullstellen von $\zeta(s)$ zwischen $\frac{1}{2}$ exkl. und 1 exkl. liegen sollte. Übrigens wäre, wie Herr von Koch damals zuerst und ich später mit meinen Methoden kürzer bewies, das x^ϵ in den beiden letzten Formeln auch durch $\log^2 x$ bezw. $\log x$ ersetzbar.

Ich benutze diese Gelegenheit, um über eine andere spezielle Dirichletsche Reihe ein paar Worte zu sagen: Für $\sigma > 1$ ist

$$\frac{1}{\zeta(s)} = \sum_{n=1}^{\infty} \frac{\mu(n)}{n^s},$$

wo $\mu(n)$ die sog. Möbiussche Funktion bezeichnet:

$\mu(1) = 1$;

$\mu(n) = 0$ für Zahlen, die mindestens eine Primzahl öfter als einmal enthalten;

$\mu(n) = (-1)^\rho$ für quadratfreie Zahlen > 1, die aus genau ρ verschiedenen Primfaktoren bestehen.

Stieltjes sprach 1885 die Behauptung

$$\sum_{n=1}^{x} \mu(n) = O(\sqrt{x})$$

aus, ohne seinen vermeintlichen Beweis mitzuteilen. Ob Stieltjes' Behauptung richtig ist, weiss ich nicht. Mit ihr wäre auch die Riemannsche Vermutung bewiesen, indem aus Stieltjes' Behauptung *a fortiori* die Konvergenz von

$$\sum_{n=1}^{\infty} \frac{\mu(n)}{n^s}$$

für $\sigma > \frac{1}{2}$, also die Riemannsche Behauptung und noch mehr—z. B. dass alle Nullstellen von $\zeta(s)$ einfache seien—folgen würde. Ich habe aus meinem oben genannten Satz über Dirichletsche Reihen, dessen Voraussetzungen bei $\frac{1}{\zeta(s)}$ verifiziert werden können, folgern können, dass umgekehrt aus der Richtigkeit der Riemannschen Vermutung die Konvergenz der Reihe

$$\sum_{n=1}^{\infty} \frac{\mu(n)}{n^s}$$

über $s = 1$ hinaus folgen würde. Aber es blieb einem jüngeren englischen Forscher aus dieser Stadt Cambridge, Herrn Littlewood, vorbehalten, in diesem Jahre 1912 zu beweisen, dass diese Reihe dann sogar für $\sigma > \frac{1}{2}$ konvergieren würde. Es gelang ihm nämlich durch scharfsinnige Schlüsse, für jedes $\delta > 0$ und jedes $\epsilon > 0$ zu beweisen, dass unter der Annahme der Richtigkeit der Riemannschen Vermutung für $\sigma \geqq \frac{1}{2} + \delta$

$$\frac{1}{\zeta(s)} = O(|t|^\epsilon)$$

wäre; daraus folgt nach dem Satz von Schnee ohne weiteres die Behauptung.

Nun will ich wieder von der Riemannschen Vermutung absehen und mich auf den festen Boden der mathematischen Wahrheiten zurückbegeben. Was weiss man über die μ-Reihe? Euler vermutete 1748, ohne es beweisen zu können, und Herr von Mangoldt bewies 1897, dass

$$\sum_{n=1}^{\infty} \frac{\mu(n)}{n}$$

konvergiert; Möbius vermutete 1832, ohne es beweisen zu können, und ich bewies 1899, dass sogar

$$\sum_{n=1}^{\infty} \frac{\mu(n) \log n}{n}$$

konvergiert. Ich bewies 1903, dass

$$\sum_{n=1}^{\infty} \frac{\mu(n)}{n^s}$$

auf der ganzen Geraden $\sigma = 1$ konvergiert und natürlich $\frac{1}{\zeta(s)}$ darstellt, ja sogar, dass

$$\sum_{n=1}^{\infty} \frac{\mu(n) \log^q n}{n^s}$$

für jedes noch so grosse feste q auf der Geraden $\sigma = 1$ konvergiert; viel mehr weiss man über diese Frage nicht. Ich habe aber 1905 die Konvergenz der soeben genannten Reihe auch mit der Modifikation bewiesen, dass n nicht alle ganzen Zahlen, sondern nur die einer arithmetischen Progression durchläuft. Desgleichen, wenn statt $\mu(n)$ die Liouvillesche Funktion $\lambda(n)$ steht, die stets $+1$ oder -1 ist, je nachdem die Anzahl der Primfaktoren von n, mehrfache mehrfach gezählt, gerade oder ungerade ist. Hieraus ergab sich das Korollar: In jeder arithmetischen Progression gibt es asymptotisch ebensoviele Zahlen, die aus einer geraden, als solche, die aus einer ungeraden Anzahl von Primfaktoren zusammengesetzt sind.

Übrigens ist es nicht ohne Interesse zu untersuchen, ob die genannten aus denselben transzendenten Quellen geschöpften Sätze über $\pi(x)$ und $\mu(n)$ aus einander direkt durch elementare Schlüsse hergeleitet werden können. Erst 1911 gelang es mir zu beweisen, dass der Primzahlsatz und der von Mangoldtsche Satz

$$\sum_{n=1}^{\infty} \frac{\mu(n)}{n} = 0$$

in diesem Sinne äquivalent sind; die eine Hälfte hiervon hatte ich schon 1899 in meiner Dissertation bewiesen.

Über Primzahlen möchte ich nur noch weniges sagen, um mich dann etwas in die Theorie der Zetafunktion zu vertiefen, deren Studium auch ohne Rücksicht auf vorläufige Anwendbarkeit in einer Reihe hervorragender Arbeiten der neueren Zeit, insbesondere meines Freundes Bohr in Kopenhagen, zum Selbstzweck geworden ist.

Die Primzahlen will ich verlassen, nachdem ich einige Fragen genannt haben werde, welche ich für unangreifbar beim gegenwärtigen Stande der Wissenschaft halte. Ich wähle Fragen mit präzisem Wortlaut, nicht so verschwommene wie: "Das Gesetz der Primzahlen zu finden" oder "$\pi(x)$ für grosse x möglichst gut abzuschätzen." Ich nenne vier Fragen und wähle in ihnen spezielle Konstanten, um den Kern deutlicher hervortreten zu lassen.

(1) Stellt die Funktion u^2+1 für ganzzahliges u unendlich viele Primzahlen dar?

(2) Hat die Gleichung $m=p+p'$ für jedes gerade $m>2$ eine Lösung in Primzahlen?

(3) Hat die Gleichung $2=p-p'$ unendlich viele Lösungen in Primzahlen?

(4) Liegt zwischen n^2 und $(n+1)^2$ für alle positiven ganzen n mindestens eine Primzahl?

Nun zur Zetafunktion zurück: Für jedes feste σ verstehe ich unter $\nu(\sigma)$ die untere Grenze der Konstanten c, für welche bei unendlich wachsendem t

$$\zeta(\sigma+ti)=O(t^c)$$

ist. Leicht beweisbar ist, dass dies ν endlich ist, und dass

$$\nu(\sigma)=0 \text{ für } \sigma \geqq 1$$

ist; auch folgt aus der Riemannschen Funktionalgleichung und einer Stieltjesschen Abschätzung der Gammafunktion leicht

$$\nu(1-\sigma)=\nu(\sigma)+\sigma-\tfrac{1}{2}$$

für jedes reelle σ; für $\sigma \leqq 0$ ist also

$$\nu(\sigma)=\tfrac{1}{2}-\sigma.$$

Wie verläuft nun die Kurve $\nu=\nu(\sigma)$ auf der Strecke $0 \leqq \sigma \leqq 1$? Herr Lindelöf hat auf Grund eines allgemeinen funktionentheoretischen Satzes von ihm selbst und Herrn Phragmén bewiesen, dass die Kurve stetig und konvex ist; daraus folgt insbesondere, dass für $0 \leqq \sigma \leqq 1$

$$\nu(\sigma) \leqq \frac{1-\sigma}{2}$$

ist. Das Lindelöfsche Endresultat ist, dass das Kurvenstück fürs Intervall $0 \leqq \sigma \leqq 1$ dem Dreieck mit den Ecken $(0, \tfrac{1}{2})$, $(\tfrac{1}{2}, 0)$, $(1, 0)$ angehört. Mehr weiss ich darüber nicht. Aber Herr Littlewood hat bewiesen, dass unter der Annahme der Richtigkeit der Riemannschen Vermutung für $\tfrac{1}{2} \leqq \sigma \leqq 1$

$$\nu(\sigma)=0,$$

also für $0 \leqq \sigma \leqq \tfrac{1}{2}$

$$\nu(\sigma)=\tfrac{1}{2}-\sigma$$

wäre. Da ich wieder einmal den festen Boden verlassen habe, füge ich hinzu, dass ich 1911 aus der Richtigkeit der Riemannschen Vermutung die Folgerung ziehen konnte, dass dann die Differenz

$$N(T)-\frac{1}{2\pi}T\log T-\frac{1+\log(2\pi)}{2\pi}T,$$

die nach Herrn von Mangoldt $O(\log T)$ ist, nicht $O(1)$ sein könnte.

Und indem ich zur Wirklichkeit zurückkehre, erwähne ich noch, dass es mir zwar nicht gelungen ist, Licht über die geheimnisvollen Zetanullstellen zu verbreiten, wohl aber ein neues Rätsel durch die Entdeckung der folgenden Tatsache (1912) aufzugeben, welche auf einen geheimnisvollen unbekannten Zusammenhang der Nullstellen mit den Primzahlen deutet. Es sei $x>0$, und ρ durchlaufe alle Nullstellen, die der oberen Halbebene angehören, nach wachsender Ordinate geordnet. Dann ist die Reihe

$$\sum_{\rho}\frac{x^\rho}{\rho}$$

divergent für $x=1$, $x=p^m$, $x=\frac{1}{p^m}$; konvergent für alle anderen $x>0$; gleichmässig konvergent in jedem Intervall $x_0 < x < x_1$, welches innen und an den Enden von jenen Divergenzpunkten frei ist; ungleichmässig konvergent in jedem Intervall $x_0 < x < x_1$, welches innen keinen Divergenzpunkt enthält, aber an mindestens einen solchen angrenzt.

Ich komme jetzt zu einigen anderen Untersuchungen über $\zeta(s)$. Es sind bei einer analytischen Funktion die Punkte, an denen sie 0 ist, zwar sehr wichtig; ebenso interessant sind aber die Punkte, an denen sie einen bestimmten Wert a annimmt. Zu beweisen, dass $\zeta(s)$ jeden Wert a annimmt, ist ein leichtes. Wo liegen aber die Wurzeln von $\zeta(s)=a$? Die erste Frage ist, welche Werte $\zeta(s)$ in der Halbebene $\sigma>1$ annimmt, aus der eine Umgebung des Poles $s=1$, z. B. durch einen Halbkreis mit dem Radius 1, herausgeschnitten ist. Herr Bohr hat 1910 die—für mich wenigstens—ganz unerwartete Tatsache bewiesen, dass $\zeta(s)$ in diesem Gebiet nicht beschränkt ist. D. h. die Ungleichung

$$|\zeta(s)|>g$$

hat bei gegebenem $g>0$ und gegebenem $t_0>0$ in der Viertelebene $\sigma>1$, $t>t_0$ eine Lösung. Daraus schlossen Bohr und ich 1910 in einer gemeinsamen Arbeit, einen Lindelöfschen Ansatz und die Arbeiten von Herrn Schottky und mir über gewisse Verallgemeinerungen des Picardschen Satzes uns zu Nutze machend: Wenn $\delta>0$ beliebig gegeben ist, so nimmt $\zeta(s)$ im Streifen $1-\delta<\sigma<1+\delta$ alle Werte mit höchstens einer Ausnahme an.

Dadurch entstand Hoffnung, die Riemannsche Vermutung zu widerlegen, indem es z. B. gelingen könnte nachzuweisen, dass $\zeta(s)$ in der Halbebene $\sigma>\frac{3}{4}$ den Wert 1 nicht annimmt; dann müsste ja $\zeta(s)$ daselbst alle übrigen Werte, insbesondere also den Wert 0 annehmen, während die Riemannsche Vermutung offenbar mit der Behauptung $\zeta(s)\neq 0$ für $\sigma>\frac{1}{2}$ identisch ist. Die Möglichkeit, auf diesem Wege das Riemannsche Problem zu lösen, verschwand aber dadurch, dass Herr Bohr 1911 die erstaunliche Tatsache nachwies, dass $\zeta(s)$ bereits im Streifen $1<\sigma<1+\delta$ jeden von 0 verschiedenen Wert annimmt, sogar unendlich oft.

Nun bewies Herr Littlewood in der schon mehrfach erwähnten Arbeit aus diesem Jahre noch den Satz: Für jedes $\delta>0$ hat in der Halbebene $\sigma>1-\delta$ mindestens eine der beiden Funktionen $\zeta(s)$ und $\zeta'(s)$ eine Nullstelle. Danach wäre also die Riemannsche Vermutung widerlegt, wenn man das Nichtverschwinden von $\zeta'(s)$ z. B. in der Halbebene $\sigma>\frac{9}{10}$ beweisen könnte. Die Untersuchung der Nullstellen in $\zeta'(s)$ ist ein schwieriges Problem, indem für $\zeta'(s)$ die schöne Produktdarstellung fehlt, welche das Studium von $\zeta(s)$ erleichtert. Aber auch hier wusste Herr Bohr Rat, und es gelang ihm in einer am 4. Juni dieses Jahres erschienenen Arbeit zu beweisen: $\zeta'(s)$ hat sogar in der Halbebene $\sigma>1$ eine Nullstelle, übrigens unendlich viele, so dass also der Littlewoodsche Satz zu keiner Lösung des Riemannschen Problems führen kann. Das fand Bohr mit dem Umwege über

$$\frac{\zeta'(s)}{\zeta(s)}=-\sum_p \frac{\log p}{p^s-1};$$

er stellte fest, dass diese Funktion in der Halbebene $\sigma>1$ den Wert 0 und sogar jeden Wert annimmt; übrigens sogar im festen Streifen $1<\sigma<1+\delta$, und zwar unendlich oft.

Zum Schluss meines Vortrags will ich erwähnen, dass meine für die Primzahltheorie geschaffenen Hilfsmittel sich auch kürzlich als geeignet erwiesen haben, andere Probleme aus der analytischen Zahlentheorie und über Abzählung von Gitterpunkten in gewissen mehrdimensionalen Bereichen zu lösen, welche vordem unerledigt geblieben waren. Ich habe dies in einer kürzlich erschienenen Abhandlung auseinandergesetzt und will hier nur einen ganz speziellen Satz daraus erwähnen. Die beiden Dirichletschen Reihen

$$f(s)=1-\frac{1}{3^s}+\frac{1}{5^s}-\frac{1}{7^s}+\frac{1}{9^s}-\frac{1}{11^s}+\dots$$

und

$$g(s)=1-\frac{1}{5^s}+\frac{1}{7^s}-\frac{1}{11^s}+\frac{1}{13^s}-\frac{1}{17^s}+\dots$$

konvergieren offenbar für $\sigma>0$. Ihr formal gebildetes Produkt ist wieder eine Dirichletsche Reihe

$$\sum_{n=1}^{\infty}\frac{c_n}{n^s};$$

dieselbe konvergiert natürlich für $\sigma>1$, wo ja die gegebenen Reihen absolut konvergieren; nach einem leicht beweisbaren Satze von Stieltjes (1885) über Dirichletsche Reihen konvergiert das formale Produkt sogar für $\sigma>\frac{1}{2}$. Andererseits hat Herr Bohr (1910) ein Beispiel zweier Dirichletscher Reihen

$$\sum_{n=1}^{\infty}\frac{a_n}{n^s}$$

und

$$\sum_{n=1}^{\infty}\frac{b_n}{n^s}$$

gebildet, die für $\sigma>0$ konvergieren, während ihr formales Produkt

$$\sum_{n=1}^{\infty}\frac{c_n}{n^s}$$

nicht über die durch den Stieltjesschen Satz gelieferte Gerade $\sigma=\frac{1}{2}$ hinaus konvergiert. Für mein obiges Beispiel $f(s)\,g(s)$ kann ich aber Konvergenz für $\sigma>\frac{1}{3}$ beweisen. Dies spezielle Beispiel repräsentiert natürlich zwei Reihen vom Typus

$$\sum_{n=1}^{\infty}\frac{\chi(n)}{n^s},$$

wo $\chi(n)$ ein sogenannter Charakter nach einem Modul k ist. Diese Reihen waren durch Dirichlet beim Beweise des Satzes von der arithmetischen Progression eingeführt, und für jedes Paar solcher Reihen, wenn nur keiner der beiden Charaktere Hauptcharakter ist (sonst konvergiert bekanntlich das Produkt überhaupt nicht einmal für $s=1$) kann ich Konvergenz für $\sigma>\frac{1}{3}$ beweisen.

Ich bitte um Entschuldigung für die Länge meines Vortrags; aber ich habe ohnehin zahlreiche unter mein Thema fallende Dinge nicht berührt. Umfasst doch allein das Litteraturverzeichnis meines 1909 erschienenen *Handbuchs der Lehre von der Verteilung der Primzahlen* mehr als 600 Abhandlungen. Und habe ich doch durch dies Handbuch erreicht, dass zahlreiche Forscher sich diesem interessanten Gebiet zuwandten, so dass seitdem viele weitere Arbeiten darüber erschienen sind. Ich würde mich freuen, wenn es mir durch diesen Vortrag gelungen sein sollte, den einen oder anderen Mitarbeiter noch hinzuzugewinnen.

THE PRINCIPLES OF INSTRUMENTAL SEISMOLOGY

BY PRINCE B. GALITZIN.

Although the accounts of great earthquakes, accompanied by severe loss of human life and property, can be traced back to the most remote historical times, it is only perhaps about 20 years that the science of earthquakes or *seismology*, considered as an exact and independent scientific discipline, exists. Former investigations concerning earthquakes, of which there were certainly no lack, and which were mainly due to the work of geologists and possessed of course their intrinsic scientific value, treated the question more from a descriptive or statistical point of view. The speedy advance and development of seismology in the last 10—20 years, of which we all are witnesses, is nearly exclusively due to the fact that seismology has seated itself down upon a sound scientific basis in adopting purely physical methods of research, based upon *instrumental observations*. This branch of seismology, i.e. *instrumental seismology* or so-called *seismometry*, in devising its instruments of research, stands in close connection with theoretical mechanics, so intimately linked with pure mathematics. In this evolution of modern seismology the lead has gradually been handed over from geologists to astronomers and physicists, who are more familiar with the wielding of the mathematical apparatus.

Pure mathematics has had such a predominant and benevolent influence in the development of all branches of natural philosophy, that it may be perhaps quite appropriate to bring forth at this mathematical congress a brief outline of the principles of instrumental seismology.

The study of the movements of the earth's surface brings forth a whole set of purely mathematical problems, to some of them I shall allude in the course of my address, and in the resolving of which seismologists are highly in need of the scientific help of pure mathematicians.

According to modern views concerning the constitution of the planet we live upon, the earth consists of an outer shell made up of more or less heterogeneous rocks under which at a depth of say 30 or 50 kilometers lies a layer of plastic magma, which forms the material that feeds volcanoes during their periodical outbursts. What lies underneath we are not very sure about. Evidently there is an interior core, which, although at a very high temperature, is subjected to such enormous pressures, quite inaccessible in our laboratory investigations, that matter in that state possesses for us as it were the properties of a solid body. According to Wiechert's investigations, based upon the velocity of propagation of seismic rays, the physical properties of this interior core, partly formed of nickel and iron, which would well account for the mean density of the earth as a whole, vary as we go deeper and deeper down, but this variation is not always a continuous one. Wiechert's latest investigations lead to the conclusion, that there are three layers, seated respectively at the depths of 1200, 1650 and 2450 kilometers, where a sudden, discontinuous change in the physical properties of the inner core takes place.

So much for the inner constitution of the earth.

As regards the causes of earthquake outbreaks, these are traced back at present to three different origins.

The first class of earthquakes is intimately connected with the activity of volcanoes; the second class is due to the yielding of underground hollows. Both types of earthquakes have mainly a local importance and are seldom felt at a great distance from their respective epicentres.

By far the most important and devastating earthquakes are the so-called tectonic earthquakes, which are due to the relative shifting of underground layers of rocks. Some of these layers exhibit a most intense folding, brought into being by the gradual cooling and shrinking of the earth and where therefore the conditions of elastic equilibrium are very unstable. This unstability is sometimes so high, that it needs only a small exterior impulse in order that the limits of elasticity should be exceeded. Then a sudden shifting of the layers takes place, accompanied by the outburst of a tectonic earthquake. As the focus or hypocentre of a tectonic earthquake lies usually comparatively deep, say 10—20 kilometers, the shock is felt at very great distances from the epicentre and sensitive seismographs record the quake all over the earth's surface. In this case we have the so-called world-recorded earthquakes, a great many of which, more or less intense, occur every year.

The original displacements engendered at the focus are transmitted through the earth's body according to the laws of propagation of elastic waves, which on reaching the earth's surface set in motion the respective seismographs installed there. These, when the problem of modern seismology is rightly understood, are always meant to give, after a shrewd analysis of the curves obtained, the *true* movement of a particle of the earth's surface at the given point.

To simplify matters the focus and epicentre are usually considered for not too small epicentral distances as being located in a point. This is certainly only a first approximation, but, considering the great dimensions of the earth, this approximation in most cases holds good.

The problem of propagation of seismic waves is therefore nothing else than a problem of the theory of elasticity, according to which in an isotropic medium two distinct types of waves are propagated independently one from another, viz. longitudinal or condensational and transverse or distortional ones.

The respective velocities of these waves V_1 and V_2 are given by the following well-known formulae:

$$V_1 = \sqrt{\frac{1-\sigma}{(1+\sigma)(1-2\sigma)} \cdot \frac{E}{\rho}},$$

$$V_2 = \sqrt{\frac{1}{2} \cdot \frac{1}{1+\sigma} \cdot \frac{E}{\rho}},$$

where E denotes Young's modulus, σ Poisson's coefficient and ρ the density of the medium.

Taking, according to Poisson, σ equal to $\frac{1}{4}$, we obtain

$$\frac{V_1}{V_2} = \sqrt{3} = 1{\cdot}732.$$

Neither of these velocities is constant, but depends upon the depth of the corresponding layer.

For the upper strata of the earth's crust these mean velocities can be easily deduced from seismometric observations in the vicinity of the epicentre of an earthquake. The observations give

$$V_1 = 7{\cdot}17 \text{ kil./sec.}$$

and

$$V_2 = 4{\cdot}01 \text{ kil./sec.},$$

therefore

$$\frac{V_1}{V_2} = 1{\cdot}788$$

or

$$\sigma = 0{\cdot}27,$$

which is very near the number adopted by Poisson for the majority of isotropic bodies.

The longitudinal waves, travelling the quickest from the focus to the station of observation, mark on their arrival the beginning of the first preliminary phase of an earthquake, usually denoted by the symbol P (undae primae). After an interval of time, usually a few seconds, the seismograms show a sudden deviation of the seismographic record. This moment corresponds to the arrival of the first transverse waves and is denoted by the symbol S (undae secundae).

The greater the distance Δ to the epicentre, reckoned along the great circle joining the epicentre with the station of observation, the greater will be the difference between S and P.

From observations made during several known earthquakes, with well-located epicentres, special tables have been computed, which give the possibility of determining the epicentral distance to a given station from the difference of time of arrival of the first longitudinal and distortional waves.

Besides these two characteristic types of seismic disturbances, which can be deduced theoretically and whose existence is confirmed by direct observation, there exists another class of waves, the so-called gravitational, long or surface waves (undae longae), which travel along the surface of the earth with a constant mean velocity V of about 3·53 kil./sec.

On searching for particular integrals of the general equations of the theory of elasticity and taking into consideration the boundary conditions for the earth's surface, Lord Rayleigh and H. Lamb have shown that the existence of these long waves can be proved theoretically. A remarkable result of this investigation is, that the velocity of these surface waves forms quite a definite fraction of the velocity V_2 of propagation of the distortional waves in the uppermost layers of the earth's crust, namely

$$V = 0{\cdot}9194\, V_2.$$

Taking $V_2 = 4{\cdot}01$, this would give $V = 3{\cdot}69$ kil./sec., which differs only slightly from the above given number, deduced from observations at Pulkovo during the great Messina earthquake on December 28, 1908.

The arrival of the long surface waves at a given station constitutes the beginning of the real maximal phase on a seismogram. At its beginning and during both preliminary phases the movement of an earth's particle in a given direction is a very intricate one, consisting of a superposition of several, rather irregular waves; but with time, in the maximal phase itself, the amplitude gets to be much larger and the movement much more regular, displaying often a whole trail of regular harmonic vibrations. The corresponding periods of these wave-form movements vary usually between 12 and 20 seconds or more.

After the maximal phase is over, the movement becomes again very irregular, but some particular periods, say 12 and 18 seconds, seem to predominate. According to Wiechert's views these movements are due to proper vibrations of the earth's outer shell or crust, a question which has not as yet been thoroughly investigated and where there is ample work for pure mathematical research, which would be most welcome to seismology. This latter part of a seismogram is known under the name of Coda, but its physical meaning is far from being clear and evident even up to the present day.

On the adjacent plate, giving the reproduction of a seismogram traced at Pulkovo by an aperiodic horizontal pendulum, with magnetic damping and galvanometric registration, during an earthquake in Asia Minor on February 9, 1909, the different above-mentioned phases can easily be traced.

This is a very typical seismogram in which all the phases are particularly well seen. Usually the first phase (P) can be easily detected, but there is sometimes much trouble about fixing the beginning of the second phase (S), when the seismographs are not of a highly sensitive type, and this difficulty may have, as we shall see later on, its theoretical reason.

The determination of the velocity of propagation of the surface waves offers no trouble, when it is possible to fix the time of arrival of one and the same maximum at stations, situated at different distances from the epicentre. But the same velocity can also be deduced during big earthquakes by comparing the times of arrival of a certain maximum in the trail of surface waves, which arrive at the point of observation by the shortest way from the epicentre (t_1) and in the trail of waves which reach the observatory from the other side, after having travelled round the earth's surface (t_2). In this case, if we denote with Δ the epicentral distance, we shall have

$$V = \frac{40000 - 2\Delta}{t_2 - t_1} \text{ kil./sec.}$$

By measuring also the corresponding parts on the seismogram and deducing therefrom the amplitudes of the true movement of an earth's particle, it is possible to calculate the coefficient of absorption of seismic surface energy. Measurements made at Pulkovo give for this coefficient $0{\cdot}00028$ when the distances are measured in kilometers. This would mean, that the energy of the surface waves at a point situated at 180° from the epicentre, i.e. in the antiepicentre, would be only the $\frac{1}{270}$th

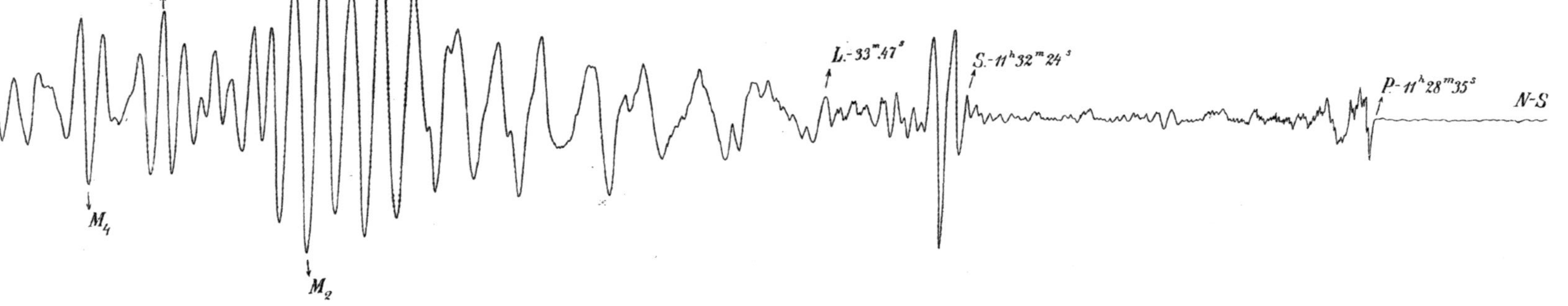
N-S
P.-11h28m35s
S.-11h32m24s
L.-33m.47s
M1
M2
M3
M4

part of the original energy in the epicentre itself. It is not to be forgotten, that in the antiepicentre the different seismic surface waves meet again together in one and the same point.

The above-mentioned theory of surface waves, based upon the general differential equations of the theory of elasticity, leads to another very important and interesting result, and that is, that when a trail of surface seismic waves passes a given place each particle of the earth's surface describes an ellipse, whose vertical axis is 1·47 times the corresponding horizontal one. This curious theoretical relation has not been as yet thoroughly tested, but the preliminary calculations based upon observations with highly damped, in fact aperiodic, seismographs made at Pulkovo show that this ratio is somewhat smaller than the theoretical value 1·47, and it varies a little from one trail of waves to another. The greatest value observed up till now did not exceed 1·28.

This fact does not mean in the least that there is a real contradiction between theory and observation. This discrepancy may be and is most probably due to the fact that the theory does not take into account the damping effect upon the waves as they travel along the surface of the earth, and this damping may be totally different for vertical and horizontal displacements; except that observations seem to point to the fact that the period of seismic surface waves augments with the path traversed by the waves along the surface of the earth. This question opens another vast field for delicate mathematical enquiry.

Returning to the two types of waves, which, starting from the focus of an earthquake, travel through the interior of the earth, it is more practical and convenient to consider, as in optics, not the seismic waves themselves but the corresponding seismic rays.

If the law giving the relationship between velocity and depth wei e exactly known, there would be no difficulty in calculating the theoretical path traversed by the seismic rays. According to Fermat's principle seismic rays must travel along brachistochronic paths. Expressing this result in mathematical language one is led to two integral expressions, giving in a general form the time T, required for a seismic ray to travel from the focus to the point of observation, as well as the epicentral distance Δ, as function of the angle of emergency e of the ray and of the depth of the focus h. By angle of emergency one understands the angle formed between the tangent to the seismic path, where the latter strikes the earth's surface, with the horizontal plane at the point of observation.

If one eliminates e from both these integral expressions, one obtains T as a function of Δ and h, which gives the form of the so-called theoretical hodograph for both types of seismic rays:

$$T = F(\Delta, h).$$

To calculate F one must make certain assumptions concerning the dependence of the velocity upon depth. Even very simple assumptions, holding good for only small intervals of depths, lead to rather awkward and clumsy mathematical expressions, which nevertheless are fairly adapted for numerical calculations.

The formula above shows distinctly that the form of the hodograph is dependent on the depth of the focus and this dependence makes itself principally evident for small epicentral distances, a fact that has not been sufficiently considered as yet by seismologists of the day. The corresponding curve has also a particular point of inflexion. From observations of the times of arrival of the longitudinal seismic rays at stations, situated at different distances Δ from the epicentre, it is possible to deduce the depth of the focus of the corresponding earthquake. This is a very delicate problem, which has seldom been treated in a satisfactory manner, the observations being usually too faulty, as this method requires very exact absolute time-reckoning, considering the comparative great velocity of propagation of the longitudinal waves in the upper layers of the earth's crust.

I have lately applied this method of determining h to the big earthquake felt in the south of Germany on November 16, 1911, and for which we have more or less reliable observations from a whole set principally of German stations, surrounding the epicentre and furnished with damped seismographs.

As result I obtained

$$h = 9{\cdot}5 \text{ klm.}$$

with a mean error of $\pm 3{\cdot}8$ klm.

Considering that h is usually small, one can as a first approximation, when Δ is not very small, suppose the focus to coincide with the epicentre; then one has simply

$$T = F(\Delta),$$

where T is the difference of times of the arrival of the first seismic waves at a given point and the outbreak of the earthquake in the epicentre itself.

The form of this mean hodograph can be deduced from direct observations of earthquakes with well-located epicentres. Seismologists have been very busy lately in trying to perfect these mean hodographs, but there is still a lot of work to do.

It is even to a certain point doubtful if such a mean hodograph really exists. Strictly speaking every earthquake has its own particular hodograph, depending upon the depth of the focus; but even if we admit that $h = 0$, still T may be dependent upon the direction from which the seismic rays reach the observatory and in general upon the geological formation of the layers of the earth's crust, through which these particular rays have travelled.

Then it is highly probable, and this fact has been emphasized by Professor Love, that there exists, in analogy with optics, a seismic dispersion. In that case the velocity of propagation of seismic waves would be dependent on their period and each particular trail of waves would have its own hodograph.

This question of seismic dispersion is of the highest importance for modern seismology, but it is a mathematical problem which has not been thoroughly investigated as yet.

When the form of the hodograph for the longitudinal waves is known, it is very easy to determine the angle of emergency e for a given seismic ray.

One has for that the following formula:

$$\cos e = V_1 \frac{dT}{d\Delta},$$

where V_1 is the velocity of the longitudinal waves in the uppermost layers of the earth's crust.

This simple and elegant formula can be deduced from the general integrals, giving T and Δ as functions of e, or through a very simple geometrical process. It holds good for any given depth of focus, if only one takes the corresponding hodograph.

We have just seen that the form of the mean hodograph depends upon the law governing the relationship between velocity and depth. This law is not known *a priori*, but plotting out the form of the hodograph from direct observations, it is possible to go backwards and deduce conclusions concerning the velocity of propagation of longitudinal and torsional waves at different depths. This is the method followed by Wiechert and his scholars and which has led to many very interesting conclusions concerning the interior constitution of the earth, whereby Wiechert has applied with success one of Abel's purely mathematical theorems.

The principal result of this investigation is, that both velocities V_1 and V_2 augment with the depth, attaining at the depth of about 1500 kilometers respectively the values of 12·7 and 6·9 kil./sec., after which they remain for some time practically constant.

In the very inner core of the earth, from a depth of about 3000 kil. inwards, these velocities seem again to diminish. The law of variation of V_1 and V_2 with depth is not, as has been stated before, always a continuous one.

The study of seismic rays opens as we see a whole new field of enquiry into the interior constitution of the earth, and at such depths which are utterly out of reach of the investigations of the geologist. These rays, travelling through the mysterious and inaccessible regions of the interior realm of our planet, bring with them some information of what nature has wrought in those profound depths. Like the optical rays, coming from the heavenly bodies, which bring with them some learning concerning the chemical constitution of the different stars and by the shifting of the spectral lines enable us to determine their translatory velocity in the line of vision, the seismic rays, when carefully studied and catalogued, will certainly enable us in future to solve the mystery of the earth's interior constitution.

The knowledge of V_1 and V_2 at different depths does not allow us to determine separately Young's modulus E and the density ρ. All that we can obtain from our theoretical formulae at present is the ratio $\frac{E}{\rho}$ and Poisson's coefficient σ, but that is already a vast achievement. These investigations show in particular that σ up to very great depths retains its same numerical value as at the earth's surface, i.e. 0·27.

A remarkable feature revealed by instrumental observations of earthquakes concerns the length of time of the earth-crust's shiverings at distant stations and near the epicentre itself. Whereas in the interior boundaries of the epicentral area nearly every earthquake is characterized by several distinct more or less intense shocks, separated by short intervals of utter calm, the whole seldom lasting for more than a few minutes, the seismograms obtained at distant stations reveal a continuous movement of the earth's surface, without any interruptions, which might correspond to the occasional lulls in the epicentral area itself, the whole lasting for ever so long, sometimes over an hour and more.

The reason of this expansion of the earth's quiverings in time may be due to three different causes.

Firstly, besides the usual types of seismic waves that reach the point of observation by brachistochronic paths, there may be a whole set of other waves, which come to the same point after having suffered different reflexions and refractions among the interior heterogeneous layers of the earth's crust, as well as one or more reflexions at the earth's surface. These latter waves have already been studied and discussed.

Another striking fact in connection with inner refraction and reflexion seems to present itself and that is, that for certain big epicentral distances the seismic rays seem not to reach the earth's surface at all, throwing on it as it were a seismic shadow of the interior core.

Secondly, it is highly probable that seismic waves of high intensity are able to start proper oscillatory movements of the earth's outer shell, characterized by special periods of vibration. The fact is known that one severe earthquake can set loose at a very considerable distance another twin-earthquake.

Thirdly, if seismic dispersion exists, and according to the general laws of physical phenomena it must exist, different types of waves with different periods would reach the point of observation at different times and this reason would also well account for the stretching out of seismograms at distant stations.

Referring to seismic dispersion, it may be stated here, that if one repeats the remarkable investigations of Lord Rayleigh and H. Lamb concerning the properties of seismic surface waves, introducing into the general equations of elasticity a frictional term, one is immediately led to a seismic dispersion, which in comparison with optics may be characterized as an anomalous dispersion, for the shorter the period is, the greater will the velocity of propagation of the corresponding wave be. The equations lead also naturally to a certain damping effect, characterized by the fact, that the shorter the period, the greater will the damping be.

Both facts seem to stand in agreement with the results of direct observations, although the question has not been studied out thoroughly as yet for lack of sufficient trustworthy and reliable data and on account of the novelty of the problem itself.

In the different theoretical investigations referred to until now it has always been implicitly admitted that the different interior layers of the earth, considered in their mean distribution in a particular shell, although varying in their physical properties with depth, still retain the properties of an isotropic medium.

Seismology at the present date, in studying the propagation of seismic waves, cannot possibly take into consideration the heterogeneity of different rocks in the earth's crust, revealed by accurate geological research, and is necessarily obliged to suppose some mean distribution of different layers to get hold of the general outlines of the phenomena to study. When these are settled, it will be a further problem to take into consideration different discrepancies, caused by various geological anomalies. This way of treating the problem seems plausible enough, especially if we go down to depths exceeding the surface of isostacy, i.e. 120 kilometers.

But are we right, in adopting even such a mean distribution of matter in each layer, to consider the corresponding medium as an isotropic body?

Rudzki finds that we are certainly wrong and in that respect I suppose he is right enough, and it is only the difficulty of the problem that has obliged seismologists to have recourse to the equations of the theory of elasticity meant for isotropic bodies.

The results obtained can be certainly of the highest importance, but nevertheless they must be considered as only a first approximation to reality.

The problem, treated in a general way for anisotropic bodies, offers certainly very great difficulties and here again seismology is very much in need of the friendly help of the pure mathematician.

Notwithstanding the difficulties of the question Rudzki set bravely to work at the problem for the case of a medium, which he defined as transverse isotropic.

Starting from the expression of the potential of the elastic forces, as given by Professor Love, Rudzki studied the form of the meridian section of the surface of a seismic wave, propagated through the interior of the earth, which led him after much trouble to a curve of the 12th degree, containing three special branches. One branch of elliptical shape corresponds to the longitudinal, whereas the two others correspond to distortional waves. One of these latter branches has a most curious shape, exhibiting four double points and eight cusps (points de rebroussement). When this particular part of the wave surface would come to strike the surface of the earth at the point of observation, the second phase S on the seismograms could not be sharply defined and under certain circumstances would correspond to three different shocks, following closely one after the other.

It would be most desirable that such investigations started by Rudzki should be taken up by mathematicians and worked out with a view of application to different seismological problems.

Modern seismology, which has led already to so many interesting and important results and which opens a vast field for further theoretical and experimental enquiry, is based upon instrumental observations, furnished by self-recording seismographs. The latter must be so adjusted as to be able to give after a careful analysis of the records the true movement of a particle of the earth's surface. What we want are not seismoscopes, which only give relative indications of the earth's quiverings, but trustworthy seismographs.

The problem, considered from a kinematical point of view, stands thus.

Taking a small element of the earth's surface and placing in it the origin of a fixed system of rectangular coordinate axes, this element may experience, as the result of a near or distant earthquake, six different movements, viz. three displacements parallel to these axes and three corresponding rotations. The rotations about horizontal axes correspond to a tilting of the ground. The complete study of the earth's movements at a given point requires therefore six different seismographs, each of which is to give the corresponding movement of the earth's surface as a function of time and this for the whole duration of the corresponding quake. This is the generalized, fundamental problem of modern seismometry which offers so many practical difficulties that at the present date it has never been treated in this exhaustive way.

Although direct, non-instrumental observations in localities which have suffered much from earthquakes prove undoubtedly that all these six movements do in reality exist, still for distant earthquakes the tilting of the ground is comparatively so small, seldom exceeding $\frac{1}{50}''$, that at the present date seismologists confine themselves only to the study of the three displacements, one in the vertical and the two others in horizontal directions at right angles to each other.

To study the *absolute* movements of a particle of the earth's surface, which alone is of real practical importance for seismological investigations, one requires a fixed point not taking part in the movements of the upper layers of the earth's crust. As no such point is to be had, seismology is obliged to have recourse to the principle of inertia and seek this immovable point in the centre of oscillation of a pendulum in its so-called *steady point.* But this point is only steady for a sudden impulse communicated to the ground. After that, the movement of the ground continuing, the pendulum begins to sway to and fro and its real movement gets very intricate, being a combination of enforced vibrations caused by the real movement of the earth's surface together with the proper movement of the pendulum itself, which is usually a harmonic, slightly damped oscillation.

In former times simple vertical pendulums were much used. But as it was found necessary, in order to augment the sensitiveness of the record obtained, to give the pendulum a very long proper period of oscillation, one passed gradually to horizontal pendulums, where the axis of rotation forms a very small angle with the vertical line and where the proper period of oscillation can be in consequence made very long. A horizontal pendulum is a very handy apparatus which can be made highly sensitive and where the movement of the pendulum can be inscribed mechanically by a pin on smoked paper fixed upon a revolving drum, or optically by means of a beam of light reflected from a small mirror attached to the pendulum and concentrated thereafter on a revolving drum covered with sensitive photographic paper. In Milne's horizontal pendulum the beam of light passes through two slits at right angles to each other, one of which is fixed to the boom of the pendulum itself. To study the true horizontal movements of the ground two such pendulums are required at right angles to each other.

There are many types of horizontal pendulums, which, although having various outward looks, differ essentially only by the mode of suspension of the heavy mass.

There are pendulums with two pivots: type Rebeur-Paschwitz; then pendulums with one pivot below and where the boom is upheld by two strings or wires going upwards and fixed to the column of the apparatus: type Omori-Bosch, Milne. When mechanical registration is used, it is necessary, in order to diminish the perturbating influence of the friction of the recording pin on the smoked paper, to use very heavy masses. But in this case the pressure sustained by the pivot is very great and it easily gets blunt, in consequence of which the sensitiveness of the apparatus will be diminished and its position of equilibrium may get shifted. To avoid this fatal effect of the blunting of the pivot one can use instead of a steel point an inverted flat steel spring; then there is no more trouble about the blunting and one can use very heavy masses. This arrangement is adopted in Mainka's heavy horizontal pendulums and in those horizontal seismographs which have been installed lately at Russian second class seismic observatories.

By far the most delicate and sensitive suspension is the Zöllner suspension, where there are no pivots and therefore scarcely any friction at all and where the boom of the pendulum with the heavy mass is supported by two inclined wires, the one going up and the other down. This type of pendulum is particularly adapted for the study of distant earthquakes and the tidal deformations of the earth under the influence of the attraction of the sun and moon. Such pendulums are used at the Russian first class seismic observatories.

Simple vertical pendulums with a comparatively short period and where the movement is divided by means of a simple mechanical contrivance into two rectangular components are still used in Italy: Vicentini's pendulum.

Another type of horizontal seismograph, installed at many seismic observatories, is the Wiechert's inverted, astatic pendulum, where the heavy mass is above and rests upon a pivot or rather on cross flat springs fixed below. In this position the pendulum is in an unsteady state of equilibrium, but is prevented from toppling over by means of crossed springs attached to the heavy mass.

There are certainly other types of horizontal seismographs, but the principal ones have been mentioned above.

For the study of the vertical component of the movement of the ground special vertical seismographs are used, although a systematic study of this component has only begun just lately and there are still very few observatories that possess a corresponding instrument. These seismographs are based upon the use of a spring.

The simplest type of a vertical seismograph is Vicentini's instrument, where the heavy mass is simply attached to a strong flat spring; but as the proper period of oscillation of such a spring is very short, the sensitiveness of the records is comparatively small.

In Wiechert's seismograph the heavy mass is suspended by several powerful vertical spiral springs, whereas in the Russian type of vertical seismograph, a copy of which is installed at the observatory of Eskdalemuir in Scotland, the boom carrying the heavy mass revolves about a horizontal axis placed at the end and is upheld in a horizontal position by only one spiral spring. By fixing the lower point of suspension of this spring under the centre of gravity, it is possible to obtain a comparatively long proper period of oscillation of the instrument, say 13 to 14 seconds, conserving at the same time the necessary stability for continuous seismometric work.

To augment the sensitiveness of all types of seismographs, especially when mechanical registration on smoked paper is used, one has recourse to magnifying levers. These latter sometimes cause a good deal of trouble, being much subjected to the influence of changes of temperature and introducing special errors in the records owing to the inevitable failures in the fixing of the joints. It is always preferable, when seismometric work of a high degree of accuracy is planned, to use the optical method of registration, which requires no levers and introduces no friction.

It is true that the sensitiveness of this optical method of registration has a certain limit not to be exceeded, as the sensitiveness of photographic paper does not admit placing the revolving drum at distances greater than 4 meters from the seismograph. In this case, when a high sensitiveness of the records is desired, it is very advantageous to use the galvanometric method of registration, which is very simple in its theory and practice, requires no levers, introduces no friction and where the magnification can be pushed to limits far exceeding those of other practical methods of registration.

This method, which can be easily applied to all types of instruments, consists in fixing to the end of the boom of the corresponding seismograph several flat induction coils, placed between the poles of a pair of strong horse-shoe magnets. Now, when the boom is set in motion, electric inducted currents are generated in the spires of the coils, their intensity being proportional to the angular velocity of the displacement of the boom. This electric current is led by two wires to a highly sensitive dead-beat galvanometer of the Deprez-D'Arsonval type, whose movements are registered on a revolving drum by the usual optical method by means of a beam of light reflected from a small mirror attached to the moveable coil of the galvanometer. In this way it is not the amplitudes of the seismograph's movements, but the corresponding velocities which are directly recorded, which makes no real practical difference for the study of seismic harmonic waves.

This galvanometric method of registration offers several important advantages.

First of all, one can place the galvanometer and registering part of the seismograph at any desirable distance from the pendulum itself in a convenient and easily accessible place and not disturb the seismograph every day or even twice a day when changing the paper on the revolving drum. This possibility of registering at a distance is certainly very convenient.

Secondly, as what one registers corresponds to the velocities and not to the amplitudes of the pendulum movement, the normal position of equilibrium of the pendulum itself is no longer of such importance; it may shift a little without causing any trouble at all. This is particularly important for the vertical seismograph, where the steel springs are so subject to the influence of a change of temperature, that they usually require a special adjustment for temperature compensation. With the galvanometric registration no such compensation is necessary.

Thirdly, the magnification obtained with this method is very great; for certain periods of waves it can be easily made to exceed 1000 and this result is attained without the use of any levers at all, which gives the possibility of using comparatively small masses of only a few kilograms, which makes the instruments compact and handy. The sensitiveness of the instrument can thereby be easily changed between very large limits by simply changing the distance of the poles of the corresponding horse-shoe magnets.

Lastly, the galvanometer being a most sensitive apparatus, it is possible to place the revolving drum comparatively near, say 1 meter from the mirror of the galvanometer, which enables one to obtain very sharp and distinct records. In fact the curves come out so clearly and of such magnification, that it is nearly always possible to fix the different phases on the seismograms. The observations made at Pulkovo show in fact, that very often one is able to locate an epicentre from one station alone by a special method, shortly to be described, when other stations fail to give even the beginning of different phases.

This method of registration has been adopted for all Russian first class seismic observatories and by some observatories abroad. This method is also used at Eskdalemuir.

It would seem desirable if possible to go yet a step further and by introducing induction currents of the second order to get at the direct registration of the acceleration of the pendulum movement, which would be of the greatest importance for the study of the true movement of the ground. This idea was suggested in a paper published by Professor Lippmann, but the experiments he and I made show that these secondary currents are so feeble that no practicable application of this principle is possible at present.

Let us now consider what conditions a seismograph must fulfil to be able to record in the most trustworthy way possible the true motion of the ground.

The motion of a horizontal or vertical seismograph subjected to displacements of the ground in a given direction, say $x=f(t)$, can be represented by a differential equation of the following form:

$$\theta'' + 2\epsilon\theta' + n^2\theta + \frac{1}{l}x'' = 0 \quad \ldots\ldots\ldots\ldots(1),$$

where θ is the angle of deviation of the instrument, ϵ a constant, which depends upon the rate of damping, $T=\frac{2\pi}{n}$ the proper period of oscillation for $\epsilon=0$ and l the reduced pendulum length.

The motion of a dead-beat galvanometer, coupled with the seismograph, is governed by the following equation:

$$\phi'' + 2n_1\phi' + n_1^2\phi + k\theta' = 0 \quad \ldots\ldots\ldots\ldots(2),$$

where ϕ is the angle of deviation of the galvanometer, $T_1=\frac{2\pi}{n_1}$ its proper normal period and k a constant, which characterises the magnification of the record.

Supposing, to simplify matters, that x corresponds to a trail of harmonic seismic waves with the period T_p; θ as a function of the time t will be represented by a curve, being a superposition of enforced oscillations with the same period T_p and the proper movement of the pendulum itself. To get rid as soon as possible of this latter movement and obtain a more truthful record of the movement of the ground it is indispensable to augment the damping of the instrument, driving it if possible to the limit of aperiodicity and making it dead-beat.

Then one obtains a very simple formula, which enables one to calculate very easily the true amplitude of the movement of the ground. The same holds good also for galvanometric registration. Special tables, lately published, make all the necessary calculations very easy and simple indeed. When damping is introduced there is always a small shifting of phase, which can easily be taken into consideration, but which unfortunately most seismologists neglect.

There is no doubt that aperiodic pendulums give much more faithful records, even if x is a complicated function of t. This has been proved by experiments with a moveable platform.

Some seismologists seem to be afraid to use dead-beat pendulums, notwithstanding all their theoretical and practical advantages, fearing that the sensitiveness of the records will be too greatly diminished. But this is, generally speaking, an utter error; all depends on the proper period of oscillation T of the seismograph itself and on the period T_p of the corresponding seismic wave. For instance, an aperiodic pendulum with a proper, normal period of 25 sec. will be, beginning with periods of seismic waves of 16 sec., *more sensitive* than a 12-second pendulum with a comparatively small damping ratio of 5 to 1. For $T_p = 40$ sec. the sensitiveness will be twice as great.

Therefore there is nothing to fear from aperiodicity, one must only take a corresponding longer proper period T. But even if the magnification came to be too short for some seismic waves, galvanometric registration would compensate tenfold all the possible loss in sensibility.

On the contrary, what one ought certainly to be afraid of and avoid is the use of wholly undamped seismographs, as their records sometimes come out very distorted and give quite an erroneous impression of the true movement of the ground. Consider, as example, only the case of a faint trail of seismic waves passing the observatory, but whose period T_p is very close to the proper period of oscillation of the pendulum itself. The corresponding seismogram would show then very large amplitudes of oscillation, which would have nevertheless nothing to do with the true motion of the ground, being only a casual effect of resonance. Of course with aperiodic instruments the magnification is not constant, but depends also on the period of the waves and this dependence must be taken into consideration when working out seismograms; but still the general outline of the record is very similar to the true motion of the ground, at all events it can be easily deduced therefrom. When undamped seismographs are used it requires a minute and tiring analysis of the curves to eliminate the proper motion of the pendulum and get at the true amplitudes and periods of the corresponding seismic waves, and no man will ever give himself the immense trouble to go over all this work for each earthquake, although the study of the true motion of the ground is the most important problem of modern seismology. Simple experiments with a moveable platform show in the most evident way how misleading undamped seismographs can be.

The advantages of damping are theoretically and practically so evident that there are few seismologists of the day who would venture to uphold a contrary opinion, but this fundamental maxim, i.e. that in order to study the true motion of the ground only damped seismographs ought to be used, has not yet found its practical realization in all countries. It is to be hoped nevertheless that with time all seismographs, of whatever type, will be furnished with some one or other damping adjustment; the scientific value of the records obtained thereby will be greatly increased.

It is also necessary, in order to be able to study the short seismic waves, which are sometimes of great importance, that the revolving drum should rotate with a sufficient velocity and as uniformly as possible. As a desirable minimum one might propose 15 m./m. in the record to the minute; in Russia the velocity for all the new instruments is double as great.

There are different means of introducing damping in a seismograph. Some instruments have air-, others oil-damping, but one of the simplest means to effect the same is to use magnetic damping. One has only to fix to the boom of the corresponding seismograph a copper plate placed between the poles of two strong horse-shoe magnets. By bringing the poles nearer to each other, the damping ratio can be augmented and the instrument easily brought to the limit of aperiodicity.

The advantages of this kind of damping is its extreme simplicity and the possibility of being easily adjusted to any type of seismograph. Moreover, it agrees absolutely with the theoretical conditions of equation (1), for according to the physical laws of electro-magnetic induction the momentum of these retarding forces is strictly proportional to the angular velocity of the corresponding instrument. No other form of damping conforms strictly to this law.

Returning now to the general problem of seismometry, i.e. to the determination of a given component of the movement of the ground, say $x=f(t)$, for a given interval of time, for instance from P up to the Coda, we meet with considerable theoretical and practical difficulties. Poincaré and Lippmann proposed in a general way to integrate the equation (1) term by term between 0 and a given interval of time t. This would lead to two squarings of the seismographic record and would seem simple enough, if only the function $x=f(t)$ had no singular points with two tangents (points angulaires). But these singular points do in reality exist and correspond to the times when a new trail of waves strikes the point of observation. Nevertheless the problem can be solved, although those who, like Pomerantzeff and Arnold, have busied themselves with the question have found it a most delicate and troublesome problem, requiring immense accuracy and circumspection.

But supposing that we have found by some means or other x as a function of t and this for all three rectangular displacements, there arises a new problem, where the aid of pure mathematicians would again be of the greatest importance.

We have supposed that we have found $x=f(t)$, but what we want to know is of what elements this curve is built up. Guided by the general laws of physical phenomena we may well admit that $f(t)$ consists of a superposition of waves, having their respective amplitudes, periods, initial phase and last but not least special damping ratio. But how are we to separate these waves one from another, even if we manage to decipher in the curve one or two particular periods? The methods of usual harmonic analysis would fail in this case, as we must necessarily admit that all these vibrations are damped. Here is a problem of seismology which will certainly stand before us in the future and it would be highly gratifying if pure mathematics would clear the way to its solution before the problem itself practically arises.

If I may allow myself a short digression and if it is not too bold of me to say so in such a distinguished assembly of pure mathematicians, I should like to express the wish that mathematicians should look down a little more condescendingly on the impending necessities of physical research. There are in fact so many different questions and problems brought up by physics and its sister sciences which remain at a standstill owing to the mathematical difficulties they involve. Either we obtain a differential equation which cannot be integrated, or we meet with some clumsy integral which does not admit of evaluation and which we are quite ready to throw away.

When it comes to numerical calculations things go on smoothly enough and one is sometimes astonished, what a simple and elegant curve a very awkward and intricate formula will give; evidently it could be expressed with quite sufficient accuracy for practical use by a much simpler function. This would be of great importance for further investigations, if we only had at the same time a criterion for the value of the error thereby involved. What we want are not absolutely strict and exhaustive, but practical, approximate solutions, which would enable us to get on with our work. If I may venture to say so, we want a whole set of different, well-tabulated functions, which may be utterly uninteresting to pure mathematicians, but may nevertheless be of much importance for physical research.

I shall quote only a very simple and trivial example.

A closer study of the proper movement of a horizontal pendulum, inscribing its corresponding curve with a pin on smoked paper, shows that the movement is no longer a harmonic damped oscillation, but is represented by the following differential equation:

$$y'' + 2\epsilon y' + n^2(y + \rho) + \xi\{y' + \nu y\}^2 = 0,$$

where ξ and ρ are two new constants, depending upon the elements of friction, whereas ν depends upon the revolving velocity of the recording drum. These latter must necessarily be taken into consideration if one wants to deduce the true movement of the earth's surface from seismographic records on smoked paper. Now what is one to do with such an equation? In our case the question is simple enough, for ξ is small and one can use the method of successive approximations, but there are corresponding physical problems where this assumption holds no longer.

Similar problems arise when treating the question of the theoretical form of the hodograph and so forth, but I shall not dwell upon this any longer.

The readings of seismograms obtained by galvanometric registration from aperiodic seismographs enable us to attack different problems which are of great importance for modern seismology.

By measuring the first maximal amplitude of the displacement of the recording light-spot directly after the arrival of the first longitudinal waves, one can deduce therefrom the true displacements of a particle of the earth's surface for two rectangular horizontal directions. This enables one to determine directly the azimuth from where the first seismic waves have come and combining this result with the epicentral distance, deduced from the difference of time of the arrival of S and P, locate the epicentre from observations made *at one seismological station alone.* The analysis of the record is certainly more complicated than in the case of the maximal phase, for the time separating the beginning of the movement till the first maximum is very short, usually 1 or 2 seconds, and one is necessarily obliged to take into consideration the influence of the proper motion of the pendulum itself, but the problem can nevertheless be mastered and auxiliary tables facilitate all the necessary calculations. If the proper periods of both pendulums are equal to one another and to those of the corresponding galvanometers, the problem gets to be exceedingly simple and the azimuth can be deduced at once. In this special case one need not suppose the presence of a harmonic seismic wave, but the true displacement of a particle of the earth's surface can be quite an arbitrary function of time.

There is only one difficulty in the question which consists in this.

Observations show that in some cases the first movement of the ground is from the epicentre and in some others towards the epicentre, the first corresponding to a condensational and the second to a dilatational wave front. There is therefore an ambiguity in the azimuth of 180°, but the use of the vertical seismograph settles the question at once. If the latter points to the fact that the first movement of the ground was upwards, the wave will be a condensational one and *vice versa*. This does away with all ambiguity and the epicentre can be fairly located.

This method of locating epicentres has the advantage over others, that it does not require as usual the observations from several other stations, but is dependent only on observations made at one and the same place, but in order to obtain good results it requires special highly sensitive and properly adjusted seismographs. This method is always used at present at the seismological observatory at Pulkovo; it has also often been tested at Eskdalemuir; the results have always been most satisfactory, if only the first phase on the seismogram is distinct enough.

This fact, namely the possibility of determining the azimuth of the epicentre by the method just described, can be considered as a direct experimental proof that the first seismic waves that reach a given point are really due to longitudinal vibrations.

This same method of determining the true direction of the movement of an earth's particle can also be applied to the second phase, where the first transverse waves strike the ground. This enables us to find the true plane of oscillation or call it plane of polarization of the corresponding waves and opens a new way to the study of the geological particularities of the upper layers of the earth's crust.

By measuring the true vertical and corresponding total horizontal displacement of an earth's particle z and h at the very beginning of the first phase of a seismogram one can deduce the visible angle of emergence $\bar{e}$ of the seismic rays:

$$tg\bar{e} = \frac{z}{h}.$$

This angle differs a little from the true angle of emergence e, as defined above, for when a seismic wave strikes the earth's surface a part of the movement is reflected inwards towards the interior of the earth. The true relationship between these two angles is a very delicate problem of the theory of elasticity, which has been treated by Wiechert and in this country by Professor Knott, although perhaps not in quite an exhaustive way. The problem I dare say requires yet further consideration being of great importance to seismology, as it is the true angle of emergence which is the most characteristic element in the study of the path traversed by the seismic rays. If Wiechert's formula stands good, the difference between both angles e and $\bar{e}$, starting from an epicentral distance $\Delta = 1500$ kil. upwards, never exceeds 2 or 3 degrees of arc. For small epicentral distances $\bar{e}$ is considerably greater than e, and for $\Delta = 0$ $\bar{e}$ would be 22°, whereas e would be 0.

The study of the angle of emergence for short epicentral distances is of very great importance. It could be conducted by means of artificial quakes, occasioned by the springing of underground mines. Such experiments would also tend to augment our very scanty knowledge concerning the laws of propagation of different types of seismic waves in the uppermost layers of the earth's crust in connection with the geological formation of the corresponding rocks. Another question would be the study of the propagation of seismic disturbances occasioned by the oscillatory movements of heavy engines and their influence upon the stability of different types of buildings. Such investigations have been conducted, but till yet on a very scanty scale, and there is still a vast field open for further scientific enquiry.

A scrupulous and systematic comparative study of records obtained at different observatories with highly sensitive instruments of the same type would open the way to the knowledge of the laws of propagation (velocity, damping) of different types of seismic waves, but all this work is yet to be done.

Seismology is still in its period of infancy and has not had time to work out all the different problems it nearly daily brings forth, some of which are of the highest importance for geophysics, but there is no doubt that it will master them with time, as it has taken the right line of work, based upon trustworthy instrumental observations.

I may just allude here in passing to the part seismology has taken in solving the great geodetical problem of the tidal deformations of the earth, about which English mathematicians have been so keen and which is so intimately connected with this town of Cambridge, owing to the work of our venerated president, Sir George Darwin. As these deformations are of the bradiseismical or slow type, they do not require damped instruments.

There are still many discrepancies which have to be explained and which the observations undertaken by the care of two international scientific bodies, namely the International Geodetical and the International Seismological Associations, will certainly clear up.

I may just mention here a new and important paper on the subject published lately by Doctor Schweydar of Potsdam, which tends to throw much light on the subject and which is in full agreement with the masterly investigations of Professor Love of Oxford.

As yet we have considered only the tachiseismic displacements of the ground, excluding the rotations or tilting, these being for distant earthquakes so very small. But when a seismic wave travels along the earth's surface, a certain tilting of the ground, however small, must necessarily take place and for comparatively small epicentral distances it certainly cannot be neglected.

But by what means are we to study this tilting?

When the tilting is slow the usual horizontal pendulum is the best adapted instrument, as it is for studying the tidal deformations of the earth, but what are we to do, when the tilting is quick and combined, as in the case of earthquakes, with corresponding displacements?

Schlüter proposed for that object a kind of balance with a very long period, which he called a klinograph, but the observations made therewith at Göttingen and Pulkovo show that it gives very scanty information and has since then been abandoned.

A remarkable instrument for its elegance and simplicity was devised, if I am not mistaken, by Mr Horace Darwin and Davison and constructed in this very town of Cambridge by the Cambridge Scientific Instrument Company. It consists of a heavy mass suspended by two vertical wires of unequal length. When a tilting takes place about an axis lying in the normal plane of the wires, the instrument turns about a small angle, which can be easily measured with great accuracy. By diminishing the distance and augmenting the difference of length of the two wires it can be made most sensitive, detecting even 0·0012″ in the tilting of the ground.

This would stand good if we had only to do with tilting alone, but in the case of earthquakes there are simultaneous displacements, which set the apparatus swinging to and fro and each such swing engenders a corresponding twisting of the instrument, so that there is no possibility of separating the two effects of displacement and tilting one from another. The mathematical theory of this instrument offers many curious and suggestive peculiarities and leads to a differential equation, also found in the theory of celestial mechanics.

The same difficulty of separating both movements arises with the horizontal pendulum.

If we denote with ψ the angle of tilting about an axis parallel to the boom and with g the acceleration of gravity, we shall have, instead of equation (1), the following generalized equation:

$$\theta'' + 2\epsilon\theta' + n^2\theta + \frac{1}{l}(x'' - g\psi) = 0 \quad\quad\quad (3).$$

Now what we measure directly is θ or, in the case of galvanometric registration, ϕ (see equation (2)). Therefore there is no possibility whatever of separating with one pendulum x from ψ.

But a special contrivance can be imagined, which will enable us to eliminate x and get at the true tilting of the ground.

Imagine quite a similar pendulum with the same proper period and same damping ratio placed at a distance s above the first. Then its differential equation will be

$$\theta_1'' + 2\epsilon\theta_1' + n^2\theta_1 + \frac{1}{l}(x'' + s\psi'' - g\psi) = 0 \quad\quad\quad (4).$$

Suppose now that both these pendulums are coupled with the same galvanometer, but so that the induction currents from each flow in opposite directions. Then the motion of the galvanometer will depend only on the *difference* of the angular velocities $(\theta_1 - \theta)'$ and x'' will be fully eliminated.

Such a double pendulum enables us therefore to study the true tilting of the ground, quite undisturbed by any displacements.

A similar apparatus has been tested on a moveable platform and has given most satisfactory results. The platform was subjected simultaneously to a tilting and shifting movement, the latter sometimes of the most irregular kind, but still the instrument recorded only the tilting, quite heedless of all displacements.

This double pendulum can be easily made extraordinarily sensitive, detecting angles of tilting down to say 0·0001″.

The practical study of tilting has not as yet been taken up systematically by seismology; this problem is reserved for the future.

We have considered until now only the small movements of the earth's surface or so-called microseisms, revealed to us by very sensitive seismographs, which, when properly adjusted and damped, enable one to deduce the true movement of the ground. The same problem arises for larger or macroseismic movements, experienced in the epicentral and surrounding areas. A profound study of the true motion of the ground in regions characterized by their seismic activity is of vital importance for elaborating practical rules for the construction of aseismic buildings able to resist in a satisfactory manner the dangerous quiverings of the ground. Nothing stands in the way of adopting for this kind of investigations the same principles of instrumental research that have been already discussed with regard to microseismic activity. It is only necessary to diminish considerably the sensitiveness of the seismographs and give them a very solid construction; simple, vertical, damped pendulums could be of great use for this kind of study. Very little has been done in this line of work up to the present day, as a systematic study of different seismic areas, requiring observations at many different points, involves a great expense and requires a trained scientific staff of observers not usually to be found. One confines oneself for the present to measuring the intensity of the seismic activity by means of purely empirical scales, like those of Rossi-Forel, Mercalli or the new scale of Sieberg, based upon personal observations, gathered through newspapers or through special enquiry. These data serve to draw the isoseists or curves of equal seismic intensity.

It is quite evident how defective such a way of proceeding is and how little dynamical value one can attribute to such experimental data.

What we require is a simple process of evaluating the intensity of the seismic movement by means of a simple, but rational, purely dynamical scale, accessible to every untrained observer, and that would give immediately the corresponding maximal acceleration of the ground's movement or at least of its two rectangular horizontal components. Such a scale can be made on the principle of overturning of rectangular blocks, subjected to a swaying movement of the base they rest upon.

The proper movement of such a block offers many interesting particularities, among which I may mention that the half-period of oscillation, when the axis of rotation passes from one rib to another, is proportional to the square root of the maximal deviation. The study of this movement is intimately connected with the problem, which is known, if I am not mistaken, in Cambridge University as the Union gate problem.

The question now arises, What are the conditions of stability of such a block, resting on a plane, subjected to harmonic oscillations, when a slipping of the block is prevented? The dynamical equations of the problem can comparatively easily be deduced, but they are not much adapted for further discussion, involving a discontinuous function of time, namely the angular velocity of the movement of the block, when the motion passes from one rib to another. Here is another case where seismology requires a helping hand from the pure mathematician.

As far as I know this problem has not as yet been mathematically solved in a general way, but experimental observations with a small moveable platform have settled the matter.

Observations show that the conditions of stability of such a block on a base, oscillating harmonically, does not depend on the period T_p or the amplitude of oscillation x_m alone, but on a certain combination of both, namely

$$w_m = \frac{4\pi^2 x_m}{T_p^2},$$

which is nothing else but the maximal acceleration of the movement of the ground.

There is an extremely simple relationship between the dimensions of this homogeneous block and the value of w_m which defines the limit of stability.

If $2h$ denotes the height and $2c$ the thickness of the block, we have simply

$$\frac{c}{h} = 0{\cdot}0012 w_m,$$

where w_m is expressed in centimeters and seconds.

When w_m is greater than this critical value the block will turn over and fall.

This equation has been established experimentally; it holds good for a wide range of values of w_m. Surely it could be deduced mathematically.

Now here is a means of establishing a rational, dynamical scale for the study of the maximal horizontal acceleration of the ground's movement, which is of all-importance to seismology.

One only has to prepare a scale of such blocks and judging after a seism which of these blocks have fallen down and which have remained upright, one can obtain two limits for the maximal acceleration of the earth's true movement. Two sets of blocks, placed at right angles to each other, will give an indication of the approximate direction of the principal seismic disturbance.

Convulsive shakings of the ground produced by earthquakes are far from being rare phenomena; on the contrary, since instrumental observations have been established, it is extraordinary what an immense number of earthquakes, large and small, have been recorded. It will not be exaggerating to state that several hundreds of earthquakes occur yearly; in fact there is scarcely a day that passes without the earth's surface shaking in some part of the world. But besides these, there is a constant pulsatory movement of the earth's surface going on, which is more prominent in the winter months and which is characterized by very regular seismic waves with short periods, varying from 4 to 10 seconds. These movements can be detected on all three components and they produce the impression as if the earth were actually breathing. A systematic study of these particular microseisms has begun only lately and their real cause is still involved in mystery. According to one hypothesis these pulsatory oscillations are due to the shocks of sea waves on steep coasts. To test this theory a special wave-counter made here in Cambridge by the Cambridge Scientific Instrument Company has been set up in this country at Newcastle-on-Tyne; the observations have been confided to Mr Morris Airey, but the results have not as yet been published.

The remarkable fact that these tremors have been detected in the most different parts of the world and even at a depth of 1000 meters under the earth's surface would seem to indicate that we have to do here with proper vibrations of the earth's outer crust. The amplitudes of these periodical movements are in reality very small, but they can easily be detected and studied by means of modern seismographs, whose sensibility is so great that displacements of the order of 0·1 of a micron can easily be measured with accuracy.

Another striking peculiarity of these oscillations is, that there seems to be a marked relation between the amplitudes and periods of oscillation, viz. the longer the period the larger will the amplitude be. This feature is not characteristic of ordinary harmonic oscillations and well deserves a dynamical and mathematical study.

Except these regular micro-tremors, there are other types of oscillatory, although less prominent, movements of the ground, with comparatively long periods, produced by winds, the shifting of barometric depressions and heat-waves. They all deserve to be thoroughly studied.

There are a great many other problems which modern seismology has brought to the front, but I will not dwell upon these questions, for lack of time, any longer.

One of these problems is of the most vital practical importance and that is, Is there any hope of being able in time to foretell the outbreak of a brewing earthquake and thereby contribute to the saving of human life and property?

The problem offers mighty difficulties, but the case is not so utterly hopeless as it might seem, for the investigations of v. Kövesligethy and others show that we may still hope in time to get at a practical solution of the question.

I should like before closing just to mention a curious relationship which has been suspected between the frequency of earthquakes and the movement of the earth's pole.

There are ample reasons to believe that some correlation between both phenomena may really exist and it would be of great importance for geophysics if we could manage to detect in some way or other the shifting of interior masses of the earth produced by severe tectonic earthquakes. Well this seems really practically possible by means of Baron Eotvös' gravimeter, an instrument of the highest sensitiveness, which enables us to detect the presence or displacements of comparatively very small masses. Such observations have been planned in Russia for the Turkestan seismic regions and two gravimeters have accordingly been ordered; one of them has already arrived and is being tested. The observations are to be conducted at particular points in the vicinity of prominent tectonic lines, indicated by geologists, and repeated again after a new, but fairly inevitable earthquake has visited those unhappy, for their unstability, but otherwise beautiful regions.

I have now come to the close of my address and I feel particularly indebted with gratitude to the organizers of this Fifth International Mathematical Congress for their kind invitation to give a lecture on the principles of instrumental seismology, the basis of all modern seismological work, at this assembly.

You will have seen during this last hour how many seismological problems require further mathematical treatment and it is useless to mention how welcome and important the friendly help of the pure mathematician would be to the seismologist, standing bewildered in front of the weird and intricate problems of modern geophysics.

I think that no place like this country and specially Cambridge could have been chosen more appropriately for the delivering of such an address, as it is particularly England that so often has had the lead in scientific seismological work. Passing over the former important work of Mallet, Oldham and others, I should just like to bring to your memory all that your countryman John Milne has done for modern seismology. He, together with Ewing, can be considered as the father of instrumental seismology; it is he who, through his remarkable activity and energy, has covered the earth's surface with a whole net of seismic observatories, gathering useful seismometric records, bearing at the same time all the burden of this vast organization nearly exclusively on his own shoulders. And not only that. I may fairly say, that there are not many questions of modern seismology that have not been attacked by Milne long before any other person had thought about them. This remarkable activity has not, notwithstanding years, relaxed in the least, a proof of which we see in the very important catalogue of earthquakes Milne has lately brought out.

Then come the theoretical and in many cases most remarkable studies of so many Englishmen like Lord Kelvin, Lord Rayleigh, Sir G. Darwin, H. Lamb, Love, Larmor, Schuster, Knott and many others on different problems of geophysics, relating to questions of the highest importance.

It is also in England that the fertile idea of creating an international seismological association was born. Since then this idea has been taken up by Rebeur-Paschwitz, Gerland and others and this scientific body brought into life, contributing by the collected work of its members to raise the veil over the mystery, that still enshrouds the natural phenomena, connected with the periodical quiverings of the earth's crust.

DÉFINITION ET DOMAINE D'EXISTENCE DES FONCTIONS MONOGÈNES UNIFORMES

Par Émile Borel.

I. *Les origines de l'idée de fonction.*

C'est l'intégration par d'Alembert de l'équation des cordes vibrantes, en 1747, qui fut l'origine d'une série de recherches desquelles se dégagea la notion de fonction arbitraire. Riemann dans l'introduction d'un mémoire célèbre a résumé les discussions auxquelles donna lieu l'importante découverte de d'Alembert: parmi les géomètres qui contribuèrent à éclaircir les idées nouvelles, on doit citer au premier rang Euler, et aussi Clairaut, Daniel Bernoulli, Lagrange.

La question qui se posait pour la première fois était celle des rapports entre la définition analytique d'une fonction et la définition en quelque sorte physique: si on écarte *arbitrairement* une corde de sa position d'équilibre, existe-t-il une formule qui représente exactement l'état initial de cette corde ?

Il était réservé à Fourier de répondre affirmativement à cette question, en faisant connaître le mode de calcul des coefficients de la série trigonométrique qui représente une fonction arbitraire. Les vues géniales de Fourier ont été entièrement confirmées par l'analyse rigoureuse que l'on doit à Lejeune-Dirichlet.

Cette découverte de Fourier bouleversait complètement les notions acquises; jusque là, on avait cru, avec Euler, qu'à une expression analytique déterminée correspondait toujours une courbe dont les portions successives dépendaient mutuellement les unes des autres: c'est pour exprimer cette interdépendance qu'Euler avait créé l'expression de fonction *continue*; le sens de cette expression a été entièrement modifié depuis. C'est aussi sous l'influence des mêmes idées que Lagrange, dans sa Théorie des fonctions analytiques, avait cherché à démontrer que toute fonction continue peut être développée en série de Taylor; ce développement en série aurait été la forme tangible de la liaison jusque là un peu mystérieuse entre les diverses portions d'une courbe continue; la connaissance d'un petit arc aurait ainsi permis de connaître toute la courbe; mais Fourier prouvait précisément que c'était là un problème illusoire et impossible, puisque le physicien qui trace une courbe arbitraire reste à chaque instant libre d'en modifier l'allure à sa guise; la courbe une fois tracée, il est toujours possible de la représenter par une expression analytique unique.

On était ainsi conduit à ce résultat paradoxal qu'il ne subsistait aucune raison logique pour regarder deux segments de la même droite ou deux arcs d'un même cercle comme correspondant à la même fonction, puisqu'il était tout aussi loisible de regarder comme une fonction unique l'ordonnée de la courbe continue formée par deux droites différentes ou par deux arcs appartenant à des cercles différents. Tout au plus aurait-on pu dire que, dans le cas de deux segments d'une même droite, la formule est *plus simple* que dans le cas de deux segments de droites différentes, mais ce criterium de simplicité ne semblait pas pouvoir être rendu bien précis, à moins que l'on ne se bornât aux fonctions algébriques, ce qui eut été exclure les développements en séries dont l'utilité apparaissait cependant chaque jour davantage.

II. *La théorie de Cauchy.*

Le paradoxe fut éclairci par l'extension du champ d'étude des fonctions; Cauchy montra que les propriétés des fonctions réelles les plus simples ne peuvent être bien connues que si l'on étudie aussi ces fonctions pour les valeurs imaginaires de la variable; la notion de fonction de variable complexe s'impose comme un auxiliaire indispensable. Cauchy basa cette notion sur la définition de la monogénéité; une fonction de la variable complexe $z = x + iy$ est dite monogène si elle admet une dérivée unique. Si une fonction est monogène en tous les points d'une région, sans aucune exception, c'est à dire si elle n'admet en cette région aucun point singulier, elle peut être développée en série de Taylor au voisinage d'un point quelconque de cette région; le rayon de convergence de la série de Taylor est d'ailleurs égal à la distance du centre du cercle de convergence au point singulier le plus voisin. C'est de ce théorème fondamental que Cauchy déduit le calcul des intégrales des équations différentielles suivant un chemin quelconque dans le plan, au moyen de séries de Taylor successives.

La théorie de Cauchy a été systématisée par Weierstrass et par Riemann. Weierstrass définit d'une manière tout à fait précise la *fonction analytique* au moyen d'un ensemble d'*éléments* ou développements de Taylor se prolongeant mutuellement; il est arrivé ainsi à la notion de domaine d'existence naturel, notion renfermée implicitement dans Cauchy, mais non explicitée par lui. Riemann a conçu la fonction monogène indépendamment de toute expression analytique *a priori* et a montré tout le parti que l'on pouvait tirer de cette conception géométrique.

En réalité, les points de vue analytique de Weierstrass et géométrique de Riemann trouvent leur synthèse la plus parfaite dans le théorème fondamental Cauchy: la monogénéité dans un cercle entraîne l'existence d'un développement de Taylor convergent dans ce cercle. C'est ce théorème qui rend si féconde la théorie des fonctions d'une variable complexe, il établit une liaison nécessaire entre les valeurs d'une même fonction comme simple conséquence de la monogénéité: il suffit donc de savoir qu'une fonction est monogène à l'intérieur d'un cercle pour que sa valeur en un point quelconque intérieur à ce cercle soit déterminée par la connaissance de ses valeurs au voisinage d'un autre point. Insistons un peu sur les théories de Cauchy-Weierstrass et de Cauchy-Riemann.

III. *Les limites de la théorie de Cauchy-Weierstrass.*

La théorie du prolongement analytique était renfermée dans les travaux de Cauchy; mais c'est à Weierstrass que l'on doit d'avoir précisé d'une manière absolument nette les limites de cette théorie. Bornons nous aux fonctions uniformes. La théorie de Weierstrass conduit à considérer des domaines que nous appellerons domaines weierstrassiens ou domaines W et qui sont caractérisés par les propriétés suivantes. Appelons cercle Γ tout cercle tel que tous les points intérieurs à Γ appartiennent à W. Tout point A de W est intérieur à un cercle Γ; les cercles Γ correspondant à deux points A et B de W peuvent être réunis par un nombre fini de cercles Γ deux à deux sécants. A toute fonction analytique uniforme correspond un domaine W; inversement, M. Runge a montré qu'à tout domaine W correspondent une infinité de fonctions analytiques uniformes, admettant précisément W comme domaine d'existence.

Si l'on admet qu'il n'y a pas d'autre procédé de prolongement que le prolongement analytique au moyen de la série de Taylor, la frontière du domaine W est une *limite naturelle* d'existence pour la fonction analytique, et les portions du plan, s'il en existe, qui n'appartiennent pas à W doivent être considérées comme un *espace lacunaire* dans lequel la fonction ne peut pas être définie. C'est là un point sur lequel Weierstrass a insisté à diverses reprises et qui a été mis en évidence de la manière la plus nette par M. Henri Poincaré. Considérons un domaine D de forme simple, tel que l'intérieur d'un cercle, et définissons une fonction $F(z)$ admettant D comme espace lacunaire et une autre fonction $F_1(z)$ définie seulement à l'intérieur de D et admettant par conséquent tout le reste du plan comme espace lacunaire. Divisons le contour de D en deux arcs A et B. M. Poincaré démontre qu'il est possible de trouver deux fonctions uniformes Φ et Φ_1 existant dans tout le plan, à l'exception de la ligne singulière A pour Φ et de la ligne B pour Φ_1, et cela de telle sorte que

$$\Phi + \Phi_1 = F \text{ à l'extérieur de } D,$$

$$\Phi + \Phi_1 = F_1 \text{ à l'intérieur de } D.$$

Si donc les fonctions Φ et Φ_1 sont regardées comme uniformes, la fonction F admet comme prolongement F_1 qui a été choisie d'une manière entièrement arbitraire; c'est donc que l'on doit écarter toute idée de prolongement à l'intérieur de l'espace lacunaire.

Ce paradoxe apparent s'éclaircit si l'on observe que, lorsqu'une fonction telle que $\Phi(z)$ possède une ligne singulière A, supposée infranchissable, cette fonction reste uniforme au sens de Weierstrass lorsqu'on lui ajoute une fonction non uniforme telle que $\log\frac{z-a}{z-b}$, a et b étant deux points de la ligne A. Le résultat remarquable dû à M. Poincaré peut donc être interprété par l'hypothèse que $\Phi(z)$ et $\Phi_1(z)$ ne sont pas véritablement uniformes: mais pour que cette hypothèse ait un sens, il est nécessaire de généraliser la définition du prolongement, de manière à pouvoir franchir en certains cas des coupures infranchissables au sens de Weierstrass; nous verrons tout à l'heure comment ce résultat peut être atteint.

IV. *La théorie de Cauchy-Riemann.*

Mais je voudrais auparavant dire quelques mots des idées de Riemann, bien que ce soit surtout dans l'étude des fonctions non uniformes, dont je n'ai pas à parler aujourd'hui, que la théorie de Riemann s'est montrée féconde.

Cauchy avait insisté à diverses reprises sur l'importance de la monogénéité. Si l'on considère une fonction élémentaire obtenue par un calcul simple effectué sur z (polynomes en z, séries toujours convergentes en z représentant la fonction exponentielle, les fonctions circulaires etc.) et si, pour une telle fonction $F(z)$, on calcule le rapport

$$\frac{F(z+\Delta z)-F(z)}{\Delta z},$$

ce rapport tend vers une limite bien déterminée lorsque Δz tend vers zéro, quelle que soit la manière dont varie son argument. Cauchy exprime ce fait important en disant que la fonction est *monogène*.

Si l'on pose

$$F(z)=P(x,y)+iQ(x,y),$$

la condition de monogénéité se traduit par les deux équations fondamentales

$$\frac{\partial P}{\partial x}=\frac{\partial Q}{\partial y},$$

$$\frac{\partial P}{\partial y}=-\frac{\partial Q}{\partial x}.$$

Cauchy a montré que ces équations, lorsqu'elles sont vérifiées en une région du plan, entraînent l'existence de la série de Taylor, c'est à dire de ce que l'on peut appeler l'analyticité au sens de Weierstrass. La démonstration de Cauchy suppose la continuité de la dérivée; M. Goursat, dans un travail bien connu, a montré que l'existence de la dérivée première suffit, et entraîne par suite la continuité et l'existence de toutes les dérivées; M. Paul Montel, suivant une voie ouverte par M. Painlevé, a étendu ce résultat aux cas où l'existence de la dérivée n'est pas supposée en certains ensembles de points. L'exposition de ces recherches est en dehors de mon cadre; je tenais cependant à les mentionner, car leurs résultats sont en quelque sorte complémentaires de ceux que j'exposerai plus loin. Ce qu'il nous suffit de retenir, c'est que, dans les domaines W, les fonctions monogènes sont analytiques; c'est pour ce motif que l'expression *fonction monogène* a cessé d'être employée par certains géomètres, l'expression *fonction analytique* étant considérée comme équivalente; comme notre but est précisément de définir des fonctions monogènes qui ne sont pas analytiques, il importait de distinguer nettement entre les deux expressions.

Il est difficile de se rendre compte si Cauchy conçut l'existence d'une fonction monogène d'une manière indépendante de toute expression analytique. En fait, il raisonna toujours sur des fonctions qui étaient définies, implicitement ou explicitement, à partir de fonctions connues, par des équations différentielles ou aux dérivées partielles; mais ses raisonnements s'appliquent sans modification à la fonction définie d'une manière purement idéale comme une correspondance entre z et $F(z)$. Cette conception fut celle de Riemann, et a certainement rendu de grands services, tant dans le champ des variables réelles que dans le champ des variables complexes, en accoutumant les mathématiciens à des raisonnements très généraux, raisonnements faits une fois pour toutes et susceptibles d'applications à des cas qui n'étaient pas prévus au moment où on faisait le raisonnement. Je dois avouer toutefois que je n'aperçois pas une différence réelle entre le point de vue de Cauchy et celui de Riemann; en fait, pour appliquer les considérations du genre de celles de Riemann à *une* fonction déterminée, il faut que cette fonction soit définie, c'est à dire puisse être distinguée des autres fonctions; et si cette définition est effective, elle rentre dans la catégorie de celles qu'eut admises Cauchy. Mais je ne veux pas développer ce point, qui se rattache aux controverses relatives à l'axiome de Zermelo: le point de vue de Riemann est certainement légitime, quelque attitude que l'on adopte dans cette controverse; pour ceux qui exigent une définition précise, il dispense de penser à tous les procédés de définition qui pourront être imaginés; pour ceux à qui la définition idéale suffit, il permet de traiter idéalement même les fonctions qui ne seront jamais définies pratiquement.

C'est à l'aide du théorème fondamental de Cauchy

$$f(\zeta)=\frac{1}{2\pi i}\int_C\frac{f(z)\,dz}{z-\zeta}$$

que l'on démontre que la monogénéité dans un domaine W entraîne l'analyticité dans ce domaine. C'est aussi à ce théorème que l'on devra avoir recours pour étudier les fonctions monogènes dans un domaine qui n'est pas W; il sera commode, pour raisonner d'une manière générale sur tous les procédés de définition possibles de ces fonctions, de les considérer comme définies à la manière de Riemann, c'est à dire en admettant que l'on ne sait rien sur une telle fonction, sinon qu'elle est monogène. Il faudra d'ailleurs montrer que la théorie ainsi construite n'est pas vide, en fournissant des exemples effectifs de fonctions définies d'une manière, non plus idéale, mais explicite. Pour plus de clarté nous donnerons d'abord ces exemples avant de développer la théorie générale.

V. *Les domaines de Cauchy.*

J'ai proposé d'appeler domaines de Cauchy, ou domaines C, les domaines les plus généraux dans lesquels peut être définie une fonction monogène uniforme, cette définition étant assujettie à la condition essentielle suivante: *La connaissance de la fonction dans une portion de* C *détermine la fonction dans tout* C. Les domaines C comprennent comme cas particulier les domaines W; il faudra en outre, en même temps que les domaines C, définir un mode de *prolongement*; ce prolongement devra coïncider avec le prolongement analytique lorsque le domaine C sera un domaine W.

Je voudrais indiquer tout d'abord un exemple aussi simple que possible de domaine C et de fonction monogène dans ce domaine. Formons la série

$$f(z)=\sum_{n=1}^{\infty}\sum_{p=0}^{n}\sum_{q=0}^{n}\frac{e^{-e^{n^4}}}{z-\dfrac{p+qi}{n}}.$$

Il est clair que cette série est convergente à l'extérieur du carré A dont les sommets sont les points $z=0,\ 1,\ i,\ 1+i$. A l'intérieur de ce carré, la série admet une infinité de pôles, à savoir tous les points dont les coordonnées sont des nombres rationnels $x=\frac{p}{n}$, $y=\frac{q}{n}$. Mais il est aisé de voir que si l'on considère les cercles ayant pour centres ces pôles et pour rayons $\frac{\epsilon}{n^4}$, la série est absolument et uniformément convergente en tous les points extérieurs à ces cercles, quelque soit le nombre fixe ϵ. Il en est de même si l'on considère les cercles ayant pour centres les points $\frac{p}{n}$, $\frac{q}{n}$, et pour rayons $\frac{1}{h}e^{-e^{n^2}}$, h étant un nombre entier fixe que nous nous réservons de faire croître indéfiniment. J'appellerai Γ_h l'ensemble de ces derniers cercles et C_h l'ensemble des points qui ne sont intérieurs à aucun des cercles Γ_h. On remarque que, pour simplifier, je désigne tous les cercles par Γ_h au lieu d'écrire $\Gamma_h^{(n)}$. Les propriétés élémentaires des nombres quadratiques permettent de prouver très aisément que toute droite telle que la suivante,

$$2x+3y-\sqrt{7}=0,$$

appartient à C_h pour une valeur finie de h.

La fonction $f(z)$ est évidemment monogène dans le domaine C_h; elle admet en effet en chaque point de ce domaine une dérivée unique bien déterminée, que l'on obtient en dérivant la série terme à terme. La valeur de cette dérivée est indépendante de la manière dont l'accroissement Δz tend vers zéro, sous la réserve bien entendu que z et $z+\Delta z$ soient intérieurs à C_h.

Nous emploierons la même notation dans la définition générale des domaines C; ces domaines seront obtenus en retranchant d'une région du plan, telle que l'intérieur d'un cercle ou d'un carré, une infinité de domaines analogues aux cercles Γ_h et que nous appellerons *domaines d'exclusion.* Ces domaines d'exclusion forment une série qui doit être supposée très rapidement convergente; je n'entre pas ici dans le détail de ces conditions de convergence. A chaque entier h on fera correspondre une suite illimitée de domaines Γ_h, dont les aires forment une série très convergente de somme σ_h; les points qui ne sont intérieurs à aucun des Γ_h forment le domaine C_h; lorsque h augmente on admettra que chaque Γ_{h+1} est intérieur au domaine Γ_h correspondant et que σ_h tend vers zéro (je laisse de côté le cas où σ_h aurait une limite différente de zéro), et l'on désignera par C la limite des domaines C_h pour h infini. Chaque domaine C_h sera dit intérieur au domaine C. Les domaines C_h seront en général parfaits, tandis que le domaine C n'est pas parfait. Il est nécessaire d'observer qu'au sens du mot intérieur adopté dans la théorie des domaines W, il n'y a pas de points intérieurs à C dans une région où les domaines d'exclusion sont partout denses. Nous donnons donc au mot *intérieur* un sens différent, défini par ce qui précède.

L'étude des fonctions monogènes dans un domaine tel que C se fait aisément par l'extension, au contour qui limite un domaine parfait convenablement choisi C_h, du théorème fondamental de Cauchy que nous rappelions tout à l'heure; mais, bien que ne présentant aucune difficulté réelle, l'exposition de cette méthode est trop longue pour que je puisse la donner ici; je préfère insister davantage, d'une part sur l'extension de la notion de prolongement, et d'autre part sur la définition des fonctions monogènes au moyen d'intégrales doubles.

VI. *Le prolongement par les séries divergentes.*

Nous venons de voir que les fonctions monogènes non analytiques se présentent naturellement comme sommes de séries dont les termes sont des fonctions analytiques. Il est donc naturel de chercher un mode de prolongement associatif applicable à de telles sommes. Le problème ainsi posé n'est autre que le problème des séries divergentes : à toute fonction analytique correspond un développement de Taylor convergent dans un cercle, mais divergent en dehors de ce cercle ; ce développement est déterminé par la connaissance de la valeur des dérivées. Si une série de fonctions analytiques est indéfiniment dérivable, ses dérivées s'expriment linéairement au moyen des dérivées des termes, et la série de Taylor qui correspond à ces dérivées est une fonction linéaire des séries de Taylor correspondant aux divers termes de la série. Mais si la fonction n'est pas analytique au point où on la développe en série, cette série de Taylor sera la somme de séries dont les rayons de convergence décroissent indéfiniment et, dans les cas que nous étudions, aura un rayon de convergence nul. Le problème des séries divergentes consiste à transformer une telle série en série convergente, de telle manière que le résultat coïncide avec le prolongement analytique dans les cas où ce prolongement est possible. C'est grâce aux beaux travaux de M. Mittag Leffler que ce problème a pu être résolu pour la première fois d'une manière entièrement satisfaisante; ces travaux et leurs relations avec les recherches antérieures ont été magistralement exposés par M. Mittag Leffler dans sa conférence du Congrès de Rome et je n'ai pas à y revenir. Je dois cependant observer que, si l'on veut utiliser ces résultats pour le prolongement des fonctions monogènes non analytiques, il est nécessaire de les interpréter, soit dans le langage des séries divergentes comme je l'ai proposé, soit dans un langage équivalent si l'on préfère ne pas parler de séries divergentes; mais en tous cas dans un langage nouveau, spécialement approprié à la nouveauté réelle des résultats, et non pas dans le langage ancien du prolongement analytique de Weierstrass; c'est là le seul langage qu'il ne soit pas permis d'employer, car il a un sens absolument précis, qui ne peut être modifié: la théorie de Weierstrass est, en quelque manière, tellement parfaite qu'on ne peut en sortir qu'en créant un langage nouveau; si, comme le proposait M. Mittag Leffler, l'on adoptait le langage de Weierstrass, la théorie des séries de M. Mittag Leffler serait un simple mode de calcul simplifié, ne renfermant rien de plus au point de vue théorique que la théorie de Weierstrass.

Je rappelle comment les résultats fondamentaux de M. Mittag Leffler peuvent être interprétés dans le langage des séries divergentes.

Appelons sommation (M) d'une série de puissances divergente l'opération qui consiste à transformer cette série en une certaine série de polynomes, dont M. Mittag Leffler a donné l'expression, expression qui est linéaire par rapport aux coefficients de la série donnée. Dans une région où cette série de polynomes est absolument et uniformément convergente la série donnée sera dite sommable (M). Il est clair que la sommation (M) est une opération distributive.

Cela posé, considérons une série de fonctions analytiques, dont chaque terme admet un seul point singulier, l'ensemble de ces points singuliers étant dense dans une certaine aire. Sous des conditions de rapidité de convergence analogues à celles dont on a déjà parlé et à celles dont il sera question plus loin, on peut tracer dans la région où les points singuliers sont denses une infinité de droites, dites droites de convergence, telles que, en chaque point de chacune de ces droites, on puisse former la série de Taylor, qui est divergente, mais sommable (M) sur les droites de convergence qui passent par ce point, et qui représente la fonction sur ces droites. On peut ainsi, les droites de convergence formant un ensemble de droites partout dense (ensemble dont le corrélatif est partout dense), arriver aussi près que l'on veut de tout point du domaine d'existence de la fonction monogène définie par la série. Il résulte manifestement de la formation même des séries (M) que si la fonction est nulle en un point, ainsi que toutes ses dérivées, elle est nulle dans tout le domaine d'existence.

Ces résultats sont déjà anciens; ils ne constituaient pas une théorie entièrement satisfaisante des fonctions monogènes non analytiques, parce qu'ils étaient établis seulement pour des fonctions définies par des séries d'une forme déterminée; il restait à prouver qu'il n'était pas possible de définir autrement d'autres fonctions monogènes coïncidant avec les premières dans une partie seulement du domaine commun d'existence, mais différentes dans d'autres parties de ce domaine. Bien évidemment, ce résultat fondamental ne pouvait être obtenu qu'en imposant à ce domaine certaines conditions, conditions comprenant comme cas particulier la définition des domaines W, mais plus larges. En d'autres termes, la définition des fonctions monogènes uniformes doit comprendre comme cas particulier la définition des fonctions analytiques.

L'étude de la question à ce point de vue exige une étude préalable de la classification des ensembles de mesure nulle, étude suivie de l'emploi de l'intégrale de Cauchy, comme je l'indiquais tout à l'heure. Je vais esquisser, en terminant, une méthode différente, dans laquelle on n'a à utiliser que des fonctions partout continues et bornées, ce qui évite beaucoup de difficultés formelles.

VII. *Les intégrales doubles analogues à l'intégrale de Cauchy.*

Considérons une fonction analytique uniforme, régulière et nulle à l'infini. Si C est un cercle tel que tous les points singuliers de la fonction soient intérieurs à C, on a, ζ étant un point quelconque extérieur à C,

$$f(\zeta)=\frac{1}{2i\pi}\int_C \frac{f(z)\,dz}{\zeta-z},$$

l'intégration étant effectuée dans le sens direct.

Soient C_1 et C_2 deux cercles concentriques extérieurs à C, a le centre de ces cercles, r_1 et r_2 leurs rayons. On a évidemment, r étant compris entre r_1 et r_2,

$$f(\zeta)=\frac{1}{2\pi}\int_0^{2\pi}\frac{f(a+re^{i\theta})\,re^{i\theta}\,d\theta}{\zeta-a-re^{i\theta}}.$$

Multiplions cette égalité par $(r_2-r)^m(r-r_1)^n$ et intégrons entre les limites r_1 et r_2; il vient

$$f(\zeta)\int_{r_1}^{r_2}(r_2-r)^m(r-r_1)^n\,dr=\frac{1}{2\pi}\int_0^{2\pi}\int_{r_1}^{r_2}\frac{f(a+re^{i\theta})\,(r_2-r)^m(r-r_1)^n\,e^{i\theta}\,r\,dr\,d\theta}{\zeta-a-re^{i\theta}}.$$

Posons
$$\int_{r_1}^{r_2}(r_2-r)^m(r-r_1)^n\,dr=\frac{1}{2\pi A_{m,n}},$$
$$a+re^{i\theta}=x+iy,$$

d'où :
$$r\,dr\,d\theta = dx\,dy,$$
il viendra
$$f(\zeta) = \iint_{C_1 C_2} \frac{A_{m,n} f(x+iy)(r_2 - r)^m (r - r_1)^n e^{i\theta}\,dx\,dy}{\zeta - x - iy},$$
ce que l'on peut écrire, en posant $\zeta = \xi + i\eta$,
$$f(\zeta) = \theta(\xi, \eta) = \iint_{C_1 C_2} \frac{\phi(x, y)\,dx\,dy}{\xi + i\eta - x - iy},$$
le domaine d'intégration étant la couronne comprise entre les cercles C_1 et C_2.

Nous définirons la fonction $\phi(x, y)$ à l'extérieur de cette couronne en lui attribuant la valeur zéro ; on peut alors prendre comme domaine d'intégration tout le plan. La fonction $\phi(x, y)$ est bornée et continue dans tout le plan ; ses dérivées sont aussi bornées, du moins jusqu'à l'ordre m sur C_1 et jusqu'à l'ordre n sur C_2 ; par un artifice analogue à celui que nous allons employer, il serait aisé de s'arranger pour que *toutes* les dérivées soient continues ; il suffit généralement de savoir que les dérivées sont continues jusqu'à un ordre fixé d'avance.

Si la fonction $f(z)$ admet un seul point singulier a, on peut faire tendre r_1 vers zéro et si, de plus, le produit $r^m f(z)$ reste fini pour $z = a$, la formule subsiste pour $r_1 = 0$; si ce produit ne restait pas fini, on remplacerait dans la formule $(r - r_1)^m$ par $e^{-\frac{1}{r}}$ ou par $e^{-e^{\frac{1}{r}}}$, etc. De plus, dans le cas d'un point singulier unique, le cercle C_2 peut être pris de rayon aussi petit que l'on veut, après que le cercle C_1 a été réduit à zéro.

On déduit aisément de là que toute fonction analytique, uniforme, régulière et nulle à l'infini peut être représentée, en tout domaine D *intérieur* à son domaine d'existence W, et aussi voisin que l'on veut de W, par une expression de la forme
$$f(\zeta) = \theta(\xi, \eta) = \int_{-\infty}^{+\infty}\int_{-\infty}^{+\infty} \frac{\phi(x, y)\,dx\,dy}{\xi + i\eta - x - iy} \quad \ldots\ldots\ldots\ldots\ldots\ldots (1),$$
la fonction $\phi(x, y)$ étant bornée et, de plus, nulle en tous les points de D (cette hypothèse entraîne que la fonction $\phi(x, y)$ est nulle à l'infini, puisque le point à l'infini appartient à D).

Inversement, toute expression de la forme (1) dans laquelle $\phi(x, y)$ est une fonction bornée, nulle à l'infini, continue dans tout le plan, ainsi que ses dérivées (au moins jusqu'à l'ordre m) représente une fonction qui est monogène en tout point où $\phi(x, y)$ est nul ; car on a, par un calcul facile,
$$\frac{\partial\theta}{\partial\xi} + i\frac{\partial\theta}{\partial\eta} = 2\pi\phi(\xi, \eta).$$
Si les points où $\phi(x, y)$ est nul forment un domaine W, la théorie des fonctions analytiques nous apprend que la fonction $\theta(\xi, \eta)$ est déterminée en tout point de W par la connaissance de ses valeurs au voisinage d'un point particulier quelconque de W. Le problème de la détermination générale du domaine d'existence des fonctions monogènes peut donc être posé comme il suit : déterminer les conditions auxquelles doit satisfaire $\phi(x, y)$ pour que cette propriété fondamentale de $\theta(\xi, \eta)$ subsiste, c'est à dire pour que la connaissance de cette fonction sur un arc de courbe où elle est monogène permette de calculer sa valeur dans tout son domaine de monogénéité.

Avant d'aborder le problème dans sa généralité, revenons pendant quelques instants sur les séries de fractions rationnelles que nous avons considérées. On a, en désignant par C_0 un cercle de centre z_0 et de rayon ρ, en supposant ζ extérieur à ce cercle, et posant $|z - z_0| = r$,

$$\frac{1}{\zeta - z_0} = \iint_{C_0} \frac{3(\rho - r)}{\pi\rho^3} \frac{dx\,dy}{\zeta - z}.$$

Lorsque le point ζ est intérieur au cercle C_0, l'intégrale se calcule aisément; si l'on pose $\left|\frac{\zeta - z_0}{\rho}\right| = \lambda$, sa valeur est

$$(3\lambda^2 - 2\lambda^3)\frac{1}{\zeta - z_0}.$$

La fonction
$$\theta_0(\xi, \eta) = \iint_{C_0} \frac{3(\rho - r)}{\pi\rho^3} \frac{dx\,dy}{\xi + i\eta - x - iy}$$

est donc bornée dans tout le plan; à l'extérieur de C_0 elle est monogène et coïncide avec la fonction analytique $\frac{1}{\zeta - z_0}$. On peut évidemment définir d'une manière analogue une infinité de fonctions $\theta_n(\xi, \eta)$, telles que l'égalité

$$\theta_n(\xi, \eta) = \frac{A_n}{\zeta - a_n}$$

ait lieu pour tout point $\zeta = \xi + i\eta$ extérieur au cercle C_n de centre a_n et de rayon ρ_n, ces fonctions étant de plus *bornées et continues* dans tout le plan; si les $|a_n|$ sont bornés et si les coefficients A_n sont tels que la série

$$\Sigma \frac{|A_n|}{\rho_n^3}$$

soit convergente, la série

$$\theta(\xi, \eta) = \Sigma\theta_n(\xi, \eta)$$

sera absolument et uniformément convergente dans tout le plan, et représentée par une intégrale de la forme

$$\theta(\xi, \eta) = \int_{-\infty}^{+\infty}\int_{-\infty}^{+\infty} \frac{\phi(x, y)\,dx\,dy}{\xi + i\eta - x - iy} \quad \text{.....................(2)},$$

la fonction $\phi(x, y)$ étant la somme d'une série partout convergente, dont les termes respectifs sont nuls à l'extérieur des divers cercles C_n; cette fonction $\phi(x, y)$ est donc nulle en tous les points extérieurs à tous ces cercles et la fonction $\theta(\xi, \eta)$ est monogène en ces points. Si les rayons ρ_n sont remplacés par $\epsilon\rho_n$, ϵ étant aussi petit qu'on veut, la fonction $\phi(x, y)$ est nulle dans une région de plus en plus étendue; elle reste bornée, mais sa borne augmente indéfiniment lorsque ϵ tend vers zéro. On est ainsi conduit à considérer *a priori* une intégrale telle que (2), et à l'étudier dans les régions où $\phi(x, y)$ est nul. Il faut évidemment partir d'une région C d'un seul tenant; nous nous bornerons au cas où cette région C se compose de domaines W (ces domaines pouvant comme cas limite se réduire à zéro) et d'un nombre fini ou d'une infinité de droites Δ, de telle manière que deux points quelconques puissent être réunis par une ligne polygonale d'un nombre fini de côtés.

Une notion importante est celle de l'ordre d'infinitude de la fonction $\phi(x, y)$ au voisinage des droites Δ. J'ai pu démontrer la convergence des développements (M) en faisant l'hypothèse que cette fonction $\phi(x, y)$ non seulement est nulle sur les droites Δ (ce qui est la condition indispensable de monogénéité), mais tend très rapidement vers zéro dans le voisinage de chaque droite. Plus précisément, σ désignant la distance du point (x, y) à la droite Δ considérée, on suppose que le produit

$$e^{e^{\frac{1}{\sigma^2}}} \phi(x, y)$$

tend uniformément vers zéro lorsque σ tend vers zéro. Moyennant cette hypothèse, on peut affirmer que la fonction $\theta(\xi, \eta)$ est déterminée dans tout son domaine d'existence par la connaissance de ses valeurs en un point quelconque de ce domaine. Cette hypothèse comprend comme cas particulier la condition vérifiée par les fonctions analytiques dans les domaines W, car si une droite est intérieure à un domaine W, la fonction $\phi(x, y)$ est identiquement nulle en tous les points dont la distance à la droite est inférieure à un nombre σ convenablement choisi.

Le domaine C peut se réduire à l'axe réel ; tel est le cas pour la fonction

$$\theta(\xi) = \int_{-\infty}^{+\infty} \int_{-\infty}^{+\infty} \frac{e^{-e^{\frac{1}{y^2}}} dx\, dy}{(x^2 + y^2)(\xi - x - iy)}.$$

Le développement de Taylor

$$\theta(\xi) = \theta(\xi_0) + (\xi - \xi_0)\,\theta'(\xi_0) + \ldots\ldots$$

diverge quelque soit ξ_0, mais est, quelque soit ξ_0, sommable (M) pour toute valeur de ξ, sa somme étant bien égale à la fonction $\theta(\xi)$.

VIII. *Les propriétés des fonctions monogènes.*

Les fonctions monogènes non analytiques possèdent les propriétés les plus importantes des fonctions analytiques ; en particulier, pour les domaines C que j'ai considérés jusqu'ici, l'existence de la dérivée première entraîne l'existence des dérivées de tous les ordres. Il n'est pas absolument évident que cette condition doive être nécessairement imposée aux généralisations ultérieures de la théorie : il est aisé de construire des fonctions monogènes dans certains domaines H, ces fonctions n'admettant pas de dérivées au delà du premier ordre ; ces domaines H ne sont pas des domaines de Cauchy, au sens que nous avons donné à ce terme ; pourront-ils être regardés comme tels grâce à une extension nouvelle de la théorie ? c'est un point que je dois laisser en suspens.

Les calculs sur les intégrales doubles de la forme que nous avons considérée conduisent aisément à des expressions de même forme ; il en est ainsi pour la dérivation, comme on le voit en transformant l'intégrale double au moyen de l'intégration par parties ; il est seulement nécessaire de supposer l'existence des dérivées de la fonction $\phi(x, y)$ précisément jusqu'à l'ordre des dérivées de $\theta(\xi, \eta)$ que l'on veut calculer. Pour calculer le produit on remarque que, si l'on pose :

$$\theta(\xi, \eta) = \iint \frac{\phi(x, y)\, dx\, dy}{\zeta - z},$$

$$\theta_1(\xi, \eta) = \iint \frac{\phi_1(x, y)\, dx\, dy}{\zeta - z}$$

il vient, les limites étant toujours $-\infty$ et $+\infty$,

$$\theta(\xi, \eta)\, \theta_1(\xi, \eta) = \iiiint \frac{\phi(x, y)\, \phi_1(x_1, y_1)\, dx\, dy\, dx_1\, dy_1}{(\zeta - z)(\zeta - z_1)};$$

or, on a
$$\frac{1}{(\zeta - z)(\zeta - z_1)} = \frac{1}{z - z_1}\left(\frac{1}{\zeta - z} - \frac{1}{\zeta - z_1}\right);$$

si donc on pose:
$$\psi(x, y) = \iint \frac{\phi(x_1, y_1)\, dx_1\, dy_1}{z - z_1},$$

$$\psi_1(x, y) = \iint \frac{\phi_1(x_1, y_1)\, dx_1\, dy_1}{z - z_1},$$

on obtient:
$$\theta(\xi, \eta)\, \theta_1(\xi, \eta) = \iint \frac{(\phi\psi_1 + \psi\phi_1)\, dx\, dy}{\zeta - z}.$$

Il est donc aisé de mettre sous la forme d'une intégrale double tout polynome P par rapport à une ou plusieurs fonctions $\theta(\xi, \eta)$ et leurs dérivées; si les domaines d'existence que nous avons définis ont des parties communes d'un seul tenant l'équation différentielle obtenue en égalant P à zéro ne peut être vérifiée en une portion de ce domaine sans être vérifiée partout.

IX. *Conclusion.*

Je dois m'excuser d'avoir été à la fois long et incomplet: le sujet que j'avais choisi était peut-être trop vaste pour une conférence. Je voudrais cependant dire au moins quelques mots des analogies, depuis longtemps remarquées, entre la théorie des fonctions d'une variable complexe et la théorie du potentiel. La transformation de l'intégrale de Cauchy en intégrale double correspond à l'hypothèse, physiquement assez naturelle, qu'il n'y a pas de masses infinies mais seulement des régions singulières dans lesquelles la densité peut être très élevée, ou, si l'on préfère, des "sphères" d'action finies attachées à chaque point singulier. Les fonctions monogènes non analytiques correspondent au cas où ces régions singulières sont à la fois extrêmement petites et extrêmement nombreuses. J'ai déjà fait observer, il y a longtemps, qu'avec certaines dispositions arithmétiquement simples de telles régions singulières, les lignes de continuité qui passent à côté de ces régions sans y pénétrer peuvent être telles que leurs propriétés soient liées très intimement avec la simplicité numérique de leurs coefficients de direction. Je ne sais si quelque analyste plus habile que moi tirera un jour de ces considérations un peu vagues des conséquences dignes d'intérêt pour les physiciens; mais il ne m'était pas possible de passer sous silence le fait que j'ai été souvent guidé par les analogies de la nouvelle théorie avec les théories de physique moléculaire aux progrès desquelles on a si puissamment contribué dans cette ville et dans ce pays.

THE PLACE OF MATHEMATICS IN ENGINEERING PRACTICE

By Sir W. H. White.

The foundations of modern engineering have been laid on mathematical and physical science; the practice of engineering is now governed by scientific methods applied to the analyses of experience and the results of experimental research. The Charter of the Institution of Civil Engineers defines engineering as the "art of directing the great sources of power in Nature for the use and convenience of man." Obviously such direction can only be accomplished by engineers who possess an adequate acquaintance with "natural knowledge"—with the laws which govern these great sources of power. Obedience to natural laws is a condition essential to the full utilisation of the great sources of power. It is true, no doubt, that notable achievements in engineering were accomplished during the last century by men whose education was imperfect, whose mathematical and scientific knowledge was small, whose appeal to past experience gave little assistance in the solution of new problems. Their successors now enjoy greatly superior educational advantages; they can profit by enormous advances made in all departments of science and manufacture; they can study and criticise works done by their predecessors in the light of long subsequent experience; but even now there is room for surprise, if not for wonder, when one realises the great success attained by these early engineers.

The advantages obtainable by the combination of scientific training with practical experience are, however, in no way depreciated because, through force of circumstances, the pioneers of engineering had to do their work as best they could. Not a few of these men recognised the serious disadvantages resulting from their lack of scientific training and gave valuable assistance to a movement which ultimately led to the existing methods of training. George Stephenson, for example, who grew to full manhood practically uneducated, took care to secure for his son Robert Stephenson the advantages of a good elementary education which was followed by a period of practical training and then by a course of scientific study at Edinburgh University. The careers of both father and son were greatly influenced by this action, and it is of interest to note that the scheme which Stephenson framed and carried out for the professional training of his son—including the alternation of scholastic and engineering work—was, in its essential features, identical with that recommended in the Report of a representative Committee of British Engineers over which I had the honour to preside eight years ago. That Report has been approved by the leading Engineering Institutions of the United Kingdom and is now largely influencing the education of engineers.

The fundamental idea underlying the accepted system of training engineers consists in the combination of an adequate knowledge of the sciences which bear upon engineering with a thorough practical training on actual engineering works. No man is now entitled to admission as a Corporate Member of the Institution of Civil Engineers unless and until he has given proof of the possession of both these qualifications. Neither kind of training standing alone, or when developed disproportionately, can be regarded as satisfactory, or as meeting the needs of engineering practice. Formerly undue prominence was given to practical training and experience; while the facilities for scientific training were at first non-existent and for a long period were inadequate. Then came a better appreciation of scientific method and a great development of technical education, departments being established for the teaching of engineering science in Universities and University Colleges. The utilisation of these opportunities for instruction by considerable numbers of young men not unnaturally brought about a swing of the pendulum which went beyond reasonable limits. For a time there was a tendency to exalt unduly scientific education, and to depreciate the value of practical training. The hard pressure of experience has done much to adjust that disproportion. University graduates when they enter upon actual work soon discover that degrees in engineering, valuable as they undoubtedly are, require to be supplemented by thorough practical training. On the other hand, men who begin their engineering careers as pupils or assistants to practising engineers or as members of engineering office-staffs, become convinced that their limit of possible attainment must be low, unless scientific knowledge is added to practical experience. Under existing conditions new and difficult problems continually arise in all branches of engineering practice, and satisfactory solutions can only be found by bringing to bear upon these problems all the resources furnished by natural knowledge, accumulated experience and experimental research.

The full equipment of an engineer must include knowledge of other sciences besides the mathematical, but our present concern is exclusively with the latter. An adequate knowledge of mathematics must be possessed by every educated engineer, because he thus acquires valuable tools, by the use of which he can overcome difficulties that would otherwise be insuperable, as well as habits of thought and methods of rigorous investigation which are invaluable when he has to deal with novel and difficult undertakings. Apart from the employment of mathematics it would not be possible for the engineer to carry out designs and construction of engineering works, of structures and machines, capable of fulfilling their intended purposes and possessing both sufficient strength and durability. The days of blind reliance upon engineering formulae and "rules of thumb" are over. Syllabuses of instruction for the guidance of engineering students, standards established for degrees and diplomas in engineering science, conditions laid down as necessary qualifications for membership of great Engineering Societies, all furnish full recognition of the fact that an adequate knowledge of mathematics is essential to the successful practice of engineering.

It may be asked what range and character of mathematical knowledge can be described as adequate? Answers to this question are to be found in Calendars of Schools of Engineering which set forth detailed courses of study considered necessary for the attainment of degrees or diplomas. Identity of conditions does not exist in these Regulations, but a closer approach to uniformity has been reached as greater experience has been gained, and it is obviously desirable that further progress should be made in that direction. Engineering Degrees ought to be based on a common standard and to represent an equal attainment. These degrees, of course, should be regarded simply as certificates of knowledge of the fundamentals of engineering science; they do not cover all the mathematical knowledge requisite for the practice of particular branches of engineering, and in most branches a greater range of mathematics is necessary. Moreover a degree-course in engineering requires to be supplemented in all cases by subsequent practical training and experience, and in many cases by advanced or specialised courses of mathematical study going beyond the standards associated with degrees. In the settlement of these advanced courses the needs of each branch of engineering must be determined on the basis of experience; and the subject is one to be dealt with satisfactorily only on the basis of conference between practising engineers and mathematicians. The former know the needs which must be met: the latter can advise as to the best methods of meeting requirements.

Differences of opinion have always prevailed, and still exist, in regard to the methods by which mathematics should be taught to engineering students. Some authorities favour the arrangement of specialised courses of instruction—"mathematics for engineers" or "practical mathematics"—and advocate the creation of separate mathematical sections for engineering schools, even when these schools form departments of Universities or Institutions which possess well-organised mathematical Departments. Other persons, whose opinions are entitled to equal respect, believe that purely mathematical instruction is best given to engineering students by mathematicians, and that a similar rule should apply to instruction in other branches of science; because that method must lead to a broader view of science and a greater capacity for original and independent investigation than can be obtained by specialised teaching narrowed down to the known requirements of previous engineering practice. Personal experience and observation—as student, teacher and practical engineer—lead me to rank myself with the supporters of the latter method of teaching mathematics. No doubt, in the actual practice of engineers, there is room for "short cuts" and special methods in the use of mathematics; but I am convinced that during the period of education it is advantageous to follow ordinary methods of teaching and to leave specialisation for the time when the performance of actual work will almost inevitably lead each individual to make his choice of the branch of engineering to be followed, and of the methods which will best economise his labour and time in doing the work of calculation. The trend of professional opinion certainly lies in the direction of utilising, as far as possible, existing mathematical departments for the instruction of engineers. During the last year the subject has been exhaustively considered by the Governors of the Imperial College of Science and Technology, a special Committee having been appointed for that purpose. The Imperial College, as is well known, has been formed by bringing together the Royal College of Science, the Royal School of Mines, and the Engineering College founded and maintained by the City and Guilds of London Institute. The last-mentioned College at the outset had to be necessarily self-contained, and had its own Mathematical Department which was admirably organised and conducted by Professor Henrici over a very long period. The College of Science and Royal School of Mines also included a Department of

Mathematics and Mechanics of which Professor Perry has been the distinguished Director for many years. Both these Departments have justified their existence and done admirable work: but the development of the scheme of the Imperial College rendered it necessary to reconsider the subject of future mathematical instruction in the College as a whole. Alternatives taken into consideration were (1) the continuance of separate provision for engineering students; (2) the creation of a single department to be presided over by a mathematician of distinction, in which engineering students would receive their fundamental instruction in mathematics. After thorough investigation the latter course was preferred and will be carried into effect. Its adoption will in no way interfere with the teaching of special applications of mathematics as parts of the courses of instruction given by professors of engineering; and no one familiar with the training of engineers would consider such a change desirable.

A second example of the opinion which now prevails respecting the teaching of mathematics to engineers may be found in a valuable Paper by Professor Hopkinson of Cambridge University, published this year as one of a Series of Special Reports to the Board of Education on the teaching of Mathematics in the United Kingdom. Professor Hopkinson states that for good reasons, and during a considerable period after the Engineering Department was established at Cambridge, its students with few exceptions "got the whole of their instruction within the walls of the Engineering Laboratory," and "had not the full advantage of their position as students of Engineering Science in a centre of mathematical learning and research." In recent years, however, "the establishment of closer relations between the two studies (Mathematics and Engineering) has made great progress, and at the present time the students of Engineering get their foundation of Mathematics and of Elementary Mechanics from College Teachers, many of whom have graduated in the Mathematical Tripos."

It may be interesting to add that in a recent "Summary Report of the Teaching of Mathematics in Japan," Mr Fujisawa has discussed the "teaching of mathematics in technical education" in a brief but interesting fashion. While advocating the "practical" method of instruction, he is careful to explain that the form of utilitarianism which he recommends "has the potentiality of manifesting its usefulness wherever there is a necessity"—a condition which obviously cannot exist unless the engineer is endowed with a good knowledge of the principles and processes of mathematics.

More than seventy years ago, men who had received a mathematical and scientific training in the first British School of Naval Architecture, wrote as follows in vindication of the necessity for the liberal education of those engaged in the designing of ships:—"The study of naval architecture brings early conviction to the mind of the constructor that he can trust little or nothing to *a priori* reasoning. He uses the exact sciences it is true, but uses them only as a means of tracing the connection between cause and effects, in order to deduce principles that may be applied to his future works, with a certainty of producing the results he contemplates." This utterance may be applied with equal force to the practice of other branches of engineering; and, amongst the exact sciences which play so important a part in successful achievement, mathematics certainly hold the first place. The "complete engineer" of the earlier and simpler periods can exist no longer under modern conditions. Even the ablest men are driven to specialise in practice: but whatever branch of engineering may be selected, the worker will need that fundamental training in mathematics to which allusion has been made. Few engineers engaged in professional work have opportunities of prosecuting mathematical studies systematically, although they are continually using mathematical tools provided during their college careers, and not infrequently have occasion to add to their mathematical equipment in order to meet new demands, or to go beyond precedent and experience. When one considers the great responsibilities which practising engineers have to bear, it is not surprising to find that they, as a class, have made comparatively few contributions to the advancement of mathematical science, although they have been well trained in mathematics and continually apply that knowledge. There are, of course, exceptions to this rule; indeed, I have known engineers who turned to mathematics as a recreation, but these men are exceptional. Another group whose members have done notable work of a mathematical nature have been trained as engineers, but have either passed out of practice to an extent which left them ample leisure or have become professors of engineering. The names and work of engineers like Rankine, Froude and John Hopkinson will always be held in honour by mathematicians as well as by members of the profession which they adorned. The labours of many mathematicians who have devoted themselves to the tuition of engineers, and after becoming acquainted with the problems of engineering have done splendid work in the formulation of mathematical theories on which have been based valuable rules of a practical nature, also deserve and always receive the grateful appreciation of engineers. But, speaking broadly, there is a real and abiding distinction between the engineer, however accomplished he may be in mathematical science, and the mathematician however well informed he may be in regard to engineering practice. The mathematician necessarily regards engineering chiefly from the scientific point of view, and although he may aim at advancing engineering science, he is primarily concerned with the bearing of mathematics thereon. The engineer, being charged with actual design and construction of efficient and permanent works in the most economical manner possible, must put considerations of a practical and utilitarian character in the forefront; and while he seeks to utilise the aids which mathematical and physical science can render, his chief aim must always be to achieve practical and commercial success. There is obviously room for both classes, and their close and friendly collaboration in modern times has produced wonderful results. Fortunately the conditions which formerly prevailed have ceased to exist. Much less is said about the alleged distinction between pure and applied science, or about the comparative untrustworthiness of theory as compared with practical experience. Fuller knowledge has led to a better understanding of what is needed to secure complete success in carrying out great engineering works on which the comfort, safety, economical transport and easy communications of the civilized world so largely depend. The true place of mathematics in engineering practice is now better understood, and it is recognised to be an important place, although not so important as was formerly claimed for it by mathematicians.

The character of the change which has taken place in the use of mathematics in connection with engineering practice may be illustrated by reference to that branch of engineering with which my life-work has been connected. It is probably true to say that no branch of engineering has benefited more from mathematical assistance than naval architecture has done, and naval architects undoubtedly require to have at least as intimate a knowledge of mathematics as any other class of engineers. Moreover it was one of the first branches of engineering for which the foundation of a mathematical theory was attempted in modern times; and these attempts were made by men who in their day and generation were recognised to be in the first rank of mathematicians. The work which they did is now almost forgotten, but it laid the foundation for the science of naval architecture as it exists to-day. To France belongs the honour of having given most encouragement to men of science to attack these problems, and the Academy of Sciences aided the movement greatly by offering prizes which brought into the field not a few of the ablest European mathematicians during the latter half of the eighteenth century. Few of these mathematicians had personal knowledge of the sea or ships, and their investigations were influenced by these limitations. Others had made long voyages; like Bouguer, who (in 1735) proceeded to Peru *pour la mesure de la Terre*, and as a consequence of that experience a very practical tone was given to his famous *Traité du Navire*, which was published in 1745. It would be interesting to sketch the valuable work done by this single mathematician, but time does not allow me to do so. My main purpose at present is to illustrate the change that has since taken place in the use of mathematics in attacking engineering problems; and this may be done better by taking a single problem and showing how it was dealt with in the eighteenth and the nineteenth centuries.

Daniel Bernoulli in 1757 won the prize offered for the second time by the Académie Royale des Sciences for an answer to the question:—What is the best means of diminishing the rolling and pitching of a ship without thereby sensibly depriving her of any of the good qualities which she should possess? His *Mémoire* was published subsequently; it is an admirable piece of work, and deals thoroughly with the stability of ships; but here Bernoulli had been anticipated by Bouguer and made acknowledgment of the fact. Greater originality was shown in a mathematical investigation of the behaviour of ships in a seaway; and in a consideration of the influence of wave-motion upon the conditions of fluid pressure, as well as the determination of the instantaneous position of equilibrium for a ship floating amongst waves. Bernoulli recognised that the particles of water in a wave must be subjected to horizontal as well as vertical accelerations, although in his mathematical expressions he took account of the latter only. He emphasised the important influence which the relative sizes of waves and ships must have upon rolling and pitching motions, and advised that attention should be mainly devoted to cases where ships were small in proportion to the waves they encountered. In this particular he departed from assumptions usually made by his contemporaries and anticipated modern views. Bernoulli also dwelt upon the critical case wherein ocean waves, forming a regular series, have a period synchronising with the period of still-water rolling of the ship which they meet. A gradual accumulation of angular motion was shown to be inevitable in such circumstances, and it was remarked that the consequent rolling motions must be considerable and might possibly become dangerous in their extent. Bernoulli recommended the conduct of experiments to determine the periods of oscillation of ships in still water, and described methods of conducting these experiments. He also insisted upon the necessity for making accurate observations of the rolling of ships when amongst waves, and made other suggestions of much practical value, which have since been repeated by writers unfamiliar with Bernoulli's work and have been practically applied. Unfortunately the neglect by Bernoulli in his mathematical investigations of the horizontal accelerations of particles of water, which he recognised as existing in waves, led to erroneous conclusions in regard to the instantaneous position of equilibrium for a ship when floating amongst large waves and the best means for securing steadiness. Bernoulli considered that when a ship was in instantaneous equilibrium, her centre of gravity and the centre of buoyancy—i.e. the centre of gravity of the volume of water instantaneously displaced by the ship—must lie in the same vertical line. This condition of course holds good for a ship floating at rest in still water, but not for a ship floating amongst waves of large relative dimensions. Bernoulli deduced from his investigations a practical rule for the guidance of naval architects: viz. that in order to minimise rolling, ships should be designed so that their centres of buoyancy when they were upright and at rest should be made to coincide with the centre of gravity. He considered that ships should be made deep, that large quantities of ballast should be used, and that the cross-sections should be approximately triangular in form. This practical rule was misleading, and if applied in a design might be exceedingly mischievous in its effect on the behaviour of ships. Bernoulli himself foresaw that, in certain cases, his rule would work badly, but he considered that these would but rarely occur. It is now known that this view was mistaken.

The detailed mathematical investigations contained in Bernoulli's *Mémoire* are still of much interest; they included the examination of cases in which were assumed widely differing ratios of the natural periods of ships to the period of the waves producing rolling motion. Throughout, the motions of ships were supposed to be unaffected by the resistance of the surrounding water, but Bernoulli did not overlook the steadying effect which water-resistance would exercise on a ship in a seaway; on the contrary he recognised the influence which changes in the underwater forms of ships must have upon the amount and steadying effect of that resistance, and he recommended the use of side-keels in order to minimise rolling. Having regard to the state of knowledge at the time this *Mémoire* appeared, it was undoubtedly a remarkable piece of work and it well deserved the reward bestowed by the Academy. It contained many practical suggestions for experimental enquiry and for guidance in the preparation of designs for ships; but it was essentially a mathematical study and had little influence on the work of naval architects.

A century later the same problems were attacked by William Froude, a graduate of Oxford University and an engineer of experience in constructional work. As an assistant to Isambard Brunel, the attention of Froude had been directed to these subjects in connection with the design and construction of the *Great Eastern*, a ship of relatively enormous dimensions and novel type, respecting whose safety, manageability and behaviour in heavy seas serious doubts had been expressed. Like Bernoulli, of whose work I feel confident Froude had no knowledge, the modern investigator perceived that, amongst waves, there must be considerable variations in the direction and magnitude of the pressure delivered by the surrounding water on the surface of a ship's bottom; and that the instantaneous position of equilibrium for a ship exposed to the action of waves of large relative dimensions must be discovered if a theory of rolling was to be framed. Froude worked out a complete theory of trochoidal wave-motion and enunciated the principle of an "effective wave-slope." In his investigations it was assumed that the resultant water-pressure on the ship at each instant acted through the centre of buoyancy and normally to the effective wave-slope. In the differential equation framed for unresisted rolling, Froude took a curve of sines for the effective wave-slope instead of a trochoid. Having obtained the general solution of that equation, he proceeded to consider the behaviour of ships as influenced by variations in the ratios of their still-water periods of rolling oscillation to the relative periods of the waves encountered. In this manner the particular cases considered by Bernoulli were readily investigated, and many of the broad deductions made a century before were amended. The critical case of synchronism of ship-period and wave-period which Bernoulli had brought into prominence was shown to be that requiring most consideration. For that case the increment of oscillation produced by the passage of each wave of a regular series was determined on the hypothesis of unresisted rolling, and was shown to be about three times as great as the maximum inclination to the horizontal of the effective wave-slope. It was also made clear that apart from the influence of water-resistance, such synchronism of periods must lead to a ship being capsized by the passage of comparatively few waves. Up to this point, the investigation made by Froude was strictly mathematical, and the modern engineer who had received a thorough mathematical training had reached results superior to those obtained by the famous mathematician a century before; becoming, in fact, the founder of a theory for the oscillations of ships amongst waves which has been universally accepted. Like Bernoulli, Froude became impressed with the necessity for experiments which would determine the periods of still-water rolling for ships; and with the desirability for making observations of the rolling of ships in a seaway. In addition he emphasised the necessity for more extensive observations on the dimensions and periods of sea-waves, a subject which had been investigated to some extent by Dr Scoresby and other observers, but had been left in an incomplete state. One great generalisation was made by Froude at an early period in this important work, and it has since become a fundamental rule in the practice of naval architects; viz. that freedom from heavy rolling under the conditions usually met with at sea was likely to be favoured by making the period of still-water rolling of ships as large as was possible under the conditions governing the designs. This rule was the exact converse of that laid down by Bernoulli, as the effect of the latter rule by increasing the stability would have lessened the period. The explanation of this simple rule is to be found in the consideration that the longer the natural period of a ship is, the less likely is she to encounter waves whose period will synchronise with her own.

Purely mathematical treatment of the subject did not satisfy the mind of a trained engineer like Froude. For practical purposes it was essential that the effect of water-resistance to rolling should be determined and brought into the account. Here purely mathematical investigation could not possibly provide solutions; experimental research, conducted in accordance with scientific methods, became necessary. Aided by the Admiralty, Froude embarked upon a series of experiments which extended over several years. Most of these experiments were made on actual ships, but models were employed in special cases. In the analysis of experimental results, mathematics necessarily played a great part; indeed without their employment, the proper deductions could not have been made. On the basis of these analyses, Froude obtained valuable data and determined experimentally "coefficients" of resistance to rolling experienced respectively by the flat and curved portions of the immersed surfaces of ships. Furthermore he demonstrated the fact that the surface disturbance produced by the rolling of ships in still-water accounted for a large part of the extinctive effect which was produced when a ship which had been set rolling in still water was allowed to come to rest. In this way, and step by step, Froude devised methods by means of which naval architects can now calculate with close approximation the extinctive effect of water-resistance for a new design. Finally, Froude produced a method of "graphic integration," the application of which in association with the calculation of the effect of water-resistance, enables a graphic record to be constructed showing the probable behaviour of a ship when exposed to the action of successive waves, not merely when they form a regular series, but when they are parts of an irregular sea. Subsequent investigators have devised amendments or extensions of Froude's methods, but in all essentials they stand to-day as he left them—a monument of his conspicuous ability, and an illustration of the modern method in which mathematics and experimental research are associated in the solution of engineering problems which would otherwise remain unsolved.

In tracing as has been done the contrast between the methods of Bernoulli and Froude, an indirect answer has been given to the question—What is the true place of mathematics in engineering practice? It has been shown that even in the hands of a great mathematician, purely mathematical investigation cannot suffice, and that Bernoulli became convinced, in the course of his study of the behaviour of ships in a seaway, that no complete or trustworthy solution could be found apart from experimental research, as well as careful observations of ocean waves and the rolling of actual ships. Bernoulli was not in a position to undertake, or to lead others to undertake, these experiments and observations. In his mathematical investigations he made, and necessarily made, certain assumptions which are now known to have been incorrect. Even the most accurate mathematical processes, when applied to equations which were framed on imperfect or incorrect assumptions, could not produce trustworthy results; and consequently the main deductions made by Bernoulli, and the rules recommended by him for the guidance of naval architects, would have led to disappointment if they had been applied in practice. On the other hand, Froude, himself a great experimentalist, was fortunately able to impress upon the British Admiralty through the Constructive Staff the importance of making experiments and extensive observations of wave-phenomena and the behaviour of ships. Not merely did Froude devote many years of personal attention to these subjects, but he was aided over a long period by the large resources of the Royal

Navy. Similar work on a very large scale was also done simultaneously by the French Navy. Some of my earliest experiences at the Admiralty forty-five years ago were gained in connection with these observations and experiments, so that I speak from personal knowledge of the influence which Froude exercised, the inspiration of his great devotion and wonderful initiative. As a result of all these efforts, a great mass of experimental data was accumulated; the results of a large number of observations were summarised and analysed; and, in the end, the soundness of the modern theory was established, and the future practice of naval architects was made more certain in their attempts to produce designs for ships which should be steady and well-behaved at sea.

At the risk of making this lecture appear to be chiefly a notice of work done by William Froude, or a summary of the advances made in the science of naval architecture, another illustration will be given of the general principle laid down in regard to the place of mathematics in engineering practice.

Mathematicians, from an early date, were attracted by the subject of the resistance offered by water to the motion of ships and made many attempts to frame satisfactory theories. The earliest investigations were based upon the assumption that the immersed surface of a ship's skin could be treated as if it consisted of an aggregation of elements, each of which was a very small plane area, set at a known angle of obliquity to the direction of motion through the water. For each elementary plane area it was proposed to estimate the resistance independently of the others, and as if it were a small isolated flat plate. The integration of such resistances over the whole surface was supposed to represent the total resistance of the water to the motion of the ship at a given speed. Certain further assumptions were made in regard to the laws connecting the resistance of each unit of area with its angle of obliquity to the direction of motion and with the speed of advance. The effect of friction was, in most cases, neglected; nor was any account taken of surface disturbance produced by the motion of the ship. It is unnecessary to dwell upon the errors and incompleteness of these assumptions. So long as ships were propelled by sails little practical importance attached to an exact determination of the resistance experienced at a certain speed. When steamships came into use it was of primary importance to have the power of making close approximations to that resistance because estimates for the engine-power required to attain a given speed had to be based thereon. The subject received great consideration, as the result of which certain simple rules were framed and commonly employed in making estimates for the engine-power to be provided in new ships. These rules were mainly based on the results obtained by trials of existing vessels; and these trial-results, of course, included not merely the effect of water-resistance—as influenced by the form and condition of the immersed surface of a ship—but were also affected by the varying efficiency of the propelling apparatus and propellers. Many attempts were made to separate these items of performance and to determine the actual amount of the resistance for a ship and the separate efficiencies of her propellers and machinery. Little progress was secured until 1868. Mathematical theories were framed, it is true, for estimating the efficiency of propellers; but while these theories were accurate enough if the assumptions underlying them had been complete and representative of actual phenomena, there was no possibility of fulfilling those conditions since the phenomena were neither fully known nor understood.

In 1868 a special and representative Committee—including Rankine and Kelvin—appointed by the British Association, made a Report on this subject and recommended that towing experiments should be made on full-sized ships. The Committee was almost unanimously of opinion that the only method which would give trustworthy information in regard to the resistances experienced at various speeds was to tow actual ships and not to depend upon models. William Froude dissented from this conclusion and recommended model experiments. Accepting the stream-line theory of resistance which Rankine had introduced, Froude based upon it a system of experiment which dealt separately with frictional resistance and applied to the residual resistance—after friction had been allowed for and deducted—a law of "corresponding speeds" between models and full-sized ships which he had worked out independently. That law had been previously recognised in a more general form by mathematicians, and had been investigated for this particular case by a French mathematician, M. Reech, of whose work Froude was then ignorant. By this happy association of mathematical theory with experimental research, Froude placed in the hands of naval architects the means of solving problems which could not be dealt with either by purely mathematical investigation, or by experience with actual ships. Experimental tanks of the character devised by Froude have now been multiplied in all maritime countries. The latest and in many respects the best of these tanks, which is due to the generosity of Mr Alfred Yarrow, is a Department of the British National Physical Laboratory. The operations of these tanks have resulted in a great addition to natural knowledge and have secured enormous economies of fuel. The success achieved in connection with modern developments of steam navigation and the attainment of very high speeds is chiefly due to tank experiments which have involved relatively small cost, and enabled naval architects to choose for every design the form which gives the least resistance possible under the conditions laid down for a new ship, even when the size or speed required go beyond all precedent. Considerations of stability, carrying capacity, available depths of water, dimensions of dock entrances and other matters, as well as speed and fuel consumption, may limit the designer and narrow the alternatives at his disposal; but ordinarily there is room for considerable variations of form in a new design, and in making the final selection of form it is essential that the designer should know how the resistances of these permissible alternatives compare. Naval architects throughout the world enjoy great advantages in this respect over their predecessors, and owe their position entirely to the genius and persistence of William Froude.

Since the work of Froude in this direction was done, model experiments have become the rule in many departments of engineering and the scientific interpretation of the results has greatly influenced the designs of structures and machines. Prominent amongst these recent applications of experimental research on models stand those relating to air-resistance and wind-pressure on bridges and other structures. In regard to the laws of wind-pressure much has been discovered in recent years, and in connection with the effects of wind-pressure on engineering structures especial reference ought to be made to the work done by Dr Stanton at the National

Physical Laboratory. All engineers owe a debt of gratitude to that distinguished experimentalist and to the Institution where he works; and they recognise the fact that he has demonstrated the trustworthiness of deductions made from tests with small models exposed to the action of air-currents when applied on the full scale to complicated structures for which independent mathematical calculations of the effect of wind-pressure could not be made. The late Sir Benjamin Baker, who was chiefly responsible for the design of the Forth Bridge, was one of the first to appreciate and make use of this experimental system, and no engineer of his time more frankly admitted than he did what a debt engineering practice owed to mathematics when used in the proper manner.

The proper use of mathematics in engineering is now generally admitted to include the following steps. First comes the development of a mathematical theory, based on assumptions which are thought to represent and embody known conditions disclosed by past practice and observation. From these theoretical investigations there originate valuable suggestions for experimental enquiries or for careful and extensive investigations. The results obtained by experimental research or from observation and experience must be subjected to mathematical analysis: and the deductions made therefrom usually lead to amendments or extensions of the original theory and to the device of useful rules for guidance in practice. Purely mathematical theories have served and still serve a useful purpose in engineering; but it is now universally agreed that the chief services of mathematics to engineering are rendered in framing schemes for experimental research, in analysing results, in directing the conduct of observations on the behaviour of existing engineering works, and in the establishment of general principles and practical rules which engineers can utilise in their daily professional employment.

One of the most recent examples of this procedure is to be found in the constitution and proceedings of the Advisory Committee appointed in 1909 by the British Government in connection with the study of Aeronautics. Its membership includes distinguished mathematicians, physicists, engineers and officers of the army and navy, and its President is Lord Rayleigh, Chancellor of the University of Cambridge. The declared intention in establishing this Committee was to bring the highest scientific talent "to bear on the problems which have to be solved" in order to endow the military and naval forces of the British Empire with efficient aerial machines. The Reports published during the last two years are of great value; the work done by the Committee—described as "the scientific study of the problems of flight with a view to their practical solution"—has been accompanied and supplemented by research and experiment carried out by the Director of the National Physical Laboratory (Dr Glazebrook) and his staff in accordance with a definite programme approved by the Committee. These investigations necessarily cover a very wide field in which there is ample room for the operations of all the branches of science and engineering represented on the Committee, and there can be no doubt that already the influence of the work done has been felt in practice. No one who has followed the progress made in aerial navigation, however, can fail to be convinced that although a considerable amount of purely mathematical investigation has been devoted to the problems of flight, it has hitherto had but little influence on practice, in comparison with that exercised by improvements due to mechanical engineering—tending to greater lightness of the engines in relation to their power—and by actual experiments made with models and full-sized flying machines. A stage has been reached, no doubt, where the interpretation by mathematicians of the experimental results available and their suggestions as to the direction in which fuller experimental research can best be carried out are of great importance, and that fact is universally recognised by engineers.

Even when the fullest use has been made of mathematical science applied in the best way and of experimental research there still remain problems which have hitherto defied all efforts at their complete solution, and engineers have to be content with provisional hypotheses. Of the James Forrest Lectures given annually at the Institution of Civil Engineers a long series has been devoted to the description of "Unsolved Problems in Engineering." Mathematicians seeking fresh fields to conquer might profitably study these utterances of practising engineers of repute. On this occasion it must suffice to mention two classes of subjects on which additional light is still needed, although they are now less obscure than they have been in the past, thanks to long years of work and experiment.

The first group has relation to the laws which govern the efficiency of screw-propellers when applied to steamships, and has long engaged the attention of a multitude of writers in all maritime countries. Many mathematical theories have been published, which are of interest and value as mathematics, and are sound if the fundamental assumptions made could be accepted. It is, however, no exaggeration to say that at the present time there exists no mathematical theory which has any considerable influence on the design of screw-propellers and the determination of the form, area and pitch. Experience and experiment are still mainly depended upon when work of that kind has to be undertaken. Of course certain mathematical principles underlie all propeller designs, but the phenomena attending the operation of a screw-propeller at the stern of a ship in motion are too variable and complex to be represented by any mathematical equation even if they were fully known and understood—which they are not. The water in which a propeller works has already been set in motion by the ship before it reaches the propeller, and the "wake" of a ship in motion is in a very confused state. The action of a propeller upon the water "passing through it" and the manner in which its effective thrust is obtained still remain subjects for discussion and for wide differences of opinion between mathematicians and experimentalists who have seriously studied them. Froude initiated a system of model-experiments for propellers, both when working in open water and when attached to and propelling ships or developing an equivalent thrust. His son, Mr R. E. Froude, has done remarkable work in the same direction, and many other experimentalists have engaged in the task: but after more than seventy years of experience in the practical use of screw-propellers we remain without complete or accurate knowledge which would enable the designs of propellers for new ships, of novel types, or of very high speed, to be prepared with a certainty of success. On the whole, naval architects and marine engineers depend largely upon the results of experience with other ships. Although model-experiments are also utilised, there is not the same confidence in passing from results obtained with model propellers to full-sized propellers as there is in passing from model ships to full-sized ships. Probably this fact is chiefly due to essential differences in the reactions between the water and the surfaces of models and the surfaces of full-sized screws moving at the rates of revolution appropriate to each. These matters are receiving and have already received careful study not merely by the Superintendents of Experimental Tanks, but by practising engineers like Sir Charles Parsons and Sir John Thornycroft. The phenomenon of "cavitation"—which has been described as the breaking-away of water from the screw surfaces when the rate of revolution of the screw exceeds certain limits, and when the thrust per unit of area on the screw exceeds certain values—is one which has given much trouble in the cases of vessels of exceedingly high speed such as destroyers and swift cruisers. It is being investigated experimentally, but up to the present time no general solution has been found. In existing conditions surprising differences in the efficiency of propellers have been produced by what appeared to be small changes in design. On the whole the largest improvements have been obtained as the result of full-scale trials made in ships, although model-experiments have been of service in suggesting the direction in which improvements were probable. In my own experience very remarkable cases have occurred, and not infrequently it has been difficult even after the event to explain the results obtained. One such case may be mentioned as an illustration. A large cruiser obtained the guaranteed speed of 23 knots on trial with a development of about 30,000 horse-power. I had anticipated a higher speed. Progressive trials made at various speeds showed that the "slip" of the propellers became excessive as the maximum speed was approached, although the blade area given to the propellers on the basis of past experience was adequate for the power and thrust. The blade-area was increased about 20 per cent., the diameter and pitch of the screw-propellers being but little changed. With these new propellers the maximum speed became 24 knots and 23 knots was obtained with a development of 27,000 horse-power, as against 30,000 horse-power required with the original screws. The increase of blade area necessarily involved greater frictional losses on the screws, yet the effective thrust was increased, a higher maximum speed was attained, and the power required at all speeds became less than in the earlier trials down to 15 knots. This incident could be paralleled from the experience of many naval architects, and it illustrates the uncertainties which still have to be faced in steamship-design when unprecedented speeds have to be guaranteed.

This open confession of lack of complete knowledge, made in the presence of the professors of an exact science such as mathematics, may be thought singular. It is the fashion to criticise, if not to condemn, designers of ships and their propelling apparatus on the ground that after long experience there ought to be a complete mastery of these problems. That criticism, however, is hardly fair; because it overlooks the fact that throughout the period of steam navigation there has been incessant change in the dimensions, forms and speeds of ships and in the character of the propelling apparatus.

Knowledge is also still incomplete, and possibly complete knowledge will never be attained, in regard to the stresses experienced by the structures of ships at sea, when driven through waves and made to perform rolling, pitching and heaving movements simultaneously. The subject has long engaged the attention of mathematicians and naval architects. Early in the last century Dr Young made a study of the causes of longitudinal bending in wood-built ships, and presented a Memoir to the Royal Society. The eminent French geometrician Charles Dupin also dealt with the subject; which had great practical interest at a date when serious "hogging" or "arching" of ships was a common occurrence. Since iron and steel have been available for ship-construction—and as a consequence the dimensions, speeds and carrying powers of ships have been enormously increased—questions of a similar character have arisen on a larger scale, and have been carefully studied. There is much in common between ship construction and bridge construction under modern conditions; and because of this resemblance engineers practising in works of a constructional nature on land have been brought into close relation with the structural arrangements of ships. Sir William Fairbairn, who was associated with the younger Stephenson in the construction of the Menai and Conway tubular bridges, and Isambard Brunel, whose chief work was on railways, but who designed the famous steamships *Great Western*, *Great Britain* and *Great Eastern*, are amongst the men of this class who have most influenced shipbuilding. There are, however, obvious and fundamental differences between the conditions of even the greatest bridge founded on the solid earth, and those holding good in the case of self-contained floating structures carrying great loads across the sea, containing powerful propelling apparatus, and necessarily exposed to the action of winds and waves. In the former case bending moments and shearing stresses which must be provided against can be closely estimated, and ample strength can be secured by adopting proper "factors of safety." In the case of ships no similar approximations are possible; because their structures are stressed not only by the unequal distribution of weight and buoyancy, but have to bear rapidly varying and compound stresses produced by rolling and pitching motions, by external water-pressure and by the action of the propelling apparatus, as well as to resist heavy blows of the sea. Inevitably, therefore, the naval architect has to face the unknown when deciding on the "scantlings" of various parts of the structure of a new ship the design of which goes beyond precedent.

Mathematicians have had the courage to attack these problems and to propound theories respecting them. Professor Kriloff of the Imperial University, St Petersburg, has been one of the latest workers in this field, and has probably carried the mathematical theory furthest: but his work, like that of his predecessors, has had little effect on the practice of naval architects. Indeed it seems too much to expect that even the most accomplished mathematician can deal satisfactorily with the complex conditions which influence the variable stresses acting, from moment to moment, on the structure of a ship at sea. In these circumstances naval architects have been compelled to fall back upon experience with ships which have been long in service at sea, and to obtain the best guidance possible from the application of mathematics to the analysis of that experience and to the device of rules of a comparative nature. In general this procedure has led to the construction of ships which have possessed ample strength, although the actual margin of strength in excess of the permissible minimum has not been ascertained. In the comparatively few cases wherein weaknesses have been brought to light on service, scientific analysis has enabled even more valuable lessons to be deduced. But it cannot be said that purely mathematical investigation has been of great service to this branch of engineering.

Rankine many years ago proposed to base comparisons of the longitudinal bending moments and shearing stresses of ships amongst waves on the hypothesis that the distribution of weight and buoyancy should be determined for two extreme cases: first when a ship was momentarily resting in equilibrium on the crest of a wave having a length equal to her own length, and a height (hollow to crest) as great as was likely to occur in a seaway—say one-twentieth of the length of the wave. Second when she was momentarily floating astride a hollow of such waves. It was recognised, of course, that these cases were purely hypothetical, but the hypothesis has proved of great value in practice. Attempts have been made to carry the calculations further, and to take account of the effects of rolling, pitching and heaving motions, and of variations in the direction and amount of water-pressure consequent on wave-motion. Practice has been influenced but little by these attempts, which have necessarily been based on more or less arbitrary assumptions themselves not free from doubt. On the other hand Rankine's method has been widely used by naval architects; and the accumulated results of calculations obtained for ships whose reputations for strength and seaworthiness were good, are now available for reference. For new designs calculations of a similar character are made, and by comparison of results with those obtained for completed ships, most closely approaching the new design in type and dimensions, the principal scantlings are determined for various parts of the structure. In calculating the strengths of ships, they are usually treated as hollow girders exposed to the action of forces tending to cause longitudinal bending. The bending moments and shearing stresses calculated for the two extreme hypothetical conditions above described, are used in order to estimate the maximum stresses corresponding thereto in any members of the new ship's structure. A comparison of these maximum stresses (per unit of area of material) with the corresponding figures for successful ships is taken as a guide for determining the sufficiency or insufficiency of the scantlings proposed for the new vessel. In providing for adequate transverse strength and for margins of strength to meet local requirements, naval architects make separate calculations, but in these cases also are guided chiefly by comparisons based on actual experience with other ships. Mathematicians may regard this procedure as unsatisfactory: but they may be assured that any suggestions for improved or more exact methods which may be made will be welcomed by naval architects provided they command confidence and are capable of practical application.

In considering the relation of mathematics to engineering practice one important fact should always be borne in mind. The mission of engineers as a class is to produce results, "to do things," which shall be of practical service to humanity, and shall ensure safety of life and property. Complete solutions of problems, in the mathematical sense, are not usually to be found by engineers; they have to be content, in many cases, with partial solutions and fairly close approximations. It may be taken for granted that engineers desire to perform efficiently the duties laid upon them and that they are ready to avail themselves of all assistance which can be rendered by contemporary science, and by mathematicians in particular. On their behalf it has been my endeavour on this occasion to make suitable acknowledgment of the debt which engineering already owes to mathematics, and to indicate a few of the many problems in which further assistance is needed. All members of the engineering profession will endorse my expression of the hope that the close and friendly relations which have long existed between mathematicians and engineers, and which have yielded excellent results during the past century, will always continue and in future be productive of even greater benefits.

BOUNDARY PROBLEMS IN ONE DIMENSION

By Maxime Bôcher.

§ 1. *Introduction.*

By a boundary problem in one dimension I understand primarily the following question:

To determine whether an ordinary differential equation has one or more solutions which satisfy certain terminal or boundary conditions, and, if so, what the character of these solutions is and how their character changes when the differential equation or the boundary conditions change*. This is the central problem, of which various modifications are possible. In its simplest forms this question is as old as the subject of differential equations itself. By the end of the nineteenth century it already had a considerable literature, which since that time has expanded rapidly. I shall try during the present hour to indicate some of the greatest advances made both as to results attained and methods used. In thus trying to get a brief and yet comprehensive survey of a large subject, the desirability of a thorough correlation of the parts becomes doubly apparent, and I trust that you will find that in this respect I have succeeded at a few points in adding something to what was to be found in the literature. The older results will be discussed in detail only so far as may seem necessary to make the scope and importance of the more recent ones intelligible.

The subject is so large that I must limit myself to certain central aspects of it by leaving out of consideration almost entirely

(1) Non-linear boundary problems, that is cases in which the differential equation or the boundary conditions or both are non-linear.

(2) Cases in which two or more parameters enter. (Klein's theorem of oscillation with its extensions.)

(3) Cases where we have to deal not with a single differential equation but with systems of differential equations.

(4) Cases in which the differential equation has singular points in or at the ends of the interval with which we deal, or, what is essentially the same thing, cases in which this interval extends to infinity.

All of these cases are of the highest importance.

* The question of finding effective means for computing the solutions in question is also one which might well be considered here.

An even more sweeping restriction than any of these is indicated by the very title of the lecture. This restriction to one dimension, i.e. to ordinary rather than partial differential equations, is made absolutely necessary by the time at my disposal if we are actually to reach the deeper lying parts of the subject. Fortunately the one-dimensional case may be regarded to a very large extent as the prototype of the higher cases; but in this simple case methods are available which enable us to go far beyond the point which we can hope at present to reach for partial differential equations.

The problem with which we deal is, then, this:

A linear differential equation which, for the sake of simplicity, I write as of the second order,

$$P(u) \equiv \frac{d^2u}{dx^2} + p_1 \frac{du}{dx} + p_2 u = r \quad \text{.......................(1)},$$

has coefficients p_1, p_2, r which are continuous functions of the real variable x in the finite interval

$$a \leqq x \leqq b \quad \text{..................................(X)}.$$

We wish to solve this equation subject to the linear boundary conditions

$$\left.\begin{aligned} W_1(u) &\equiv \alpha_1 u(a) + \alpha_1' u'(a) + \beta_1 u(b) + \beta_1' u'(b) = \gamma_1 \\ W_2(u) &\equiv \alpha_2 u(a) + \alpha_2' u'(a) + \beta_2 u(b) + \beta_2' u'(b) = \gamma_2 \end{aligned}\right\} \quad \text{.........(2)},$$

where the α's, β's, γ's are constants.

Why is this problem an important one? The most obvious answer is that it is one of which special cases come up constantly in applied mathematics; that even its special cases are of sufficient difficulty to have demanded the serious attention of the best mathematicians for nearly two hundred years; that in connection with this problem methods and results of large scope have been developed. From another and more abstract point of view also this problem may claim importance: it is one of the simplest and most natural generalizations of that most central of all subjects, the theory of a system of linear algebraic equations. This is a fact which has been known ever since John and Daniel Bernoulli in their treatment of vibrating strings replaced the uniform string by a massless one weighted at equal intervals by heavy particles. The effect of this was to replace the differential equation for determining the simple harmonic vibrations of the string, which is a special case of (1), and the boundary conditions, which come under (2), by a system of linear algebraic equations.

The idea involved in this physical example may be formulated more generally as follows:

We may replace (1) by a *difference* equation of the second order:

$$L_i u_{i+1} + M_i u_i + N_i u_{i-1} = R_i \qquad (i = 1, 2, \ldots n-1) \quad \text{...........}(\bar{1}),$$

and the boundary conditions by

$$\left.\begin{aligned} A_1 u_0 + A_1' u_1 + B_1 u_{n-1} + B_1' u_n = C_1 \\ A_2 u_0 + A_2' u_1 + B_2 u_{n-1} + B_2' u_n = C_2 \end{aligned}\right\} \quad \text{.................}(\bar{2}).$$

The equations $(\bar{1})$ and $(\bar{2})$ taken together form a system of $n+1$ linear algebraic equations for determining $u_0, u_1, \ldots u_n$. If now we allow n to become infinite, causing the coefficients of ($\bar{1}$) and ($\bar{2}$) to vary in the proper way, we easily obtain the system (1), (2) as the limiting form.

In the same way the linear boundary problem for a differential equation of the nth order may be regarded as the limit of a linear boundary problem for a difference equation of the nth order, that is, again, of a system of linear algebraic equations.

It goes without saying that this relation yields a fertile source of suggestions both as to the facts in the transcendental case and as to possible methods of proof. It was indeed the unpublished method which Sturm originally used in his fundamental investigations*. On the other hand, the passage to the limit may be rigorously carried through, as was done by Cauchy in his proof of the fundamental existence-theorem for differential equations (not merely in the linear case). This proof was completed in 1899 by Picard and Painlevé by showing that the solution of the difference equation approaches that of the differential equation uniformly not only in a certain small neighbourhood of the point where the initial conditions are given, but throughout any closed interval about this point in which the solution in question of the differential equation is continuous. With this fact at our disposal there is no longer any difficulty in carrying through rigorously the passage to the limit from the difference equation to the differential equation in other cases of boundary problems, as was shown in a sufficiently general case by Porter† more than ten years ago. Thus we may regard this method of passage to the limit as one of the well-established methods, both heuristic and otherwise, of approaching boundary problems.

This linear boundary problem for difference equations has, however, also distinct interest in itself apart from any assistance it may give us in the transcendental case. During the last few years great interest has been awakened in the theory of difference equations from a very different side by the remarkable work of Galbrun, Birkhoff, and Nörlund. It seems therefore an opportune time that this side of the subject should be also further developed. I shall return to this matter presently.

§ 2. *Generalities. Green's Function.*

A special case of the general linear boundary problem (1), (2) is the *homogeneous* boundary problem in which $r \equiv 0$, $\gamma_1 = \gamma_2 = 0$:

$$P(u) = 0 \qquad \text{(1')},$$

$$W_1(u) = 0, \quad W_2(u) = 0 \qquad \text{(2')}.$$

This system we shall call the *reduced system* of (1), (2). If this system has no solution except the trivial solution $u = 0$, I call it incompatible. If it has essentially only one solution, I call it simply compatible; if it has two linearly independent solutions, I call it doubly compatible. If we have to deal with a differential equation of the nth order, we may have compatibility of order as high as n. One of the most fundamental theorems here, and yet one which, I believe, has been enunciated and proved

* For a reconstruction of this work see the paper by Porter cited below and Bôcher, *Bull. Amer. Math. Soc.* vol. 18 (1911), p. 1.

† *Annals of Mathematics*, 2nd series, vol. 3 (1902), p. 55. This was more than two years before Hilbert, in 1904, took a similar step for integral equations.

only within the last few years*, is that *a necessary and sufficient condition that the general boundary problem* (1), (2) *have one and only one solution is that the reduced problem be incompatible.* It should be noticed that this is the direct analogue of a familiar theorem concerning linear algebraic equations.

It is readily seen that the general case is that in which the reduced system is incompatible. The case in which the reduced system is compatible, so that the complete system has either no solution or an infinite number of solutions, we may therefore speak of, for brevity, simply as the *exceptional case.* This exceptional case will always occur when the boundary conditions (2′) are linearly dependent. It may however occur in other cases too, and it is from this fact that the most interesting and important questions relating to boundary problems arise.

Of all the boundary problems by far the simplest and most important is what we may call the *one-point problem* in which all the β's or all the α's are zero, so that the boundary conditions (2) involve only one of the end-points of (X). If in this case conditions (2′) are linearly independent, equations (2) may be solved for $u(a)$ and $u'(a)$ (we assume for definiteness that the β's are zero) and thus be written in the form

$$u(a)=\delta_1, \quad u'(a)=\delta_2.$$

Now the most fundamental existence-theorem in the theory of differential equations tells us that there always exists one and only one solution of (1) which satisfies these conditions. This existence-theorem may then be regarded as the answer to our boundary problem in this case, and phrased as follows: *In the one-point boundary problem the exceptional case can occur only when conditions* (2′) *are linearly dependent.*

So far as we have yet gone there is no necessity for the two points which enter the boundary conditions (2) to be precisely the end-points a, b of (X); they may instead be any two points x_1, x_2 of this interval. Moreover, we may make a further generalization by considering in place of (2) conditions of the form

$$\alpha_1 u(x_1)+\alpha_1' u'(x_1)+\alpha_2 u(x_2)+\alpha_2' u'(x_2)+\ldots+\alpha_k u(x_k)+\alpha_k' u'(x_k)=\gamma,$$

which involve not two points but k; and we may at the same time consider differential equations of the nth order. This is a subject which has hardly been touched upon in the literature so far, but which seems likely to become of importance. The one result which I find in the literature is that if the boundary conditions consist in giving at each of the k points the value of u and of a certain number of its earliest derivatives, *and if the k points are sufficiently near together* the problem always has one and only one solution. This fact was established (not merely in the linear case) by Niccoletti† as a generalization of some methods and results of Picard for certain non-linear differential equations of the second order.

Still another direction in which we may generalize the boundary problem, either in connection with the generalization last mentioned or independently of it, is to admit in connection with the equation of the nth order more than n boundary conditions. We shall have occasion to mention some cases of this sort later.

* Cf. Mason, *Math. Ann.* vol. 58 (1904), p. 532; *Trans. Amer. Math. Soc.* vol. 7 (1906), p. 340; and Bôcher, *Annals of Math.* ser. 2, vol. 13 (1911), p. 71.

† *Turin Atti*, vol. 33 (1898), p. 746.

One occasionally finds the boundary conditions (2) replaced by conditions which involve definite integrals and which, on their face, are not boundary conditions at all*. Such conditions may however often be reduced to precisely the form (2). As an example of this we mention the problem of solving the equation

$$\frac{d^2u}{dx^2} - Gu = r$$

subject to the condition $$\int_a^b \Phi(x)\,u(x)\,dx = C,$$

where Φ is a given continuous function, and C a given constant.

Let $\phi(x)$ be any solution of the equation

$$\frac{d^2\phi}{dx^2} - G\phi = \Phi.$$

By combining this with the equation for u we readily find the formula

$$\int_a^b \Phi u\,dx = \Big[\phi' u - u'\phi\Big]_a^b + \int_a^b r\phi\,dx.$$

The above integral condition may therefore be replaced by

$$-\phi'(a)\,u(a) + \phi(a)\,u'(a) + \phi'(b)\,u(b) - \phi(b)\,u'(b) = C - \int_a^b r\phi\,dx,$$

a condition of precisely the form (2).

If we approach the subject from the point of view of difference equations, this simply means that if we have in place of the boundary conditions *general* linear equations between $u_0, u_1, \ldots u_n$, these conditions can by using the difference equation be reduced to the ordinary four (or if we prefer three) term boundary condition form,—an obvious algebraic fact.

Let us leave these generalizations, however, and return to the case in which the conditions (2) involve merely two points, the end-points of the interval (X). While what I am about to say may readily be extended to equations of the nth order, I will again, for the sake of simplicity, speak merely of the equation of the second order, i.e. of the system (1), (2) in precisely the form in which we wrote it at first.

If, as is in general the case, the reduced system (1′), (2′) is incompatible, we are led to the important conception of the Green's Function by trying to find a function not identically zero satisfying (2′) and which comes as near as possible to being a solution of (1′)—it is to fail in this only through a finite jump of magnitude 1 at a point ξ of (X) in its first (or in the case of equations of the nth order in its $(n-1)$th) derivative. Such a function, $G(x, \xi)$, always exists and is uniquely determined when (1′) and (2′) are incompatible. A characteristic property of this function and one upon which its importance depends is that when (1′), (2′) are incompatible, the solution of the semi-homogeneous problem (1), (2′), which then exists and is uniquely determined, is given by the formula

$$u = \int_a^b G(x, \xi)\,r(\xi)\,d\xi \quad\ldots\ldots\ldots\ldots\ldots\ldots(3),$$

* Cf Picone, *Annali della R. Scuola Normale Superiore di Pisa*, vol. 11 (1909), p. 8; and v. Mises, *Heinrich Weber Festschrift* (1912), p. 252.

which, as we mention in passing, includes as a special case (viz. when conditions (2) involve only one of the points a or b) the formula for the solution of (1) obtained by the method of variation of constants.

These Green's Functions may also be regarded, if we wish, as the limits of the Green's Functions for the difference equation, i.e. the solution of the reduced system corresponding to $(\bar{1})$, $(\bar{2})$, except that *for a single value of i* the second member of $(\bar{1})$ is to be taken not as zero but as 1*. The formula (3) then becomes a special case of the obvious one for building up the solution of a general system of non-homogeneous linear algebraic equations of non-vanishing determinant from the solutions of the special non-homogeneous system obtained by replacing one of the second members by 1 while all the other second members are replaced by zero.

So far we have demanded merely the continuity of the coefficients of (1). If in addition we demand the existence and continuity of the first derivative of p_1†, we can add considerably to the properties of the Green's function. When regarded as a function of ξ, it then satisfies the differential equation adjoint to (1′)

$$Q(v) \equiv \frac{d^2v}{d\xi^2} - \frac{d(p_1v)}{d\xi} + p_2v = 0 \qquad \text{.................(1″)},$$

except when $\xi = x$. Moreover, still regarding it as a function of ξ, we find that it satisfies a system of homogeneous boundary conditions precisely analogous to (2′) but with different coefficients, these coefficients being however independent of the parameter x just as the coefficients of (2′) are independent of ξ:

$$\left.\begin{aligned}\overline{W}_1(v) \equiv \bar{\alpha}_1 v(a) + \bar{\alpha}_1' v'(a) + \bar{\beta}_1 v(b) + \bar{\beta}_1' v'(b) = 0\\ \overline{W}_2(v) \equiv \bar{\alpha}_2 v(a) + \bar{\alpha}_2' v'(a) + \bar{\beta}_2 v(b) + \bar{\beta}_2' v'(b) = 0\end{aligned}\right\} \qquad \text{.........(2″)}.$$

The system (1″), (2″) is of fundamental importance in the whole theory of linear boundary problems and is called the system adjoint to (1′), (2′). A special case of it was used by Liouville‡ but the general formulation and application of the conception was made for the first time by Birkhoff§ less than five years ago. The reason why even now this conception is not as well known as it deserves to be is that the special cases which have almost exclusively absorbed the attention of mathematicians belong to the class of *self-adjoint* systems where not only the equation (1′) is self-adjoint but the boundary conditions (2″) are also identical with (2′). It is true that a somewhat more general case than this has received a little attention from Hilbert and his pupils‖, namely the case which they call that of "Greenian boundary conditions"¶ where (2′) and (2″) are identical without (1′) being self-

* Cf. Bôcher, *Annals of Math.* 2nd series, vol. 13 (1911), p. 71, where other references for the literature of Green's Functions will be found.

† For the equation of the nth order

$$\frac{d^nu}{dx^n} + p_1\frac{d^{n-1}u}{dx^{n-1}} + \ldots + p_nu = 0,$$

the requirement would be the existence and continuity of the first $n-i$ derivatives of p_i.

‡ Liouville's *Journal*, vol. 3 (1838), p. 604.

§ *Trans. Amer. Math. Soc.* vol. 9 (1908), p. 373. See also for the relation to Green's functions, Bôcher, *Bull. Amer. Math. Soc.* vol. 7 (1901), p. 297 and *Annals of Math.* vol. 13 (1911), p. 81.

‖ See for instance Westfall, *Zur Theorie der Integralgleichungen* (dissertation), Göttingen, 1905, p. 19.

¶ It remains to be seen whether this case is really of sufficient importance to deserve a name.

adjoint. The general case, however, in which (2′) and (2″) are different is, apart from Birkhoff's fundamental paper, only just beginning to receive attention.

Here too the analogies for difference equations are interesting and simple. In place of the adjoint system (differential equation and boundary conditions) we now have the system of homogeneous linear algebraic equations whose matrix is the conjugate (transposed) of the original system; it is this system which the Green's function of the difference equation satisfies when regarded as a function of its second argument. The self-adjoint case now becomes the case in which the matrix of the system of linear equations is symmetric or can be made symmetric by a combination of rows and columns. Such expressions as

$$\int_a^b vP(u)\,dx$$

which occur in Green's Theorem

$$\int_a^b \Big[vP(u) - uQ(v)\Big]\,dx = \Big[T(u, v)\Big]_a^b \quad \text{.................(4)}$$

(where T is a homogeneous bilinear differential expression of order one less than P), have as their analogues, in the case of difference equations, bilinear forms. I shall not go into these analogies in detail, since they have become very familiar during the last eight years in the similar case of linear integral equations as developed by Hilbert and his pupils. I wished however to say enough to make it clear that we can get to a large extent the satisfaction and the benefit of these analogies in the case of linear differential equations, without going to the subject of integral equations, by simply regarding the differential equation (of any order) as the limit of a difference equation. This same remark applies equally well to those parts of the subject upon which I have not yet touched, and I shall not in general think it necessary to repeat it.

§ 3. *Small Variations of the Coefficients.*

All the deeper lying parts of the theory of boundary problems depend directly or indirectly on the effect produced by changes in the coefficients of the differential equation or of the boundary conditions or of both. Such changes are frequently, indeed usually as the literature of the subject now stands, produced by supposing these coefficients to depend on one or more parameters. The more general point of view, however, is to consider arbitrary variations in these coefficients; and here, before coming to the deeper lying questions, it is essential to know under what conditions *small* variations of this sort will produce a small variation in the solution of the problem. The fundamental fact here is*

I. *If the reduced system* (1′), (2′) *is incompatible, it remains incompatible after a variation of the coefficients of* (1) *and* (2) *which is uniformly sufficiently small; and*

* I have not found this fact in the literature. In the special case in which only one of the end-points appears in the boundary conditions I proved it in *Trans. Amer. Math. Soc.* vol. 3 (1902), p. 208 and *Amer. Journ. of Math.* vol. 24 (1902), p. 315. The general theorem may be deduced from this special case by following the general lines of the reasoning given by me in *Annals of Math.* vol. 13 (1911), p. 74. Indeed the case of the equation of the nth order where the boundary conditions involve k points (cf. § 2) presents no difficulty here.

such a variation produces a variation in the solution of (1), (2) *and in its first two derivatives which is uniformly small throughout* (X).

It is merely a special case of this if we assume the coefficients of (1) to be continuous functions of (x, λ) when x is in (X) and the parameter λ lies in any one or two dimensional region Λ of the complex λ-plane. The coefficients of (2) we then also assume to be continuous functions of λ in Λ. An immediate corollary of the above theorem is then:

II. *If for a certain point λ_0 of Λ the system* (1′), (2′) *is incompatible, the same will be true throughout a certain neighbourhood of λ_0, and throughout this neighbourhood the solution of* (1), (2) *and its first two derivatives are continuous functions of* (x, λ).

Something essentially new is, however, added if we demand that the coefficients be analytic functions of λ and wish to infer the analytic character of the solution. Here the facts are these:

III. *If when x lies in* (X) *and λ in a certain two-dimensional continuum Λ of the λ-plane the coefficients of* (1) *are continuous functions of* (x, λ) *and analytic functions of λ, and if the coefficients of* (2) *are analytic in λ throughout Λ, and if λ_0 is a point in Λ such that when $\lambda = \lambda_0$,* (1′) *and* (2′) *are incompatible, then the same will be true throughout a certain neighbourhood of λ_0 and the solution of* (1), (2) *throughout this neighbourhood is, together with its first two derivatives with regard to x, continuous in* (x, λ) *and analytic in λ**.

If the coefficients depend on a parameter λ, as in cases II and III, the values of λ for which (1′), (2′) are compatible are readily seen to be precisely the roots of the equation

$$\begin{vmatrix} W_1(y_1) & W_1(y_2) \\ W_2(y_1) & W_2(y_2) \end{vmatrix} = 0 \quad \text{.........................(5)},$$

where y_1 and y_2 are any pair of solutions of (1′) which do not become linearly dependent for any value of λ with which we are concerned. This equation we call the characteristic equation and its roots the *characteristic parameter values* (Eigenwerte), or characteristic numbers. In case III it is clear that (5) may be taken as analytic in λ, so that in this case, provided (5) is not identically satisfied, the characteristic numbers are all isolated though there may be an infinite number of them with cluster-points on the boundary of Λ. These characteristic numbers are the only singularities of the solution of (1), (2) regarded as a function of λ, and also of the Green's function of (1′), (2′), and it may be shown that these functions can have no other singularities there than poles. In special cases the solution of (1), (2) may have no singularity at some of these points. Those characteristic numbers for which (1′), (2′) become simply compatible we call simple characteristic numbers, those for which they become doubly compatible, double characteristic numbers, and so on in the higher cases when we are dealing with equations of higher order than the second.

In all that has been said so far no restrictions have been made concerning the

* In the special case in which (2) involves only one of the points a or b the proof of this theorem follows from the uniform convergence of the method of successive approximations. The general case may be inferred from this as indicated in a similar case in the preceding foot-note.

reality of the quantities used except that x be real. In particular the coefficients of (1) may be complex. If the system (1), (2) is real, then when (1'), (2') is incompatible, the solution of (1), (2) is real; while if (1'), (2') is compatible, it has a *real* solution not identically zero. In this case we can add various further facts to those already mentioned in this section, of which I mention the following immediate consequence of II and III.

IV. *If for a certain real range* Λ *of values of* λ *the coefficients of* (1) *are real continuous functions of* (x, λ) *and the coefficients of* (2) *real continuous functions of* λ; *if there is no characteristic value of* λ *in* Λ; *and if for no point in* Λ *the solution* u *of* (1), (2) *vanishes at* a *or* b, *or at any interior point where its derivative also vanishes; then* u *has the same number of roots in* (X) *for all values of* λ *in* Λ *and these roots are continuous functions of* λ.

If we add to our hypothesis that the coefficients of (1), (2) *be analytic in* λ, *we may add to the conclusion that the roots are analytic functions of* λ.

It must not be inferred from what I have said so far that the theory of boundary problems consists wholly, or even chiefly, in establishing existence-theorems or in proving by the exact methods of modern analysis facts which a hundred years ago would have seemed self-evident to any mathematician. Some applied mathematicians make it a reproach to pure mathematics that it has come now to a state where it is interested solely in questions of this sort. If this were so it would indeed be a cause for reproach; but it should perhaps rather be regarded as a warning of whither certain extreme tendencies in modern pure mathematics might lead us if allowed to get too much the upper hand. The good old-fashioned view that it is the main object of mathematics to discover essentially new facts is, however, hardly in danger of becoming obsolete in a generation which has just witnessed the splendid achievements of Poincaré. In the subject of boundary problems, while we need as a foundation the existence-theorems, and exact proofs of facts which in themselves are quite to be expected, these are *only* a foundation. We wish not merely to be able to say: under such and such conditions there exists a solution of the boundary problem which is continuous (or analytic) but also to be able to say what this solution is like and what can be done with it. However incomplete the theory still is, we can make important statements of this sort, as we shall now see.

§ 4. *Sturm's Fundamental Results and their Recent Extensions.*

Sturm's great memoir of 1836, which forms to a certain extent the foundation of our whole subject, produces on most superficial readers the effect of being complicated and diffuse. Nothing could be a greater mistake. The paper is very rich in content, and, while it would no doubt be possible to present the material more compactly than Sturm has done, there is by no means the repetition of which one gets the impression on a first reading owing to similarity in appearance of theorems which are really very different. It must be confessed, however, that Sturm does fail to emphasize sufficiently his really fundamental results.

Sturm, throughout whose work all quantities used are assumed real, takes the differential equation in the self-adjoint form

$$\frac{d}{dx}\left(K\frac{du}{dx}\right) - Gu = 0, \quad (K > 0) \qquad (6),$$

where K has a continuous first derivative, to which any homogeneous linear differential equation of the second order can readily be reduced. Perhaps the most fundamental result of the whole paper is the one which when stated roughly says that if the solutions of (6) oscillate in (X), they will oscillate more rapidly when G or K is decreased. The precise statement is this:

If we consider the two differential equations

$$\left.\begin{aligned}\frac{d}{dx}\left(K_1\frac{du}{dx}\right)-G_1u=0\\ \frac{d}{dx}\left(K_2\frac{du}{dx}\right)-G_2u=0\end{aligned}\right\}\quad\ldots\ldots\ldots\ldots\ldots\ldots\ldots\ldots(7),$$

where throughout (X)

$$0<K_2\leqq K_1,\qquad G_2\leqq G_1\quad\ldots\ldots\ldots\ldots\ldots\ldots\ldots\ldots(8),$$

and if a solution u_1 of the first equation has two successive roots at x_1 and x_2, then every solution u_2 of the second will vanish at least once in the interval $x_1<x<x_2$ provided both equality signs in (8) *do not hold at every point of this interval.*

If we note that this theorem tells us that if u_2 is a solution which vanishes with u_1 at x_1, then it vanishes again *before* u_2 vanishes for the first time, we see the appropriateness of the statement that the solutions of the second equation oscillate more rapidly than those of the first.

The proof of this theorem is made to depend by Sturm on the formula

$$[K_2u_1u_2'-K_1u_2u_1']_{c_1}^{c_2}+\int_{c_1}^{c_2}(G_1-G_2)\,u_1u_2dx+\int_{c_1}^{c_2}(K_1-K_2)\,u_1'u_2'dx=0\ldots(9),$$

where u_1 and u_2 are any solutions of the first and second equations (7) respectively and c_1, c_2 any points of (X). This formula, which may be regarded as merely a special application of Green's Theorem*, yields an immediate and extremely brief proof of the theorem we are considering in the special, but very important, case $K_1\equiv K_2$. In the general case the proof is by no means so easy, it being necessary then to introduce a parameter so as to pass over continuously from K_1, G_1 to K_2, G_2, and to consider carefully the effect of small changes of this parameter. Simpler methods have therefore since been devised for treating the general case, of which I will mention the extremely elegant one recently given by Picone†. This consists in using in place of (9) the formula

$$\int_{c_1}^{c_2}(K_1-K_2)\,u_1'^2dx+\int_{c_1}^{c_2}(G_1-G_2)\,u_1^2dx+\int_{c_1}^{c_2}K_2\left(u_1'-u_2'\frac{u_1}{u_2}\right)^2dx$$
$$+\left[\frac{u_1}{u_2}(K_2u_1u_2'-K_1u_2u_1')\right]_{c_1}^{c_2}=0\quad\ldots\ldots(10),$$

which may be deduced without difficulty from the differential equations. In applying this formula we must assume that u_2 does not vanish between c_1 and c_2, and vanishes at one or both of these points only if u_1 vanishes there. By means of this formula the proof of Sturm's theorem is immediate.

* Cf. Dunkel, *Bull. Amer. Math. Soc.* vol. 8 (1902), p. 288.

† *Annali della R. Scuola Normale Superiore di Pisa*, vol. 11 (1909), p. 1, where however only special cases of (10) are used. Another brief proof, based on the use of Riccati's resolvent of (6), had been previously given by me: *Trans. Amer. Math. Soc.* vol. 1 (1900), p. 414.

I have insisted somewhat at length on this one simple result of Sturm both on account of its great importance and because it represents a direction for investigation which, I believe, might well be pursued farther. The question is: What changes in K and G will cause the solutions of (6) to oscillate more rapidly? Sturm's theorem gives one answer to this question. There are, however, many other changes in K and G besides a decrease in one or both which will have this same effect. Further theorems can of course be obtained by multiplying (6) before and after the change by different constants, or by making a change of independent or of dependent variable. All these results, while they may be formally more general, may be said not to go *essentially* beyond Sturm's classical theorem. An illustration of this which will be of some importance for us is the following:

The special case of equation (6) where $G = l - \lambda g$, $K = k$, where g, l, k are continuous functions of x independent of the parameter λ,

$$\frac{d}{dx}\left(k\frac{du}{dx}\right) + (\lambda g - l)\,u = 0 \quad \text{.......................(11)},$$

has been much considered ever since Sturm's time. If $g \geqq 0$, the equality sign not holding at all points with which we are concerned, an increase of λ will produce a decrease of G and consequently it is merely a special case of Sturm's theorem in its simplest form to infer that if for one value of λ a solution oscillates, the solutions will oscillate more rapidly for a larger value of λ. Precisely the reverse is clearly true if $g \leqq 0$. During the last few years, however, another case of (11) has also been considered by several authors using various methods, namely the case $l \geqq 0$, while g changes sign. An increase in λ then causes G to decrease for some parts of (X) and to increase for others. It looks as though we had here a case going decidedly beyond that of Sturm. If, however, we divide (11) by $|\lambda|$, we get an equation in which

$$K = \frac{k}{|\lambda|}, \quad G = \frac{l}{|\lambda|} - (\operatorname{sgn} \lambda)\, g.$$

Consequently an increase in $|\lambda|$ (λ retaining one sign) produces a decrease in K while G either decreases or remains constant, and we see that we have precisely Sturm's case.

I know of no published result* which goes in this direction, and in the sense I have explained, essentially beyond Sturm's.

By the side of this theorem I will recall to you another one even simpler and better known and which Sturm proved by the same methods. It may indeed be regarded as a limiting case of the above theorem.

The roots of two linearly independent real solutions of a real homogeneous linear differential equation of the second order separate each other.

These theorems perhaps hardly come within the subject of boundary problems if we take the term in a strict sense, since no particular boundary conditions are laid

* From a verbal communication of Professor R. G. D. Richardson I understand that in a paper shortly to appear in the *Mathematische Annalen* he has made progress in this direction in the case of (11) when g changes sign and l is negative at some or all points of (X). This would appear to be a case really different from Sturm's.

down, but they are so fundamental for all work whose object is to determine the nature of the solutions of boundary problems that they could not be omitted here. Other theorems of the same sort contained in Sturm's memoir refer to the roots of u' or more generally of functions of the form $\phi_1 u - \phi_2 u'$, where ϕ_1 and ϕ_2 are given functions satisfying certain conditions*.

To the same category of theorems, preliminary, so to speak, to true boundary problems, are the various tests which have been given, some of which are contained in or follow readily from Sturm's memoir, for the equation (1) being oscillatory in (X), that is possessing solutions which vanish more than once there†.

In all of these cases we have theorems whose extension to equations of higher order is by no means easy, not merely because of essentially new difficulties which may be and doubtless are involved in the proofs, but still more because it is not easy to surmise what the character of the analogous theorems will be. The only investigation in this direction with which I am acquainted is a recent paper by Birkhoff‡ in which theorems concerning the roots of the real solutions of real homogeneous linear differential equations of the third order are obtained. The method used is one which, while familiar in other parts of the theory of linear differential equations, had never, I think, been used in treating boundary problems or questions relating to them. It consists in interpreting a fundamental system of solutions, u_1, u_2, u_3, as the homogeneous coordinates of a point in a plane. As x varies, this point traces out a curve whose shape is characteristic for the oscillatory properties of the solutions. I mention as a sample one of the simpler results obtained, from which it will be evident that we really have to deal with an extension of the results of Sturm mentioned above. Birkhoff proves that in an interval (X), where q and its derivative q' are real and continuous, the equation

$$u''' + qu' + \tfrac{1}{2}q'u = 0,$$

to which every self-adjoint equation of the third order may be reduced, always has real solutions which do not vanish, but that if two real solutions do vanish, their roots separate each other either singly or in pairs. Moreover, if q is increased, the maximum number of roots in (X) increases.

We have here a field worthy of further cultivation.

§ 5. *Boundary Problems as Treated by Sturm.*

Sturm's memoir may perhaps best be divided from a logical point of view into three parts, though this division is by no means followed out by the author in his method of exposition. We have

First those parts of the memoir which do not involve any boundary conditions. These we have already sufficiently considered.

* This part of Sturm's memoir, while extensive, is rather incomplete. Much more general results have been obtained by another method by Bôcher, *Trans. Amer. Math. Soc.* vol. 2 (1901), p. 428.

† Bôcher, *Bull. Amer. Math. Soc.* vol. 7 (1901), p. 333. Of a somewhat different character is Kneser's paper, *Math. Ann.* vol. 42 (1893), p. 409, since it deals with an infinite interval. The question there is essentially the behaviour of solutions in the neighbourhood of a singular point.

‡ *Annals of Math.* vol. 12 (1911), p. 103.

Secondly those theorems that refer to what we have called one-point boundary conditions, viz. $u(a) = \gamma_1$, $u'(a) = \gamma_2$. Since the existence-theorem here was well known, being merely the fundamental existence-theorem for differential equations, the theorems concern (a) the *character* of the solution of the boundary problem and (b) the *changes* produced in it by changes in the differential equation or in the boundary conditions. What is most essential here is contained in what I have called Sturm's two Theorems of Comparison.

Thirdly there comes a special kind of two-point boundary problem, the boundary conditions being the so-called Sturmian Conditions:

$$\left.\begin{array}{ll} \alpha u(a) + \alpha' u'(a) = 0, & |\alpha| + |\alpha'| \neq 0 \\ \beta u(b) + \beta' u'(b) = 0, & |\beta| + |\beta'| \neq 0 \end{array}\right\} \quad \ldots\ldots\ldots\ldots\ldots(12),$$

characterised by the fact that each involves only one end-point of the interval. Here all three aspects of boundary problems are considered: (a) the existence of characteristic numbers; (b) the nature of the characteristic functions; (c) the changes produced in the characteristic numbers and functions by changes in the differential equation or the boundary conditions. The main result here is the Theorem of Oscillation, or perhaps it would be more correct to say the Theorems of Oscillation, since a variety of these may be formulated.

The first theorem of comparison may be roughly but sufficiently characterised by saying that it tells us that a decrease of G, or K, or $K(a)\,u'(a)/u(a)$ causes all the roots of u in (X) to decrease; while the second theorem of comparison tells us that under the same conditions the value of $K(b)\,u'(b)/u(b)$ will decrease provided the number of roots of u has not been changed. Both of these theorems are proved by Sturm by means of formula (9), which may, when K_1 and K_2 are not identically equal, be advantageously replaced by (10).

I shall not stop to enunciate Sturm's theorem of oscillation in any very general form. The general case would be that in which K or G or both are functions of (x, λ) which decrease as λ increases, while the ratio $K(a)\,u'(a)/u(a)$ may also decrease with λ. I enunciate, however, merely two special cases in which λ does not enter the boundary conditions and where the differential equation has the form (11).

I. *If* $g \geqq 0$, *the equality sign not holding throughout* (X), *and if* $\alpha, \alpha', \beta, \beta'$ *are constants, there exist an infinite number of real characteristic numbers for the system* (11), (12). *These are all simple and have no cluster-point except* $+\infty$. *If, when arranged in order of increasing magnitude, they are denoted by* $\lambda_0, \lambda_1, \lambda_2, \ldots$ *and the corresponding characteristic functions by* $u_0, u_1, u_2, \ldots$, *then* u_n *has exactly* n *roots in the interval* $a < x < b$.

This is the best known special case of the theorem of oscillation. Another special case which, after division by $|\lambda|$, follows with exactly the same ease is this:

II. *If* g *changes sign in* (X) *and*

$$l \geqq 0, \qquad \alpha\alpha' \leqq 0, \qquad \beta\beta' \geqq 0,$$

there exist an infinite number of real characteristic numbers for the system (11), (12). *These are all simple and have* $+\infty$ *and* $-\infty$ *as cluster-points. If the positive and negative characteristic numbers arranged each in order of increasing numerical value are denoted by*

$$\lambda_0^+, \quad \lambda_1^+, \quad \lambda_2^+, \ldots$$

and

$$\lambda_0^-, \quad \lambda_1^-, \quad \lambda_2^-, \ldots$$

and the corresponding characteristic functions by

$$u_0^+, \quad u_1^+, \quad u_2^+, \ldots$$

and

$$u_0^-, \quad u_1^-, \quad u_2^-, \ldots,$$

then u_n^+ *and* u_n^- *have exactly* n *roots in the interval* $a < x < b$.

I doubt if it has been noticed before that this theorem is substantially contained in Sturm's results. It has been re-discovered three times during the last few years*.

That in the first of these cases there can be no imaginary characteristic numbers had been shown by Poisson by means of a special case of (9). A slight modification of this reasoning establishes this same fact for the second case†.

Sturm thus had both existence-theorems for the characteristic numbers and, in the theorems of oscillation, some rather specific information as to the nature of the characteristic functions. The next thing was to consider the changes produced in the characteristic numbers and functions by changes in the coefficients of the equation or of the boundary conditions. Such questions are also touched upon by Sturm, but we will not enter upon their consideration here.

As has already been said, all of these results including the theorems of oscillation, have their counterparts in the theory of linear difference equations, and it was from this side that the subject was first approached by Sturm. However, these oscillation properties will not hold for all equations of the form

$$L_i u_{i+1} + M_i u_i + N_i u_{i-1} = 0 \quad \ldots\ldots\ldots\ldots\ldots\ldots\ldots\ldots(13),$$

but only for those for which $L_i N_i > 0$ for all values of i with which we are concerned. As an illustration let us take the theorem that the roots of two linearly independent solutions of (1′) separate each other. In order to get the analogous theorem for (13) we must introduce the conception of *nodes* as follows: Corresponding to the values $i = 1, 2, \ldots n$ let us mark points $x_1, x_2, \ldots x_n$ on the axis of x, whether equally spaced or not is for our present purpose of no consequence. At the point x_i we erect an ordinate equal to u_i and we join the successive points thus obtained by straight lines. We regard the broken line thus formed as representing the solution u_i of (13), and the points where this line meets the axis of x we call the nodes of u_i. If the condition $L_i N_i > 0$ is fulfilled, it is readily seen that a solution of (13) not identically zero corresponds to a broken line which crosses the axis of x at each of its nodes, and here the theorem holds that the nodes of any two linearly independent

* Sanielevici, *Ann. de l'École Normale Supérieure*, 3rd ser. vol. 26 (1909), p. 19; Picone, *loc. cit.* (1909), and Richardson, *Math. Ann.* vol. 68 (1910), p. 279. The mere fact of the existence of an infinite number of positive and also of negative characteristic numbers (proved for instance under certain restrictions in Hilbert's 5th *Mitteilung*) is an even more obvious corollary of Sturm's work, even if no restriction is placed on the sign of l.

† Picone, *loc. cit.* p. 16.

solutions of (13) separate each other*. Without the restriction in question the theorem is false as the example†

$$u_{i+1} - u_i - u_{i-1} = 0$$

shows. Here the solution determined by the initial conditions $u_0 = 0$, $u_1 = 1$ gives for positive values of i Fibonacci's numbers 0, 1, 1, 2, 3, 5, 8, 13, ... with no node; while the solutions determined by $u_0 = -10$, $u_1 = 6$ and by $u_0 = -10$, $u_1 = 7$ both have several positive nodes, but these nodes do not separate each other. In the same way the other more complicated theorems of Sturm are, for the case of difference equations, essentially bound to the inequality in question.

This apparent failure of the analogy is less surprising when we notice that every linear differential equation of the second order may be obtained as the limit of an equation of the form ($\bar{1}$) in which *after a certain point in the limiting process* the inequality in question holds. It is therefore only those difference equations that come nearest to the differential equations, so to speak, which share with them the simple oscillation properties. Difference equations of the form (13) in general will have oscillation properties of a very different character concerning which, so far as I know, nothing has been published, though from Sturm's brief remarks it seems possible that he had developed this theory also.

The results of Sturm concerning the oscillatory properties of the solutions of differential equations and the existence of characteristic values have been carried forward in various directions since his time, partly by methods more or less closely related to his own and partly by a number of essentially different methods. Of these there are four which we may describe briefly as

(1) Liouville's method of asymptotic expressions.

(2) The method of successive approximations.

(3) The minimum principle.

(4) Integral equations.

It will be well for us to glance briefly at these methods in succession before proceeding to consider the present state of knowledge of the theory of one-dimensional boundary problems.

§ 6. *Asymptotic Expressions.*

Liouville's greatest contribution to the theory of boundary problems, which had been so brilliantly inaugurated by his friend Sturm a few years before, was first the discovery of asymptotic expressions for the large characteristic values and the corresponding characteristic functions, and secondly the application of these expressions in the theory of the development of arbitrary functions‡. It is the first of these questions which we must now consider.

* E. J. Moulton, *Annals of Math.* vol. 13 (1912), p. 137.

† Or, more generally, the difference equation satisfied by Gauss's symbols $[a_1, a_2, \ldots a_n]$.

‡ Liouville's *Journal*, vol. 2 (1837), p. 16 and p. 418.

Liouville begins by reducing equation (11), in which he assumes $k>0$, $g>0$, by a change of both independent and dependent variable to the normal form*

$$\frac{d^2u}{dx^2}+(\mu^2-\bar{l})\,u=0, \qquad (\mu^2=c^2\lambda) \quad \dots\dots\dots\dots\dots\dots(14),$$

where for the sake of simplicity we may suppose that the transformation has been so made that the interval (X) goes over into the interval $(0, \pi)$. It is then sufficient to consider this simpler equation. The boundary conditions (12) may be written

$$\left.\begin{aligned} u'(0)-hu(0)&=0\\ u'(\pi)+Hu(\pi)&=0\end{aligned}\right\} \quad \dots\dots\dots\dots\dots\dots(12'),$$

provided we assume $\alpha'\neq 0$, $\beta'\neq 0$. If we suppose u multiplied by a suitable constant, the first equation (12′) may be replaced by the two non-homogeneous conditions

$$u(0)=1, \qquad u'(0)=h \quad \dots\dots\dots\dots\dots\dots(15),$$

and it is the non-homogeneous boundary problem (14), (15) which Liouville first considers. He shows that its solution satisfies the relation

$$u=\cos\mu x+\frac{h}{\mu}\sin\mu x+\frac{1}{\mu}\int_0^x \bar{l}(\xi)\,u(\xi)\sin\mu(x-\xi)\,d\xi \quad \dots\dots\dots(16).$$

This is of interest as being the first occurrence, so far as is known, of an integral equation of the second kind, and also because it is the first appearance of an integral equation as the equivalent of the system consisting of a differential equation and boundary conditions†.

By means of (16) Liouville readily infers that u and u' may be written

$$\left.\begin{aligned} u&=\cos\mu x+\frac{\psi_1(x,\mu)}{\mu}\\ u'&=-\mu\sin\mu x+\psi_2(x,\mu)\end{aligned}\right\} \quad \dots\dots\dots\dots\dots\dots(17),$$

where ψ_1 and ψ_2 (and all functions which in this section are denoted by ψ) are continuous functions of (x, μ) which for all real values of μ and all values of x in $(0, \pi)$ remain in absolute value less than a certain constant. From (17) we see that u differs when μ is large only in unessential ways from $\cos\mu x$, so that the large characteristic values of μ may be approximately obtained, as is readily shown with entire rigour, by substituting $\cos\mu x$ in the second condition (12′) in place of u. If then we denote the squares of the characteristic numbers arranged in order of increasing magnitude by $\mu_0^2, \mu_1^2, \mu_2^2, \dots$, we may write for the positive values μ_i the expression $n+i+\gamma_i$, where γ_i approaches zero as i becomes infinite, and n denotes an integer independent of i whose value is as yet unknown. It is worth while to notice that we thus get a new proof, quite independent of Sturm's, of the existence of an infinite number of positive characteristic values, and of a part of the theorem of oscillation, namely that *at least after a certain point* each characteristic function

* Here and in what follows certain conditions of differentiability etc. must be satisfied by the coefficients of (11). Concerning the possibility of removing these restrictions cf. A. C. Dixon, *Phil. Trans.* vol. 211 (1911), p. 411.

† I.e. not only is (16) a consequence of (14), (15), but conversely (14), (15) is a consequence of (16). This last fact, it is true, is not brought out by Liouville.

has just one more root in $(0, \pi)$ than the preceding one. We see also that there are at most a finite number of negative or imaginary values for μ^2.

If, however, we are willing to make use of Sturm's theorem of oscillation, we may readily infer that $n=0$ and thus get the more specific asymptotic formula

$$\mu_i = i + \gamma_i, \qquad \left(\lim_{i=\infty} \gamma_i = 0\right) \quad \text{.......................(18)}.$$

The formulae (17), (18) are merely the roughest kind of asymptotic formulae, and Liouville proceeded to sharpen them by a further application of the integral equation (16). This process was carried a little farther by the same method by Hobson* whose results we record

$$u = \cos \mu x \left[1 + \frac{\psi_3(x, \mu)}{\mu^2}\right] + \sin \mu x \left[\frac{\phi(x)}{\mu} + \frac{\psi_4(x, \mu)}{\mu^2}\right] \quad \text{......(17')},$$

where ϕ is continuous in $(0, \pi)$;

$$\mu_i = i + \frac{c}{i} + \frac{\delta_i}{i^2}, \qquad \pi c = h + H + \frac{1}{2}\int_0^\pi \bar{l}(x)\, dx \quad \text{...........(18')},$$

and the constants δ_i are all in absolute value less than a certain constant.

By substituting these asymptotic expressions for μ_i in the asymptotic expressions for u, Liouville and Hobson readily obtain, after certain reductions, asymptotic expressions for the characteristic functions, which need not be here recorded.

All of these formulae, even the simple one (18), require certain modifications† when in the boundary conditions (12) $\alpha' = 0$ or $\beta' = 0$; the method to be used, however, remains the same.

These asymptotic expressions can be indefinitely sharpened. Thus Horn‡, using another method, obtains expressions of the form (17'), (18') except that instead of containing only the first and second powers of $1/\mu$ and $1/i$, the powers from 1 to n enter, where n is an arbitrary positive integer.

This paper of Horn was the starting point for the modern developments of this subject of asymptotic expressions. In a second paper by Horn§ and in papers by Schlesinger|| and Birkhoff¶ similar asymptotic expressions are obtained not only for equations of the second order in which the parameter enters in a more general way and in which the coefficients of the equation are not all assumed real, but also for equations of higher order in similarly general cases. These investigations refer, however, merely to asymptotic expressions of solutions of a differential equation either without special reference to the boundary conditions or else in the case where

* *Proc. London Math. Soc.* 2nd ser. vol. 6 (1908), p. 349.

† This was mentioned on p. 445 of my article II A 7*a* in the *Encyclopädie* and the formulae corresponding to (18) when one but not both of the quantities α', β' are zero were given. By an oversight the case $\alpha' = \beta' = 0$, where (18) must be replaced by

$$\mu_i = 1 + i + \gamma_i,$$

was not mentioned. These cases are considered by Kneser, *Math. Ann.* vol. 58 (1903), p. 136.

‡ *Math. Ann.* vol. 52 (1899), p. 271.

§ *Math. Ann.* vol. 52 (1899), p. 340.

|| *Math. Ann.* vol. 63 (1907), p. 277.

¶ *Trans. Amer. Math. Soc.* vol. 9 (1908), p. 219. The method used in this paper was rediscovered by Blumenthal, *Archiv d. Math. u. Phys.* 3rd ser. vol. 19 (1912), p. 136.

the boundary conditions refer to a single point. The question of characteristic values does not present itself. This question, however, was taken up by Birkhoff in a second paper* in the case of the equation

$$\frac{d^n u}{dx^n} + p_1 \frac{d^{n-1}u}{dx^{n-1}} + \ldots + p_{n-1}\frac{du}{dx} + (p_n + \lambda g)\, u = 0 \quad \text{..........(19)},$$

where g and the p_i's are continuous functions of x of which all except g may be complex while g is assumed to be real and not to vanish in (X). General linear homogeneous boundary conditions are considered, certain special cases merely (so-called *irregular cases*) being excluded. Under these very general conditions Birkhoff establishes the existence of an infinite number of characteristic values, of which when n is odd all but a finite number are simple, while when n is even an infinite number of multiple characteristic values can occur only in very special cases; and at the same time he obtains an asymptotic expression for them. By means of this result an asymptotic expression for the characteristic functions is obtained.

The question of the reality of the characteristic numbers, even when the coefficients of (19) are real, is not touched upon. Professor Birkhoff, however, calls my attention to the fact that it is possible to treat questions of this sort by the methods there given. For instance, to mention only an obvious case, one sees that if n is odd there can, apart from the irregular cases†, be at most a finite number of real characteristic values.

§ 7. *The Method of Successive Approximations.*

Still another method which goes back to Liouville is the method of successive approximations. Although in his published papers he used this method only in very special cases, it is certain that he was familiar with it in more general forms, though it is impossible now to say to what extent of generality he had carried it. The method may be formulated as follows in order to include the special cases to be found in the literature and many others:

Let us write the homogeneous linear differential expression

$$P(u) = \frac{d^n u}{dx^n} + p_1 \frac{d^{n-1}u}{dx^{n-1}} + \ldots + p_n u$$

in the form

$$P(u) \equiv L(u) - M(u),$$

where L, M are homogeneous linear differential expressions, whose coefficients we assume to be continuous, of orders n and $m < n$. The differential equation (1) may then be written

$$L(u) = M(u) + r \quad \text{.........................(20)}.$$

We wish to solve this equation subject to a system of linear boundary conditions which we will write in the form

$$U_i(u) = V_i(u) + \gamma_i, \qquad (i = 1, 2, \ldots n) \quad \text{..............(21)},$$

where U_i and V_i are homogeneous linear expressions in $u(a)$, $u'(a)$, ... $u^{[n-1]}(a)$,

* *Trans. Amer. Math. Soc.* vol. 9 (1908), p. 373.

† To these irregular cases belongs the one treated by Liouville in Liouville's *Journal*, vol. 3 (1838), p. 561.

$u(b), \dots u^{[n-1]}(b)$. We have thus transposed part of equation (20) and conditions (21) to the second member, and we will suppose this so done that the auxiliary homogeneous system

$$\left.\begin{aligned} &L(u)=0, \\ &U_i(u)=0, \qquad (i=1,2,\dots n) \end{aligned}\right\} \quad \dots\dots\dots\dots\dots\dots\dots(22)$$

is incompatible. We may then start from any function u_0 for which $M(u_0)$ is continuous in (X) and $V_i(u_0)$ $(i=1,\dots n)$ are defined, and determine a succession of functions $u_1, u_2, \dots$ by means of the equations

$$\begin{aligned} &L(u_{j+1})=M(u_j)+r, \\ &U_i(u_{j+1})=V_i(u_j)+\gamma_i, \qquad (i=1,2,\dots n). \end{aligned}$$

If u_j and its first $n-1$ derivatives converge uniformly throughout (X), and this is what we shall mean when we say the process converges, the limit of u_j is precisely the solution of the problem (20), (21). The question whether this process converges or not depends, as was noticed by Liouville in some special cases*, on the characteristic values for the problem

$$\left.\begin{aligned} &L(u)=\lambda M(u), \\ &U_i(u)=\lambda V_i(u), \qquad (i=1,2,\dots n) \end{aligned}\right\} \quad \dots\dots\dots\dots\dots\dots(23).$$

This connection can best be stated by considering the system

$$\left.\begin{aligned} &L(u)=\lambda[M(u)+r_1]+r_2, \\ &U_i(u)=\lambda[V_i(u)+\gamma_i']+\gamma_i'', \qquad (i=1,2,\dots n) \end{aligned}\right\} \quad \dots\dots\dots(24),$$

where $r_1+r_2\equiv r$, $\gamma_i'+\gamma_i''=\gamma_i$, $M(u_0)+r_1=0$, and $V_i(u_0)+\gamma_i'=0$, so that when $\lambda=1$ (24) reduces to (20), (21). The fact then is this:

The method of successive approximations applied to (24) (*in the same way in which it was applied above to* (20), (21)) *converges for values of* λ *which lie in a certain circle (finite or infinite) described about* $\lambda=0$ *as centre and diverges outside. If this circle is not infinite, its radius is precisely the absolute value of one of the characteristic values of* (23). *All the characteristic values of* (23) *which lie within this circle are such that for them the system* (24) *has solutions (necessarily in infinite number), while if all the characteristic values on its circumference are simple roots of the equation* (5), *there is at least one of them for which the system* (24) *has no solution*†.

If there exist no characteristic values, it follows that the method of successive approximations will always converge; and this will, in particular, be the case if only one of the end-points a or b enter in the boundary conditions. The well-known fact that in this case the method of successive approximations surely converges‡ appears thus as a special case of the above general theorem.

In other cases, in which characteristic values do exist, it will be important in applying the theorem to know whether for a given characteristic value of λ the system (24) has solutions or not. Necessary and sufficient conditions of this sort

* Liouville's *Journal*, vol. 5 (1840), p. 356.

† The statement here made goes far beyond anything I have found in the literature, and is sufficient for our purposes, although a considerable generalization is possible. I expect to take up this matter in detail on another occasion.

‡ Cf. Fuchs, *Annali di Matematica*, ser. 2, vol. 4 (1870), p. 36.

have been given by Mason* in fairly general cases for differential equations of the second order. By means of such conditions the above theorem can of course be thrown into other, but equivalent, forms. In all investigations with which I am acquainted where the method of successive approximations is used a special case of this theorem in some of its forms plays a central part. Thus in Picard's well-known application of the method† to the semi-homogeneous problem

$$\frac{d^2u}{dx^2} + \lambda A(x)\, u = 0, \qquad A > 0$$

$$u(a) = u(b) = 1,$$

the fact upon which the possibility of applying the method depends is that the successive approximations converge or diverge according as $|\lambda|$ is less or greater than the smallest characteristic value; and this is readily seen to be substantially a special case of the above theorem. Again, although the term successive approximation is not used, §§ 9, 10 of Kneser's paper of 1903‡ are in substance an application of this method to the semi-homogeneous problem

$$\frac{d}{dx}\left(k\frac{du}{dx}\right) + (\lambda g - l)\, u + f = 0,$$

$$\alpha u(a) + \alpha' u'(a) = 0, \qquad |\alpha| + |\alpha'| \neq 0,$$

$$\beta u(b) + \beta' u'(b) = 0, \qquad |\beta| + |\beta'| \neq 0,$$

and the first of these sections may be regarded as a proof of the above theorem so far as it refers to this special case.

A second essential element in almost all applications of the method of successive approximations is constituted by Schwarz's constants which serve the purpose of giving a second test of the range of convergence of the process. It is by a comparison of the inferences drawn from these two methods that the final result is deduced. For details we refer here to the work of Picard and Kneser already cited.

§ 8. *The Minimum Principle.*

That linear boundary problems can be brought into intimate connection with the calculus of variations was first noticed and is still best known in connection with Laplace's equation, where the method involved has received the now almost universally accepted misnomer Dirichlet's Principle. It was pointed out by Weierstrass some fifty years ago that the existence of a minimum is here by no means obvious and that Dirichlet's Principle does not establish rigorously the existence of a solution of the boundary problem in question. This criticism was, however, not generally known in 1868 when H. Weber§ applied a similar method to establish the existence of characteristic numbers for the partial differential equation

$$\frac{\partial^2 u}{\partial x^2} + \frac{\partial^2 u}{\partial y^2} + k^2 u = 0$$

* *Trans. Amer. Math. Soc.* vol. 7 (1906), p. 337. See also *C. R.* vol. 140 (1905), p. 1086. Cf. also for equations of higher order, Dini, *Annali di Mat.* vol. 12 (1906), pp. 240 ff.

† *Traité d'Analyse*, vol. 3, 2nd edition, p. 100.

‡ *Math. Ann.* vol. 58, p. 81. See also for more general cases the paper of Dini just cited.

§ *Math. Ann.* vol. 1, p. 1.

subject to the boundary condition $u = 0$. While this work of Weber thus remained inconclusive, it at least made it clear that *granting the existence of a minimum* a precisely similar method could be carried through in the similar one-dimensional case*. The facts here are these:

Consider the problem of determining the function $u(x)$ with continuous first and second derivatives in (X) which satisfies the conditions

$$u(a) = u(b) = 0,$$

$$\int_a^b Au^2 dx = 1$$

(A being a given function which is everywhere positive), and which minimizes the integral

$$J = \int_a^b u'^2 dx.$$

If we admit that such a function, u_0, exists and call the corresponding (minimum) value of J, λ_0, it is readily proved that λ_0 is a characteristic number for the differential equation

$$\frac{d^2u}{dx^2} + \lambda A u = 0$$

with the boundary conditions $\quad u(a) = u(b) = 0,$

and that u_0 is the corresponding characteristic function. Moreover it is shown that λ_0 is the smallest characteristic number.

To get the next characteristic number we add to the conditions imposed above on u the further one

$$\int_a^b Au_0 u\, dx = 0.$$

The function u_1 satisfying this condition as well as those stated above and minimizing J is the second characteristic function and this minimum value of J is the second smallest characteristic number, λ_1.

By adding to the conditions already imposed the further one

$$\int_a^b Au_1 u\, dx = 0,$$

we get the third characteristic value and function, etc.

After Hilbert's brilliant achievement in 1899 of inventing a method by which in many cases the existence of a minimizing function in problems of the calculus of variations may be established, it was natural to hope that this method might be applied successfully to this problem also. This was in fact done by Holmgren†, but a far simpler method of accomplishing the same result for this special problem, as well as for certain other boundary conditions, had been invented a little earlier by Mason‡ to whom the problem had been proposed by Hilbert. As first given, this

* Cf. Picard, *Traité d'Analyse*, 1st edition, vol. 3 (1896), p. 117, where only a partial account of the matter is given.

† *Arkiv för Mat., Astr. och Fysik*, vol. 1 (1904), p. 401.

‡ Dissertation, Göttingen, 1903. Some serious mistakes contained here were corrected in the abridged version, *Math. Ann.* vol. 58 (1904), p. 528.

method involved the use of some of Fredholm's results in the newly developed theory of integral equations, but it has been subsequently modified by Mason so as to be entirely independent of this theory and at the same time extended to much more general cases*. Still another method of establishing the existence of a minimum was given by Richardson†. This method depends essentially on development theorems in the theory of integral equations.

§ 9. *The Method of Integral Equations.*

We come finally to the method of integral equations which has held such a prominent place in the mathematical literature of the last few years. The central fact here is that a linear differential equation, whether ordinary or partial, together with a system of linear boundary conditions can be replaced by a single integral equation of the second kind. We have already seen how this fact presented itself in a very special case in the early work of Liouville. In the case of the fundamental boundary problem for Laplace's equation it formed the starting point for Fredholm's epoch-making investigations. It was however reserved for Hilbert‡ to bring out this relation clearly in more general cases, and to make use of it in the theory of characteristic numbers of differential equations and of the developments according to their characteristic functions.

The relation of the linear boundary problem for ordinary differential equations to the subject of integral equations is actually established by formula (3) above which may be regarded as an integral equation of the first kind for the function $r(x)$. The integral equation of the second kind originally used by Hilbert in the case of certain self-adjoint systems was

$$f(x) = u(x) + \lambda \int_a^b G(x, \xi)\, u(\xi)\, d\xi \quad \text{.....................(25)},$$

where G is the Green's function of a certain homogeneous system, the differential equation of which we will denote by $L(u) = 0$. Hilbert shows that the reciprocal, $\bar{G}(x, \xi, \lambda)$, of the kernel $\lambda G(x, \xi)$ of this equation (the "solving function") is precisely the Green's function of the equation

$$L(u) + \lambda u = 0 \quad \text{..............................(26)},$$

with the same boundary conditions as before. Since the characteristic numbers for this last system are the poles of its Green's function, we see from one of the most fundamental of Fredholm's results that these characteristic numbers are the values of λ for which the determinant of equation (25) vanishes; that is they are, according to Hilbert's terminology, the characteristic numbers of the homogeneous integral equation

$$u(x) + \lambda \int_a^b G(x, \xi)\, u(\xi)\, d\xi = 0 \quad \text{.....................(27)}.$$

A comparison of this equation with the equation of the first kind (3) shows that

* *Trans. Amer. Math. Soc.* vol. 7 (1906), p. 337, also vol. 13, p. 516. For the treatment of the special case here mentioned see *The New Haven Colloquium* of the American Mathematical Society, 1910, p. 210.

† *Math. Ann.* vol. 68 (1910), p. 279.

‡ *Göttinger Nachrichten*, 1904, Zweite Mitteilung, p. 213.

if λ is a characteristic value, every solution of (27) is a solution of (26) which satisfies the given boundary conditions, and *vice versa*. Consequently we have in (27) a homogeneous integral equation of the second kind equivalent to the homogeneous system consisting of (26) and a set of homogeneous boundary conditions independent of λ.

In the cases considered by Hilbert, G is a real symmetric function of (x, ξ), that is we have to deal here with what we have called a real self-adjoint system. Here Hilbert's beautiful theory of integral equations with real symmetric kernels comes into play*, the fundamental theorem in which is that such a kernel always has at least one characteristic number† and can have no imaginary characteristic numbers. It was possible for Hilbert to go at once farther since the kernel G was readily shown to be *closed*, that is to be such that the equation

$$\int_a^b G(x, \xi)\, u(\xi)\, d\xi = 0$$

is satisfied by no continuous function u except zero. For such kernels he had established the existence of an infinite number of characteristic numbers. He thus obtained at one stroke the theorem: Every real self-adjoint system in which the parameter λ does not enter the boundary conditions, and enters the differential equation only in the form (26), has an infinite number of real and no imaginary characteristic numbers‡.

Other applications made by Hilbert, including the theorems concerning the developments according to characteristic functions, will be mentioned later.

Hilbert has sketched at the close of his fifth and in his sixth *Mitteilung*§ still another method for reducing a boundary problem to an integral equation of the second kind. This method, which he carried through in detail only in the case of a special partial differential equation, leads us to a kernel which is not a Green's function, but is formed by means of a *parametrix*, that is a function of (x, ξ) which satisfies the same boundary conditions as the Green's function, and whose $(n-1)$th derivative has the same discontinuity, but which does not satisfy the differential equation.

A linear integral equation may be regarded as the limiting form of a system of linear algebraic equations. This fact, which had been noticed by Volterra and put to essential use by Fredholm, as the very names determinant and minor sufficiently indicate, was made by Hilbert in his first paper the foundation, not merely

* *Göttinger Nachrichten*, 1904, Erste Mitteilung, p. 49. This theory was subsequently put into still more elegant and complete form by E. Schmidt, Göttingen dissertation, 1905, *Math. Ann.* vol. 63 (1907), p. 433.

† In my Tract: *Introduction to the Study of Integral Equations*, Cambridge, England, 1909, p. 47, I erroneously attributed this theorem to Schmidt. This mistake will shortly be corrected in a second edition.

‡ It is no essential generalization, as Hilbert himself points out, to consider the differential equation

$$L(u) + \lambda g u = 0,$$

where g is continuous and does not vanish. The general conception of a self-adjoint system is not formulated by Hilbert, but his work evidently applies to this case.

§ *Göttinger Nachrichten*, 1906, p. 480; 1910, p. 8.

heuristically but also in the way of rigorous deduction, of the theory of integral equations of the second kind. We thus have two methods of treating a boundary problem in one dimension as the limit of an algebraic problem concerning linear equations; first the direct method of difference equations described near the beginning of this lecture, and secondly the indirect method of replacing the boundary problem by an integral equation and regarding this as the limit of a linear algebraic system. Not only do these two methods look very unlike when superficially considered, but they present also a deeper lying difference: the determinant and its minors of the linear algebraic system whose limit is the integral equation approach definite limits, namely the Fredholm determinant and the Fredholm minors of the integral equation; whereas the determinant and its minors of the system of difference equations do not approach any limits as we pass over to the transcendental case. In spite of this apparently essential difference, there is the very closest relation between these two methods of obtaining the transcendental problem as the limit of an algebraic one. This relation was pointed out to me a few days ago in conversation by Dr Toeplitz of Göttingen, and may be briefly stated as follows:

If we use Hilbert's original method of passing from the differential to the integral equation by means of the Green's function $G(x, \xi)$, as explained above, the connection with the system of difference equations may be established by considering the homogeneous linear algebraic system *reciprocal* to the system of difference equations of which $L(u) = 0$ is the limit. This system has as its matrix, as is readily seen, precisely the Green's function of the difference equation, and if we add to the terms in the principal diagonal the quantities

$$\frac{u_0}{\lambda}, \frac{u_1}{\lambda}, \dots \frac{u_n}{\lambda},$$

we get the linear algebraic system of which the integral equation (27) is the limit.

On the other hand, if we use the parametrix to pass from the differential to the integral equation, the connection with the difference equation is even more direct. In order to make the determinant and its minors of the difference equation converge when we pass to the limit, it is sufficient to combine the linear algebraic equations into an equivalent system by taking suitable linear combinations with constant coefficients of the equations, and this can be done in an infinite number of ways. The limit of the algebraic system as thus modified is precisely the integral equation of the second kind yielded by the use of the parametrix.

These relations will be explained in detail in Dr Toeplitz's forthcoming book on integral equations.

All the methods which have been devised to treat linear integral equations, for instance Hilbert's method of infinitely many variables, may be regarded as being indirect methods for the treatment of linear boundary problems; but any discussion of such questions would obviously be beyond the scope of this lecture.

§ 10. *The Present State of the Problem.*

The methods discussed in the last three sections have in common the very important advantage that they are capable of generalization without serious difficulty to the case of partial differential equations. It was therefore well worth while for their inventors and others to apply them somewhat systematically to the proofs of theorems in the case of the one-dimensional problem which had been already proved by other methods. The fact that the proofs by the newer methods were almost invariably both less direct and less simple than the earlier proofs leaves these applications of the newer methods still of decided interest, since they pointed the way to be followed in deducing really new results for partial differential equations. As an example in point I mention Richardson's use of the calculus of variations in proving Sturm's theorem of oscillation*.

I wish now, however, to indicate the stage which has been reached in *results* rather than in methods, and in doing this we begin with the case of the differential equation of the second order.

Twelve years ago in writing the article on boundary problems in one dimension for the mathematical Encyclopaedia I was obliged to present as an unsolved, and indeed until then almost unformulated, problem the question of solving the real homogeneous equation of the second order

$$\frac{d^2u}{dx^2} + p(x)\frac{du}{dx} + q(x, \lambda) u = 0 \quad \text{.......................(28)},$$

subject to the "periodic" boundary conditions

$$u(a) = u(b), \quad u'(a) = u'(b) \quad \text{.......................(29)}.$$

If we assume that as λ increases through the interval

$$l < \lambda < L \quad \text{....................................}(\Lambda),$$

q constantly increases from negative or zero values to values which at least for some part of (X) become positively infinite, and that

$$\int_a^b p\, dx = 0,$$

this problem has since been answered by the following theorem of oscillation which I quote in detail because it really goes beyond Sturm's results and is at the same time simple†:

The problem (28), (29) *has an infinite number of characteristic numbers in the interval* (Λ), *and these have L as their only cluster point. If we indicate these characteristic numbers in order of increasing magnitude by* $\lambda_0, \lambda_1, \lambda_2, \ldots$, *each double characteristic number being repeated, and the corresponding characteristic functions by* $u_0, u_1, u_2, \ldots$, *then* u_n *vanishes an even number of times, namely n or n + 1 times.*

This theorem was not completely proved until Birkhoff‡ in 1909 established it as a special case of a much more general theorem of oscillation referring to the

* *Math. Ann.* vol. 68 (1910), p. 279.

† Given in my Encyclopaedia article for the special case in which q has the same value at $a+\xi$ as at $b-\xi$ while the values of p at these two points are the negatives of each other (the statement as to p is there incorrectly given); for the case where $q=\lambda g - l\ (g>0)$, by Mason, *C. R.* vol. 140 (1905), p. 1086 (see also Tzitzéica, *ibid.* p. 492); as here stated, by Bôcher, *C. R.* vol. 140 (1905), p. 928, except that it was not there proved that *only two* u's have the same number of roots.

‡ *Trans. Amer. Math. Soc.* vol. 10, p. 259. Special cases of these results were subsequently deduced by another method by Haupt, Dissertation, Würzburg, 1911.

general real self-adjoint homogeneous problem for the differential equation of the second order*. These results of Birkhoff, which he obtains by a natural extension of Sturm's methods, may be regarded as on the whole the high-water-mark of our subject so far as theorems of oscillation are concerned. They do not, however, at present include Sturm's theorems of oscillation when K depends on λ or even the special case mentioned above where q is replaced by $\lambda g - l$, g changing sign, but $l \geqq 0$.

It is perhaps of interest to return for a moment to the periodic case in order to remark that if we seek there not the truly periodic solutions (on the supposition that p and q have the period $b-a$) but periodic solutions of the second kind, i.e. if we seek to make $u(b)$ and $u'(b)$ merely proportional to $u(a)$ and $u'(a)$, we are imposing not a linear but a quadratic homogeneous boundary condition, viz.

$$u(a)\,u'(b) - u'(a)\,u(b) = 0.$$

This example of a quadratic boundary problem is interesting because of its relative simplicity—the problem always has one, and in general two, linearly independent solutions. It was considered explicitly by Floquet† in 1883, but is essentially the problem of Riemann and Fuchs concerning the existence of solutions of an analytic linear differential equation which behave multiplicatively when we go around a singular point. Concerning this quadratic problem and its relations to the linear problem (28), (29) reference should also be made to the work of Liapounoff‡.

We come next to a series of interesting but rather special investigations concerning the equation of the fourth order. The equations here considered are of the self-adjoint form

$$\frac{d^2}{dx^2}\left(k\frac{d^2u}{dx^2}\right) + \lambda g u = 0, \qquad k > 0, \quad g > 0.$$

In 1900, and more generally in 1905, Davidoglou§ treated this equation by the method of successive approximations, the boundary conditions being the very special ones which present themselves in the theory of the vibrating rod. By using Picard's methods it was shown that Sturm's theorem of oscillation may be transferred without change to this case, multiple roots for the characteristic functions never occurring *between* the points a and b. This same differential equation has since been treated by Haupt (*loc. cit.*) subject to more general, but still very special, real homogeneous self-adjoint boundary conditions; the method used being to consider the effect on the characteristic numbers and functions of continuous changes in the differential equation—a method, it will be seen, not unlike in spirit, however it may differ in detail, from the methods used by Sturm.

In all the cases mentioned so far only self-adjoint problems have been considered. Liouville‖, in 1838, considered a special real but not self-adjoint homogeneous

* This requires $p \equiv 0$ in (28), but this is no essential restriction.

† *Annales de l'École Normale Supérieure*, 2nd ser. vol. 12, p. 47.

‡ *Memoirs of the Academy of St Petersburg*, 8th ser. vol. 13 (1902), No. 2, where references to some earlier work by the same mathematician will be found.

§ *Annales de l'École Normale Supérieure*, 3rd ser. vols. 17 and 22, pages 359 and 539.

‖ Liouville's *Journal*, vol. 3, p. 561.

equation of the nth order with boundary conditions of a rather special form* to which special methods were applicable resembling those used in establishing Fourier's theorem concerning the number of real roots of algebraic equations. In this way a theorem of oscillation precisely like Sturm's was established. Liouville noticed that the characteristic values were the same for this problem and its adjoint, and that the corresponding characteristic functions for these two problems have the same number of roots.

Finally we note a very recent paper by v. Mises† who reverts to Sturm's original method of obtaining the differential equation as the limiting form of a difference equation to treat the equation (11) either under the assumption $g > 0$ or $l > 0$ and with the boundary conditions

$$\int_a^b Au\,dx = 0, \qquad \int_a^b Bu\,dx = 0,$$

where A and B are given functions. From what was said in § 2 it will be seen that these are equivalent to conditions of the form (2′), where, however, the coefficients are in general functions of λ of a special kind.

The only other result of a general character which has been obtained is Birkhoff's proof, already mentioned, of the existence of an infinite number of characteristic numbers for the general (not necessarily real or self-adjoint) boundary problem in which the parameter does not enter the boundary conditions, and enters the differential equation only in the form indicated in (19), g being real and positive, and his asymptotic expressions in this case. A similar result of Hilb‡ deserves notice, although it refers only to special equations of the first and second orders, because it involves non-homogeneous differential equations with $n+1$ instead of n non-homogeneous boundary conditions; a case, however, which may readily be reduced to the type of problem we have been considering (i.e. a homogeneous system involving n boundary conditions) provided we are willing to admit the parameter into the coefficients of one of the boundary conditions.

§ 11. *The Sturm-Liouville Developments of Arbitrary Functions.*

Almost as old as linear boundary problems themselves, and indeed one of the chief causes for the importance of and continued interest in these problems, is the question of developing a more or less arbitrarily given function $f(x)$ in the form of a series whose terms are the characteristic functions of such a problem. The simplest case here is that of the system (11), (12), with which alone we shall be concerned in this section§. Moreover we assume $g > 0$. Denoting the characteristic functions by $u_0, u_1, \ldots$, we have the problem of determining the coefficients $c_0, c_1, \ldots$ so that the development

$$f(x) = c_0 u_0 + c_1 u_1 + \ldots \quad \text{.........................(30)}$$

* Namely $n-1$ homogeneous conditions involving a and one homogeneous condition involving b. Liouville writes, to be sure, n non-homogeneous conditions at a, but they are, for his purposes, equivalent to $n-1$ homogeneous ones.

† H. Weber, *Festschrift*, 1912, p. 252.

‡ Crelle's *Journal*, vol. 140 (1911), p. 205.

§ We will assume that neither α' nor β' is zero. These are exceptional cases which require a separate treatment which presents no difficulty.

shall be valid. By means of (9) we readily see that the u_i's satisfy the relation

$$\int_a^b g u_i u_j dx = 0, \qquad (i \neq j) \quad \text{......................(31)},$$

by means of which the formal determination of the coefficients of (30), precisely as in the case of Fourier's series, is effected, namely

$$c_i = \frac{\int_a^b g f u_i dx}{\int_a^b g u_i^2 dx}, \qquad (i = 0, 1, \ldots) \quad \text{......................(32)}.$$

Liouville* set himself the problem of considering this formal development of Sturm and proving first that it converges, and secondly that its value is $f(x)$, but though he invented methods of great importance and got some valuable results, he did not succeed in carrying his treatment even for the simplest functions $f(x)$ to a successful conclusion.

Let us first consider the question of showing that if $f(x)$ is continuous and the series (30) with coefficients (32) converges uniformly in (X), its value must be precisely $f(x)$. Liouville by a simple and ingenious process showed that under these conditions the function represented by the series coincides with the function $f(x)$ for an infinite number of values of x in (X), but did not perceive that this was not sufficient. A rigorous proof was first given by Stekloff† in 1901 (modified and simplified in 1903 by Kneser‡) by the method of successive approximations. Further proofs have since been given, namely one by Hilbert§ completed by Kneser|| by means of integral equations, and a very simple one by Mason¶ by means of the calculus of variations.

If we turn to the question of the convergence of the series, we find that Liouville accomplished decidedly more than in the matter just considered, since he proved by a method, which when examined in the light of our modern knowledge proves to be essentially rigorous, that if $f(x)$ is continuous and consists of a finite number of pieces each of which has a continuous derivative, the series will converge uniformly. This he did by means of the asymptotic expressions of § 6. Finally Kneser** in his remarkable papers of 1903 and 1905, which so far as we have not already described them depend essentially on the use of asymptotic values, gave a comprehensive, rigorous, and simple treatment of this whole subject which applies to functions satisfying Dirichlet's conditions throughout the region (X), and even establishes the uniform convergence of the development in any portion of (X) where $f(x)$ is continuous. Thus, with Kneser's papers, all the more fundamental questions concerning the development of an arbitrary function in a Sturm-Liouville series were completely and satisfactorily settled.

* Liouville's *Journal*, vol. 1 (1836), p. 253; vol. 2 (1837), p. 16 and p. 418.

† *Ann. de la Faculté des sciences de Toulouse*, ser. 2, vol. 3, p. 281.

‡ *Math. Ann.* vol. 58, p. 81.

§ *Göttinger Nachrichten*, 1904, 2te Mitteilung, p. 213.

|| *Math. Ann.* vol. 63 (1907), p. 477.

¶ *Trans. Amer. Math. Soc.* vol. 8 (1907), p. 431.

** *Math. Ann.* vol. 58, p. 81 and vol. 60, p. 402.

It was however of interest to accomplish the same thing in other ways, and two other methods essentially distinct from Kneser's and from each other have since been developed. The first of these was Hilbert's remarkable application of integral equations to this development problem*, while the second by A. C. Dixon† involved Cauchy's method of residues.

The subject was not however hereby exhausted. There remained, for instance, the question of showing that, as in the case of Fourier's series, the convergence of the development at a particular point depends, roughly speaking, only on the behaviour of $f(x)$ in the neighbourhood of this point, a question which was successfully treated by Hobson‡. One could, however, hardly have anticipated that there was still room for such an extensive advance as was to be made by Haar§ in two papers which seem to have such a degree of finality that we must consider them in some detail.

Haar's work, like almost all other work on this subject, involves the reduction of the differential equation to the normal form (14) by means of Liouville's transformation, and, for the sake of simplicity, it is only of this normal form I shall speak. Moreover we will assume that the characteristic functions have been multiplied by such constants as to make the denominators of the coefficients (32) have the value 1.

From the earlier work on the development of functions we need merely assume as known that the very simplest kind of functions, say analytic functions, are represented uniformly by their Sturm-Liouville development.

Let us now denote by $s_n(x)$ and $\sigma_n(x)$ the sums of the first $n+1$ terms of the Sturm-Liouville and of the cosine development of $f(x)$ respectively:

$$s_n(x) = \int_0^\pi f(\alpha) \sum_{i=0}^n u_i(\alpha)\, u_i(x)\, d\alpha,$$

$$\sigma_n(x) = \int_0^\pi f(\alpha) \left[\frac{1}{\pi} + \frac{2}{\pi} \sum_{i=1}^n \cos i\alpha \cos ix\right] d\alpha.$$

We have then

$$s_n(x) - \sigma_n(x) = \int_0^\pi f(\alpha)\, \Phi_n(\alpha, x)\, d\alpha \quad \text{..................(33)},$$

where

$$\Phi_n(\alpha, x) = \sum_{i=0}^n u_i(\alpha)\, u_i(x) - \frac{1}{\pi} - \frac{2}{\pi} \sum_{i=1}^n \cos i\alpha \cos ix.$$

Now the central fact discovered by Haar, from which everything else flows with the greatest ease, is that *whatever continuous function $f(x)$ represents*

$$\lim_{n=\infty} [s_n(x) - \sigma_n(x)] = 0 \quad \textit{uniformly}.$$

* *Göttinger Nachrichten*, 1904, 2te Mitteilung, p. 213, where, however, the conditions imposed on $f(x)$ were extremely restrictive. The matter was treated more generally by Kneser, *Math. Ann.* vol. 63 (1907), p. 477.

† *Proc. London Math. Soc.* ser. 2, vol. 3 (1905), p. 83.

‡ *Proc. London Math. Soc.* ser. 2, vol. 6 (1908), p. 349.

§ *Zur Theorie der orthogonalen Funktionensysteme.* Göttingen dissertation (1909). Reprinted *Math. Ann.* vol. 69 (1910), p. 331. Also a second paper, *Math. Ann.* vol. 71 (1911), p. 38. See also Mercer, *Phil. Trans.* vol. 211 (1911), p. 111.

The proof consists of three steps of which I give all but the first completely:

(*a*) By means of the asymptotic expressions for u_i it is shown that there exists a constant M (independent of α, x, n) such that

$$|\Phi_n(\alpha, x)| < M.$$

(*b*) If $f(x)$ is analytic we know that $s_n(x)$ and $\sigma_n(x)$ both approach $f(x)$ uniformly. Consequently in this case, by (33),

$$\lim_{n=\infty} \int_0^\pi f(\alpha)\,\Phi_n(\alpha, x)\,d\alpha = 0 \quad \textit{uniformly}.$$

(*c*) Whatever be the continuous function $f(x)$, form a sequence $\phi_1(x)$, $\phi_2(x)$, ... of analytic functions which approach $f(x)$ uniformly. We may write

$$s_n(x) - \sigma_n(x) = \int_0^\pi [f(\alpha) - \phi_m(\alpha)]\,\Phi_n(\alpha, x)\,d\alpha + \int_0^\pi \phi_m(\alpha)\,\Phi_n(\alpha, x)\,d\alpha.$$

Since ϕ_m approaches f uniformly, we see by (*a*) that m may be so chosen that for all n's and x's the first of these integrals is in absolute value less than $\frac{1}{2}\epsilon$. Having thus fixed m, we see by (*b*) that the second integral can be made in absolute value less than $\frac{1}{2}\epsilon$ by taking n sufficiently large. This completes the proof.

It is now merely restating a *part* of what we have just proved if we say:

The Sturm-Liouville development of any continuous function $f(x)$ in the case of the normal system (14), (12′) *converges or diverges at any point of* (X) *according as the cosine development of $f(x)$ converges or diverges there. It diverges to $+\infty$ $(-\infty)$ when and only when the cosine development does this. It converges uniformly through a portion of* (X) *when and only when this is true of the cosine development.*

If we now denote by $S_n(x)$ and $\Sigma_n(x)$ the arithmetic means of the first n s's and σ's respectively, we may infer easily from the fact that $s_n - \sigma_n$ approaches zero uniformly, the further fact that

$$\lim_{n=\infty} [S_n(x) - \Sigma_n(x)] = 0 \quad \textit{uniformly}.$$

Consequently, since Fejér has proved that the cosine development of a continuous function of x is always uniformly summable by the method of the arithmetic mean to the value of the function, it follows that the same is true of the Sturm-Liouville development of any continuous function.

The extension to the development of discontinuous functions is not at all difficult and leads, as is indicated by Haar, to analogous results.

Finally in his second paper Haar shows how still other theorems concerning trigonometric series, namely those established by Riemann and his followers, can be carried over to the Sturm-Liouville developments with only very slight changes.

§ 12. *Other Developments.*

The most immediate and natural extension of the Sturm-Liouville developments is to the development according to the characteristic functions of a system which consists of the differential equation (11), in which $g > 0$, and in place of the Sturmian conditions (12) a more general pair of real self-adjoint conditions, thus including, for instance, the periodic conditions (29). The formal work in these cases is the same as before, since the relation (31) is still satisfied. Some cases of this sort were treated by Hilbert in his second *Mitteilung* (1904) by the method of integral equations but only under very restrictive conditions on the function $f(x)$ to be developed, namely the continuity of its first and second derivatives, besides the further fact that $f(x)$ must satisfy the same boundary conditions as the characteristic functions in terms of which it is to be developed. Shortly after, the general case here described was treated by A. C. Dixon, in the paper referred to above, by Cauchy's method of residues, the restrictions to be placed upon $f(x)$ being very much less restrictive.

Here again an essential advance was made by Birkhoff* in 1908. Even more significant here than the generalization to equations of the nth order of the form (19)† is the fact that the condition of reality is dropped and that the system considered is no longer required to be self-adjoint. This last generalization makes, as Liouville had already noticed in a special case‡, an essential change even in the formal work of expansion, since formula (31) is no longer valid. It is desirable now to consider by the side of the given problem the adjoint problem. This has, as we know, the same characteristic values as the original system, and if we denote the corresponding characteristic functions first of the original system and then of the adjoint system by

$$u_0,\ u_1,\ u_2,\ \dots$$

$$v_0,\ v_1,\ v_2,\ \dots$$

respectively, we have the relation

$$\int_a^b g u_i v_j dx = 0, \qquad (i \neq j) \quad \dots\dots\dots\dots(34),$$

which reduces to (31) when the system is self-adjoint. We have then essentially not an orthogonal but what is known as a biorthogonal system. By means of this equation the coefficients may be formally determined by the expression

$$c_i = \frac{\int_a^b g f v_i dx}{\int_a^b g u_i v_i dx} \quad \dots\dots\dots\dots(35),$$

where, however, the question of the possible vanishing of the denominator must be further considered. This formal work, which had been given by Liouville in a special case, is the basis of Birkhoff's paper.

At a characteristic number λ_i the Green's function $G(x, \xi)$ has in general a pole of the first order whose residue Birkhoff finds to be given by the formula

$$\frac{u_i(x)\, v_i(\xi)}{\int_a^b g u_i v_i dx}.$$

* *Trans. Amer. Math. Soc.* vol. 9, p. 373. A very special case of Birkhoff's result was subsequently obtained by essentially the same method by Hilb, *Math. Ann.* vol. 71 (1911), p. 76.

† Westfall had in 1905 (Göttingen dissertation) considered the real self-adjoint case where the equation is of even order, where, however, no essentially new features occur. The method used was Hilbert's and the restrictions imposed on f were correspondingly great.

‡ Liouville's *Journal*, vol. 3 (1838), p. 561.

This when multiplied by $f(\xi)$ and integrated from a to b is precisely the general term of the formal development of $f(x)$ according to the functions u_i. Consequently the sum of the first $n+1$ terms of this formal development may readily be expressed as a contour integral in the λ-plane whose path surrounds the first $n+1$ characteristic numbers $\lambda_0, \lambda_1, \ldots \lambda_n$. Birkhoff then evaluates the limit of this contour integral as n becomes infinite by means of the asymptotic expressions for the characteristic functions u_i, v_i, and thus establishes at one stroke in fairly general cases both the convergence of the series and the fact that it represents the function $f(x)$.

A similar treatment has since been given by Hilb in the case of two special non-homogeneous systems mentioned at the end of § 10.

The Sturm-Liouville developments have also been generalized in one other direction, namely to the case where in the equation of the second order (11) the function g changes sign while $l \geqq 0$. The results here are still very incomplete, only the real case with certain special self-adjoint boundary conditions having been so far treated. The first treatment was by Hilbert* in 1906, when by means of his theory of polar integral equations he succeeded in establishing the validity of the development under very special restrictions including the continuity of the first four derivatives of the function to be developed. Mason's proof by means of the calculus of variations, referred to above, that if f is continuous and the series converges uniformly, the development represents the function, is valid in this case also.

The numerous important contributions which have been made during the last few years to the theory of series of orthogonal or biorthogonal functions in general all have a direct bearing on the questions here considered, and some of them give, even in the special cases we are here concerned with, essentially new results. It would, however, lead us too far if we should attempt to follow up these more general investigations.

§ 13. *Conclusion.*

The questions we have been considering may be classified roughly as (a) Existence Theorems, (b) Oscillation Properties, (c) Asymptotic Expressions, (d) Development Theorems. For the Sturm-Liouville system (11), (12) the investigation of all of these questions has been carried to a high degree of perfection, although even here the field is not yet exhausted. In the real self-adjoint case for the equation of the second order (11) where $g > 0$ results of a fair degree of completeness in all these directions have also been attained. In most other cases, however, the ground has only just been broken and nearly everything is still to be done.

Of the methods invented during the last few years undoubtedly that of integral equations is the most far-reaching and powerful. This method would seem however to be chiefly valuable in the cases of two or more dimensions where many of the simplest questions are still to be treated. In the case of one dimension where we now have to deal with finer or more remote questions other, in the main older, methods have so far usually proved to be more serviceable. It is only fair to mention

* *Göttinger Nachrichten*, 5te Mitteilung, p. 473. Cf. also Fubini, *Annali di Mat.* ser. 3, vol. 17 (1910), p. 111, where Hilbert's restriction that g vanish only a finite number of times in (X) is removed.

here the very important treatment given by Weyl of cases in which singular points occur at *a* or *b*. The development theorems here, where we have frequently not series but definite integrals, or even mixed forms, have so far been handled only by the use of integral equations. Apart from this, it may fairly be said that the greatest advances of recent years in the theory of boundary problems in one dimension, I recall for instance Birkhoff's three important contributions, have been made by other methods, largely indeed by methods more or less closely analogous to the original methods of Sturm and of Liouville. If my lecture to-day can serve to emphasize not the historical importance but the present vitality of these methods it will have served one of its main purposes.

ON THE DYNAMICS OF RADIATION

By Sir Joseph Larmor.

The subject of this title is coextensive with the whole range of the physics of imponderable agencies. For if it is correct to say with Maxwell that all radiation is an electrodynamic phenomenon, it is equally correct to say with him that all electrodynamic relations between material bodies are established by the operation, on the molecules of those bodies, of fields of force which are propagated in free space as radiation and in accordance with the laws of radiation, from the one body to the other. It is not intended to add to the number of recent general surveys of this great domain. The remarks here offered follow up some special points: they are in part in illustration of the general principle just stated: and in part they discuss, by way of analogy with cognate phenomena now better understood, the still obscure problem of the mode of establishment of the mechanical forces between electric systems.

The essential characteristic of an electrodynamic system is the existence of the correlated fields, electric and magnetic, which occupy the space surrounding the central body, and which are an essential part of the system; to the presence of this pervading aethereal field, intrinsic to the system, all other systems situated in that space have to adapt themselves. When a material electric system is disturbed, its electrodynamic field becomes modified, by a process which consists in propagation of change outward, after the manner of radiation, from the disturbance of electrons that is occurring in the core. When however we are dealing with electric changes which are, in duration, slow compared with the time that radiation would require to travel across a distance of the order of the greatest diameter of the system—in fact in all electric manifestations except those bearing directly on optical or radiant phenomena—complexities arising from the finite rate of propagation of the fields of force across space are not sensibly involved: the adjustment of the field surrounding the interacting systems can be taken as virtually instantaneous, so that the operative fields of force, though in essence propagated, are sensibly statical fields. The practical problems of electrodynamics are of this nature—how does the modified field of force, transmitted through the aether from a disturbed electric system, and thus established in the space around and alongside the neighbouring conductors which alone are amenable to our observation, penetrate into these conductors and thereby set up electric disturbance in them also? and how does the field emitted in turn by these new disturbances interact with the original exciting field and with its core? For example, if we are dealing with a circuit of good conducting quality and finite cross section, situated in an alternating field of fairly rapid frequency, we know that the penetration of the arriving field into the conductor is counteracted by the mobility of its electrons, whose motion, by obeying the force, in so far annuls it by Newtonian kinetic reaction; so that instead of being propagated, the field soaks in by diffusion, and it does not get very deep even when adjustment is delayed by the friction of the vast numbers of ions which it starts into motion, and which have to push their way through the crowd of material molecules; and the phenomena of surface currents thus arise. If (by a figure of speech) we abolish the aether in which both the generating circuit and the secondary circuit which it excites are immersed, in which they in fact subsist, the changing phases of the generator could not thus establish, from instant to instant, by almost instantaneous radiant transmission, their changing fields of force in the ambient region extending across to the secondary circuit, and the ions in and along that circuit would remain undisturbed, having no stimulus to respond to. The aethereal phenomenon, viz., the radiant propagation of the fields of force, and the material phenomenon, viz., the response of the ions of material bodies to those fields, involving the establishment of currents with new fields of their own, are the two interacting factors. The excitation of an alternating current in a wire, and the mode of distribution of the current across its section, depend on the continued establishment in the region around the wire, by processes of the nature of radiation, of the changing electromagnetic field that seizes hold on the ions and so excites the current; and the question how deep this influence can soak into the wire is the object of investigation. The aspect of the subject which is thus illustrated, finds in the surrounding region, in the aether, the seat of all electrodynamic action, and in the motions of electrons its exciting cause. The energies required to propel the ions, and so establish an induced current, are radiant energies which penetrate into the conductor from its sides, being transmitted there elastically through the aether; and these energies are thereby ultimately in part degraded into the heat arising from fortuitous ionic motions, and in part transformed to available energy of mechanical forces between the conductors. The idea—introduced by Faraday, developed into precision by Maxwell, expounded and illustrated in various ways by Heaviside, Poynting, Hertz—of radiant fields of force, in which all the material electric circuits are immersed, and by which all currents and electric distributions are dominated, is the root of the modern exact analysis of all electric activity.

The elementary phenomena of steady currents, including Ohm's law and all the rest of the relations which are so easily formulated directly, are the simple synthesis to which this scheme of activity leads, when the changes of the controlling fields are slow enough to be considered as derived from statical potentials. In the electric force integrated round a circuit, viz., the electromotive force so called, the undetermined portion of the electric field, that arises from electric distributions which adjust the current to flow 'full-bore,' after the manner of a stream, is eliminated by the integration; and the fully developed current, thus adjusted to be the same at all sections of the wire, is of necessity proportional to the impressed electromotive force, as Ohm postulated.

The principle of the controlling influence of the activities in the intervening aether emerges in the strongest light when we recall the mechanical illustrations of the Kelvin period; we may say that the disturbance of the field of aether exerts influence on the conductors in the same general kind of way as the pressures involved in the inertia of moving fluid control the motion of the vortices or solid bodies which are immersed in the fluid.

As regards the mode of establishment of these fields of aethereal activity, we here merely recall that the phenomena of free propagation in space are covered by Hertz's brilliant analysis for the case of a simple electric vibrator or dipole, that being the type of element out of which all more complex sources of radiation, including the radiation of moving ions, may be built up by superposition. More remarkable, on account of its sharp contrast with the familiar phenomena of light and sound, is the guidance of electric radiation by a wire, which was explored experimentally by Hertz, in full touch with the mathematical theory, but with so much trouble and vexation arising from the influence of casual conductors serving as the 'return circuits' on which the issuing lines of electric force must find their terminations. It was not long however until the conditions were made manageable, first by Lecher, by the simple device of introducing a parallel wire, on which as return circuit the lines of force from the other wire could again converge, thus restricting the propagated field of force to the region extending between the original and this return circuit; this arrangement, by preventing lateral spreading of the radiation, thus avoided all effects both of moderate curvature of the guiding conductor and of disturbance by neighbouring bodies. The problem of conducting free radiant energy to a distance in the open space around the guiding conductor, without lateral loss any more than there is when electric undulations travel along the interior dielectric region between the coatings of a submarine cable, thus became securely realized and understood.

The equations of propagation of radiant effects in free space are, as we have seen, the foundation of all electric phenomena, whether static or kinetic. In the case of slow changes, both the electric field (PQR) and the magnetic field $(\alpha\beta\gamma)$ * have their potentials; namely we have

$$\operatorname{curl}(PQR)=0, \qquad \operatorname{curl}(\alpha\beta\gamma)=0.$$

In the case of rapid alternations these relations have to become adapted to the transmission of undulations; which we know in advance must be of necessity transverse, owing to the absence of divergence of the vectors concerned in them, as expressed by the equations

$$\operatorname{div}(PQR)=0, \qquad \operatorname{div}(\alpha\beta\gamma)=0.$$

When this order of ideas is pursued, as it was in a way by Hertz, the appropriate kind of modification of the statical equations can hardly be missed: it must in fact be expressed by relations of the type

$$\operatorname{curl}(PQR)=-A\frac{d}{dt}(\alpha\beta\gamma), \qquad \operatorname{curl}(\alpha\beta\gamma)=A\frac{d}{dt}(PQR).$$

* A British writer will be pardoned for retaining the commodious and classical notation of Maxwell, in which he has been educated, until there is some consensus of opinion as to what other notation, if any, is to replace it.

We are bound to recall here that precisely equivalent relations had been laid down, in brilliant fashion, as the result of a tentative process of adaptation of analytical theory to optical phenomena, by MacCullagh* as far back as 1838, as a scheme consistently covering the whole ground of physical optics: and especially that their form was elucidated, and their evidence fortified, by him in the following year, by showing that they fitted into the Lagrangian algorithm of 'Least Action,' which was thus already recognized in physics as the generalized compact expression and criterion of the relations appropriate to a dynamical system. The analysis proceeded on the basis of the wider range of indications afforded by the study of optical phenomena in crystals, and accordingly the result was reached in the generalized form appropriate to aeolotropic media. But MacCullagh could not construct any model of the dynamical operation of his analytical equations, on the lines of the properties of ordinary material bodies, with which alone we are familiar through experiment—a task which indeed is now widely recognized to be an unreasonable one, though at that time it largely dominated all problems of physical interpretation.

Guidance of electric radiation by wires.

In the arrival of a magnetic field and its related electric field, in the space around the conducting circuit, the field being transmitted there by radiant processes, we have recognized the essential cause of the excitation of an alternating current in the circuit and of its location, when the changes are rapid, in the outer part. An electric field so transmitted through space will not, of course, be along the length of the wire; but the component that is oblique will be at once compensated and annulled statically by the electric separation (of amount in other respects negligible) which it produces across the section of the conductor. The longitudinal force that remains may be treated as the 'induced electromotive force per unit length' at the place in question. When this adjustment of the electric field has been effected by excitation of free charge, the total force becomes longitudinal, and its distribution in the cross-section of the wire is necessarily restricted to those solutions of its cylindrical harmonic characteristic equation which remain finite across the section; in the simplest and usual case of axial symmetry the distribution is represented as *infra* by the Bessel function of complex argument and zero order.

It is perhaps worth while formally to set out the analysis from this point of view†, when this longitudinal electric field, arising from the transmitted field and the transverse electric distribution induced by it—which is the field propelling the current—is a uniform force $R_0 e^{ipt}$, operating on a long conducting cylinder from outside and all along its surface.

The equations of the field are, in Maxwell's notations,

$$-\frac{da}{dt}=\frac{dR}{dy},\quad \frac{db}{dt}=\frac{dR}{dx},\quad \frac{dc}{dt}=0,$$

* *Collected Works of James MacCullagh* (1880), p. 145.

† Cf. the rather different analysis in Maxwell, *Phil. Trans.* 1865, *Elec. and Mag.* vol. ii. § 690, developed by Rayleigh, *Phil. Mag.* vol. 21 (1886), p. 387.

where $$4\pi w = \frac{d\beta}{dx} - \frac{d\alpha}{dy}, \text{ where } w = \left(\frac{K}{4\pi C^2}\frac{d}{dt} + \sigma\right) R.$$

Thus $$\frac{d^2R}{dx^2} + \frac{d^2R}{dy^2} = -m^2R, \text{ and } w = -\frac{m^2}{4\pi\mu p}\iota R,$$

where $$-m^2 = -\frac{K\mu}{C^2}p^2 + 4\pi\sigma p\mu\iota, \qquad \iota p = \frac{d}{dt}.$$

Thus, for symmetrical distribution in a circular section,

$$\frac{d^2R}{dr^2} + \frac{1}{r}\frac{dR}{dr} + m^2R = 0;$$

so that inside the section the distribution is determined as

$$R_{\text{in}} = AJ_0(mr),$$

m being a complex quantity; while outside the section, in free space,

$$R_{\text{out}} = R_0 + BI_0\left(\frac{p}{C}r\right),$$

where*, if $S_n = 1 + \frac{1}{2} + \frac{1}{3} + \ldots + \frac{1}{n}$, and γ represents Euler's constant ·577,

$$I_0(x) = (\gamma + \log \tfrac{1}{2}\iota x)\left(1 - \frac{x^2}{2^2} + \frac{x^4}{2^2 . 4^2} - \ldots\right)$$
$$+ \frac{x^2}{2^2}S_1 - \frac{x^4}{2^2 . 4^2}S_2 + \frac{x^6}{2^2 . 4^2 . 6^2}S_3 - \ldots,$$

$$J_0(x) = 1 - \frac{x^2}{2^2} + \frac{x^4}{2^2 . 4^2} - \ldots.$$

The continuity of the field across the boundary of the conductor requires that R and $\frac{1}{\mu}\frac{dR}{dr}$ are there continuous. Thus for a wire of radius a,

$$AJ_0(ma) = R_0 + BI_0\left(\frac{p}{C}a\right) \quad\ldots\ldots(1),$$

$$A\frac{m}{\mu}J_0'(ma) = B\frac{p}{C}I_0'\left(\frac{p}{C}a\right) \quad\ldots\ldots(2).$$

Also, the total current C is determined by

$$C = \int_0^a w . 2\pi r\,dr = -\frac{m^2}{2\mu p}\iota A\int_0^a J_0(mr)\,r\,dr$$
$$= \frac{1}{2\mu p}\iota A \Big| mrJ_0'(mr)\Big|_0^a,$$

giving finally $$C = \frac{ma}{2\mu p}\iota AJ_0'(ma) \quad\ldots\ldots(3).$$

The elimination of the undetermined constants A and B between the equations (1), (2), (3) will lead to the circumstances of propagation of this total current C excited in the infinite cylinder.

* Cf. e.g. Rayleigh's *Theory of Sound*, vol. ii. § 341; referring back to cognate physical applications by Stokes, *Phil. Trans.* 1868, or *Collected Papers*, vol. iv. p. 321.

As pa/c is usually very small, we may write approximately

$$I_0(x) = \gamma + \log \tfrac{1}{2}\iota x, \qquad I_0'(x) = \frac{1}{x};$$

and after some reductions we have

$$\frac{R_0}{C} = -\frac{2\mu p}{ma}\iota\frac{J_0(ma)}{J_0'(ma)} - 2p\iota\left(\gamma + \tfrac{1}{2}\pi + \log\frac{p}{2c}a\right);$$

and finally, eliminating the complex character of this equation by writing dC/dt for $pC\iota$, bearing in mind that m is complex, we arrive at the usual form of relation

$$R_0 = L\frac{dC}{dt} + \rho C,$$

and thus obtain expressions for the inductance L and the resistance ρ of a wire per unit length when it is so nearly straight that the radius of curvature is many times its diameter, and when no disturbing conductor is near.

When the conductivity σ is very large, and the frequency $p/2\pi$ is not excessive so that the second term in the expression for m^2 is preponderant, we have

$$m = (2\pi\sigma\mu p)^{\frac{1}{2}}(1 - \iota).$$

Also when, as here, the real part of the argument z is very great and positive,

$$J_0(z)/J_0'(z) \text{ becomes equal to } \iota.$$

Thus in this case

$$\frac{R_0}{C} = \left(\frac{\mu p}{2\pi\sigma}\right)^{\frac{1}{2}} - \iota p\,.\,2\left\{\gamma + \tfrac{1}{2}\pi + \log\frac{pa}{2c} - \tfrac{1}{2}\left(\frac{\mu}{2\pi p\sigma}\right)^{\frac{1}{2}}\right\},$$

viz.

$$R_0 = \left(\frac{\mu p}{2\pi\sigma}\right)^{\frac{1}{2}} C + \left\{2\log\frac{2c}{pa} - \pi - 2\gamma + \left(\frac{\mu}{2\pi p\sigma}\right)^{\frac{1}{2}}\right\}\frac{dC}{dt}.$$

As R_0 is the electric force impressed along the conductor just outside it, the coefficient of C is the effective resistance ρ of the conductor per unit length, and that of dC/dt is the effective inductance L per unit length, in this limiting case. These expressions agree with Rayleigh's results*, except that in this mode of formulation the value of L is definite and does not retain any undetermined constant.

The transmission, along a circular wire, of electric waves which maintain and propagate their own field subject to the inevitable damping, involves a different point of view; as R_0 is now absent, a velocity of propagation is determined by equating the values of A/B derived from equations (1) and (2).

General theory of pressure exerted by waves.

If a perfectly reflecting structure has the property of being able to advance through an elastic medium, the seat of free undulations, without producing disturbance of structure in that medium, then it follows from the principle of energy alone that these waves must exert forces against such a reflector, constituting a pressure equal in intensity at each point to the energy of the waves per unit volume. Cf. p. 208, *infra*. The only hypothesis, required in order to justify this general result, is that the velocity of the undulations in the medium must be independent

* Cf. *Theory of Sound*, ed. 2, § 235*v*.

of their wave-length; viz., the medium is to be non-dispersive, as is the free aether of space.

This proposition, being derived solely from consideration of conservation of the energy, must hold good whatever be the character of the mechanism of propagation that is concerned in the waves. But the elucidation of the nature of the pressure of the waves, of its mode of operation, is of course concerned with the constitution of the medium. The way to enlarge ideas on such matters is by study of special cases: and the simplest cases will be the most instructive.

Let us consider then transverse undulations travelling on a cord of linear density ρ_0, which is stretched to tension T_0. Waves of all lengths will travel with the same velocity, namely $C = (T_0/\rho_0)^{\frac{1}{2}}$, so that the condition of absence of dispersion is satisfied. A solitary wave of limited length, in its transmission along the cord, deflects each straight portion of it in succession into a curved arc. This process implies increase in length, and therefore increased tension, at first locally. But we adhere for the present to the simplest case, where the cord is inextensible or rather the elastic modulus of extension is indefinitely great. The very beginnings of a local disturbance of tension will then be equalized along the cord with speed practically infinite; and we may therefore take it that at each instant the tension stands adjusted to be the same (T_0) all along it. The pressure or pull of the undulations at any point is concerned only with the component of this tension in the direction of the cord; this is

$$T_0\left(1 + \frac{d\eta^2}{dx^2}\right)^{-\frac{1}{2}},$$

where η is the transverse displacement of the part of the cord at distance x measured along it; thus, up to the second order of approximation, the pull of the cord is

$$T_0 - \tfrac{1}{2}T_0\left(\frac{d\eta}{dx}\right)^2.$$

The tension of the cord therefore gives rise statically to an undulation pressure

$$\tfrac{1}{2}T_0\left(\frac{d\eta}{dx}\right)^2, \quad \text{or } \tfrac{1}{2}T_0C^{-2}\left(\frac{d\eta}{dt}\right)^2, \quad \text{or } \tfrac{1}{2}\rho_0\left(\frac{d\eta}{dt}\right)^2.$$

The first of these three equivalent expressions can be interpreted as the potential energy per unit length arising from the gathering up of the extra length in the curved arc of the cord, against the operation of the tension T_0; the last of them represents the kinetic energy per unit length of the undulations. Thus there is a pressure in the wave, arising from this statical cause, which is at each point equal to half its total energy per unit length.

There is the other half of the total pressure still to be accounted for. That part has a very different origin. As the tension is instantaneously adjusted to the same value all along, because the cord is taken to be inextensible, there must be extra mass gathered up into the curved segment which travels along it as the undulation. The mass in this arc is

$$\int \rho_0\left(1 + \frac{d\eta^2}{dx^2}\right)^{\frac{1}{2}} dx,$$

or to the second order is approximately

$$\rho_0 l + \int \tfrac{1}{2}\rho_0 \left(\frac{d\eta}{dx}\right)^2 dx$$

In the element δx there is extra mass of amount

$$\tfrac{1}{2}\rho_0 \left(\frac{d\eta}{dx}\right)^2 \delta x,$$

which is carried along with the velocity c of the undulatory propagation. This implies momentum associated with the undulation, and of amount at each point equal to $\frac{1}{2}\rho_0 c \left(\frac{d\eta}{dx}\right)^2$ per unit length. Another portion of the undulation pressure is here revealed, equal to the rate at which the momentum is transmitted past a given point of the cord; this part is represented by $\frac{1}{2}\rho_0 c^2 \left(\frac{d\eta}{dx}\right)^2$ or $\frac{1}{2}\rho_0 \left(\frac{d\eta}{dt}\right)^2$, and so is equal to the component previously determined.

In our case of undulations travelling on a stretched cord, the pressure exerted by the waves arises therefore as to one half from transmitted intrinsic stress and as to the other half from transmitted momentum.

The kinetic energy of the cord can be considered either to be energy belonging to the transverse vibration, viz., $\int \frac{1}{2}\rho \left(\frac{d\eta}{dt}\right)^2 ds$, or to be the energy of the convected excess of mass moving with the velocity of propagation c*, viz., $\int \frac{1}{2}\rho \left(\frac{d\eta}{dx}\right)^2 c^2 dx$; for these quantities are equal by virtue of the condition of steady propagation $\frac{d\eta}{dt} = c\frac{d\eta}{dx}$.

On the other hand the momentum that propagates the waves is transverse, of amount $\rho\frac{d\eta}{dt}$ per unit length; it is the rate of change of this momentum that appears in the equation of propagation

$$\frac{d}{dt}\left(\rho\frac{d\eta}{dt}\right) = \frac{d}{dx}\left(T\frac{d\eta}{dx}\right).$$

But the longitudinal momentum with which we have been here specially concerned is $\frac{1}{2}\rho\left(\frac{d\eta}{dx}\right)^2 c$ per unit length, which is $\frac{1}{2}\frac{d\eta}{dx}\cdot\rho\frac{d\eta}{dt}$. Its ratio to the transverse momentum is very small, being $\frac{1}{2}\frac{d\eta}{dx}$; it is a second-order phenomenon and is not essential to the propagation of the waves. It is in fact a special feature, and there are types of wave motion in which it does not occur. The criterion for its presence is that the medium must be such that the reflector on which the pressure is exerted can advance through it, sweeping the radiation along in front of it, but not disturbing the structure; possibly intrinsic strain, typified

[* This specification is fictitious; indeed a factor $\frac{1}{2}$ has been dropped in its expression just following. There is however *actual* energy of longitudinal motion; as it belongs to the whole mass of the cord, which moves together, it is very small in amount, its ratio to the energy of transverse vibration being $\frac{1}{4}(d\eta/dx)^2$.]

by the tension of the cord, may be an essential feature in the structure of such a medium.

If we derive the dynamical equation of propagation along the cord from the Principle of Action $\delta\int(T-W)dt=0$, where $T=\int\frac{1}{2}\rho\left(\frac{d\eta}{dt}\right)^2 ds$ and $W=\int\frac{1}{2}T_0\left(\frac{d\eta}{dx}\right)^2 dx$, the existence of the pressure of the undulations escapes our analysis. A corresponding remark applies to the deduction of the equations of the electrodynamic field from the Principle of Action*. In that mode of analysis the forces constituting the pressure of radiation are not in evidence throughout the medium; they are revealed only at the place where the field of the waves affects the electrons belonging to the reflector. Problems connected with the Faraday-Maxwell stress lie deeper; they involve the structure of the medium to a degree which the propagation of disturbance by radiation does not by itself give us means to determine.

We therefore proceed to look into that problem more closely. We now postulate Maxwell's statical stress system; also Maxwell's magnetic stress system, which is, presumably, to be taken as of the nature of a kinetic reaction. But when we assert the existence of these stresses, there remain over uncompensated terms in the mechanical forcive on the electrons which may be interpreted as due to a distribution of momentum in the medium†. The pressure of a train of radiation is, on this hypothetical synthesis of stress and momentum, due entirely (p. 207) to the advancing momentum that is absorbed by the surface pressed, for here also the momentum travels with the waves. This is in contrast with the case of the cord analysed above, in which only half of the pressure is due to that momentum.

The pressure of radiation against a material body, of amount given by the law specified by Maxwell for free space, is demonstrably included in the Maxwellian scheme of electrodynamics, when that scheme is expanded so as to recognize the electrons with their fields of force as the link of communication between aether and matter. But the illustration of the stretched cord may be held to indicate that it is not yet secure to travel further along with Maxwell, and accept as realities the Faraday-Maxwell stress in the electric field, and the momentum which necessarily accompanies it; it shows that other dynamical possibilities of explanation are not yet excluded. And, viewing the subject from the other side, we recognize how important have been the experimental verifications of the law of pressure of radiation which we owe to Lebedew, too early lost to science, to Nichols and Hull, and to Poynting and Barlow. The law of radiation pressure in free space is not a necessary one for all types of wave-motion; on the other hand if it had not been verified in fact, the theory of electrons could not have stood without modification.

The pressure of radiation, according to Maxwell's law, enters fundamentally in the Bartoli-Boltzmann deduction of the fourth power law of connexion between total radiation in an enclosure and temperature. Thus in this domain also, when we pass beyond the generalities of thermodynamics, we may expect to find that the laws of distribution of natural radiant energy depend on structure which is deeper

* Cf. Larmor, *Trans. Camb. Phil. Soc.* vol. xviii. (1900), p. 318; or *Aether and Matter*, Chapter vi.

† For the extension to the most general case of material media cf. *Phil. Trans.* vol. 190 (1897), p. 253.

seated than anything expressed in the Maxwellian equations of propagation. The other definitely secure relation in this field, the displacement theorem of Wien, involves nothing additional as regards structure, except the principle that operations of compression of a field of natural radiation in free space are reversible. The most pressing present problem of mathematical physics is to ascertain whether we can evade this further investigation into aethereal structure, for purposes of determination of average distribution of radiant energy, by help of the Boltzmann-Planck expansion of thermodynamic principles, which proceeds by comparison of the probabilities of the various distributions of energy that are formally conceivable among the parts of the material system which is its receptacle.

Momentum intrinsically associated with Radiation.

We will now follow up, after Poynting*, the hypothesis thus implied in modern statements of the Maxwellian formula for electric stress, namely that the pressure of radiation arises wholly from momentum carried along by the waves. Consider an

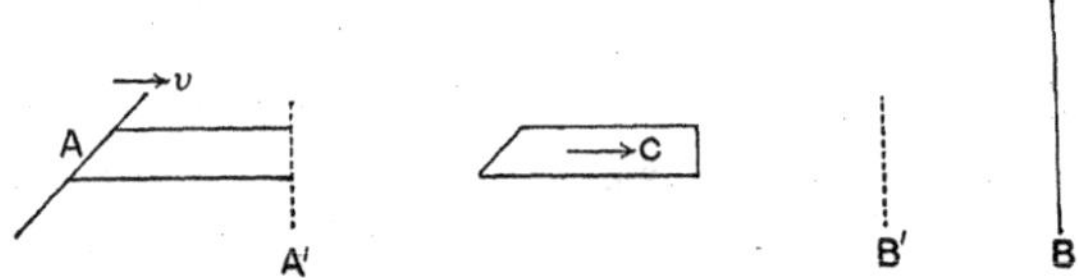

isolated beam of definite length emitted obliquely from a definite area of surface A and absorbed completely by another area B. The automatic arrangements that are necessary to ensure this operation are easily specified, and need not detain us. In fact by drawing aside an impervious screen from A we can fill a chamber AA' with radiation; and then closing A and opening A', it can emerge and travel along to B, where it can be absorbed without other disturbance, by aid of a pair of screens B and B' in like manner. Let the emitting surface A be travelling in any direction while the absorber B is at rest. What is emitted by A is wholly gained by B, for the surrounding aether is quiescent both before and after the operation. Also, the system is not subject to external influences; therefore its total momentum must be conserved, what is lost by A being transferred ultimately to B, but by the special hypothesis now under consideration, existing meantime as momentum in the beam of radiation as it travels across. If v be the component of the velocity of A in the direction of the beam, the duration of emission of the beam from A is $(1 - v/C)^{-1}$ times the duration of its absorption by the fixed absorber B. Hence the intensity of pressure of a beam of issuing radiation on the moving radiator must be affected by a factor $(1 - v/C)$ multiplying its density of energy; for pressure multiplied by time is the momentum which is transferred unchanged by the beam to the absorber for which v is null. We can verify readily that the pressure of a beam against a moving absorber involves the same factor $(1 - v/C)$. If the emitter were advancing with the velocity of light this factor would make the pressure vanish, because the emitter would keep permanently in touch with the beam: if the absorber were receding with the velocity of light there would be no pressure on it, because it would just keep ahead of the beam.

* Cf. *Phil. Trans.* vol. 202, A (1903).

There seems to be no manner other than these two, by altered intrinsic stress or by convected momentum, in which a beam of limited length can exert pressure while it remains in contact with the obstacle and no longer. In the illustration of the stretched cord the intrinsic stress is transmitted and adjusted by tensional waves which travel with velocity assumed to be practically infinite. If we look closer into the mode of this adjustment of tension, it proves to be by the transmission of longitudinal momentum; though in order that the pressure may keep in step, the momentum must travel with a much greater velocity, proper to tensional waves. In fact longitudinal stress cannot be altered except by fulfilling itself through the transfer of momentum, and it is merely a question of what speeds of transference come into operation.

In the general problem of aethereal propagation, the analogy of the cord suggests that we must be careful to avoid undue restriction of ideas, so as, for example, not to exclude the operation, in a way similar to this adjustment of tension by longitudinal propagation, of the immense but unknown speed of propagation of gravitation. We shall find presently that the phenomena of absorption lead to another complication.

So long, however, as we hold to the theory of Maxwellian electric stress with associated momentum, there can be no doubt as to the validity of Poynting's modification of the pressure formula for a moving reflector, from which he has derived such interesting consequences in cosmical astronomy. To confirm this, we have only to contemplate a beam of radiation of finite length l advancing upon an obstacle A in which it is absorbed. The rear of it moves on with velocity C; hence if the body A is in motion with velocity whose component along the beam is v, the beam will be absorbed or passed on, at any rate removed, in a time $l/(C-v)$. But by electron theory the beam possesses a distribution of at any rate *quasi*-momentum identical with the distribution of its energy, and this has disappeared or has passed on in this time. There must therefore be a thrust on the obstructing body, directed along the beam and equal to $\epsilon(1-v/C)$, where ϵ is the energy of the beam per unit length which is also the distribution of the *quasi*-momentum along the free beam.

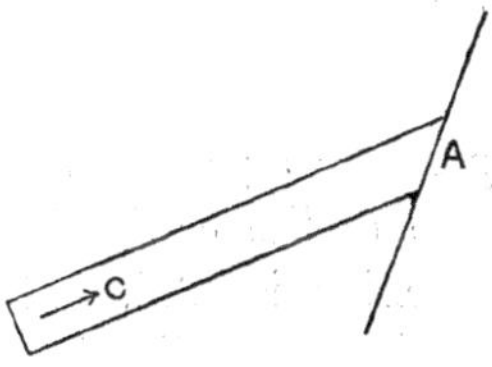

The back pressure on a radiating body travelling through free space, which is exerted by a given stream of radiation, is by this formula smaller on its front than on its rear; so that if its radiation were unaffected by its motion, the body would be subject to acceleration at the expense of its internal thermal energy. This of course could not be the actual case.

The modifying feature is that the intensity of radiation, which corresponds to a given temperature, is greater in front than in rear. The temperature determines the amplitude and velocity of the ionic motions in the radiator, which are the same whether it be in rest or in uniform motion: thus it determines the amplitude of the oscillation in the waves of aethereal radiation that are excited by them and travel out from them. Of this oscillation the intensity of the magnetic field represents the velocity. If the radiator is advancing with velocity v in a direction inclined at an angle θ to an emitted ray, the wave length in free aether is shortened in the ratio $1-\frac{v}{c}\cos\theta$; thus the period of the radiation is shortened in the same ratio; thus the velocity of vibration, which represents the magnetic field, is altered in the inverse ratio, and the energy per unit volume in the square of that ratio, viz., that energy is now $\epsilon\left(1-\frac{v}{c}\cos\theta\right)^{-2}$; and the back pressure it exerts involves a further factor $1-\frac{v}{c}\cos\theta$ owing to the convection; so that that pressure is $\epsilon\left(1-\frac{v}{c}\cos\theta\right)^{-1}$, where ϵ is the energy per unit volume of the natural radiation emitted from the body when at rest. The pressural reaction on the source is in fact E'/c, where E' is the actual energy emitted in the ray per unit time.

Limitation of the analogy of a stretched cord.

In the case of the inextensible stretched cord, the extra length due to the curved arc in the undulation is proportional to the energy of the motion. The loss of energy by absorption would imply slackening of the tension; and the propositions as to pressure of the waves, including Poynting's modification for a moving source, would not hold good unless there were some device at the fixed ends of the cord for restoring the tension. The hypothesis of convected momentum would imply something of the same kind in electron structure.

It is therefore worth while to verify directly that the modified formula for pressure against a moving total reflector holds good in the case of the cord, when there is no absorption so that the reflexion is total. This analysis will also contain the proof of the generalization of the formula for radiant pressure that was enunciated on p. 202 *supra**.

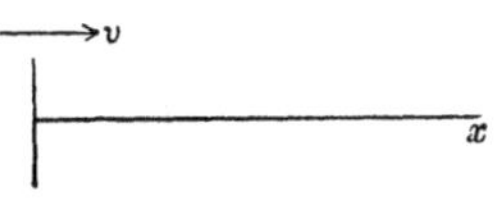

Let the wave-train advancing to the reflector and the reflected wave-train be represented respectively by

$$\dot\eta_1 = A_1\cos m_1(x+ct),$$

$$\dot\eta_2 = A_2\cos m_2(x-ct).$$

At the reflector, where $x=vt$, we must have

$$\int\dot\eta_1 dt = \int\dot\eta_2 dt;$$

this involves two conditions,

$$\frac{A_1}{m_1}=\frac{A_2}{m_2} \text{ and } m_1(c+v)=m_2(c-v).$$

Now the energies *per unit length* in these two simple wave-trains are

$$\tfrac{1}{2}\rho A_1^2 \text{ and } \tfrac{1}{2}\rho A_2^2;$$

* See Larmor, *Brit. Assoc. Report*, 1900. [The statement that follows here is too brief, unless reference is made back to the original, especially as a *minus* sign has fallen out on the right of the third formula below. The reflector consists of a disc with a small hole in it through which the cord passes; this disc can move along the cord sweeping the waves in front of it while the cord and its tension remain continuous through the hole—the condition of reflexion being thus $\eta_1+\eta_2=0$ when $x=vt$. In like manner a material perfect reflector sweeps the radiation in front of it, but its molecular constitution is to be such that it allows the aether and its structure to penetrate across it unchanged. For a fuller statement, see *Encyclopaedia Britannica*, ed. 9 or 10, article Radiation.]

thus the gain of energy *per unit time* due to the reflexion is

$$\delta E = (c-v)\tfrac{1}{2}\rho A_2^2 - (c+v)\tfrac{1}{2}\rho A_1^2$$
$$= \tfrac{1}{2}\rho A_1^2 \left\{(c-v)\left(\frac{c+v}{c-v}\right)^2 - (c+v)\right\}$$
$$= \tfrac{1}{2}\rho A_1^2 . 2\frac{c+v}{c-v}v.$$

This change of energy must arise as the work of a pressure P exerted by the moving reflector, namely it is Pv; hence

$$P = \tfrac{1}{2}\rho A_1^2 . 2\frac{c+v}{c-v}.$$

The total energy per unit length, incident and reflected, existing in front of the reflector is

$$E_1 + E_2 = \tfrac{1}{2}\rho A_1^2 + \tfrac{1}{2}\rho A_2^2$$
$$= \tfrac{1}{2}\rho A_1^2 . 2\frac{c^2+v^2}{(c-v)^2}.$$

Hence finally

$$P = (E_1 + E_2)\frac{c^2 - v^2}{c^2 + v^2},$$

becoming equal to the total density of energy $E_1 + E_2$, in accordance with Maxwell's law, when v is small.

If we assume Poynting's modified formula for the pressure of a wave-train against a travelling obstacle, the value ought to be

$$P = E_1\left(1 + \frac{v}{c}\right) + E_2\left(1 - \frac{v}{c}\right);$$

and the truth of this is readily verified.

It may be remarked that, if the relation connecting strain with stress contained quadratic terms, pressural forces such as we are examining would arise in a simple wave-train*. But such a medium would be dispersive, so that a simple train of waves would not travel without change, in contrast to what we know of transmission by the aether of space.

Momentum in convected aethereal fields.

If any transfer of momentum, analogous to what has been here described for the case of a stretched cord, is operative in free aether, the concentration of inertia on which it depends must, as in that case, be determined by and involved in the nature of the strain-system. Now this strain is expressed by the electric field, and therefore by the tubes of electric force. Thus we have to consider in what cases the change of the electric field can be supposed to be produced by convection of the tubes of force†.

* Cf. Poynting, *Roy. Soc. Proc.* vol. 86, A (1912), pp. 534—562, where the pressure exerted by torsional waves in an elastic medium, such as steel, is exhaustively investigated on both the experimental and the mathematical side.

† The considerations advanced in this section were suggested by the study of a passage in J. J. Thomson's *Recent Researches* (1893), § 9 *seq.*

Let the scheme of tubes of force be in motion with velocity varying from point to point, equal at the point (xyz) to (pqr), but without other change. If N be the number of electric tubes enclosed by a fixed circuit, then by Ampère's circuital electrodynamic relation

$$\frac{1}{c^2}\frac{dN}{dt} = \int(\alpha dx + \beta dy + \gamma dz),$$

for the left-hand side is equal to 4π times the total (in this case aethereal) current through the circuit. But if the tubes of the current (uvw) all enter the circuit by cutting across its contour with the velocity (pqr), i.e. if none of them originate *de novo* during the operation, the rate of gain of total current (uvw) is expressed kinematically by

$$-\int\{(qw - rv)\,dx + (ru - pw)\,dy + (pv - qu)\,dz\}.$$

And as the current is here wholly aethereal, $(u, v, w) = (\dot{f}, \dot{g}, \dot{h})$.

The equivalence of these two line integrals, as it holds good for all circuits, requires that

$$-\frac{\alpha}{4\pi} = qh - rg - \frac{d\psi}{dx},$$

$$-\frac{\beta}{4\pi} = rf - ph - \frac{d\psi}{dy},$$

$$-\frac{\gamma}{4\pi} = pg - qf - \frac{d\psi}{dz}.$$

These relations must in fact be satisfied for every field of aethereal strain (fgh) whose changes occur by pure convection.

If permanent magnets are absent, the potential ψ will not enter, and we have then the relations

$$\alpha f + \beta g + \gamma h = 0 \quad \text{and} \quad p\alpha + q\beta + r\gamma = 0.$$

Thus in a convected field the magnetic vector must be at right angles to the electric vector and to the velocity of convection, is in fact 4π times their vector product.

In the case of the stretched cord, the kinetic energy is expressible as that of the convected concentrated mass on the cord. Following that analogy, the kinetic energy is here to be expressed in terms of the velocity of convection (pqr) and the electric field; viz.

$$T = \frac{1}{8\pi}\int(\alpha^2 + \beta^2 + \gamma^2)\,d\tau$$

$$= 2\pi\int\{(qh - rg)^2 + (rf - ph)^2 + (pg - qf)^2\}\,d\tau.$$

Now generally in a dynamical change which occurs impulsively, so that the position of the system is not sensibly altered during the change, if Φ is the component of the impulse corresponding to the coordinate ϕ, the corresponding applied force is $\dot{\Phi}$, and the increase of kinetic energy is equal to the work of this force, viz., to

$$\Sigma\int\dot{\Phi}\,d\phi=\Sigma\int\dot{\Phi}\dot{\phi}\,dt=\Sigma\int\dot{\phi}\,d\Phi;$$

thus
$$\delta T_{\Phi}=\Sigma\dot{\phi}\,\delta\Phi.$$

But $T_{\dot{\phi}}$ is a quadratic function of the velocities: thus T_{Φ} must be a quadratic function of the momenta, and therefore

$$2T=\Sigma\Phi\frac{dT}{d\Phi}=\Sigma\Phi\dot{\phi}.$$

Hence
$$\delta T_{\dot{\phi}}=\Sigma\Phi\delta\dot{\phi}.$$

This argument applies for example in the field of hydrodynamics.

In the present case the Cartesian components of momentum would therefore be $\dfrac{dT}{dp},\dfrac{dT}{dq},\dfrac{dT}{dr}$; so that in volume $\delta\tau$ they are

$$\begin{vmatrix} f & g & h \\ a & b & c \end{vmatrix}\delta\tau.$$

We thus arrive at the same distribution of momentum as the one that has to be associated with the Maxwellian stress system. In this case of supposed pure convection, that momentum is of type having for its x component

$$4\pi p\,(f^2+g^2+h^2)-4\pi f\,(pf+qg+rh).$$

If the motion of the tube of force is wholly transverse, as symmetry would seem to demand, we have the condition

$$pf+qg+rh=0;$$

thus the momentum is now along the direction of the motion (pqr) and belongs to a simple travelling inertia $4\pi\,(f^2+g^2+h^2)$ per unit volume. The convected inertia here suggested is equal to twice the strain energy multiplied by c^{-2}, double what it was in the case of the stretched cord.

The two conditions above introduced

$$pf+qg+rh=0 \text{ and } \alpha f+\beta g+\gamma h=0$$

are equivalent to

$$\frac{f}{q\gamma-r\beta}=\frac{g}{r\alpha-p\gamma}=\frac{h}{p\beta-q\alpha}$$

which, as we shall see presently, may be treated as an indication that the magnetic tubes are convected as well as the electric tubes. Under these circumstances of complete convection, electric and magnetic, it is thus suggested that there is a mechanical momentum in the field, which arises from convection of inertia and is of amount equal to twice the energy of strain multiplied by c^{-2}. But here again restrictions will arise.

Meantime we have to supply the condition that the magnetic lines are simply convected. If n denote the number of unit magnetic tubes that are enclosed by a circuit, then by Faraday's circuital law

$$-\frac{dn}{dt}=\int(Pdx+Qdy+Rdz),$$

where (PQR) is at each point the force exerted per unit charge on an electric particle moving along with the circuit.

Now when the change of the tubes is due to convection solely we must have

$$-\frac{dn}{dt} = \int \{(qc - rb)\,dx + (ra - pc)\,dy + (pb - qa)\,dz\}.$$

These line integrals are therefore equivalent for all circuits fixed in the aether: hence

$$P = qc - rb - \frac{dV}{dx},$$

$$Q = ra - pc - \frac{dV}{dy},$$

$$R = pb - qa - \frac{dV}{dz}.$$

If there are no free charges and the potential V is thus null,

$$4\pi c^2 (fgh) = \begin{vmatrix} p & q & r \\ a & b & c \end{vmatrix}.$$

Thus when the system is completely convected, as regards its electromagnetic activity, and it has no charges, the velocity of convection is at each point at right angles to both the electric and magnetic lines which are at right angles to each other: this velocity is everywhere c, that of radiation, and the ratio of magnetic to electric induction is $4\pi c$. But these are precisely the characteristics of a field of pure radiation, to which alone therefore the preceding argument can possibly apply: and the general argument even in this case is destroyed by the circumstance that (pqr) is restricted to the constant value c, and so is not amenable to variation. But as already seen, if we accept on other grounds this convection of momentum by radiation, the validity of the Maxwell stress will follow.

In the wider case when charges are operative, both these circuital relations for the convective system are satisfied, if $v^2 = p^2 + q^2 + r^2$, by

$$\left(1 - \frac{v^2}{c^2}\right)(P, Q, R) = -\left(\frac{d}{dx}, \frac{d}{dy}, \frac{d}{dz}\right) V,$$

$$(\alpha, \beta, \gamma) = -4\pi\,(qh - rg,\ rf - ph,\ pg - qf).$$

When (pqr) is constant, this is the well known field of a uniformly convected electrostatic system specified by the potential V. But to no other system is it applicable unless it can satisfy the necessary conditions of absence of divergence in (PQR) and $(\alpha\beta\gamma)$.

Frictional resistance to the motion of a radiating body.*

We proceed to examine the retarding force exerted on a body translated through the aether with uniform velocity v, arising from its own radiation. A ray transmitting energy E per unit time pushes backward with a force

* It appears to have escaped notice that Balfour Stewart definitely established, by a qualitative thermodynamic argument based on the disturbance of compensation in the exchanges, as early as 1871, that a moving body must be subject to retardation owing to its own radiation; and that we should 'expect some loss of visible energy in the case of cosmical bodies approaching or receding from one another.' See *Brit. Assoc. Report*, 1871, p. 45.

$\frac{E}{C}\left(1-\frac{v}{C}\cos\theta\right)$. Consider first by themselves the sheaf of nearly parallel rays of natural radiation emitted from all parts of the surface whose directions are

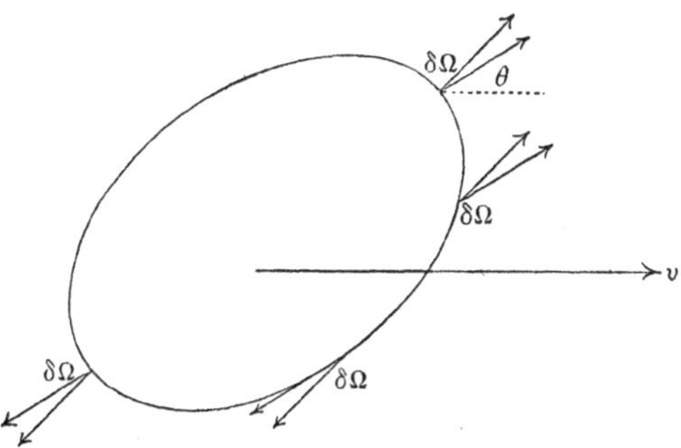

included within the same cone of infinitesimal angle $\delta\Omega$. Their energy (E'), emitted per unit time, as regards the part issuing from an element of surface δS is (p. 207), for a perfect radiator, if ϵ *now* represents natural radiation per unit time,

$$\epsilon\delta\Omega\left(1-\frac{v}{C}\cos\theta\right)^{-1}\times\text{projection of } \delta S \text{ on } \delta\Omega\,;$$

and for the whole surface, taking the front and rear parts separately, they give a back pressure along the rays

$$\frac{\epsilon\delta\Omega}{C}\left(1-\frac{v}{C}\cos\theta\right)^{-1}S_\theta-\frac{\epsilon\delta\Omega}{C}\left(1+\frac{v}{C}\cos\theta\right)^{-1}S_\theta,$$

where S_θ is the projection of the surface on the plane perpendicular to the rays: neglecting $(v/C)^2$ this is

$$2\delta\Omega\,\frac{\epsilon v}{C^2}\cos\theta\,.\,S_\theta\,.$$

If the body is symmetrical around the direction of its translatory convection v, so that S_θ is independent of azimuthal angle, we have $\delta\Omega=-2\pi\,.\,d\cos\theta$; and the aggregate backward pressure opposing the motion of the body is

$$\int_0^1 2\,\frac{\epsilon v}{C^2}\cos^2\theta\,.\,2\pi\,.\,d\cos\theta\,.\,S_\theta\,,$$

that is

$$4\pi\,\frac{v}{C^2}\,\epsilon\int_0^1 S_\theta\cos^2\theta\,.\,d\cos\theta.$$

If the travelling perfect radiator is a sphere of radius a, we have $S_\theta=\pi a^2$, and the force resisting its motion is

$$\tfrac{1}{3}\pi\epsilon\,.\,\frac{v}{C^2}\,.\,4\pi a^2\,*.$$

For a plane radiating disc of area S advancing broadside on, the resisting force is

$$4\pi\,\frac{v}{C^2}\,\epsilon\int S\cos^3\theta\,d\cos\theta,\ \text{which is}\ \pi\,\frac{v}{C^2}\,\epsilon S,$$

namely is $\frac{3}{4}$ of the value for a sphere of the same radius.

* Agreeing with Poynting, *Phil. Trans.* vol. 202, A (1903), p. 551, where important applications in cosmical astronomy are developed.

The natural radiation is more usually defined by R the total radiation per unit area in all directions per unit time: then

$$R = \int \epsilon \cos\phi \, d\Omega \text{ where } d\Omega = -2\pi d\cos\phi,$$

so that

$$\epsilon = \pi^{-1} R.$$

If the radiating body is in an enclosed region whose walls are also convected with the same uniform velocity v, the radiation contained in the region will attain to a steady state. Then the density travelling in the region in each direction (θ) will be equal to that emitted in that direction from a complete radiator; thus it will involve a factor $\left(1 + \frac{v}{C^2}\cos\theta\right)^{-2}$, and so be an aeolotropic distribution.

The separate elements of surface of a perfect radiator will not now maintain a balance of exchanges of radiant energy in their emission and absorption. It may be calculated* that the extra pressure, due to its own radiation, on an element of area δS whose normal makes an angle β with the direction of convection v, is $\delta S \frac{v}{C^2} \epsilon \frac{1}{4}\pi(1 + \cos^2\beta)$, agreeing with the two special cases above. Also the extra radiation emitted by it is $\delta S \, . \, \epsilon \frac{v}{C} \pi \cos\beta$ and the extra radiation absorbed by it is $-\delta S \, . \, \epsilon \frac{v}{C} \pi \cos\beta$; the equilibrium of exchanges is thus vitiated and there is either a compensating flux of heat in the radiator from rear to front, or, if an adiabatic partition is inserted, there is a diminished temperature of the part in front. The same statements apply as regards the front and rear walls of the enclosure itself.

Generalization to forcive on any convected system.

It is of interest to attempt to extend this analysis so as to include the resistance to the motion through aether of any electrodynamic system whatever. The transformation of Lorentz is appropriate to effect this generalization. When that transformation is extended to include the second power of v/C†, the phenomena in any system of electrons at rest or in motion, uniformly convected with velocity v

* As follows:—The extra pressure resolved along v is

$$\delta S \iint \epsilon \frac{v}{C^2} \cos\phi \, (dA \, . \, d\cos\theta) \cos\phi,$$

and the extra radiation of the surface is

$$\delta S \iint \epsilon \frac{v}{C} \cos\phi \, (dA \, . \, d\cos\theta),$$

where A is the azimuth of the ray, and $dA \, . \, d\cos\theta$ is an element of its solid angle referred to the normal n to δS, while

$$\cos\phi = \cos\theta\cos\beta + \sin\theta\sin\beta\cos A.$$

Integrating for A from 0 to 2π and for $\cos\theta$ from 0 to 1, the expressions in the text are obtained.

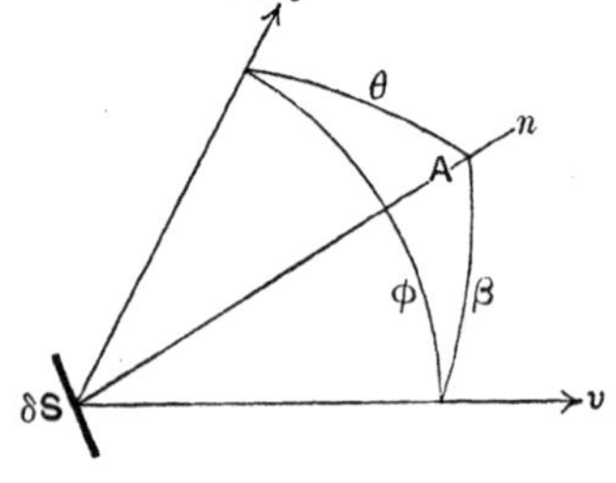

† See *Aether and Matter* (1900), Chapter xi. p. 176; or *Phil. Mag.* June, 1904. That the transformation, with ϵ thus inserted in it, so as to include the second order, is in fact exact so far as regards free space, not merely to the second order which is as far as experiment can go, but to *all* orders of v/C, was pointed out by Lorentz later, thus opening a way for the recent discussions on absolute relativity, an idea which involves of course, on the very threshold, complete negation of any aethereal medium.

parallel to x, are the same as for that system unconvected, provided only that (in electromagnetic units)

$$(f, g, h) \text{ and } (a, b, c) \quad \text{.................................(1)}$$

for the convected system are put equal to the values of

$$\epsilon^{\frac{1}{2}}\left(\epsilon^{-\frac{1}{2}}f,\ g - \frac{v}{4\pi C^2}c,\ h + \frac{v}{4\pi C^2}b\right) \text{ and } \epsilon^{\frac{1}{2}}(\epsilon^{-\frac{1}{2}}a,\ b + 4\pi vh,\ c - 4\pi vg) \ldots(2)$$

for the stationary system, and this change is accompanied by a shrinkage of space and local time involving $\epsilon^{\frac{1}{2}}$ and so of the second order, where $\epsilon = (1 - v^2/C^2)^{-1}$.

Now the force exerted on any system of electrons is determined by and statically equivalent to the Maxwell *quasi*-stress over a boundary enclosing the system, diminished by what would be needed to maintain in the region a distribution of momentum of density $\begin{vmatrix} f & g & h \\ a & b & c \end{vmatrix}$. In free space (abc) and $(\alpha\beta\gamma)$ are the same. The x component of this stress yields a force

$$2\pi C^2 \int \{(f^2 - g^2 - h^2)\, l + 2fg\,.\,m + 2fh\,.\,n\}\, dS$$
$$+ \frac{1}{8\pi}\int \{(\alpha^2 - \beta^2 - \gamma^2)\, l + 2\alpha\beta\,.\,m + 2\alpha\gamma\,.\,n\}\, dS.$$

The effect on this force of the translatory velocity is found by substituting (2) in place of (1). When we neglect as usual terms in $(v/C)^2$, it is (expressed in terms of the stationary system)

$$v\int \{(g\gamma - h\beta)\, l - f\gamma\,.\,m + f\beta\,.\,n\}\, dS + v\int \{(g\gamma - h\beta)\, l + h\alpha\,.\,m - g\alpha\,.\,n\}\, dS,$$

that is,

$$v\int \{2\,(g\gamma - h\beta)\, l + (h\alpha - f\gamma)\, m + (f\beta - g\alpha)\, n\}\, dS^*.$$

Now the rate of increase of the x component of the *quasi*-momentum inside this fixed surface, which (p. 207) represents the pressure of radiation on the system, is, so far as it arises from convection only,

$$v\int (g\gamma - h\beta)\, l\, dS.$$

When this is subtracted there still remains, for this case of a steady convected system, a force of the same order of magnitude parallel to v of amount

$$\frac{v}{C^2}R,$$

where R is the loss of energy per unit time by flux across this fixed surface, which if the body is symmetrical fore and aft can arise only from excess of radiation emitted over radiation received.

* It may be shown that owing to the steady convection the Poynting flux of energy in the field is modified by the following additions:—twice the total density of energy is carried on with velocity v, from which is subtracted twice the electrostatic energy W carried along the electric field with velocity equal to the component of v in that direction, and twice the magnetic energy T carried along the magnetic field with like velocity. And similarly the *quasi*-momentum is altered by that of a mass $2E/C$ convected with the system, of a mass $-2W/C$ convected along the electric field with velocity the component of v, and of a mass $-2T/C$ convected along the magnetic field with like velocity.

It may perhaps be suggested that this excess of the total force on the system, beyond pressure connected with the radiation, arises from increase of effective mass (δm) of the source itself, owing to loss of radiant energy from it; for this would involve an increase of momentum $v\delta m$, to be supplied by impressed force inside the radiator itself, if the velocity is to be maintained. A loss of energy E would thus increase the effective mass of the system by E/C^2*, owing presumably to resulting change in minute internal configuration. But Poynting's astronomical calculation as to the time of clearance of cosmical dust from celestial spaces would stand, as the changes of thermal content that are there possible cannot introduce any sensible change of effective inertia†.

But these second order structural phenomena appear to be still obscure.

In final illustration of the principle that *quasi*-momentum is somehow transmitted along the rays, I take this opportunity formally to correct two statements occurring in a paper 'On the Intensity of the Natural Radiation from moving bodies and its Mechanical Reaction'; *Boltzmann Festschrift*, 1904, p. 591, or *Phil. Mag.*, May 1904, p. 578.

It was shown there that when radiation is incident directly on a reflector it exerts no force on the surface layer, unless indeed the conductivity is so great that we can regard this layer as containing a current sheet. But this does not prove absence of radiation pressure unless there is conduction. The argument from a beam of limited length, on p. 207, is decisive on that point, so long as the principles of the theory of electrons are maintained. The correct inference is that the radiation pressure in a parallel beam is *transmitted* without change, unless where the existence of conductivity gives rise to electric flow. Thus it is the Ampèrean force acting on the current sheet, induced in the surface of a good conductor, that *equilibrates* the radiant pressure advancing with the incident beam, and prevents its transmission beyond that surface.

The attempted disproof, in a postscript, of Poynting's modification of the law of pressure of radiation for a moving body, is also at fault, and must be replaced by the general discussion just given (p. 215).

* The writer is reminded (by a summary in W. Schottky, 'Zur Relativtheoretischen Energetik und Dynamik,' *Berlin Dissertation*, 1912, p. 6) that this relation, obtained by Einstein in 1905, expanded in 1907 by Planck, Hasenöhrl, Laue, *etc.*, is fundamental in the generalized relativity theory. It may be permitted to mention that the analysis in the text is taken, modified by various corrections, from private correspondence with Poynting relating to his *Phil. Trans.* memoir of 1903 on resistance due to radiation, which has been already referred to.

† On similar principles, the pressure of solar radiation on the earth would produce, owing to its orbital motion, an increase of the secular retardation of the Earth's diurnal rotation, which however proves to be only about a [ten] thousandth part of the amount astronomically in question.

APPENDIX

I have received from Prof. Levi-Civita a more general discussion of the mechanical side of the problem of radiant pressure, which I have his permission to append here.

SIR J. LARMOR'S MECHANICAL MODEL OF THE PRESSURE OF RADIATION

[From a letter of Prof. T. LEVI-CIVITA to Prof. SIR J. LARMOR.]

. . . .

I have perused your beautiful lecture "On the dynamics of radiation" which I was fortunate to hear in Cambridge. Will you allow me to present in a little more general aspect your idea leading to a mechanical model of the pressure of waves? I shall refer myself, as you do, to a vibrating string.

1. *Specification of the assumptions.—Flow of energy.—Pressure on moving end.*

The small transverse oscillations of a stretched cord in absence of bodily forces, whatever the initial and boundary conditions may be, follow the equation

$$\rho \frac{\partial^2 \eta}{\partial t^2} = T \frac{\partial^2 \eta}{\partial x^2} \quad \text{..............................(1)},$$

where $\eta(x, t)$ is the displacement, and ρ and T are constants of well-known signification. As usual I shall write c^2 for T/ρ, c thus designating the velocity of propagation of transverse waves along the string.

Let us suppose that the undulations η extend only to a finite (variable) portion

of our string: from a moving end (reflector) R, at which $x = vt$, to a fixed S, at which $x = b$ (v, b positive constants).

The condition of (perfect) reflexion at R shall be further introduced. Independently from it, we may specify in usual way the energetic point of view.

The density of energy, both kinetic and potential (at any place x, between R and S, and time t) is

$$e = \tfrac{1}{2}\rho\dot{\eta}^2 + \tfrac{1}{2}T\left(\frac{\partial \eta}{\partial x}\right)^2 = \tfrac{1}{2}\rho\left\{\dot{\eta}^2 + c^2\left(\frac{\partial \eta}{\partial x}\right)^2\right\} \quad \text{.................(2)},$$

where the dot stands for $\partial/\partial t$. Especially, in front of the reflector, it becomes

$$e_R = [e]_{x=vt} \quad \text{..................................(3)}.$$

The energy stored in the whole extent of the disturbed string at the time t amounts therefore to

$$E = \int_{vt}^{b} e\,dx \quad \text{..................................(4)},$$

from which we easily get an expression of dE/dt convenient for our aim. It is in fact

$$\frac{dE}{dt} = - e_R v + \int_{vt}^{b} \frac{\partial e}{\partial t}\, dx.$$

But, by (2),
$$\frac{\partial e}{\partial t} = \rho \left\{ \dot{\eta}\, \frac{\partial^2 \eta}{\partial t^2} + \frac{\partial \eta}{\partial x}\frac{\partial \dot{\eta}}{\partial x} \right\}.$$

Since, by partial integration,

$$\int_{vt}^{b} \frac{\partial \eta}{\partial x}\frac{\partial \dot{\eta}}{\partial x}\, dx = \left[\frac{\partial \eta}{\partial x}\, \dot{\eta}\right]_{vt}^{b} - \int_{vt}^{b} \dot{\eta}\, \frac{\partial^2 \eta}{\partial x^2}\, dx,$$

and $\dot{\eta}$ vanishes at the fixed end $S\,(x=b)$, it remains

$$\int_{vt}^{b} \frac{\partial e}{\partial t}\, dx = \rho \int_{vt}^{b} \dot{\eta} \left\{ \frac{\partial^2 \eta}{\partial t^2} - c^2 \frac{\partial^2 \eta}{\partial x^2} \right\} dx - \rho c^2 \left[\frac{\partial \eta}{\partial x}\, \dot{\eta}\right]_{x=vt}.$$

On account of the fundamental equation (1), this reduces to the last term, and gives

$$\frac{dE}{dt} = - e_R v - \rho c^2 \left[\frac{\partial \eta}{\partial x}\, \dot{\eta}\right]_{x=vt}.$$

Putting
$$f_R = - \rho c \left[\frac{\partial \eta}{\partial x}\, \dot{\eta}\right]_{x=vt} \quad \text{.............................(5)},$$

the formula may be written

$$\frac{dE}{dt} = - e_R v + f_R c, \quad \text{...........................(6)},$$

or also
$$\frac{dE}{dt} = P v \quad \text{..................................(6')},$$

where
$$P = - e_R + f_R \cdot \frac{c}{v} \quad \text{..............................(7)}.$$

The formulae (6) and (6′) are capable of expressive interpretations.

Let us firstly pay attention to the formula (6), supposing $v=0$ (R fixed). It means that the exchanges of energy between RS and the outside take place as if a flow f_R (directed inward if positive) passed through R with the wave-velocity c. We recognize obviously the one-dimensional form of the Poynting-Volterra investigations.

In the general case where v is not zero, the wave-flow f_R must be increased by the convection-flow $-e_R$, travelling with the velocity v, that is—we may say—convected by the moving end R.

To get the interpretation of the (equivalent) formula (6′), we have only to recall the principle of conservation of energy in its pure mechanical form. It states that dE/dt, for any material system (the string in our case) must be equal to the time-rate of doing work of all external forces. At the present no external forces act on the system, except at the ends of the disturbed portion, arising from the connections: with the reflector at R, with some fixed body at S. But the last does not do work because S is at rest.

Hence, in the equation (6′), P means the force exerted *on* the considered system by the reflector, the positive sense being of course that of the increasing x. Reversing the positive sense and availing ourselves of the principle of reaction, we may also regard P *as the pressure supported by the advancing reflector* (traction if P should result negative).

2. *Adiabatic arrangement.*

The formula (6) and its consequences have been deduced on the hypothesis that the cord is fixed at S, so that $\frac{\partial \eta}{\partial x}\dot{\eta}$ vanishes for $x=b$. If it be not so, we have in the second member of (6) a further term $-cf_S$, where

$$-f_S = \rho c \left[\frac{\partial \eta}{\partial x}\dot{\eta}\right]_{x=b}.$$

This would introduce a flow of energy across S, to be considered together with the flow across R; the preceding argument would therefore be altered.

There is however an obvious arrangement for which the formula equally holds: it consists in admitting a proper supply of energy at S, just as it is required to compensate the flow $-cf_S$. The connection at S, between RS and the outside, may then be called adiabatic. We shall henceforth adopt this assumption, getting thus free from the more restrictive one of a fixed end.

3. *Decomposition of the disturbance in two wave-trains.—Perfect reflexion.*

If the solution η of (1) is a function (any whatever) of $x+ct$, we have the case of waves advancing to the reflector R. Then $c\frac{\partial \eta}{\partial x}=\dot{\eta}$, and f_R becomes identical with $-e_R=\rho\dot{\eta}^2$, giving to (6) the form

$$\frac{dE}{dt} = -e_R(c+v):$$

the flow of energy occurs as if the waves were carrying their energy with the (absolute) velocity c, i.e. $c+v$ relative to the reflector.

For a train $\eta(x-ct)$ (reflected from R), we find in analogous way the flow $e_R(c-v)$.

Now any solution of (1) has the form

$$\eta = \eta_1(x+ct) + \eta_2(x-ct) \quad \text{.........................(8)},$$

η_1, η_2 being arbitrary functions of their respective arguments. We get, accordingly,

$$\dot{\eta} = \dot{\eta}_1 + \dot{\eta}_2,$$

$$c\frac{\partial \eta}{\partial x} = c\left(\frac{\partial \eta_1}{\partial x}+\frac{\partial \eta_2}{\partial x}\right) = \dot{\eta}_1 - \dot{\eta}_2;$$

therefore, from (2),
$$e = \rho(\dot{\eta}_1^2+\dot{\eta}_2^2) \quad \text{..............................(2')},$$

and, from (5)
$$f_R = \rho[\dot{\eta}_2^2-\dot{\eta}_1^2]_{x=vt} \quad \text{............................(5')}.$$

With these values the expression (6) of dE/dt may be written

$$\frac{dE}{dt} = \rho\dot{\eta}_2^2(c-v) - \rho\dot{\eta}_1^2(c+v) \quad \text{......................(6'')}.$$

Thus the gain of energy appears caused by flows (relative to R) of the energies carried by the two opposite wave-trains. It is your favourite point of view.

Now we proceed to the condition of perfect reflexion at R. You properly conceive the reflector to be realised by a plate with a hole through which the cord passes. As the plate advances along the cord, it sweeps the waves in front, restoring behind it the resting straight configuration of the cord. Under these circumstances the condition at R is obviously that the total displacement shall be annulled, that is

$$\eta_1 + \eta_2 = 0 \quad \text{for} \quad x = vt \qquad (9).$$

Having thus achieved the general premisses, a mathematical observation may find place, viz., that it would not be difficult to determine functions η_1, η_2 of their respective arguments $x + ct$, $x - ct$, satisfying rigorously to the nodal condition $\eta_1 + \eta_2 = 0$ as well for $x = vt$ as for $x = b$. But I propose only to apply the above to your particular solution.

4. *Case of simple wave-trains.—Mean pressure.*

You assume

$$\left.\begin{aligned} \eta_1 &= \frac{A_1}{m_1 c} \sin m_1 (x + ct) \\ \eta_2 &= -\frac{A_2}{m_2 c} \sin m_2 (x - ct) \end{aligned}\right\} \qquad (10),$$

and consequently

$$\left.\begin{aligned} \dot{\eta}_1 &= A_1 \cos m_1 (x + ct) \\ \dot{\eta}_2 &= A_2 \cos m_2 (x - ct) \end{aligned}\right\} \qquad (11),$$

A_1, m_1, A_2, m_2 being constants to be disposed with regard to (9).

It requires

$$A_2 (c - v) = -A_1 (c + v) \qquad (12),$$

$$m_2 (c - v) = \pm m_1 (c + v) \qquad (13).$$

With the determination (10) of $\dot{\eta}_1$, $\dot{\eta}_2$, the expression (2′) of e becomes a sum of two periodic functions of x, and of t. Its average value $\bar{e}$, with respect to x as well as to t, is

$$\bar{e} = \tfrac{1}{2} \rho (A_1^2 + A_2^2) \qquad (14).$$

The same value belongs to the time average of $e_R = (e)_{x=vt}$. On the other hand, averaging the expression (5′) of f_R, we have

$$\bar{f}_R = \tfrac{1}{2} \rho (A_1^2 - A_2^2) \qquad (15).$$

But, by (12),

$$A_2^2 = \frac{(c + v)^2}{(c - v)^2} A_1^2;$$

hence, from (14),

$$\bar{e} = \rho \frac{c^2 + v^2}{(c - v)^2} A_1^2,$$

and, from (15),

$$\bar{f}_R = -2\rho \frac{cv}{(c - v)^2} A_1^2 = -\frac{2cv}{c^2 + v^2} \bar{e}.$$

We finally arrive at the pressure P defined by (7). Its mean value (e_R being identical with e) becomes

$$\bar{P} = -\bar{e} - \frac{c}{v} \bar{f}_R = \frac{c^2 - v^2}{c^2 + v^2} \bar{e},$$

which is your result.

. . . .

UNIVERSITY OF PADUA, *Oct.* 9.

COMMUNICATIONS

SECTION I

(ARITHMETIC, ALGEBRA, ANALYSIS)

SOME PROBLEMS OF DIOPHANTINE APPROXIMATION

By G. H. Hardy and J. E. Littlewood.

1. Let us denote by $[x]$ and (x) the integral and fractional parts of the real number x, so that

$$(x) = x - [x], \quad 0 \leqq (x) < 1.$$

Let θ be an irrational number, and α any number between 0 and 1 (0 included). Then it is well known that it is possible to find a sequence of positive integers $n_1, n_2, n_3, \ldots$ such that

$$(n_r \theta) \to \alpha$$

as $r \to \infty$. Now let $f(n)$ denote a positive increasing function of n, integral when n is integral, such as

$$n, n^2, n^3, \ldots, 2^n, 3^n, \ldots, n!, 2^{n^2}, \ldots, 2^{2^n}, \ldots,$$

and let f_r denote the value of $f(n)$ for $n = n_r$. The result just stated suggests the following question, which seems to be of considerable interest:—*For what forms of $f(n)$ is it true that, for any irrational value of θ, and any value of α such that $0 \leqq \alpha < 1$, a sequence n_r can be found such that*

$$(f_r \theta) \to \alpha\,?$$

It is easy to see that, when the increase of $f(n)$ is sufficiently rapid, the result suggested will not generally be true. Thus, if $f(n) = 2^n$, and θ is a number which, when expressed in the binary scale, shows at least k 0's following upon every 1, it is plain that

$$(2^n \theta) < \tfrac{1}{2} + \lambda_k,$$

where λ_k is a number which can be made as small as we please by increasing k sufficiently. There is thus an "excluded interval" of values of α, the length of which can be made as near to $\frac{1}{2}$ as we please. If $f(n) = 3^n$ we can obtain an excluded interval whose length is as near $\frac{2}{3}$ as we please, and so on, while if $f(n) = n!$ it is (as is well known) possible to choose θ so that $(n!\,\theta)$ has a unique limit. Thus

$$(n!\,e) \to 0.$$

2. The first object of this investigation has been to prove the following theorem:—

Theorem 1. *If $f(n)$ is a polynomial in n, with integral coefficients, then a sequence can be found for which $(f_r \theta) \to \alpha$.*

We shall give the proof in the simple case in which

$$f(n) = n^2,$$

a case which is sufficient to exhibit clearly the fundamental ideas of our analysis. Our argument is based on the following general principle, which results from the work of Pringsheim and London on double sequences and series*:

If $$f_{r,s},\quad \phi_{r,s},\quad \dots$$

are a finite number of functions of the positive integral variables r, s; and if

$$\lim_{s\to\infty}\lim_{r\to\infty} f_{r,s} = a,\quad \lim_{s\to\infty}\lim_{r\to\infty} \phi_{r,s} = b,\ \dots;$$

then we can find a sequence of pairs of numbers

$$(r_1, s_1),\quad (r_2, s_2),\quad (r_3, s_3)\ \dots$$

such that $r_i \to \infty$, $s_i \to \infty$ *and* $f_{r_i, s_i} \to a$, $\phi_{r_i, s_i} \to b, \dots,$ *as* $i \to \infty$.

We shall first apply this principle to prove that a sequence n_r can be found so that

$$(n_r\theta) \to 0,\quad (n^2{}_r\theta) \to 0$$

simultaneously. We shall, in the argument which follows, omit the brackets in $(n_r\theta)$, etc., it being understood always that integers are to be ignored.

We can choose a sequence n_r so that $n_r\theta \to 0$. The corresponding values $n^2{}_r\theta$ are infinite in number, and so have at least one limiting point ξ; ξ may be positive or zero, rational or irrational. We can (by restricting ourselves to a subsequence of the n_r's) suppose that

$$n_r\theta \to 0,\quad n^2{}_r\theta \to \xi.$$

If $\xi = 0$, we have what we want. If not we write

$$f_{r,s} = (n_r + n_s)\,\theta,\quad \phi_{r,s} = (n_r + n_s)^2\,\theta.$$

Then
$$\lim_{s\to\infty}\lim_{r\to\infty} f_{r,s} = \lim_{s\to\infty} n_s\theta = 0,$$

$$\lim_{s\to\infty}\lim_{r\to\infty} \phi_{r,s} = \lim_{s\to\infty} (\xi + n^2{}_s\theta) = 2\xi.$$

Hence, by the general principle, we can pick out a new sequence p_r such that

$$p_r\theta \to 0,\quad p^2{}_r\theta \to 2\xi.$$

Repeating the argument, with $n_r + p_s$ in the place of $n_r + n_s$, we are led to a sequence q_r such that

$$q_r\theta \to 0,\quad q^2{}_r\theta \to 3\xi;$$

and it is plain that by proceeding in this way sufficiently often we can arrive at a sequence $n_{r,k}$ such that

$$n_{r,k}\,\theta \to 0,\quad n^2{}_{r,k}\,\theta \to k\xi,$$

for any integral value of k.

Now whatever number ξ is, rational or irrational, we can find a sequence k_s such that

$$k_s\xi \to 0$$

as $s \to \infty$. Then

$$\lim_{s\to\infty}\lim_{r\to\infty} n_{r,k_s}\,\theta = \lim_{s\to\infty} 0 = 0,$$

$$\lim_{s\to\infty}\lim_{r\to\infty} n^2{}_{r,k_s}\,\theta = \lim_{s\to\infty} k_s\xi = 0.$$

* Pringsheim, *Sitzungsberichte der k. b. Akademie der Wiss. zu München*, vol. 27, p. 101, and *Math. Annalen*, vol. 53, p. 289; London, *Math. Annalen*, *ibid.*, p. 322.

Applying the general principle once more we deduce a sequence of values of n for which $(n\theta) \to 0$, $(n^2\theta) \to 0$ simultaneously.

When we have proved that there is a sequence n_r for which $n^2{}_r\theta \to 0$, it is very easy to define a sequence $\nu_r n_r$, where ν_r is an integer depending on r, which gives any arbitrary α as a limit. We thus complete the proof of Theorem 1 in the case $f(n) = n^2$. An analogous method may be applied in the case of the general power n^k. As in the course of this proof we obtain a sequence for which

$$n\theta \to 0, \quad n^2\theta \to 0, \quad \ldots, \quad n^k\theta \to 0$$

simultaneously, we thus prove the theorem when $\alpha = 0$ for the general polynomial $f(n)$. The extension to the case $\alpha > 0$ may be effected on the same lines as in the case $f(n) = n^k$, but it is more elegant to complete the proof by means of the theorems of the next section.

It may be observed that the relation

$$n\theta \to 0$$

may be satisfied *uniformly* for all values of θ, rational or irrational; that is to say, given any positive ϵ, a number $N(\epsilon)$ can be found such that

$$n\theta < \epsilon$$

for every θ and some n, which depends on ϵ and θ but is less that $N(\epsilon)$. Similar results may be established for $n^2\theta$, $n^3\theta$, ... The chief interest of this result lies in the fact that it shows that there must be *some* function $\phi(n)$, independent of θ, which tends to zero as $n \to \infty$ and is such that for every θ there is an infinity of values of n for which

$$n^2\theta < \phi(n)^*.$$

3. The following generalisation of the theorem quoted at the beginning of § 1 was first proved by Kronecker†:—

If $\theta, \phi, \psi, \ldots$ are any number of linearly independent irrationals (i.e. if no relation of the type

$$a\theta + b\phi + c\psi + \ldots = 0,$$

where $a, b, c, \ldots$ are integers, not all zero, holds between $\theta, \phi, \psi, \ldots$), and if $\alpha, \beta, \gamma, \ldots$ are any numbers between 0 and 1 (0 included), then a sequence n_r can be found such that

$$n_r\theta \to \alpha, \quad n_r\phi \to \beta, \quad n_r\psi \to \gamma, \quad \ldots.$$

This theorem, together with the results of § 2, at once suggest the truth of the following theorem:—

Theorem 2. *If $\theta, \phi, \psi, \ldots$ are linearly independent irrationals, and*

$$\alpha_l, \beta_l, \gamma_l, \ldots \; (l = 1, 2, \ldots, k)$$

* It is well known that, in the case of $n\theta$, $\phi(n)$ may be taken to be $1/n$. No such simple result holds when $\alpha > 0$: exception has to be made of certain aggregates of values of θ. On the other hand, if θ is a fixed *irrational*, the relation $n\theta \to \alpha$ holds uniformly *with respect to* α. All these results suggest numerous generalisations.

† *Werke*, vol. 3, p. 31. The theorem has been rediscovered independently by various authors, e.g. by Borel, F. Riesz, and Bohr (see for example Borel, *Leçons sur les séries divergentes*, p. 135, and F. Riesz, *Comptes Rendus*, vol. 139, p. 459).

k sets of numbers all lying between 0 and 1 (0 included), then it is possible to find a sequence of values of n for which

$$n\theta \to \alpha_1,\quad n\phi \to \beta_1,\quad n\psi \to \gamma_1,\quad \ldots,$$
$$n^2\theta \to \alpha_2,\quad n^2\phi \to \beta_2,\quad n^2\psi \to \gamma_2,\quad \ldots,$$
$$\ldots\ldots\ldots\ldots\ldots\ldots\ldots\ldots\ldots$$
$$n^k\theta \to \alpha_k,\quad n^k\phi \to \beta_k,\quad n^k\psi \to \gamma_k,\quad \ldots.$$

This theorem we prove by means of two inductions, the first from the case of k sets $\alpha_l, \beta_l, \gamma_l, \ldots$ to the case of $k+1$ sets *in which the numbers of the last set are all zero*, the second from this last case to the general case of $k+1$ sets. The principles which we employ do not differ from those used in the proof of the simpler propositions discussed in § 2.

4. The investigations whose results are summarised in the preceding sections were originally begun with the idea of obtaining further light as to the behaviour of the series

$$\Sigma e^{n^2\theta\pi i},\quad \Sigma e^{n^3\theta\pi i},\quad \ldots$$

from the point of view of convergence, summability, and so forth. If we write*

$$s_n^{(2)} = \sum_{\nu \leqq n} e^{(\nu-\frac{1}{2})^2\theta\pi i},\quad s_n^{(3)} = \sum_{\nu \leqq n} e^{\nu^2\theta\pi i},\quad s_n^{(4)} = \sum_{\nu \leqq n} (-1)^{\nu-1} e^{\nu^2\theta\pi i}$$

it is obvious that, if s_n is any one of $s_n^{(2)}, \ldots$, then $s_n = O(n)$. If θ is *rational*, either $s_n = O(1)$ or $s_n = A_n + O(1)$, where A is a constant: the cases may be differentiated by means of the well known formulae for "Gauss's sums." Similar remarks apply to the higher series in which (e.g.) ν^2 is replaced by $\nu^3, \nu^4, \ldots$. The results of the preceding sections have led us to a proof of

Theorem 3. *If θ is irrational, then $s_n = o(n)$: the same result is true for the corresponding higher sums.*

The argument by which we prove this theorem has a curious and unexpected application to the theory of the Riemann ζ-function; it enables us to replace Mellin's result $\zeta(1+ti) = O(\log|t|)$† by

$$\zeta(1+ti) = o(\log|t|).$$

Theorem 4. *Theorem 3 is the best possible theorem of its kind, that is to say the $o(n)$ which occurs in it cannot be replaced by $O(n\phi)$, where ϕ is any definite function of n, the same for all θ's, which tends to zero as $n \to \infty$.*

But although Theorem 3 contains the most that is true for *all* irrational θ's, it is possible to prove much more precise results for special classes of θ's. Here we use methods of a less elementary (though in reality much easier) type than are required for Theorem 3, the proof of which is intricate.

In Chap. 3 of his *Calcul des Résidus*‡ M. Lindelöf gives a very elegant proof of the formula

$$\sum_0^{q-1} e^{\nu^2 p\pi i/q} = \sqrt{\left(\frac{iq}{p}\right)} \sum_0^{p-1} e^{-\nu^2 q\pi i/p}$$

* The notation is chosen so as to run parallel with Tannery and Molk's notation for the ϑ-functions: n is not necessarily an integer.

† Landau, *Handbuch der Lehre von der Verteilung der Primzahlen*, p. 167.

‡ pp. 73 *et seq.*

of Genocchi and Schaar. Here p and q are integers of which one is even and the other odd. By a suitable modification of Lindelöf's argument, we establish the formula

$$s_n^{(\lambda)}(\theta)=\sqrt{\left(\frac{i}{\theta}\right)}\,s_{n\theta}^{(\lambda_1)}\left(-\frac{1}{\theta}\right)+\frac{O(1)}{\sqrt{\theta}},$$

where θ is an irrational number, which we may suppose to lie between -1 and 1, λ is one of 2, 3, 4, λ_1 a corresponding one of the same numbers, and $O(1)$ stands for a function of n and θ less in numerical value than an absolute constant.

We observe also that the substitution of $\theta+1$ for θ merely permutes the indices 2, 3, 4, and that the substitution of $-\theta$ for θ changes s_n into its conjugate. If now we write θ in the form of a simple continued fraction

$$\frac{1}{a_1+}\,\frac{1}{a_2+}\,\frac{1}{a_3+}\dots,$$

and put

$$\theta=\frac{1}{a_1+\theta_1},\quad \theta_1=\frac{1}{a_2+\theta_2},\quad \dots$$

we obtain

$$\begin{aligned}s_n^{(\lambda)}(\theta)&=\sqrt{\left(\frac{i}{\theta}\right)}\,s_{n\theta}^{(\lambda_1)}(-\theta_1)+\frac{O(1)}{\sqrt{\theta}}\\&=\sqrt{\left(\frac{1}{\theta\theta_1}\right)}\,s_{n\theta\theta_1}^{(\lambda_2)}(\theta_2)+O(1)\left\{\frac{1}{\sqrt{\theta}}+\frac{1}{\sqrt{(\theta\theta_1)}}\right\}\\&=\sqrt{\left(\frac{i}{\theta\theta_1\theta_2}\right)}\,s_{n\theta\theta_1\theta_2}^{(\lambda_3)}(-\theta_3)+O(1)\left\{\frac{1}{\sqrt{\theta}}+\frac{1}{\sqrt{(\theta\theta_1)}}+\frac{1}{\sqrt{(\theta\theta_1\theta_2)}}\right\},\end{aligned}$$

and so on. We can continue this process until $n\theta\theta_1\theta_2\dots<1$, when the first term vanishes, and we are left with an upper limit for $|s_n|$ the further study of which depends merely on an analysis of the continued fraction.

We thus arrive at easy proofs of Theorems 3 and 4 for $k=2$. We can also prove

Theorem 5. *If the partial quotients a_n of the continued fraction for θ are limited* then $s_n(\theta)=O(\sqrt{n})$. In particular this is true if θ is a quadratic surd, pure or mixed.*

5. The question naturally arises whether Theorem 5 is the best possible of its kind. The answer to this question is given by

Theorem 6. *If θ is any irrational number, it is possible to find a constant H and an infinity of values of n such that*

$$|s_n(\theta)|>H\sqrt{n}.$$

The same is true of all Cesàro's means formed from the series.

The attempt to prove this theorem leads us to a problem which is very interesting in itself, namely that of the behaviour of the modular functions

$$\Sigma q^{(n-\frac{1}{2})^2},\quad \Sigma q^{n^2},\quad \Sigma(-1)^{n-1}q^{n^2}$$

as q tends along a radius vector to an "irrational place" $e^{\theta\pi i}$ on the unit circle. If $f(q)$ denotes any one of these functions, it is trivial that

$$f(q)=O\{(1-|q|)^{-\frac{1}{2}}\}.$$

* This hypothesis may be generalised widely.

If q tends to a *rational* place, it is known that $f(q)$ tends to a limit or becomes definitely infinite of order $\frac{1}{2}$. By arguments depending upon the formulae of transformation of the ϑ-functions, and similar in principle to, though simpler than, those of § 4, we prove

Theorem 7. *When q tends to any irrational place on the circle of convergence,*

$$f(q) = o\{(1-|q|)^{-\frac{1}{2}}\}.$$

No better result than this is true in general. If $q \to e^{\theta\pi i}$, where θ is one of the irrationals defined in Theorem 5, then

$$f(q) = O\{(1-|q|)^{-\frac{1}{4}}\}.$$

Further, whatever be the value of θ, we can find a constant H and an infinity of values of $|q|$, tending to unity, such that

$$|f(q)| > H\{(1-|q|)^{-\frac{1}{4}}\}.$$

In so far as these results assign *upper* limits for $|f(q)|$, they could be deduced from our previous theorems. But the remaining results are new, and Theorem 6 is a corollary of the last of them. Another interesting corollary is

Theorem 8. *The series*

$$\Sigma n^{-\alpha} e^{(n-\frac{1}{2})^2\theta\pi i}, \quad \Sigma n^{-\alpha} e^{n^2\theta\pi i}, \quad \Sigma (-1)^n n^{-\alpha} e^{n^2\theta\pi i},$$

where θ is irrational, and $\alpha \leqq \frac{1}{2}$, can never be convergent, or summable by any of Cesàro's means.

On the other hand, if $\alpha > \frac{1}{2}$, these series are each certainly convergent for an everywhere dense set of values of θ. They are connected with definite integrals of an interesting type: for example

$$\sum_1^\infty \frac{(-1)^{n-1}}{n} e^{n^2\theta\pi i} = \sqrt{\left(\frac{i}{\pi}\right)} \int_0^\infty e^{-ix^2} \log(4\cos^2 \varpi x)\, dx,$$

where $\varpi = \sqrt{(\theta\pi)}$, whenever the series is convergent.

6. We have also considered series of the types $\Sigma(n\theta)$, $\Sigma(n^2\theta)$, ... It is convenient to write

$$\{n\theta\} = (n\theta) - \tfrac{1}{2}, \quad s_n = \sum_{\nu \leqq n} \{\nu\theta\}.$$

Arithmetic arguments analogous to those used in proving Theorems 3 and 4 lead to

Theorem 9. *If θ is any irrational number, then $s_n = o(n)$. The same result holds for the series in which ν is replaced by $\nu^2, \nu^3, \ldots, \nu^k, \ldots$*. Further, this result is the best possible of its kind.*

* This result, in the case $k=1$, has (as was kindly pointed out to us by Prof. Landau) been given by Sierpinski (see the *Jahrbuch über die Fortschritte der Math.*, 1909, p. 221). Similar results hold for the function

$$x + a - [x + a] - \tfrac{1}{2}$$

which reduces to $\{x\}$ for $a = 0$.

When $k=1$, we can obtain more precise results analogous to those of §§ 4, 5. The series $\Sigma\{n\theta\}$ behaves, in many ways, like the series $\Sigma e^{n^2\theta\pi i}$. The *rôle* of the formula of Genocchi and Schaar is now assumed by Gauss's formula

$$\sum_{1}^{\frac{1}{2}(q-1)}\left[\frac{\nu p}{q}\right]+\sum_{1}^{\frac{1}{2}(p-1)}\left[\frac{\nu q}{p}\right]=\tfrac{1}{4}(p-1)(q-1),$$

where p, q are odd integers. Taking this formula as our starting point we easily prove Theorem 9 in the case $k=1$. Further, we obtain

Theorem 10. *If θ is an irrational number of the type defined in Theorem* 5, *then $s_n=O(\log n)$.*

This corresponds to Theorem 5. When we come to Theorem 6 the analogy begins to fail. We are not able to show that, for every irrational θ (or even for every θ of the special class of Theorem 5), s_n is sometimes effectively of the order of $\log n$. The class in question includes values of θ for which this is so, but, for anything we have proved to the contrary, there may be values of θ for which $s_n=O(1)$. And when we consider, instead of s_n, the corresponding Cesàro mean of order 1, this phenomenon does actually occur. While engaged on the attempt to elucidate these questions we have found a curious result which seems of sufficient interest to be mentioned separately. It is that

$$\sum_{\nu\leqq n}\{\nu\theta\}^2=\tfrac{1}{12}n+O(1)$$

for all irrational values of θ. When we consider the great irregularity and obscurity of the behaviour of $\Sigma\{\nu\theta\}$, it is not a little surprising that $\Sigma\{\nu\theta\}^2$ (and presumably the corresponding sums with higher *even* powers) should behave with such marked regularity.

7. The exceedingly curious results given by the transformation formulae for the series $\Sigma e^{n^2\theta\pi i}$, $\Sigma\{n\theta\}$ suggest naturally the attempt to find similar formulae for the higher series. It is possible, by a further modification of Lindelöf's argument, to obtain a relation between the two sums

$$\sum^{n} e^{\nu^3\theta\pi i},\quad \sum^{m}\mu^{-\frac{1}{4}}e^{-K\mu^{3/2}\pi i},$$

where $K=\sqrt{(32/27\theta)}$. The relation thus obtained gives no information about the first series that is not trivial. We can however deduce the non-trivial result

$$\sum_{1}^{n} e^{\nu^{3/2}\theta\pi i}=O(n^{\frac{3}{4}}).$$

Similar remarks apply to the higher series $\Sigma e^{n^k\theta\pi i}$ and to the series $\Sigma\{n^k\theta\}$, where $k>1$. But it does not seem probable that we can make much progress on these lines with any of our main problems.

In conclusion we may say that (with the kind assistance of Dr W. W. Greg, Librarian of Trinity College, and Mr J. T. Dufton, of Trinity College) we have tabulated the values of $(n^2\theta)$ for the first 500 values of n, in the cases

$$\theta=\frac{1}{\sqrt{10}}=\cdot 31622776\ldots,\ \theta=e.$$

The distribution of these values shows striking irregularities which encourage a closer scrutiny.

ON THE FUNDAMENTAL FUNCTIONAL OPERATION OF A GENERAL THEORY OF LINEAR INTEGRAL EQUATIONS

By Eliakim Hastings Moore.

1. The General Linear Integral Equation G. The Foundation Σ_5 of the General Theory.

In a memoir "On the Foundations of the Theory of Linear Integral Equations," published in the April 1912 number of the *Bulletin of the American Mathematical Society* (ser. 2, vol. 18, pp. 334—362), I have indicated the foundations, that is, the terminology or basis, and the system of postulates, of a theory of the general linear integral equation

$$\xi = \eta - zJ\kappa\eta \quad \text{..............................(G)},$$

including, as special instances, the regular cases of the equations

$$\xi(s) = \eta(s) - z\sum_{t=1}^{n} \kappa(s, t)\,\eta(t) \quad (s = 1, \ldots, n) \text{...........}(\mathrm{II}_n);$$

$$\xi(s) = \eta(s) - z\sum_{t=1}^{\infty} \kappa(s, t)\,\eta(t) \quad (s = 1, 2, \ldots) \text{...........(III)};$$

$$\xi(s) = \eta(s) - z\int_a^b \kappa(s, t)\,\eta(t)\,dt \quad (a \leqq s \leqq b) \text{...........(IV)},$$

which are various analogues of the simple equation

$$x = y - zky \text{....................................(I)}.$$

In the memoir are references to earlier papers of my own on general analysis and to the current literature of the linear integral equation IV and the infinite system III of linear equations.

The general equation G we term Fredholm's equation in general analysis. We write this equation explicitly, in the first place, in the form

$$\xi(s) = \eta(s) - zJ_t\kappa(s, t)\,\eta(t) \quad (s) \text{.....................(G)},$$

with the basis $(\mathfrak{A};\ \mathfrak{P};\ \mathfrak{M};\ \mathfrak{N};\ \mathfrak{K};\ J)$.

Here $\mathfrak{A} \equiv [a]$ denotes the class of real, or, more generally, of complex numbers a, and $\mathfrak{P} \equiv [p]$ denotes a class of elements p; $\mathfrak{M} \equiv [\mu]$ and $\mathfrak{N} \equiv [\nu]$ are two classes of functions μ and ν respectively on $\mathfrak{P}$ to $\mathfrak{A}$, and $\mathfrak{K} \equiv [\kappa]$ is a class of functions κ on $\mathfrak{P}\mathfrak{P}$ to $\mathfrak{A}$, that is, μ or $\mu(p)$, ν or $\nu(p)$, κ or $\kappa(p_1, p_2)$ or $\kappa(s, t)$ are single-valued real- or complex-valued functions respectively of one, one, two arguments ranging (in the case of κ independently) over the class or range $\mathfrak{P}$; and J is a function on $\mathfrak{N}$ to $\mathfrak{A}$, that is, a functional operation turning a function ν of the class $\mathfrak{N}$ into a number of the class $\mathfrak{A}$ denoted by $J\nu$. We may conveniently indicate the character just described of the basis by writing it as follows:

$$(\mathfrak{A};\ \mathfrak{P};\ \mathfrak{M}^{\text{on } \mathfrak{P} \text{ to } \mathfrak{A}};\ \mathfrak{N}^{\text{on } \mathfrak{P} \text{ to } \mathfrak{A}};\ \mathfrak{K}^{\text{on } \mathfrak{P}\mathfrak{P} \text{ to } \mathfrak{A}};\ J^{\text{on } \mathfrak{N} \text{ to } \mathfrak{A}}).$$

In the equation G ξ and η are functions of the class $\mathfrak{M}$; κ is a function of the class $\mathfrak{K}$; z is a parameter of the class $\mathfrak{A}$; for every value of s of the class $\mathfrak{P}$ $\kappa(s, t)\,\eta(t)$, considered as a function of t, is of the class $\mathfrak{N}$ or $\mathfrak{N}_t$, so that $J_t\kappa(s, t)\,\eta(t)$ is a function of s on $\mathfrak{P}$; and the equation G is to hold for every value of s on $\mathfrak{P}$; the functions ξ, κ and the parameter z are given; the function η is to be determined.

For the respective instances II_n; III; IV the range $\mathfrak{P}$ is

$$p=1, \ldots, n; \quad p=1, 2, 3, \ldots; \quad a \leqq p \leqq b;$$

the functional operation J_t is

$$\sum_{t=1}^{n}\ ; \quad \sum_{t=1}^{\infty}\ ; \quad \int_a^b dt;$$

and, for the regular cases, the classes $\mathfrak{M}$, $\mathfrak{N}$, $\mathfrak{K}$ are the classes of all functions μ, ν, κ on the respective ranges $\mathfrak{P}$, $\mathfrak{P}$, $\mathfrak{P}\mathfrak{P}$ which in the instances III and IV satisfy certain conditions, viz., in the instance IV the functions μ, ν, κ are continuous over their respective ranges, while in the instance III the functions μ, ν give rise to absolutely convergent series $\sum_{p=1}^{\infty} \mu(p)^2$, $\sum_{p=1}^{\infty} \nu(p)$, and for every function κ there exists a function μ of the class $\mathfrak{M}$ such that for every pair of arguments s, t on the range $\mathfrak{P}$

$$|\,\kappa(s, t)\,| \leqq |\,\mu(s)\,\mu(t)\,|.$$

In the memoir cited above I have explained how the analogy of the sphere and the ellipsoid leads to the determination of a foundation, at the same time simpler and more general, although less obvious, than that just described, of a theory of the general equation G. To this end we write the equation G in the modified explicit form

$$\xi(s) = \eta(s) - zJ_{(tu)}\,\kappa(s, t)\,\eta(u) \quad (s)\ldots\ldots\ldots\ldots\ldots\ldots\ldots(\text{G}).$$

On the understanding that the variables s and u range over a class $\tilde{\mathfrak{P}}$ while the variable t ranges over a class $\hat{\mathfrak{P}}$, we have a basis

$$(\mathfrak{A};\ \tilde{\mathfrak{P}};\ \hat{\mathfrak{P}};\ \mathfrak{M}^{\text{on } \tilde{\mathfrak{P}} \text{ to } \mathfrak{A}};\ \mathfrak{N}^{\text{on } \hat{\mathfrak{P}}\tilde{\mathfrak{P}} \text{ to } \mathfrak{A}};\ \mathfrak{K}^{\text{on } \tilde{\mathfrak{P}}\hat{\mathfrak{P}} \text{ to } \mathfrak{A}};\ J^{\text{on } \mathfrak{N} \text{ to } \mathfrak{A}}),$$

for the determination, by means of suitable postulates, of a general Fredholm theory of this equation G.

For the purpose of obtaining a general Hilbert-Schmidt theory of the equation G with symmetric real-valued kernel κ ($\kappa(s, t) \equiv \kappa(t, s)$), or, more generally, with hermitian complex-valued kernel κ ($\overline{\kappa(s, t)} \equiv \kappa(t, s)$) we must identify the classes $\tilde{\mathfrak{P}}$, $\hat{\mathfrak{P}}$ and increase the system of postulates. The basis is then

$$(\mathfrak{A};\ \mathfrak{P};\ \mathfrak{M}^{\text{on } \mathfrak{P} \text{ to } \mathfrak{A}};\ \mathfrak{N}^{\text{on } \mathfrak{P}\mathfrak{P} \text{ to } \mathfrak{A}};\ \mathfrak{K}^{\text{on } \mathfrak{P}\mathfrak{P} \text{ to } \mathfrak{A}};\ J^{\text{on } \mathfrak{N} \text{ to } \mathfrak{A}}).$$

We obtain perhaps the simplest general theory including the Fredholm and Hilbert-Schmidt theories of the regular cases of the classical instances II_n, III, IV by identifying the classes $\mathfrak{N}$ and $\mathfrak{K}$, indeed by defining them as the $*$-composite $(\mathfrak{M}\mathfrak{M})_*$ of the class $\mathfrak{M}$ with itself, and by imposing upon the class $\mathfrak{M}$ certain five conditions $LCDD_0R$ and upon the functional operational J certain five conditions $LMHPP_0$. Thus, as foundation of the general theory of linear integral equations here in question we have

$$(\mathfrak{A};\ \mathfrak{P};\ \mathfrak{M}^{\text{on } \mathfrak{P} \text{ to } \mathfrak{A}\,.\,LCDD_0R};\ \mathfrak{K} \equiv (\mathfrak{M}\mathfrak{M})_*;\ J^{\text{on } \mathfrak{K} \text{ to } \mathfrak{A}\,.\,LMHPP_0}).$$

This is the foundation, designated as Σ_5, defined in the memoir cited above.

The simple general theory with the foundation Σ_5 occupies a position within the complex of general theories of the equation G in a measure analogous to the position occupied by the theory of continuous functions within the complex of various types of functions.

It should be added that this foundation Σ_5 is effective for the development of a theory of the linear integral equation G in general analysis, including, for the regular cases of the classical instances II_n, III, IV and other classical instances, not only the Fredholm and Hilbert-Schmidt theories in the narrower sense, but also at least the principal developments of those theories due amongst others to Plemelj, Goursat, I. Schur and Landsberg. With additional postulates it is effective moreover for the development of a theory of non-linear integral equations in general analysis along the lines laid down by Schmidt in instance IV.

It is the purpose of this paper to consider more closely the fundamental functional operation J of the foundation Σ_5, by indicating its principal properties deducible from the fundamental postulates and useful throughout the development of the general theory of the linear integral equation G in its second explicit form. For the convenience of the reader the requisite fundamental definitions of the memoir cited above are briefly repeated here. In so far as the foundation Σ_5 itself, rather than the theory erected on it, is immediately in question we speak of the system Σ_5 of basal terms and postulates, and similarly of systems Σ of type Σ_5.

2. On Classes of Functions on General Ranges.

We consider a class (multiplicity, *Menge*) $\mathfrak{M} \equiv [\mu]$ of functions μ on a general range $\mathfrak{P} \equiv [p]$ to the class $\mathfrak{A} \equiv [a]$ of all real or complex numbers. The range $\mathfrak{P}$ is a general class of elements p; this *general* is the *true general* in the sense of *any particular*; for the purposes of the theory there are no restrictions on the character of the elements p or of the class $\mathfrak{P}$, while for the purposes of illustration or of application any restrictions or specifications are permissible. It is the presence in a theory of a class of functions on a general range which constitutes the theory a doctrine of the form of general analysis which we are developing. For a definite function μ of the class $\mathfrak{M}$ and element p of the range $\mathfrak{P}$ $\mu(p)$ denotes a definite real or complex number of the class $\mathfrak{A}$.

(*a*) **Relative uniformity of convergence.**—A sequence $\{\phi_n\}$ of functions $\phi_n (n = 1, 2, \ldots)$ is said to converge to a function θ on the range $\mathfrak{P}$ relative to a function σ as scale of uniformity, in notation

$$\underset{n}{L}\phi_n = \theta \quad (\mathfrak{P};\ \sigma),$$

in case for every real positive number e there is an index n_e (depending on e alone) such that for every index $n \geqq n_e$ the relation

$$|\phi_n(p) - \theta(p)| \leqq e|\sigma(p)|$$

holds for every argument p of the range $\mathfrak{P}$. This relative uniformity of convergence is a fundamentally important generalization of the classical uniformity, to which it reduces in case the scale function σ is the constant function $\sigma \equiv 1$.

A sequence $\{\phi_n\}$ is said to converge to a function θ relative to a class $\mathfrak{S} \equiv [\sigma]$ of functions σ as scale class of uniformity, in notation

$$\underset{n}{L}\phi_n = \theta \quad (\mathfrak{P};\ \mathfrak{S}),$$

in case there is a function σ of the class $\mathfrak{S}$ effective as scale of relatively uniform convergence: $L_n\phi_n = \theta \quad (\mathfrak{P};\ \sigma)$.

Similar definitions hold for the relative uniformity of convergence of a series $\Sigma_n\phi_n$ of functions as to a function σ or a class $\mathfrak{S}$ of functions on $\mathfrak{P}$ as scale of uniformity, in notation

$$\sum_n \phi_n = \theta\ (\mathfrak{P};\ \sigma); \quad \sum_n \phi_n = \theta\ (\mathfrak{P};\ \mathfrak{S}).$$

The series $\Sigma_n\phi_n$ is said to converge absolutely-uniformly in case the series $\Sigma_n|\phi_n|$ converges uniformly as to σ or as to $\mathfrak{S}$.

As necessary and sufficient condition for the relative uniformity of convergence of sequence or of series we have the usual Cauchy condition taken in the sense of relative uniformity as to scale function or scale class of functions.

(*b*) **The classes $\overline{\mathfrak{M}}$, $\mathfrak{M}_{\mathfrak{S}}$, $\mathfrak{M}_L$, $\mathfrak{M}_*$.**—A real or complex number a has a (definite) conjugate real or complex number $\bar{a}$. A function θ on $\mathfrak{P}$ has a (definite) conjugate function $\bar{\theta}$ such that for every argument p of the range $\mathfrak{P}$ $\bar{\theta}(p) = \overline{\theta(p)}$. A class $\mathfrak{M} \equiv [\mu]$ of functions μ on $\mathfrak{P}$ has a (definite) conjugate class $\overline{\mathfrak{M}} \equiv [\bar{\mu}]$ consisting of the conjugate functions $\bar{\mu}$ of the functions μ of the class $\mathfrak{M}$.

Relative to two classes $\mathfrak{M} \equiv [\mu]$, $\mathfrak{S} \equiv [\sigma]$ of functions on $\mathfrak{P}$, the extension of $\mathfrak{M}$ as to $\mathfrak{S}$ is the class $\mathfrak{M}_{\mathfrak{S}} \equiv [\mu_{\mathfrak{S}}]$ of all functions of the form

$$\mu_{\mathfrak{S}} = \underset{n}{L}\mu_n\ (\mathfrak{P};\ \mathfrak{S}), \quad \text{or} \quad \mu_{\mathfrak{S}} = \underset{n}{L}\mu_n\ (\mathfrak{P};\ \sigma),$$

viz., of all limit functions of sequences $\{\mu_n\}$ of functions of the class $\mathfrak{M}$ converging uniformly as to (some function σ of) the class $\mathfrak{S}$.

The linear extension $\mathfrak{M}_L \equiv [\mu_L]$ of a class $\mathfrak{M} \equiv [\mu]$ of functions on $\mathfrak{P}$ is the class of all functions of the form $\mu_L \equiv a_1\mu_1 + \ldots + a_n\mu_n$, viz., of all linear homogeneous combinations of a finite number of functions of the class $\mathfrak{M}$ with numerical coefficients belonging to the class $\mathfrak{A}$.

The $*$-extension (read *star-extension*) $\mathfrak{M}_* \equiv [\mu_*]$ of a class $\mathfrak{M} \equiv [\mu]$ of functions on $\mathfrak{P}$ is the class $(\mathfrak{M}_L)_{\mathfrak{M}}$, the extension as to $\mathfrak{M}$ of the linear extension of the class $\mathfrak{M}$, viz., the class of all functions μ_* of the form

$$\mu_* = \underset{n}{L}\mu_{Ln}\ (\mathfrak{P};\ \mu),$$

i.e., of the form

$$\mu_* = \underset{n}{L}\sum_{g=1}^{m_n} a_{ng}\mu_{ng}\ (\mathfrak{P};\ \mu).$$

(*c*) **The closure properties** L, C, R.—A class $\mathfrak{M}$ of functions on $\mathfrak{P}$ is linear (L); closed (C); real (R),—in notation

$$\mathfrak{M}^L;\quad \mathfrak{M}^C;\quad \mathfrak{M}^R,\text{—}$$

in case respectively

$$\mathfrak{M} = \mathfrak{M}_L;\quad \mathfrak{M} = \mathfrak{M}_{\mathfrak{M}};\quad \mathfrak{M} = \overline{\overline{\mathfrak{M}}},$$

or, what is equivalent, in case $\mathfrak{M}$ contains respectively $\mathfrak{M}_L$; $\mathfrak{M}_{\mathfrak{M}}$; $\overline{\overline{\mathfrak{M}}}$.

(*d*) **The dominance properties** D, D_0.—A function ϕ on $\mathfrak{P}$ is dominated by a function θ on $\mathfrak{P}$ in case for every argument p of $\mathfrak{P}$ $|\phi(p)| \leqq |\theta(p)|$. A class $\mathfrak{M} \equiv [\mu]$ of functions on $\mathfrak{P}$ is dominated by a class $\mathfrak{S} \equiv [\sigma]$ of functions on $\mathfrak{P}$ in case every function μ of $\mathfrak{M}$ is dominated by some function σ of $\mathfrak{S}$.

A class $\mathfrak{M} \equiv [\mu]$ of functions on $\mathfrak{P}$ has the dominance property D, in notation $\mathfrak{M}^D$, in case for every sequence $\{\mu_n\}$ of functions of $\mathfrak{M}$ there is a function μ_0 of $\mathfrak{M}$ (variable with the sequence) such that every function μ_n of the sequence is dominated by some numerical multiple $a_n\mu_0$ of the function μ_0, so that for every n and p $|\mu_n(p)| \leqq |a_n\mu_0(p)|$. The class $\mathfrak{M}$ has the dominance property D_0, in notation $\mathfrak{M}^{D_0}$, in case every function μ of $\mathfrak{M}$ is dominated by some nowhere negative real-valued function μ_0 of $\mathfrak{M}$, so that for every p $|\mu(p)| \leqq \mu_0(p)$.

(*e*) **Composition of ranges and classes of functions.**—Consider two general ranges $\mathfrak{P}' \equiv [p']$, $\mathfrak{P}'' \equiv [p'']$ conceptually but not necessarily actually distinct. The product range $\mathfrak{P}'\mathfrak{P}''$ is the class of all composite elements (p', p'') or $p'p''$, the first constituent being an element p' of $\mathfrak{P}'$ and the second constituent being an element p'' of $\mathfrak{P}''$. The product class of two classes $\mathfrak{M}'$, $\mathfrak{M}''$ of functions on the respective ranges $\mathfrak{P}'$, $\mathfrak{P}''$ consists of all product functions $\mu'\mu''$ or $\mu'(p')\mu''(p'')$ of a function μ' of the class $\mathfrak{M}'$ on the range $\mathfrak{P}'$ and a function μ'' of the class $\mathfrak{M}''$ on the range $\mathfrak{P}''$. The $*$-composite $(\mathfrak{M}'\mathfrak{M}'')_*$ of the two classes $\mathfrak{M}'$, $\mathfrak{M}''$ is, as indicated by the notation, the $*$-extension of the product class, viz., the class $((\mathfrak{M}'\mathfrak{M}'')_L)_{(\mathfrak{M}'\mathfrak{M}'')}$, the extension as to the product class $\mathfrak{M}'\mathfrak{M}''$ of the linear extension $(\mathfrak{M}'\mathfrak{M}'')_L$ of the product class.

The classes $\mathfrak{M}'\mathfrak{M}''$, $(\mathfrak{M}'\mathfrak{M}'')_*$ are on the product range $\mathfrak{P}'\mathfrak{P}''$.

These definitions are extensible in an obvious sense to a finite number of ranges $\mathfrak{P}^1, \ldots, \mathfrak{P}^n$ and of classes $\mathfrak{M}^1, \ldots, \mathfrak{M}^n$ of functions on the respective ranges.

(*f*) **Illustrations.**—The availability of these definitions for a general theory of linear integral equations is indicated by the fact that the classes $\mathfrak{M}$, $\mathfrak{K}$ for the regular cases of the classical instances II_n, III, IV (defined in § 1) have the properties $LCDD_0R$ and the class $\mathfrak{K}$ is the $*$-composite $(\mathfrak{M}\mathfrak{M})_*$ of the class $\mathfrak{M}$ with itself.

The notion of relative uniformity enters into the definition of closure (C) and of $*$-extension and $*$-composition. Were we to use classical uniformity instead of relative uniformity, we should not succeed in unifying the instances II_n, III, IV for the purpose of a general theory. In fact, let the class $\mathfrak{M}$ be said to have the closure property C_0 in case it contains the limit function of every sequence $\{\mu_n\}$ of its functions which converges, in the classical sense, uniformly on the range $\mathfrak{P}$. Then for the cases II_n, IV the class $\mathfrak{M}$ has, while for the case III the class $\mathfrak{M}$ has not, this closure property C_0.

(*g*) **Theorems concerning $*$-extension and $*$-composition of classes of functions.**—$\mathfrak{M} \equiv [\mu]$ being a class of functions μ on $\mathfrak{P}$, $\mathfrak{A}\mathfrak{M} \equiv [a\mu]$ is the class of all numerical multiples $a\mu$ of the functions μ of $\mathfrak{M}$ by the numbers a of $\mathfrak{A}$. For brevity let B_1 denote "is dominated by," so that $\mathfrak{M}^{B_1\mathfrak{S}}$ denotes "class $\mathfrak{M}$ of functions is dominated by class $\mathfrak{S}$ of functions." Further after Peano "$.$" denotes "and" and "$.\supset.$" denotes "implies" or "if ( ) then ( )", so that "$A.B.\supset.C.D$" denotes "if A and B, then C and D." The systematic use of these and a few other abbreviations will be found to tend to perspicuity as well as to brevity of statement of theorems and proofs.

We notice the following fundamental theorems.

$$(1)\quad \mathfrak{P}.\mathfrak{M}^{LC}.\supset.\mathfrak{M}_* = \mathfrak{M},$$

viz., if the class $\mathfrak{M}$ on the range $\mathfrak{P}$ is linear and closed, then it is identical with its $*$-extension.—The converse is also true.

$$(2)\quad \mathfrak{P}.\mathfrak{M}^{D}.\supset.\mathfrak{M}_*{}^{B_1\mathfrak{A}\mathfrak{M}.LCD},$$

viz., if the class $\mathfrak{M}$ of functions on the range $\mathfrak{P}$ has the dominance property D, then its $*$-extension $\mathfrak{M}_*$ is dominated by the class $\mathfrak{A}\mathfrak{M}$ ($B_1\mathfrak{A}\mathfrak{M}$), is linear ($L$) and closed ($C$) and has the dominance property D.

$$(3)\quad \mathfrak{P}'.\mathfrak{M}'^{D}.\mathfrak{P}''.\mathfrak{M}''^{D}.\supset.(\mathfrak{M}'\mathfrak{M}'')^{D},$$

viz., if the classes $\mathfrak{M}'$, $\mathfrak{M}''$ of functions on the respective ranges $\mathfrak{P}'$, $\mathfrak{P}''$ have the dominance property D, so has their product class $\mathfrak{M}'\mathfrak{M}''$ (of functions on the product range $\mathfrak{P}'\mathfrak{P}''$).

$$(4)\quad \mathfrak{P}'.\mathfrak{M}'^{D}.\mathfrak{P}''.\mathfrak{M}''^{D}.\supset.(\mathfrak{M}'\mathfrak{M}'')_*{}^{B_1\mathfrak{A}\mathfrak{M}'\mathfrak{M}''.LCD}.(\mathfrak{M}'\mathfrak{M}'')_* = (\mathfrak{M}'{}_*\mathfrak{M}'')_*$$
$$= (\mathfrak{M}'\mathfrak{M}''{}_*)_* = (\mathfrak{M}'{}_*\mathfrak{M}''{}_*)_*.$$

$$(5)\quad \mathfrak{P}'.\mathfrak{M}'^{LCDD_0R}.\mathfrak{P}''.\mathfrak{M}''^{LCDD_0R}.\supset.(\mathfrak{M}'\mathfrak{M}'')_*{}^{B_1\mathfrak{M}'\mathfrak{M}''.LCDD_0R}.$$

The theorems 3, 4, 5 are extensible to any finite number of classes $\mathfrak{M}^1, \ldots, \mathfrak{M}^n$ of functions on the respective ranges $\mathfrak{P}^1, \ldots, \mathfrak{P}^n$ (conceptually but not necessarily actually distinct) having the properties D, D, $LCDD_0R$; the product class $\mathfrak{M}^1\mathfrak{M}^2\ldots\mathfrak{M}^n$ replaces the product class $\mathfrak{M}'\mathfrak{M}''$, and the second part of the conclusion of theorem 4 has, for the instance $n=3$, the form

$$(\mathfrak{M}^1\mathfrak{M}^2\mathfrak{M}^3)_* = (\mathfrak{M}^1{}_*\mathfrak{M}^2\mathfrak{M}^3)_* = \ldots = (\mathfrak{M}^1{}_*\mathfrak{M}^2{}_*\mathfrak{M}^3)_* = \ldots$$
$$= (\mathfrak{M}^1{}_*\mathfrak{M}^2{}_*\mathfrak{M}^3{}_*)_* = ((\mathfrak{M}^1\mathfrak{M}^2)_*\mathfrak{M}^3)_* = \ldots = ((\mathfrak{M}^1\mathfrak{M}^2)_*\mathfrak{M}^3{}_*)_*;$$

in permutation of the classes the understanding is that the arguments permute with the functions.

As to theorem 5 we remark that owing to the linearity (L) of $\mathfrak{M}'$ (or $\mathfrak{M}''$) the classes $\mathfrak{M}'\mathfrak{M}''$ and $\mathfrak{A}\mathfrak{M}'\mathfrak{M}''$ are identical, so that the properties $B_1\mathfrak{M}'\mathfrak{M}''$ and $B_1\mathfrak{A}\mathfrak{M}'\mathfrak{M}''$ are equivalent, and further that owing to the property D_0 of $\mathfrak{M}'$ and $\mathfrak{M}''$ we have for every function θ of $(\mathfrak{M}'\mathfrak{M}'')_*$ a relation of the form $|\theta| \leqq \mu_0'\mu_0''$, where μ_0', μ_0'' are suitably chosen real-valued nowhere negative functions of the respective classes $\mathfrak{M}'$, $\mathfrak{M}''$.

(*h*) **Theorems involving the notion of relative uniformity of convergence of series.**—The following theorems illustrate the notion of relative uniformity. Theorem 8 is a corollary of theorem 7, and itself enters as a lemma in the proof of the fundamental convergence theorem of § 5 *l*. The functions entering the theorems are functions of a common argument p on a general range $\mathfrak{P}$, and the inequalities involving functions hold identically in the common argument p.

(6) The uniform convergence of the series $\Sigma_n \bar{\phi}_n \phi_n$, $\Sigma_n \bar{\psi}_n \psi_n$ as to the respective scale functions σ, τ implies the uniform convergence of the series $\Sigma_n |\phi_n \psi_n|$ as to the scale function $\sqrt{|\sigma\tau|}$.

(7) The uniform convergence of the series $\Sigma_n \bar{\phi}_n \phi_n$ as to the scale function σ and the convergence of the series $\Sigma_n \bar{\psi}_n \psi_n$ to a sum-function $\leqq \tau$ imply the uniform convergence of the series $\Sigma_n |\phi_n \psi_n|$ as to the scale function $\sqrt{|\sigma\tau|}$.

(8) The convergence of the numerical series $\Sigma_n \bar{a}_n a_n$ and the convergence of the series $\Sigma_n \bar{\psi}_n \psi_n$ to a sum-function $\leqq \tau$ imply the uniform convergence of the series $\Sigma_n |a_n \psi_n|$ as to the scale function $\sqrt{\tau}$.

3. On the Fundamental Functional Operation J.

Of the system Σ_5:

$$(\mathfrak{A};\ \mathfrak{P};\ \mathfrak{M}^{\text{on } \mathfrak{P} \text{ to } \mathfrak{A} . LCDD_0R};\ \mathfrak{K} \equiv (\mathfrak{M}\mathfrak{M})_*;\ J^{\text{on } \mathfrak{K} \text{ to } \mathfrak{A} . LMHPP_0}),$$

we have in § 2 considered those features not involving the functional operation J. In § 3 we define the fundamental properties $LMHPP_0$ and certain additional properties $CC_*H_*P_*P_{0*}$ and then formulate a number of theorems involving these properties. The operation J is a function on $\mathfrak{K}$ to $\mathfrak{A}$, where $\mathfrak{K}$ is the class $(\mathfrak{M}\mathfrak{M})_*$ of functions on $\mathfrak{P}\mathfrak{P}$ to $\mathfrak{A}$ derived from the class $\mathfrak{M}$ of functions on $\mathfrak{P}$ to $\mathfrak{A}$.

(*a*) **The properties L, C, C_*, M of the operation J on the class $\mathfrak{K}$.**—These properties are definable for a functional operation J on $\mathfrak{K}$ to $\mathfrak{A}$, where $\mathfrak{K} \equiv [\kappa]$ is a class of functions κ on a general range $\mathfrak{Q} \equiv [q]$ to $\mathfrak{A}$. Thus for every function κ or $\kappa(q)$ of the class $\mathfrak{K}$ or $\mathfrak{K}_q$ $J\kappa$ or $J_q \kappa(q)$ denotes a definite number of the class $\mathfrak{A}$, denotable likewise by $J_r \kappa(r)$ where r is any notation for a variable element of the class $\mathfrak{Q}$.

The operation J is linear (L), in notation J^L, in case

$$a_1\kappa_1 + a_2\kappa_2 = \kappa \text{ implies } a_1 J\kappa_1 + a_2 J\kappa_2 = J\kappa.$$

The operation J is continuous (C), in notation J^C, in case

$$\underset{n}{L}\kappa_n = \kappa \quad (\mathfrak{Q};\ \mathfrak{K}), \text{ or } \underset{n}{L}\kappa_n = \kappa \quad (\mathfrak{Q};\ \kappa_0),$$

implies

$$\underset{n}{L} J\kappa_n = J\kappa.$$

The operation J is ultracontinuous (C_*), in notation J^{C_*}, in case for every range $\mathfrak{U} \equiv [u]$ (conceptually but not necessarily actually distinct from the range $\mathfrak{Q}$)

$$\underset{n}{L}\phi_n = \phi \quad (\mathfrak{Q}\mathfrak{U};\ \kappa v) \text{ implies } \underset{n}{L} J\phi_n = J\phi \quad (\mathfrak{U};\ v),$$

where υ or $\upsilon(u)$ is a function on $\mathfrak{U}$ to $\mathfrak{A}$, κ or $\kappa(q)$ is a function of the class $\mathfrak{K}$ on $\mathfrak{Q}$ (so that the product $\kappa\upsilon$ or $\kappa(q)\upsilon(u)$ is a function on the product range $\mathfrak{Q}\mathfrak{U}$), and the functions ϕ or $\phi(q, u)$ and ϕ_n or $\phi_n(q, u)$ $(n=1, 2, \ldots)$ are functions on the product range $\mathfrak{Q}\mathfrak{U}$, which for every value of u of the range $\mathfrak{U}$, quâ functions of the argument q on the range $\mathfrak{Q}$, belong to the class $\mathfrak{K}$, and where further $J\phi$ or $J_q\phi(q, u)$ and $J\phi_n$ or $J_q\phi_n(q, u)$ $(n=1, 2, \ldots)$, having as to the operation J or J_q the parameter u, are functions of u on the range $\mathfrak{U}$.

The operation J has the modular property (M), in notation J^M, in case there exists an operation (called the modulus or modular operation) M on the class $\mathfrak{K}_{\geqq 0}$ (consisting of all real-valued nowhere negative functions κ of the class $\mathfrak{K}$) to $\mathfrak{A}$ so related to the operation J that

$$|\kappa| \leqq \kappa_0 \text{ implies } |J\kappa| \leqq M\kappa_0.$$

If κ belongs to $\mathfrak{K}_{\geqq 0}$ we have in particular $|J\kappa| \leqq M\kappa$. If the class $\mathfrak{K}_{\geqq 0}$ is the null-class, the operation J is said to have the modular property vacuously. In the classical instances the modular operation M is the operation J itself. It is to be noted that the modular operation M is not supposed to be linear.

(*b*) **Theorems involving the properties L, C, C_*, M of the operation J of the class $\mathfrak{K}$ of functions on the range $\mathfrak{Q}$.**—

(1)* $\mathfrak{K}^L \,.\, J^L :\supset: \begin{matrix} a_1, \ldots, a_n \\ \kappa_1, \ldots, \kappa_n \end{matrix} \,.\supset.\, J(a_1\kappa_1 + \ldots + a_n\kappa_n) = a_1 J\kappa_1 + \ldots + a_n J\kappa_n.$

(2)† $\mathfrak{K}^C \,.\, J^C :\supset: \underset{n}{L}\kappa_n \; (\mathfrak{Q}; \mathfrak{K}) \,.\supset.\, J\underset{n}{L}\kappa_n = \underset{n}{L}J\kappa_n.$

(3)‡ $\mathfrak{K}^{LC} \,.\, J^{LC} :\supset: \underset{n}{\Sigma}\kappa_n \; (\mathfrak{Q}; \mathfrak{K}) \,.\supset.\, J\underset{n}{\Sigma}\kappa_n = \underset{n}{\Sigma}J\kappa_n.$

(4) $\mathfrak{K} \,.\, J^{C_*} \,.\supset.\, J^C.$

(5) $\mathfrak{K}^{LD_0} \,.\, J^{LM} \,.\supset.\, J^{CC_*}.$

(*c*) **The transpose function $\breve{\omega}$ and the hermitian property (H) of a function ω on the product range $\mathfrak{P}\mathfrak{P}$.**—Consider a function ω or $\omega(s, t)$ on the product range $\mathfrak{P}\mathfrak{P}$, the variable s, t ranging independently over the class $\mathfrak{P}$. The transpose function $\breve{\omega}$ of ω is the function for which $\breve{\omega}(s, t) = \omega(t, s)$ identically in s and t on $\mathfrak{P}$. The operation of transposition is linear, i.e., the transpose of $a_1\omega_1 + a_2\omega_2$ is $a_1\breve{\omega}_1 + a_2\breve{\omega}_2$.

The function ω on $\mathfrak{P}\mathfrak{P}$ is hermitian (H), in notation ω^H, in case its conjugate function $\bar{\omega}$ and its transpose function $\breve{\omega}$ are identical: $\bar{\omega} = \breve{\omega}$, that is, in case $\overline{\omega(s, t)} = \omega(t, s)$ identically in s and t on $\mathfrak{P}$. Otherwise expressed, a hermitian

* For a linear operation J on a linear class $\mathfrak{K}$ with any n numbers $a_1, \ldots, a_n$ of $\mathfrak{A}$ and n functions $\kappa_1, \ldots, \kappa_n$ of $\mathfrak{K}$ we have $J(a_1\kappa_1 + \ldots + a_n\kappa_n) = a_1 J\kappa_1 + \ldots + a_n J\kappa_n$. (The implication "$:\supset:$" has as its conclusion the implication "$.\supset.$")

† For a continuous operation J on a closed class $\mathfrak{K}$ on a range $\mathfrak{Q}$ if a sequence $\{\kappa_n\}$ of functions of $\mathfrak{K}$ converges on $\mathfrak{Q}$ uniformly as to $\mathfrak{K}$ then $JL_n\kappa_n = L_nJ\kappa_n$.

‡ For a continuous linear operation J on a closed linear class $\mathfrak{K}$ on a range $\mathfrak{Q}$ if a series $\Sigma_n\kappa_n$ of functions of $\mathfrak{K}$ converges on $\mathfrak{Q}$ uniformly as to $\mathfrak{K}$ then $J\Sigma_n\kappa_n = \Sigma_nJ\kappa_n$.

function ω is the transpose of its conjugate: $\omega = \breve{\bar{\omega}}$, or the conjugate of its transpose: $\omega = \bar{\breve{\omega}}$.

The function ω or $\omega(s, t)$ on $\mathfrak{P}\mathfrak{P}$ may be the product $\alpha(s)\,\beta(t)$ of a function α of s on $\mathfrak{P}$ by a function β of t on $\mathfrak{P}$. In this case we write $\omega = \alpha\beta$, *the order being of importance*, the first or left argument (s) of ω being the argument of the first or left factor α of the product $\alpha\beta$ and the second or right argument (t) of ω being the argument of the second or right factor β of the product $\alpha\beta$. Thus the product function $\alpha\beta$ has as conjugate $\bar{\alpha}\bar{\beta}$, as transpose $\beta\alpha$, and as transpose of conjugate $\bar{\beta}\bar{\alpha}$. Hermitian product functions are of the form $\bar{\alpha}\alpha$ or $\alpha\bar{\alpha}$.

(*d*) **The properties H, H_*, P, P_0, P_*, P_{0*} of the operation J on the class $\mathfrak{K} \equiv (\mathfrak{M}\mathfrak{M})_*$ derived from a real class $\mathfrak{M}$ on the range $\mathfrak{P}$.**—We consider a real class $\mathfrak{M}$ on the range $\mathfrak{P}$, that is, a class $\mathfrak{M}$ containing the conjugate $\bar{\mu}$ of its every function μ. Then the class $\mathfrak{K} \equiv (\mathfrak{M}\mathfrak{M})_*$ on the product range $\mathfrak{P}\mathfrak{P}$ is real and self-transpose, that is, the class $\mathfrak{K}$ contains the conjugate $\bar{\kappa}$ and the transpose $\breve{\kappa}$, and accordingly the transpose of the conjugate $\breve{\bar{\kappa}}$, of its every function κ. The class $\mathfrak{K}$ contains the class $\mathfrak{M}\mathfrak{M}$ of product functions $\mu_1\mu_2$ of two functions μ_1, μ_2 (of independent arguments) of the class $\mathfrak{M}$.

The operation J is hermitian (H), in notation J^H, in case for every two functions μ_1, μ_2 of $\mathfrak{M}$

$$\overline{J\mu_1\mu_2} = J\bar{\mu}_2\bar{\mu}_1.$$

It is ultrahermitian (H_*), in notation J^{H_*}, in case for every function κ of $\mathfrak{K}$

$$\overline{J\kappa} = J\breve{\bar{\kappa}}.$$

Hence, if J is ultrahermitian and κ is hermitian, $J\kappa$ is a real number.

The operation J is positive (P), in notation J^P, in case for every function μ of $\mathfrak{M}$

$$J\bar{\mu}\mu \geqq 0\,*,$$

and definitely (P_0) so, in notation J^{P_0}, in case moreover

$$J\bar{\mu}\mu = 0 \text{ implies } \mu = 0.$$

The operation J is ultrapositive (P_*), in notation J^{P_*}, in case for every function κ of $\mathfrak{K}$ having the form

$$\kappa = \underset{n}{L} \sum_{g=1}^{m_n} \bar{\mu}_{ng}\mu_{ng} \quad (\mathfrak{P}\mathfrak{P};\ \mathfrak{M}\mathfrak{M}),$$

we have

$$J\kappa \geqq 0,$$

and definitely (P_{0*}) so in case moreover for such a function

$$J\kappa = 0 \quad \text{implies} \quad \kappa = 0.$$

Functions κ of the form specified above are hermitian and of positive type (cf. § 4 *c*).

(*e*) **Theorems concerning the operation J on the class $\mathfrak{K} \equiv (\mathfrak{M}\mathfrak{M})_*$.**—

$$(6) \quad \mathfrak{M}^{D_0} \,.\, J^{LM} :\supset:$$

$$\kappa = \underset{n}{L} \sum_{g=1}^{m_n} a_{ng}\,\beta_{ng}\,\gamma_{ng} \quad (\mathfrak{P}\mathfrak{P};\ \mathfrak{M}\mathfrak{M}) \,.\supset.\, J\kappa = \underset{n}{L} \sum_{g=1}^{m_n} a_{ng}\,J\beta_{ng}\,\gamma_{ng},$$

* $J\bar{\mu}\mu$ is a real non-negative number.

where the functions β_{ng}, γ_{ng} $(g=1, \ldots, m_n;\ n=1, 2, \ldots)$ belong to the class $\mathfrak{M}$.—Every function κ of the class $\mathfrak{K} \equiv (\mathfrak{M}\mathfrak{M})_*$ has the form just indicated, and, conversely, every function of this form is a function κ.

$$(7)\quad \mathfrak{M}^{R} . J^{H_*} . \supset . J^{H}.$$

$$(8)\quad \mathfrak{M}^{D_0 R} . J^{LMH} . \supset . J^{H_*}.$$

$$(9)\quad \mathfrak{M}^{R} . J^{P_*} . \supset . J^{P}.$$

$$(10)\quad \mathfrak{M}^{D_0 R} . J^{LMP} . \supset . J^{P_*}.$$

$$(11)\quad \mathfrak{M}^{R} . J^{P_{0*}} . \supset . J^{P_0}.$$

$$(12)\quad \mathfrak{M}^{LCDD_0 R} . J^{LMPP_0} . \supset . J^{P_{0*}}.$$

The proof of theorem 12 is of interesting nature. The conclusion, $\kappa=0$, of the implicational property P_{0*} of the operation J is to be proved. One proves indirectly that $\kappa(p, p)=0$ identically in p on $\mathfrak{P}$, and then directly that $\kappa(s, t)$ is on $\mathfrak{P}\mathfrak{P}$ everywhere purely imaginary and finally null.

As a corollary of theorems stated above we have

$$(13)\quad \mathfrak{M}^{LCDD_0 R} . J^{LMHPP_0} . \supset . J^{CC_* H_* P_* P_{0*}},$$

that is, the

THEOREM. *The fundamental functional operation J of the system Σ_5 has the properties $LMCC_*HH_*PP_*P_0P_{0*}$.*

4. Instances of Systems Σ_5.

The instances, now to be indicated, of systems Σ_5 serve to indicate to some extent the scope of the general theory of the linear integral equation G or of other theories developable on the foundation Σ_5. Of importance for the development of such general theories are the $*$-composition of systems Σ_5, explained in (*b*), and the abbreviational notations in connection with iterated operations, explained in (*c*).

(*a*) **The classical instances.**—The regular cases of the classical instances II_n, III, IV may be treated on the foundation Σ_5, the operation J or $J_{(st)}$ on the class $\mathfrak{K}$ or $\mathfrak{K}_{st}$ having the definition

$$J_{(st)}\kappa(s, t) = J_p\kappa(p, p)$$

for every function κ of $\mathfrak{K}$, the operation J_p being in the respective instances

$$\sum_{p=1}^{n};\quad \sum_{p=1}^{\infty};\quad \int_a^b dp\ .$$

(*b*) **$*$-composition of systems** Σ_5.—Consider two systems Σ', Σ'' of type Σ_5, viz.,

$$\Sigma' \equiv \left(\mathfrak{A};\ \mathfrak{P}';\ \mathfrak{M}'^{\text{ on } \mathfrak{P}' \text{ to } \mathfrak{A}};\ \mathfrak{K}' \equiv (\mathfrak{M}'\mathfrak{M}')_*;\ J'^{\text{ on } \mathfrak{K}' \text{ to } \mathfrak{A}}\right);$$

$$\Sigma'' \equiv \left(\mathfrak{A};\ \mathfrak{P}'';\ \mathfrak{M}''^{\text{ on } \mathfrak{P}'' \text{ to } \mathfrak{A}};\ \mathfrak{K}'' \equiv (\mathfrak{M}''\mathfrak{M}'')_*;\ J''^{\text{ on } \mathfrak{K}'' \text{ to } \mathfrak{A}}\right),$$

where the classes $\mathfrak{A}$ are identical, the ranges $\mathfrak{P}'$, $\mathfrak{P}''$ are conceptually distinct, and the classes $\mathfrak{M}'$, $\mathfrak{M}''$ have the properties $LCDD_0R$ while the operations J', J'' have the properties $LMHPP_0$. The $*$-composite $(\Sigma'\Sigma'')_*$ of these two systems Σ', Σ'' is the system Σ:

$$\Sigma \equiv (\Sigma'\Sigma'')_* \equiv (\mathfrak{A};\ \mathfrak{P} \equiv \mathfrak{P}'\mathfrak{P}'';\ \mathfrak{M} \equiv (\mathfrak{M}'\mathfrak{M}'')_*;\ \mathfrak{K} \equiv (\mathfrak{K}'\mathfrak{K}'')_*;\ J \equiv J'J'').$$

This system Σ is of type Σ_5, for with use of theorems of §§ 2, 3 it is provable that the class $\mathfrak{M} \equiv (\mathfrak{M}'\mathfrak{M}'')_*$ on the range $\mathfrak{P} \equiv \mathfrak{P}'\mathfrak{P}''$ has the properties $LCDD_0R$, that the class $\mathfrak{K} \equiv (\mathfrak{K}'\mathfrak{K}'')_*$ on $\mathfrak{P}\mathfrak{P}$ is the class $(\mathfrak{M}'\mathfrak{M}'\mathfrak{M}''\mathfrak{M}'')_*$, viz. the class $(\mathfrak{M}\mathfrak{M})_*$, and that the operation J, defined as the iteration of the operations J', J'' is applicable to functions κ of the class $\mathfrak{K}$ and, as an operation on $\mathfrak{K}$ to $\mathfrak{A}$, has the properties $LMHPP_0$.

The two operations J', J'' on a function κ are commutative: $J'J''\kappa = J''J'\kappa$. In the proof that $J \equiv J'J''$ is definitely positive (P_0P) it is convenient to use the notion of orthogonality (cf. § 5).

The $*$-composite $(\Sigma^1\Sigma^2 \ldots \Sigma^n)_*$, similarly defined, of a finite number of systems $\Sigma^1, \Sigma^2, \ldots, \Sigma^n$ of type Σ_5 is likewise a system of type Σ_5.

For the classes $\mathfrak{M}'$ of the classical instances we have simple functional characterizations* of the functions μ of the $*$-composite class $\mathfrak{M} \equiv (\mathfrak{M}'\mathfrak{M}'')_*$. Thus

If $\mathfrak{M}'$ is the class $\mathfrak{M}'^{\text{III}}$ or $\mathfrak{M}'^{\text{IV}}$ of the regular case of the classical instance III or IV (defined in § 1) and $\mathfrak{M}''$ is any linear closed class with the dominance property D, then $\mathfrak{M} \equiv (\mathfrak{M}'\mathfrak{M}'')_*$ consists of all functions μ on $\mathfrak{P} \equiv \mathfrak{P}'\mathfrak{P}''$ satisfying the two conditions: (1) for every p' the function μ or $\mu\,(p', p'')$ quâ function of p'' belongs to the class $\mathfrak{M}''$; (2 III) the function μ is dominated by some product function of the form $\mu'\mu''$; (2 IV) the function μ or $\mu\,(p', p'')$ is for every p'', quâ function of p', uniformly continuous on the range p'^{IV}, and this uniform continuity is uniform as to the parameter p'', viz., for every positive number e there exists a positive number d_e such that $|p_1' - p_2'| \leqq d_e$ implies for every p'' $|\mu\,(p_1', p'') - \mu\,(p_2', p'')| \leqq e$.

Hence (cf. § 1) the class $\mathfrak{K} \equiv (\mathfrak{M}\mathfrak{M})_*$ consists of the functions κ which are, in instance III, dominated each by some product function of the form $\mu\mu$, and, in instance IV, continuous over the square $\mathfrak{P}^{\text{IV}}\mathfrak{P}^{\text{IV}}$, while for the class $\mathfrak{M} \equiv (\mathfrak{M}'^{\text{III}}\mathfrak{M}''^{\text{IV}})_*$ we have two distinct but equivalent characterizations of the constituent functions.

The linear integral equation G for the system $\Sigma \equiv (\Sigma'^{\text{III}}\Sigma''^{\text{IV}})_*$ has the form

$$\xi\,(s', s'') = \eta\,(s', s'') - z \sum_{t'=1}^{\infty} \int_a^b \kappa\,(s', s'', t', t'')\,\eta\,(t', t'')\,dt'' \quad (s' = 1, 2, \ldots\ a \leqq t' \leqq b).$$

Here ξ and η belong to the class $\mathfrak{M} \equiv (\mathfrak{M}'^{\text{III}}\mathfrak{M}''^{\text{IV}})_*$ while κ belongs to the class

$$\mathfrak{K} \equiv (\mathfrak{K}'^{\text{III}}\mathfrak{K}''^{\text{IV}}) = (\mathfrak{M}\mathfrak{M})_* = (\mathfrak{M}'^{\text{III}}\mathfrak{M}''^{\text{IV}}\mathfrak{M}'^{\text{III}}\mathfrak{M}''^{\text{IV}})_*.$$

To the general question of characterization of the constituent functions of classes arising by $*$-composition of two or more classes of functions are devoted §§ 56—84, pp. 96—149, of my memoir: "Introduction to a Form of General Analysis," cited in the footnote.

* Cf. §§ 57 *a*, 60, 62, 65, 66 of my memoir: "Introduction to a Form of General Analysis," pp. 1—150 of *The New Haven Mathematical Colloquium*, Yale University Press, New Haven, 1910. The proofs there specified for the real number system $\mathfrak{A}$ are effective without change for the complex number system $\mathfrak{A}$.

(*c*) **The instance suggested by the analogy of the sphere and the ellipsoid.**—Consider a system Σ of type Σ_5 and in it two functions ω, κ of the class $\mathfrak{K}$ and two functions ξ, η of the class $\mathfrak{M}$. Then, as phenomena of the $*$-composite system $(\Sigma\,\Sigma)_*$, the iterations $J_{(st)} J_{(uv)}$, $J_{(uv)} J_{(st)}$ are applicable to the functions $\xi(s)\,\omega(t, u)\,\eta(v)$, $\kappa(s, v)\,\omega(t, u)$ and we have

$$J_{(st)} J_{(uv)}\, \xi(s)\, \omega(t, u)\, \eta(v) = J_{(uv)} J_{(st)}\, \xi(s)\, \omega(t, u)\, \eta(v),$$

$$J_{(st)} J_{(uv)}\, \kappa(s, v)\, \omega(t, u) = J_{(uv)} J_{(st)}\, \kappa(s, v)\, \omega(t, u).$$

Here s, t, u, v are four independent variables on the range $\mathfrak{P}$.

Having regard to the order of the arguments of the operand functions of such iterational operations we may conveniently abbreviate the notations. For instance, in $\kappa(s, v)\,\omega(t, u)$ the arguments are in the order *svtu* and we write indifferently

$$J_{(st)} J_{(uv)}\, \kappa(s, v)\, \omega(t, u)\,; \quad J_{(13)} J_{(42)}\, \kappa\omega,$$

so that, for instance, $$J_{(13)} J_{(42)}\, \kappa\omega \equiv J_{(31)} J_{(24)}\, \omega\kappa.$$

In this notation, the two equalities first written are

$$J_{(12)} J_{(34)}\, \xi\omega\eta = J_{(34)} J_{(12)}\, \xi\omega\eta,$$

$$J_{(13)} J_{(42)}\, \kappa\omega = J_{(42)} J_{(13)}\, \kappa\omega.$$

As to the operation J the function ω of the class $\mathfrak{K}$ is of positive type (P), in notation ω^P, in case for every function μ of the class $\mathfrak{M}$

$$J_{(12)} J_{(34)}\, \bar{\mu}\omega\mu \geqq 0,$$

and it is definitely (P_0) so, in notation: ω^{P_0}, in case moreover

$$J_{(12)} J_{(34)}\, \bar{\mu}\omega\mu = 0 \ \text{ implies } \ \mu = 0.$$

Now consider a hermitian function $\omega : \bar{\omega} = \breve{\omega}$, of definitely positive type as to the operation J. Then from the system Σ we obtain a system Σ_ω by replacing the operation J on $\mathfrak{K}$ by the operation J_ω on $\mathfrak{K}$ where for every function κ of $\mathfrak{K}$

$$J_\omega \kappa = J_{(13)} J_{(42)}\, \kappa\omega.$$

This system Σ_ω is of type Σ_5, for the operation J_ω has the properties $LMHPP_0$.

It was the desire to secure a theorem of this character, expressing in a sense the analogy between, let us say, the sphere and the ellipsoid, which led to the second explicit formulation of the general linear integral equation G and to the replacement of the unary operation J_p, as fundamental functional operation of the general theory, by the binary operation $J_{(st)}$.

It may be remarked that, if the hermitian function ω is of positive type, although not of definitely positive type, the operation J_ω has the properties $LMHP$ but not the property P_0.

(*d*) **Adjunctional composition of systems** Σ_5.—Consider n systems $\Sigma^1, \ldots, \Sigma^n$ of the type Σ_5 conceptually, but not necessarily actually, distinct. Let the classes $\mathfrak{A}$ be identical. Further let no two of the n ranges $\mathfrak{P}^i$ have elements in common; this state of affairs, being always securable by transformation, involves no essential restriction of generality. The n systems $\Sigma^1, \ldots, \Sigma^n$ give rise to a system $\Sigma^{1\ldots n}$, their adjunctional composite, which is likewise of type Σ_5, as follows.

The class $\mathfrak{A}$ of the system $\Sigma^{1\ldots n}$ is the common class $\mathfrak{A}$ of the systems $\Sigma^1, \ldots, \Sigma^n$.

The range $\mathfrak{P}$ of $\Sigma^{1\ldots n}$ is the adjunctional composite or logical sum or aggregate of elements of the n ranges $\mathfrak{P}^i$ of Σ^i, and the range $\mathfrak{P}^i$ is the ith component of the range $\mathfrak{P}$. Then the product range $\mathfrak{P}\mathfrak{P}$ is the adjunctional composite of the n^2 product ranges $\mathfrak{P}^i\mathfrak{P}^j$.

The class $\mathfrak{M}$ of functions μ on the composite range $\mathfrak{P}$ is the adjunctional composite of the n classes $\mathfrak{M}^i$ of functions μ^i on the respective component ranges $\mathfrak{P}^i$, viz., every combination of functions $\mu^1, \ldots, \mu^n$ on the respective ranges $\mathfrak{P}^1, \ldots, \mathfrak{P}^n$ determines a function μ on $\mathfrak{P}$ by the stipulation that on every component range $\mathfrak{P}^i$ the function μ is identical with the corresponding function μ^i, and every function μ is so determinable; the function μ is the adjunctional composite of the n functions μ^i, and the function μ^i is the ith component of the function μ; we write $\mu = (\mu^1, \ldots, \mu^n)$ or $\mu = (\mu^i)$.

Then the class $\mathfrak{K} \equiv (\mathfrak{M}\mathfrak{M})_*$ of functions κ on the product range is the adjunctional composite of the n^2 classes $\mathfrak{K}^{ij} \equiv (\mathfrak{M}^i\mathfrak{M}^j)_*$ of functions κ^{ij} on the respective component ranges $\mathfrak{P}^i\mathfrak{P}^j$; we write $\kappa = (\kappa^{ij})$.

The functional operation J on $\mathfrak{K}$ is the adjunctional composite of the n operations J^i on the respective classes $\mathfrak{K}^i \equiv (\mathfrak{M}^i\mathfrak{M}^i)_* \equiv \mathfrak{K}^{ii}$, viz., if the function κ is the adjunctional composite of the n^2 functions κ^{ij}: $\kappa = (\kappa^{ij})$,

$$J\kappa \equiv J(\kappa^{ij}) \equiv \sum_{i=1}^{n} J^i \kappa^{ii}.$$

Consider the linear equation

$$\xi = \eta - zJ\kappa\eta \quad \ldots\ldots\ldots\ldots\ldots\ldots\ldots\ldots\ldots\ldots(\mathrm{G})$$

for this basal system $\Sigma^{1\ldots n}$. Setting

$$\xi = (\xi^i), \quad \eta = (\eta^i), \quad \kappa = (\kappa^{ij})$$

we have

$$(\xi^i) = (\eta^i) - zJ(\kappa^{ij})(\eta^g) = (\eta^i) - z\left(\sum_{j=1}^{n} J^j \kappa^{ij} \eta^j\right) \quad \ldots\ldots\ldots(\mathrm{G}),$$

that is, equating the components, we have the system

$$\xi^i = \eta^i - z\sum_{j=1}^{n} J^j \kappa^{ij} \eta^j \qquad (i = 1, \ldots, n) \quad \ldots\ldots\ldots\ldots(\mathrm{G}_n{}^{1\ldots n})$$

of n simultaneous equations for the set of component systems $\Sigma^1, \ldots, \Sigma^n$; and conversely, every such system $\mathrm{G}_n{}^{1\ldots n}$ of equations is expressible as a single equation G for the composite basis $\Sigma^{1\ldots n}$.

Accordingly the general theory of the simple equation G with a basis Σ of type Σ_5 covers by specialization the general theory of the more general system $\mathrm{G}_n{}^{1\ldots n}$ of n simultaneous equations based on n systems $\Sigma^1, \ldots, \Sigma^n$ of type Σ_5.

(*e*) **The transpose system $\breve{\Sigma}$ of the system Σ.**—Consider a system Σ of type Σ_5. The transpose system $\breve{\Sigma}$ likewise of type Σ_5 is obtained by replacing the operation J by the transpose operation $\breve{J}$. Here $\breve{J}$ is the operation such that for every function κ of $\mathfrak{K}$

$$\breve{J}\kappa = J\breve{\kappa},$$

that is, by the definition (§ 3 c) of the transpose $\check{\kappa}$ of the function κ,

$$\check{J}_{(st)}\,\kappa(s,t) = J_{(st)}\,\kappa(t,s) = J_{(ts)}\,\kappa(s,t),$$

so that we have

$$\check{J}_{(st)} = J_{(ts)}.$$

Here one must recall, as noted in § 3 d, that the class $\mathfrak{K}$ contains the transpose $\check{\kappa}$ of its every function κ.

(*f*) **A class $\mathfrak{M}^{LCDD_0R}$ and the operations J^{LMHPP_0} on $\mathfrak{K} \equiv (\mathfrak{M}\mathfrak{M})_*$.**— We consider a range $\mathfrak{P}$ and a class $\mathfrak{M}^{LCDD_0R}$ of functions on $\mathfrak{P}$ to $\mathfrak{A}$ and the associated class $\mathfrak{J}$ of all operations J^{LMHPP_0} on $\mathfrak{K} \equiv (\mathfrak{M}\mathfrak{M})_*$ to $\mathfrak{A}$.

The class $\mathfrak{J}$ is self-transpose, as we have seen in § 4 e, viz., if the operation J belongs to the class $\mathfrak{J}$, so does the transpose operation $\check{J}$: $\check{J}\kappa = J\check{\kappa}$.

The class $\mathfrak{J}$ is positively-linear, viz., if the two (possibly identical) operations J_1, J_2 belong to the class $\mathfrak{J}$, so does every operation of the form $a_1J_1 + a_2J_2$:

$$(a_1J_1 + a_2J_2)\,\kappa = a_1J_1\kappa + a_2J_2\kappa,$$

where a_1, a_2 are two positive real numbers.

Further, as we saw in § 4 c, if the operation J belongs to the class $\mathfrak{J}$, so does every operation J_ω, where ω is a hermitian function of the class $\mathfrak{K}$ of definitely positive type as regards the operation J. If ω is merely of positive type, J_ω has at least the properties $LMHP$.

These are closure properties of the class $\mathfrak{J}$. Another closure property is given by the following theorem.

If $J_1, J_2, \ldots, J_n, \ldots$ is an infinite sequence of operations on $\mathfrak{K}$ to $\mathfrak{A}$ such that (1) every operation J_n has the properties $LMHP$, while at least one operation J_n has the additional property P_0, and so belongs to the class $\mathfrak{J}$; (2) for every function κ of $\mathfrak{K}$ the infinite series $\sum_{n=1}^{\infty} J_n\kappa$ converges absolutely; (3) there exists a sequence $M_1, M_2, \ldots, M_n, \ldots$ of modular operations on $\mathfrak{K}_{\geqq 0}$ to $\mathfrak{A}$, the operation M_n being associated with the operation J_n in the sense of the modular property M of J_n, of such a nature that for every function κ_0 of $\mathfrak{K}_{\geqq 0}$ the infinite series $\sum_{n=1}^{\infty} M_n\kappa_0$ converges, then the operation $J \equiv \sum_{n=1}^{\infty} J_n$: $J\kappa = \sum_{n=1}^{\infty} J_n\kappa$, is an operation on $\mathfrak{K}$ to $\mathfrak{A}$ with the properties $LMHPP_0$ and accordingly belongs to the class $\mathfrak{J}$.

We notice the following example. Let $p_1, \ldots, p_m$ be m fixed elements of the range $\mathfrak{P}$. The corresponding operations

$$R_{p_1}, \ldots, R_{p_m}:\ R_{p_1}\kappa = \kappa(p_1, p_1), \ldots, R_{p_m}\kappa = \kappa(p_m, p_m),$$

are operations on $\mathfrak{K}$ to $\mathfrak{A}$ with the properties $LMHP$. Accordingly, if J is an operation of the class $\mathfrak{J}$, so is the operation $J_1 \equiv R_{p_1} + \ldots + R_{p_m} + J$. For this operation J_1 the linear equation G: $\xi = \eta - zJ_1\kappa\eta$, has the form

$$\xi(s) = \eta(s) - z\left(\sum_{n=1}^{m} \kappa(s, p_n)\,\eta(p_n) + J_{(tu)}\,\kappa(s,t)\,\eta(u)\right) \qquad (s),$$

and to this equation the general theory applies. For the classical instance IV this equation has the form

$$\xi(s)=\eta(s)-z\left(\sum_{n=1}^{m}\kappa(s,p_n)\eta(p_n)+\int_a^b\kappa(s,t)\eta(t)\,dt\right)\qquad(a\leqq s\leqq b),$$

where $p_1, \ldots, p_m$ are m arguments of the range $a\leqq p\leqq b$.

5. The Function Space $\mathfrak{M}$. Unitary Orthogonal Sets of Functions. The Pythagoras and Schwarz Inequalities.

We are now to consider certain geometric properties of the class $\mathfrak{M}$ of functions relative to the operation J of the system Σ_5, to the extent requisite for the theory of the general linear integral equation G. For the algebraic instance II_n a function ξ of the class $\mathfrak{M}$ is interpretable as a point or a vector in space of n (real or complex) dimensions, and this instance is to be borne in mind in the sequel.

(*a*) **Modulus of a function. Orthogonality of a pair of functions.**— A function ξ of the class $\mathfrak{M}$ has the real non-negative modulus $M(\xi)$, where $M^2(\xi)=J\bar{\xi}\xi$, which vanishes if, and only if, $\xi=0$. A function ξ of modulus 1 is unitary or a unit function. If $M(\xi)\neq 0$, $\xi/M(\xi)$ is unitary.

Two functions ξ, η are orthogonal, in notation: $(\xi,\eta)^O$, in case $J\bar{\xi}\eta=0$. The relation of orthogonality is symmetric.

For two orthogonal functions ξ, η we have the Pythagoras equality

$$M^2(\xi+\eta)=M^2(\xi)+M^2(\eta).$$

The modulus M and orthogonality O are relative to the operation J. Similarly we define relative to the transpose operation $\breve{J}$ the transpose modulus $\breve{M}$ and orthogonality $\breve{O}$, and have the relations

$$\breve{M}(\xi)=M(\bar{\xi});\ (\xi,\eta)^{\breve{O}}\equiv(\bar{\xi},\bar{\eta})^O.$$

It should be noted that in general we do not have the relations

$$M(\xi)=M(\bar{\xi});\ (\xi,\eta)^O\equiv(\bar{\xi},\bar{\eta})^O,$$

that is, modulus and orthogonality are in general not self-conjugate. For example, consider the algebraic instance II_2 derived from the classical instance II_2 by the utilization (in the sense of § 4 *c*) of the hermitian function ω:

$$\omega(1,1)=\omega(2,2)=5,\quad \omega(1,2)=4i,\quad \omega(2,1)=-4i\qquad(i=\sqrt{-1}),$$

of definitely positive type. Then, setting $\kappa(g,h)=k_{gh}\,(g,h=1,2)$, we have

$$J\kappa=5(k_{11}+k_{22})+4i(k_{12}-k_{21}),$$

and accordingly for two functions ξ, η:

$$\xi(g)=x_g,\ \eta(g)=y_g\,(g=1,2),\ \text{ or }\ \xi=(x_1,x_2),\ \eta=(y_1,y_2),$$

we have
$$J\bar{\xi}\eta=5(\bar{x}_1y_1+\bar{x}_2y_2)+4i(\bar{x}_1y_2-\bar{x}_2y_1).$$

Then the moduli of the conjugate functions $(1, i)$, $(1, -i)$ are respectively $\sqrt{2}$, $\sqrt{18}$; and the functions $(1, 0)$, $(4, 5i)$ are orthogonal, while the conjugate functions $(1, 0)$, $(4, -5i)$ are not orthogonal.

(*b*) **The combinatorial product of m functions. The multiplication theorem for integral determinants.**—Consider a set $(\xi_1, \ldots, \xi_m)$ of m functions of the class $\mathfrak{M}$. By multiplication of these functions in independent arguments $p_1, \ldots, p_m$ of the range $\mathfrak{P}$ we have the product function

$$\xi_1\xi_2\ldots\xi_m \quad \text{or} \quad \xi_1(p_1)\,\xi_2(p_2)\ldots\xi_m(p_m)$$

of the class $\mathfrak{M}\mathfrak{M}\ldots\mathfrak{M}$ or $\mathfrak{M}_{p_1}\mathfrak{M}_{p_2}\ldots\mathfrak{M}_{p_m}$; in the notation $\xi_1\xi_2\ldots\xi_m$ the order is of importance (cf. §§ 3 *c*, 4 *c*). The combinatorial product $(\xi_1\xi_2\ldots\xi_m)$ of the m functions is the sum

$$(\xi_1\xi_2\ldots\xi_m) \equiv \sum_{f_1, \ldots, f_m}^{1, m} \pm_{f_1\ldots f_m}\xi_{f_1}\xi_{f_2}\ldots\xi_{f_m},$$

where the symbol $\pm_{f_1\ldots f_m}$ is 0 if two indices are equal, and otherwise $+1$ or -1, according as $f_1, \ldots, f_m$ is an even or odd permutation of $1, \ldots, m$. Thus, the combinatorial product $(\xi_1\ldots\xi_m)$ in the arguments $p_1, \ldots, p_m$ is the determinant

$$(\xi_1\xi_2\ldots\xi_m)(p_1, \ldots, p_m) \equiv |\,\xi_f(p_h)\,| \qquad (f, h = 1, \ldots, m).$$

If the functions $\xi_1, \ldots, \xi_m$ are linear combinations of functions $\eta_1, \ldots, \eta_m$:

$$\xi_f = \sum_{g=1}^{m} a_{fg}\eta_g \qquad (f = 1, \ldots, m),$$

we have $$(\xi_1\xi_2\ldots\xi_m) = |\,a_{fg}\,|\,(\eta_1\eta_2\ldots\eta_m) \qquad (f, g = 1, \ldots, m).$$

The combinatorial product $(\xi_1\xi_2\ldots\xi_m)$ changes sign on the interchange of two of the m functions ξ_f, and it vanishes identically if and only if the m functions ξ_f are linearly related.

From two sets $(\xi_1, \ldots, \xi_m)$, $(\eta_1, \ldots, \eta_m)$, each of m functions, arises the integral determinant

$$|\,J\xi_f\eta_g\,| \qquad (f, g = 1, \ldots, m),$$

in which the m^2 elements are the integrals $J\xi_f\eta_g$ of the m^2 products $\xi_f\eta_g$. This determinant for $m = 1$ is the integral $J\xi\eta$; we denote it by the notation*

$$J^m(\xi_1, \ldots, \xi_m)(\eta_1, \ldots, \eta_m).$$

This determinant vanishes if the functions of either set are linearly related and it is an invariant of weight 1 as to linear homogeneous transformations of either set of functions.

Multiplication theorem.—For two sets $(\xi_1, \ldots, \xi_m)$, $(\eta_1, \ldots, \eta_m)$ of functions of the class $\mathfrak{M}$ the integral determinant multiplied by $m!$ is equal to a certain m-fold integral of the product of their combinatorial products, viz.,

$$m!\,J^m(\xi_1, \ldots, \xi_m)(\eta_1, \ldots, \eta_m) = J^m(\xi_1\xi_2\ldots\xi_m)(\eta_1\eta_2\ldots\eta_m),$$

where on the right J^m denotes the m-fold operation

$$J_{(1\ m+1)}\,J_{(2\ m+2)}\ldots J_{(m\ 2m)},$$

in the notation of § 4 *c*.

* Here J^m is an operation (associated with the operation J) whose operand is always of the form: $(\xi_1, \ldots, \xi_m)(\eta_1, \ldots, \eta_m)$. If we multiply the two sets to form the matrix $(\xi_f\eta_g)$ $(f, g = 1, \ldots, m)$, we may write $J^m(\xi_f\eta_g) \equiv |\,J\xi_f\eta_g\,|$ $(f, g = 1, \ldots, m)$, and accordingly, more generally, for a matrix (κ_{fg}) of functions of the class $\mathfrak{K}$, we may write $J^m(\kappa_{fg}) \equiv |\,J\kappa_{fg}\,|$ $(f, g = 1, \ldots, m)$.

Proof.—The right side has the value

$$\sum_{\substack{f_1, \dots, f_m \\ g_1, \dots, g_m}}^{1, m} \pm_{f_1 \dots f_m} \pm_{g_1 \dots g_m} J\xi_{f_1}\eta_{g_1} J\xi_{f_2}\eta_{g_2} \dots J\xi_{f_m}\eta_{g_m},$$

or

$$m! \sum_{f_1, \dots, f_m}^{1, m} \pm_{f_1 \dots f_m} J\xi_{f_1}\eta_1 J\xi_{f_2}\eta_2 \dots J\xi_{f_m}\eta_m,$$

which is an expansion of the left side.

It is to be noted that the factor $m!$ on the left may be omitted if we replace either combinatorial product on the right by the corresponding simple product, viz.,

$$J^m(\xi_1, \dots, \xi_m)(\eta_1, \dots, \eta_m) = J^m(\xi_1\xi_2 \dots \xi_m)\eta_1\eta_2 \dots \eta_m = J^m\xi_1\xi_2 \dots \xi_m(\eta_1\eta_2 \dots \eta_m).$$

(*c*) **Modulus of a set of functions.**—The modulus $M(\xi)$: $M^2(\xi) = J\bar{\xi}\xi$, of a function ξ of the class $\mathfrak{M}$ is a real non-negative number which vanishes if and only if $\mu = 0$. Similarly, the modulus $M(\xi_1, \dots, \xi_m)$ of a set $(\xi_1, \dots, \xi_m)$ of functions of the class $\mathfrak{M}$, suitably defined by the equation

$$M^2(\xi_1, \dots, \xi_m) = J^m(\bar{\xi}_1, \dots, \bar{\xi}_m)(\xi_1, \dots, \xi_m) = |J\bar{\xi}_f\xi_g|,$$

where the integral determinant on the right has, by § 5 *b*, the expressions as m-fold integrals

$$\frac{1}{m!} J^m(\bar{\xi}_1\bar{\xi}_2 \dots \bar{\xi}_m)(\xi_1\xi_2 \dots \xi_m), \quad J^m(\bar{\xi}_1\bar{\xi}_2 \dots \bar{\xi}_m)\xi_1\xi_2 \dots \xi_m, \quad J^m\bar{\xi}_1\bar{\xi}_2 \dots \bar{\xi}_m(\xi_1\xi_2 \dots \xi_m),$$

is a real non-negative number which vanishes if and only if the functions ξ_f are linearly related.

The proof by induction of this fact, true for $m = 1$, depends upon the identity

$$M^2(\theta) = M^2(\xi_1, \dots, \xi_m) M^2(\xi_1, \dots, \xi_m, \xi_{m+1}),$$

where the function θ is a certain function of the form

$$\theta = M^2(\xi_1, \dots, \xi_m)\xi_{m+1} + a_1\xi_1 + \dots + a_m\xi_m,$$

viz., the function $J^m\bar{\xi}_1\bar{\xi}_2 \dots \bar{\xi}_m(\xi_1\xi_2 \dots \xi_m\xi_{m+1})$. $M^2(\theta)$ or $J\bar{\theta}\theta$ is readily put in the form

$$\sum_{\substack{f_1, \dots, f_m, f_{m+1} \\ g_1, \dots, g_m, g_{m+1}}}^{1, m+1} \pm_{f_1 \dots f_m f_{m+1}} \pm_{g_1 \dots g_m g_{m+1}} J\bar{\xi}_{f_1}\xi_1 \dots J\bar{\xi}_{f_m}\xi_m J\bar{\xi}_1\xi_{g_1} \dots J\bar{\xi}_m\xi_{g_m} J\bar{\xi}_{f_{m+1}}\xi_{g_{m+1}},$$

or, since the terms with $f_{m+1} = 1, \dots, m$ occur in pairs with opposite signs, in the form

$$\sum_{f_1, \dots, f_m}^{1, m} \pm_{f_1 \dots f_m} J\bar{\xi}_{f_1}\xi_1 \dots J\bar{\xi}_{f_m}\xi_m \cdot \sum_{g_1, \dots, g_m, g_{m+1}}^{1, m+1} \pm_{g_1 \dots g_m g_{m+1}} J\bar{\xi}_1\xi_{g_1} \dots J\bar{\xi}_m\xi_{g_m} J\bar{\xi}_{m+1}\xi_{g_{m+1}};$$

thus

$$\begin{aligned} M^2(\theta) &= J^m(\bar{\xi}_1\bar{\xi}_2 \dots \bar{\xi}_m)\xi_1\xi_2 \dots \xi_m J^{m+1}\bar{\xi}_1\bar{\xi}_2 \dots \bar{\xi}_m\bar{\xi}_{m+1}(\xi_1\xi_2 \dots \xi_m\xi_{m+1}) \\ &= M^2(\xi_1, \dots, \xi_m) M^2(\xi_1, \dots, \xi_m, \xi_{m+1}), \end{aligned}$$

as stated.

Now admit the statements as to the modulus of a set of m functions, and suppose that $(\xi_1, \dots, \xi_m, \xi_{m+1})$ is a set of linearly independent functions. Then $M^2(\xi_1, \dots, \xi_m)$ and $M^2(\theta)$ are real and positive, since $(\xi_1, \dots, \xi_m)$ is a set of linearly independent functions and θ is not the function 0. Hence, by the preceding remark, $M^2(\xi_1, \ldots, \xi_m, \xi_{m+1})$ is real and positive. Accordingly, the statements as to the modulus of a set of $m+1$ functions are valid, since the modulus vanishes if the functions are linearly related, as one sees from either expression of the square of the modulus as a multiple integral, the combinatorial product vanishing identically.

The definitions and theorems of § 5 *a*, *b*, *c* in essentially the present form, although for the more special integral operation J_p and with different postulates on the class $\mathfrak{M}$ and the operation J, have been in my possession since 1906, and I gave the multiplication theorem for integral determinants in a course of lectures on the general theory of determinants at Chicago in the summer of 1907. The equivalence of the identical vanishing of the combinatorial product and the linear dependence of the functions of a set of functions was, for the instance IV, perhaps first published by Goursat (*Annales de Toulouse*, ser. 2, vol. 10, p. 80, 1908). The multiplication theorem for instance IV was first published by I. Schur (*Mathematische Annalen*, vol. 67, p. 319, 1909), while the names "integral determinant" and "multiplication theorem" are due to Landsberg (*ibid.*, vol. 69, p. 231, 1910).

(*d*) **Hadamard's theorem in general analysis.**—For functions ξ of the class $\mathfrak{M}$ we have

$$M(\xi_1, \ldots, \xi_m, \xi_{m+1}) \leqq M(\xi_1, \ldots, \xi_m)\, M(\xi_{m+1}),$$

and accordingly the general Hadamard theorem:

$$M(\xi_1, \ldots, \xi_m) \leqq M(\xi_1)\, M(\xi_2) \ldots M(\xi_m).$$

The first inequality hold as equality if and only if either the functions $(\xi_1, \ldots, \xi_m)$ are linearly related or the function ξ_{m+1} is orthogonal to the various functions $(\xi_1, \ldots, \xi_m)$. Accordingly, the second inequality hold as equality if and only if the functions $(\xi_1, \ldots, \xi_m)$ are in pairs orthogonal*.

We prove the first inequality and added remark in case the functions $(\xi_1, \ldots, \xi_m)$ are linearly independent, so that $M(\xi_1, \ldots, \xi_m) \neq 0$. Set

$$\chi_{m+1} \equiv J^m \bar{\xi}_1 \bar{\xi}_2 \ldots \bar{\xi}_m (\xi_1 \xi_2 \ldots \xi_m \xi_{m+1})\,; \quad \psi_{m+1} \equiv \chi_{m+1}/M^2(\xi_1, \ldots, \xi_m).$$

Then $$M(\chi_{m+1}) = M^2(\xi_1, \ldots, \xi_m)\, M(\psi_{m+1}),$$

while, as proved in § 5 *c*,

$$M(\xi_1, \ldots, \xi_m)\, M(\xi_1, \ldots, \xi_m, \xi_{m+1}).$$

Hence $$M(\xi_1, \ldots, \xi_m, \xi_{m+1}) = M(\xi_1, \ldots, \xi_m)\, M(\psi_{m+1}).$$

We readily see that $\psi_{m+1} = \xi_{m+1} - \eta_{m+1}$, where η_{m+1} is of the form $b_1\xi_1 + \ldots + b_m\xi_m$, and that χ_{m+1} and ψ_{m+1} are orthogonal to the various functions $(\xi_1, \ldots, \xi_m)$, since

$$J\bar{\xi}_f \chi_{m+1} = J^{m+1} \bar{\xi}_1 \bar{\xi}_2 \ldots \bar{\xi}_m \bar{\xi}_f (\xi_1 \xi_2 \ldots \xi_m \xi_{m+1}) = J^{m+1} (\bar{\xi}_1 \bar{\xi}_2 \ldots \bar{\xi}_m \bar{\xi}_f)\, \xi_1 \xi_2 \ldots \xi_m \xi_{m+1} = 0,$$

the combinatorial product $(\bar{\xi}_1 \bar{\xi}_2 \ldots \bar{\xi}_m \bar{\xi}_f)$ vanishing identically $(f = 1, \ldots, m)$. Hence ψ_{m+1} is orthogonal to η_{m+1}, and with $\xi_{m+1} = \psi_{m+1} + \eta_{m+1}$ we have the Pythagoras equality: $M^2(\xi_{m+1}) = M^2(\psi_{m+1}) + M^2(\eta_{m+1})$, and accordingly $M(\psi_{m+1}) \leqq M(\xi_{m+1})$, the equality holding if and only if $\eta_{m+1} = 0$, and so if and only if ξ_{m+1} is orthogonal to

* If $m=1$, this condition is understood to hold (as we say, vacuously).

the various functions $(\xi_1, \ldots, \xi_m)$. Hence the first inequality and added remark are proved.

We notice further that the functions $(\psi_1, \ldots, \psi_{m+1})$:

$$\psi_1 \equiv \xi_1;\ \psi_f \equiv J^{f-1}\bar{\xi}_1\bar{\xi}_2 \ldots \bar{\xi}_{f-1}(\xi_1\xi_2 \ldots \xi_{f-1}\xi_f)/M^2(\xi_1, \ldots, \xi_{f-1}) \quad (f = 2, \ldots, m+1),$$

are mutually orthogonal and connected with the functions $(\xi_1, \ldots, \xi_{m+1})$ by a linear homogeneous transformation of determinant 1, and accordingly

$$M(\xi_1, \ldots, \xi_m, \xi_{m+1}) = M(\psi_1, \ldots, \psi_{m+1}) = M(\psi_1)\,M(\psi_2) \ldots M(\psi_{m+1}).$$

Hadamard's classical theorem on determinants of order m is the instance II_m of the general Hadamard theorem stated above.

(*e*) **Equivalence of sets of functions. Unitary sets and orthogonal sets of functions.**—A set $(\phi_1, \ldots, \phi_m)$ of m functions of the class $\mathfrak{M}$ as basis gives rise to the linear class $(\phi_1, \ldots, \phi_m)_L$ consisting of all functions ξ of the form

$$x_1\phi_1 + \ldots + x_m\phi_m,$$

where the coefficients x_g are numbers of the class $\mathfrak{A}$. If the functions ϕ_g of the basis are linearly independent, a function ξ of the linear class $(\phi_1, \ldots, \phi_m)_L$ is uniquely determined by and uniquely determines the corresponding set $(x_1, \ldots, x_m)$ of numbers, the coördinates of ξ as to the basis $(\phi_1, \ldots, \phi_m)$; on occasion, we write $\xi = (x_1, \ldots, x_m)$; the linear class $(\phi_1, \ldots, \phi_m)_L$ is thus a function space of m real or complex dimensions, according as $\mathfrak{A}$ is the class of real or complex numbers.

Two sets $(\phi_1, \ldots, \phi_m)$, $(\psi_1, \ldots, \psi_n)$ of functions are linearly equivalent: in notation, $(\phi_1, \ldots, \phi_m) \sim (\psi_1, \ldots, \psi_n)$, in case the two classes $(\phi_1, \ldots, \phi_m)_L$, $(\psi_1, \ldots, \psi_n)_L$ are identical, that is, in case the functions ϕ_f are linearly expressible in the functions ψ_g and conversely.

The set $(\phi_1, \ldots, \phi_m)$ is unitary, in notation: $(\phi_1, \ldots, \phi_m)^U$, or orthogonal, in notation: $(\phi_1, \ldots, \phi_m)^O$, in case its functions are respectively individually unitary or in pairs orthogonal.

A unitary orthogonal set $(\phi_1, \ldots, \phi_m)^{UO}$ of functions, necessarily linearly independent, plays the rôle of a set of m mutually orthogonal unit vectors in geometry. Thus, consider two functions:

$$\xi = (x_1, \ldots, x_m) = \sum_{g=1}^{m} x_g\phi_g; \quad \eta = (y_1, \ldots, y_m) = \sum_{g=1}^{m} y_g\phi_g,$$

of the m dimensional function space $(\phi_1, \ldots, \phi_m)_L$. The coördinates x_g are determined by the formulae

$$x_g = J\bar{\phi}_g\xi, \quad \bar{x}_g = J\bar{\xi}\phi_g \qquad (g = 1, \ldots, m).$$

Hence we have the equations

$$\xi = \sum_{g=1}^{m} \phi_g J\bar{\phi}_g\xi; \quad \bar{\xi} = \sum_{g=1}^{m} J\bar{\xi}\phi_g\,\bar{\phi}_g;$$

$$\bar{\xi}\eta = \sum_{g=1}^{m} \bar{x}_g y_g = \sum_{g=1}^{m} J\bar{\xi}\phi_g J\bar{\phi}_g\eta.$$

For the modulus $M(\xi)$ we have

$$M^2(\xi) = J\bar{\xi}\xi = \sum_{g=1}^{m} \bar{x}_g x_g;$$

and for the modulus $M(\xi-\eta)$ or distance $\xi\eta$ between the two points ξ, η of the function space $(\phi_1, \ldots, \phi_m)_L$ we have

$$M^2(\xi-\eta) = J(\bar{\xi}-\bar{\eta})(\xi-\eta) = \sum_{g=1}^{m} (\bar{x}_g - \bar{y}_g)(x_g - y_g).$$

Accordingly, the function space $(\phi_1, \ldots, \phi_m)_L$ with unitary orthogonal basis $(\phi_1, \ldots, \phi_m)$ is, with respect to the operation J, a real m dimensional euclidean space or a complex m dimensional parabolic hermitian space in the sense of Study, according as $\mathfrak{A}$ is the real or the complex number system.

The rank of a set $(\xi_1, \ldots, \xi_m)$ of functions is the largest number of linearly independent functions contained in the set. Two equivalent sets are of the same rank. For a set of rank $n>0$ there is an equivalent unitary orthogonal set of rank n.

The infinite set $(\phi_1, \phi_2, \ldots, \phi_n, \ldots)$ of functions is unitary, or orthogonal, if its functions are, respectively, individually unitary, or in pairs orthogonal.

Consider an infinite set $(\xi_1, \xi_2, \ldots, \xi_n, \ldots)$ of functions of the class $\mathfrak{M}$. We proceed to construct an infinite orthogonal set $(\psi_1, \psi_2, \ldots, \psi_n, \ldots)$ of such a nature that for every m the sets $(\xi_1, \ldots, \xi_m)$, $(\psi_1, \ldots, \psi_m)$ are equivalent, in fact, in such a way that $\psi_1 - \xi_1 = 0$ and for every m $\psi_{m+1} - \xi_{m+1}$ is linearly expressible in $\xi_1, \ldots, \xi_m$ or in $\psi_1, \ldots, \psi_m$.

Construction.—Take $\psi_1 \equiv \xi_1$. Whenever the set $(\xi_1, \ldots, \xi_m)$ is of rank 0 take $\psi_{m+1} \equiv \xi_{m+1}$. Whenever the set $(\xi_1, \ldots, \xi_m)$ is of rank $n>0$, denote by $(\xi_{g_1}, \ldots, \xi_{g_n})$ a set of n linearly independent functions chosen from the m functions $\xi_g (g=1, \ldots, m)$. Then the function χ_{m+1}:

$$\chi_{m+1} \equiv J^n \bar{\xi}_{g_1} \cdots \bar{\xi}_{g_n} (\xi_{g_1} \cdots \xi_{g_n} \xi_{m+1}),$$

is linear in $\xi_{g_1}, \ldots, \xi_{g_n}, \xi_{m+1}$, and is orthogonal to the functions $\xi_{g_1}, \ldots, \xi_{g_n}$ and hence to the functions $\xi_1, \ldots, \xi_m$; and the coefficient of ξ_{m+1} in χ_{m+1} is $J^n \bar{\xi}_{g_1} \cdots \bar{\xi}_{g_n} (\xi_{g_1} \cdots \xi_{g_n})$, that is, $M^2(\xi_{g_1}, \ldots, \xi_{g_n})$, which is different from 0. Take $\psi_{m+1} \equiv \chi_{m+1}/M^2(\xi_{g_1}, \ldots, \xi_{g_n})$. It is readily seen that the set $(\psi_1, \ldots, \psi_n, \ldots)$ so determined satisfies the conditions specified, and that it is the only set $(\psi_1, \ldots, \psi_n, \ldots)$ which satisfies those conditions.

Consider now an infinite set $(\xi_1, \ldots, \xi_n, \ldots)$ of functions of such a nature that for every m the set $(\xi_1, \ldots, \xi_m)$ is of rank m. We proceed to construct an infinite unitary orthogonal set $(\phi_1, \ldots, \phi_n, \ldots)$ of such a nature that for every m the sets $(\xi_1, \ldots, \xi_m)$, $(\phi_1, \ldots, \phi_m)$ are equivalent.

Construction.—Take

$$\phi_1 \equiv \frac{\xi_1}{M(\xi_1)}, \quad \phi_{m+1} \equiv \frac{\chi_{m+1}}{M(\chi_{m+1})} \qquad (m=1, 2, \ldots, n, \ldots),$$

where

$$\chi_{m+1} \equiv J^m \bar{\xi}_1 \ldots \bar{\xi}_m (\xi_1 \ldots \xi_m \ \xi_{m+1}) \qquad (m=1, 2, \ldots, n, \ldots).$$

These definitions are legitimate, since, from the hypothesis, for every m $M(\xi_1, \ldots, \xi_m) \neq 0$, and, as we saw in § 5 c,

$$M^2(\chi_{m+1}) = M^2(\xi_1, \ldots, \xi_m) M^2(\xi_1, \ldots, \xi_m, \xi_{m+1}).$$

It is readily seen that the set $(\phi_1, \ldots, \phi_n, \ldots)$ so determined satisfies the conditions specified, and that these conditions determine the set $(\phi_1, \ldots, \phi_n, \ldots)$ up to numerical multipliers a_n (of the respective functions ϕ_n) belonging to the class $\mathfrak{A}$ and of unit modulus: $|a_n| = 1$.

(f) **The Pythagoras or Bessel inequality.**—Consider in the function space $\mathfrak{M}$ a unitary orthogonal set $(\phi_1, \ldots, \phi_m)$ and the corresponding m-space $(\phi_1, \ldots, \phi_m)_L$. A function ξ of $\mathfrak{M}$ is uniquely separable into two components ξ_1, ξ_2: $\xi = \xi_1 + \xi_2$, ξ_1 of $(\phi_1, \ldots, \phi_m)_L$ and ξ_2 orthogonal to $(\phi_1, \ldots, \phi_m)_L$, viz.,

$$\xi_1 \equiv \sum_{n=1}^{m} \phi_n J\bar{\phi}_n\xi; \quad \xi_2 \equiv \xi - \xi_1 \equiv \xi - \sum_{n=1}^{m} \phi_n J\bar{\phi}_n\xi.$$

The function ξ_1 is the orthogonal projection of ξ on $(\phi_1, \ldots, \phi_m)_L$; the function ξ_2 is the component of ξ orthogonal to $(\phi_1, \ldots, \phi_m)_L$. For the orthogonal functions ξ_1, ξ_2 we have the Pythagoras equality: $M^2(\xi_1) + M^2(\xi_2) = M^2(\xi_1 + \xi_2) = M^2(\xi)$.

Accordingly, for a function ξ and unitary orthogonal set $(\phi_1, \ldots, \phi_m)$ of functions of $\mathfrak{M}$ we have the Pythagoras or Bessel inequality

$$\sum_{n=1}^{m} J\bar{\xi}\phi_n J\bar{\phi}_n\xi \leqq J\bar{\xi}\xi,$$

wherein the inequality holds as equality if, and only if, ξ is a function of $(\phi_1, \ldots, \phi_m)_L$. Hence, for a function ξ and an infinite unitary orthogonal set $(\phi_1, \ldots, \phi_n, \ldots)$ the infinite series $\sum_{n=1}^{\infty} J\bar{\xi}\phi_n J\bar{\phi}_n\xi$ of real-valued non-negative terms converges to a sum at most $J\bar{\xi}\xi$, and we have the corresponding Pythagoras inequality

$$\sum_{n=1}^{\infty} J\bar{\xi}\phi_n J\bar{\phi}_n\xi \leqq J\bar{\xi}\xi.$$

Thus, for a function ξ and a finite or infinite unitary orthogonal set $(\phi_1, \ldots, \phi_n, \ldots)$ we have the Pythagoras inequality

$$\sum_{n} J\bar{\xi}\phi_n J\bar{\phi}_n\xi \leqq J\bar{\xi}\xi,$$

where $\sum_{n}$ is of the form $\sum_{n=1}^{m}$ or $\sum_{n=1}^{\infty}$, according as the set $(\phi_1, \ldots, \phi_n, \ldots)$ is finite or infinite. Since the class $\mathfrak{M}$ is real and the operation J is hermitian, this inequality occurs also in the form

$$\sum_{n} J\xi\phi_n \overline{J\xi\phi_n} \leqq J\xi\bar{\xi}.$$

(g) **The Pythagoras or Bessel inequality for a finite set of functions.**—The Pythagoras inequality, in the two forms given at the conclusion of § 5 f, may be generalized by taking instead of the function ξ a finite set $(\xi_1, \ldots, \xi_k)$ of functions of the class $\mathfrak{M}$. We have thus the two forms

$$\sum_{n_1 < \ldots < n_k} J^k(\bar{\xi}_1, \ldots, \bar{\xi}_k)(\phi_{n_1}, \ldots, \phi_{n_k}) J^k(\bar{\phi}_{n_1}, \ldots, \bar{\phi}_{n_k})(\xi_1, \ldots, \xi_k) \leqq J^k(\bar{\xi}_1, \ldots, \bar{\xi}_k)(\xi_1, \ldots, \xi_k);$$

$$\sum_{n_1 < \ldots < n_k} J^k(\xi_1, \ldots, \xi_k)(\phi_{n_1}, \ldots, \phi_{n_k}) \overline{J^k(\xi_1, \ldots, \xi_k)(\phi_{n_1}, \ldots, \phi_{n_k})} \leqq J^k(\xi_1, \ldots, \xi_k)(\bar{\xi}_1, \ldots, \bar{\xi}_k).$$

Here the terms on the right and the factors of terms on the left are integral determinants of order k, expressible in various forms as k-fold integrals, as explained in § 5 b; the summation indices $n_1, \ldots, n_k$ run from 1 to ∞ if the unitary orthogonal set $(\phi_1, \ldots, \phi_n, \ldots)$ is infinite, and from 1 to m if the set is finite: $(\phi_1, \ldots, \phi_m)$, in which case if $k > m$ the left sides are to be interpreted as having the value 0, the inequality becoming (in view of § 5 c) trivial.

This generalized Pythagoras inequality for the case of an infinite unitary orthogonal set $(\phi_1, \ldots, \phi_n, \ldots)$ follows (as in § 5 f for $k = 1$) from the generalized inequality for the case of a finite unitary orthogonal set $(\phi_1, \ldots, \phi_m)$ with $m \geqq k$.

This latter inequality holds as an equality, if the functions $\xi_1, \ldots, \xi_k$ are linearly expressible in the functions $\phi_1, \ldots, \phi_m$. This equality is a particular case of a more general equality which we state as a lemma.

Lemma 1. If the k functions $\xi_1, \ldots, \xi_k$ are linearly expressible in terms of m mutually orthogonal functions $\phi_1, \ldots, \phi_m$ $(m \geqq k)$:

$$\xi_f = \sum_{n=1}^{m} x_{fn}\phi_n \qquad (f = 1, \ldots, k),$$

then
$$J^k(\bar{\xi}_1, \ldots, \bar{\xi}_k)(\xi_1, \ldots, \xi_k) = \sum_{n_1 < \ldots < n_k}^{1,\, m} |\bar{x}_{fn_g}|\,|x_{fn_g}|\, J\bar{\phi}_{n_1}\phi_{n_1} \ldots J\bar{\phi}_{n_k}\phi_{n_k},$$

where the determinants are of order k, viz., $f, g = 1, \ldots, k$.

Proof.—By the multiplication theorem we have

$$J^k(\bar{\xi}_1, \ldots, \bar{\xi}_k)(\xi_1, \ldots, \xi_k) = \frac{1}{k!} J^k\, |\bar{\xi}_f(s_g)|\,|\xi_f(t_g)|,$$

where J^k on the right denotes the k-fold operation $J_{(s_1 t_1)} \ldots J_{(s_k t_k)}$. This expression readily takes the form

$$\frac{1}{k!} \sum_{\substack{u_1, \ldots, u_k \\ v_1, \ldots, v_k}}^{1,\, m} J^k\, |\bar{x}_{fu_g}\bar{\phi}_{u_g}(s_g)|\,|x_{fv_g}\phi_{v_g}(t_g)|,$$

that is,
$$\frac{1}{k!} \sum_{\substack{u_1, \ldots, u_k \\ v_1, \ldots, v_k}}^{1,\, m} |\bar{x}_{fu_g}|\,|x_{fv_g}|\, J\bar{\phi}_{u_1}\phi_{v_1} \ldots J\bar{\phi}_{u_k}\phi_{v_k},$$

from which, in view of the mutual orthogonality of the functions ϕ_n, the required form is readily obtained.

In case the set $(\phi_1, \ldots, \phi_m)$ is unitary orthogonal we have $x_{fn} = J\bar{\phi}_n \xi_f$ $(f=1, \ldots, k;\ n=1, \ldots, m)$, and readily see that this case of lemma 1 is the special case specified above of the generalized Pythagoras inequality.

It remains to prove the generalized inequality for the finite unitary orthogonal set $(\phi_1, \ldots, \phi_m)$ and any set $(\xi_1, \ldots, \xi_k)$ where $m \geqq k$. By means of lemma 1 this will be reduced to the special case.

Separate the functions ξ_f $(f=1, \ldots, m)$ into orthogonal components η_f, ζ_f: $\xi_f = \eta_f + \zeta_f$, with respect to the space $(\phi_1, \ldots, \phi_m)_L$, η_f being the orthogonal projection of ξ_f on the space and ζ_f being the component of ξ_f orthogonal to the space. Then the left of the Pythagoras inequality is unchanged, if we replace the functions ξ_f respectively by the functions η_f, since the elements of the determinants involved are of the forms $J\bar{\xi}_f \phi_n$, $J\bar{\phi}_n \xi_f$, and, in view of the orthogonality of ζ_f to $(\phi_1, \ldots, \phi_m)_L$, $J\bar{\zeta}_f \phi_n = J\bar{\phi}_n \zeta_f = 0$. Thus, by the preceding case, the inequality holds if on the right we replace the functions ξ_f respectively by the functions η_f, and accordingly it holds as written, in view of the following lemma.

Lemma 2. If the k functions η_f $(f=1, \ldots, k)$ are orthogonal to the k functions ζ_g $(g=1, \ldots, k)$: $J\bar{\eta}_f \zeta_g = 0$ $(f, g = 1, \ldots, k)$, then

$$J^k (\bar{\eta}_1, \ldots, \bar{\eta}_k)(\eta_1, \ldots, \eta_k) + J^k (\bar{\zeta}_1, \ldots, \bar{\zeta}_k)(\zeta_1, \ldots, \zeta_k) \leqq J^k (\bar{\xi}_1, \ldots, \bar{\xi}_k)(\xi_1, \ldots, \xi_k),$$

viz.,

$$M^2 (\eta_1, \ldots, \eta_k) + M^2 (\zeta_1, \ldots, \zeta_k) \leqq M^2 (\xi_1, \ldots, \xi_k),$$

where $\xi_f = \eta_f + \zeta_f$ $(f=1, \ldots, k)$.

For $k=1$ this inequality holds as an equality, the Pythagoras equality, while for $k>1$, on the introduction of orthogonal sets $(\psi_1, \ldots, \psi_k)$, $(\chi_1, \ldots, \chi_k)$ respectively equivalent to the sets $(\eta_1, \ldots, \eta_k)$, $(\zeta_1, \ldots, \zeta_k)$, we readily secure the inequality by three applications of lemma 1.

The generalized Pythagoras inequality is now fully established.

(*h*) **The Schwarz inequality.**—For two functions ξ, η of the class $\mathfrak{M}$ we have the Schwarz inequality

$$J\bar{\xi}\eta \, J\bar{\eta}\xi \leqq J\bar{\xi}\xi \, J\bar{\eta}\eta.$$

This inequality holds obviously if $\xi = \eta = 0$, and, if say ξ is not 0, it is the Pythagoras inequality $(m=1)$ for the function η and the unitary function $\xi / M(\xi)$. The inequality holds as an equality if and only if one of the functions is a numerical multiple of the other.

The Schwarz inequality has also the second form

$$J\xi\eta \, \overline{J\xi\eta} \leqq J\xi\bar{\xi} \, J\bar{\eta}\eta.$$

It may be remarked that the Schwarz inequality has as a consequence,—and, if $\mathfrak{A}$ is the real number system, is equivalent to—the geometric theorem for the function space $\mathfrak{M}$ that the length of one side $\xi\eta$ of a triangle $\xi\eta\zeta$ is at most the sum of the lengths of the other two sides $\xi\zeta$, $\zeta\eta$,—viz., the theorem that for three functions ξ, η, ζ of the class $\mathfrak{M}$

$$M(\xi - \eta) \leqq M(\xi - \zeta) + M(\zeta - \eta),$$

that is, the theorem that for two functions ξ, η of the class $\mathfrak{M}$

$$M(\xi+\eta) \leqq M(\xi) + M(\eta),$$

or

$$J\bar{\xi}\eta + J\bar{\eta}\xi \leqq 2\sqrt{J\bar{\xi}\xi\, J\bar{\eta}\eta}.$$

These two inequalities in the functions ξ, η hold as equalities if, and only if, one of the two functions is a positive numerical multiple of the other. The preceding inequality in the functions ξ, η, ζ holds as an equality if, and only if, ζ has the form $(x\xi + y\eta)/(x+y)$, where x and y are real non-negative numbers with $x + y \neq 0$, that is, geometrically, if the point ζ is on the interval $\xi\eta$ of the normal chain joining the points ξ, η.

(*i*) **The Schwarz inequality for two sets of functions.**—For two sets $(\xi_1, \ldots, \xi_k)$, $(\eta_1, \ldots, \eta_k)$ of functions of the class $\mathfrak{M}$ we have the Schwarz inequality in the two forms:

$$J^k(\bar{\xi}_1, \ldots, \bar{\xi}_k)(\eta_1, \ldots, \eta_k)\, J^k(\bar{\eta}_1, \ldots, \bar{\eta}_k)(\xi_1, \ldots, \xi_k)$$
$$\leqq J^k(\bar{\xi}_1, \ldots, \bar{\xi}_k)(\xi_1, \ldots, \xi_k)\, J^k(\bar{\eta}_1, \ldots, \bar{\eta}_k)(\eta_1, \ldots, \eta_k);$$

$$J^k(\xi_1, \ldots, \xi_k)(\eta_1, \ldots, \eta_k)\, \overline{J^k(\xi_1, \ldots, \xi_k)(\eta_1, \ldots, \eta_k)}$$
$$\leqq J^k(\xi_1, \ldots, \xi_k)(\bar{\xi}_1, \ldots, \bar{\xi}_k)\, J^k(\bar{\eta}_1, \ldots, \bar{\eta}_k)(\eta_1, \ldots, \eta_k).$$

This inequality is readily seen to hold as an equality if the functions ξ_f are linearly expressible in the functions $\eta_1, \ldots, \eta_k$. For the general case, separate the functions $\xi_f\ (f=1, \ldots, k)$ into orthogonal components ζ_f, ω_f: $\xi_f = \zeta_f + \omega_f$, respectively lying in and orthogonal to the function space $(\eta_1, \ldots, \eta_k)_L$. Then the left of the Schwarz inequality is unchanged, if we replace the functions ξ_f respectively by the functions ζ_f. Since, by the preceding case, the inequality holds if we make this substitution on the right, it holds as written, in view of lemma 2 of § 5 *g*.

(*j*) **The generalized Pythagoras inequality.**—The Pythagoras inequality of § 5 *f* for a single function ξ and unitary orthogonal set $(\phi_1, \ldots, \phi_n, \ldots)$ has been in § 5 *g* generalized by replacing the function ξ by a set $(\xi_1, \ldots, \xi_k)$ of functions of the class $\mathfrak{M}$. We proceed, by the suitable introduction of a parameter into the function ξ, to obtain another generalization of use in the theory of the general linear integral equation G.

Consider two systems Σ', Σ'' of type Σ_5 and their $*$-composite system $\Sigma \equiv (\Sigma'\Sigma'')_*$ defined in § 4 *b*. Then a function ξ or $\xi(p', p'')$ of the class $\mathfrak{M} \equiv (\mathfrak{M}'\mathfrak{M}'')_*$ of the system Σ is for fixed argument p' a function of the argument p'' belonging to the class $\mathfrak{M}''$ of the system Σ''. Hence, by § 5 *f*, for a function ξ of the class $\mathfrak{M}$ and a unitary orthogonal set $(\phi_1'', \ldots, \phi_n'', \ldots)$ of functions of the class $\mathfrak{M}''$ as to the operation J'' we have, as an identity over the range $\mathfrak{P}'$, the Pythagoras inequality

$$\sum_n (J''\bar{\xi}\phi_n''\, J''\bar{\phi}_n''\xi)_{R'} \leqq (J''\bar{\xi}\xi)_{R'}.$$

Here $J''\bar{\xi}\phi_n''$, $J''\bar{\phi}_n''\xi$, $J''\bar{\xi}\xi$ are functions of the respective classes $\mathfrak{M}'$, $\mathfrak{M}'$, $\mathfrak{K}' \equiv (\mathfrak{M}'\mathfrak{M}')_*$ on the ranges $\mathfrak{P}'$, $\mathfrak{P}'$, $\mathfrak{P}'\mathfrak{P}'$; and the suffix R' (of reduction on the range $\mathfrak{P}'$) denotes that the two arguments of range $\mathfrak{P}'$, whether on the left or on the right of the inequality, are to be set equal. Written explicitly, the inequality is as follows:

$$\sum_n J''_{(s''t'')} \bar{\xi}(p', s'') \phi_n''(t'') J''_{(s''t'')} \bar{\phi}_n''(s'') \xi(p', t'') \leqq J''_{(s''t'')} \bar{\xi}(p', s'') \xi(p', t'') \quad (p').$$

To this inequality we may, in the classical instances, apply the operation $J' \equiv J'_{p'}$ to secure the new inequality desired. In general we may not, however, similarly apply the operation $J' \equiv J'_{(s't')}$.

The generalized Pythagoras inequality is the following:

$$\sum_n J'(J''\bar{\xi}\phi_n'' J''\bar{\phi}_n''\xi) \leqq J'J''\bar{\xi}\xi = J\bar{\xi}\xi,$$

that is,

$$\sum_n J'_{(s't')}(J''_{(s''t'')} \bar{\xi}(s', s'') \phi_n''(t'') J''_{(s''t'')} \bar{\phi}_n''(s'') \xi(t', t''))$$
$$\leqq J'_{(s't')} J''_{(s''t'')} \bar{\xi}(s', s'') \xi(t', t'') = J_{(st)} \bar{\xi}(s) \xi(t),$$

where ξ is a function of the class $\mathfrak{M} \equiv (\mathfrak{M}'\mathfrak{M}'')_*$ on $\mathfrak{P} \equiv \mathfrak{P}'\mathfrak{P}''$ and $(\phi_1'', \ldots, \phi_n'', \ldots)$ is a unitary orthogonal set of functions of the class $\mathfrak{M}''$ as to the operation J''.

Proof.—In the system Σ the function ξ has relative to a finite unitary orthogonal set $(\phi_1'', \ldots, \phi_m'')$ of the system Σ'' the two orthogonal components

$$\xi_1 \equiv \sum_{n=1}^{m} \phi_n'' J''\bar{\phi}_n''\xi, \quad \xi_2 \equiv \xi - \xi_1,$$

while $$M^2(\xi_1) = J\bar{\xi}_1\xi_1 = J'J''\bar{\xi}_1\xi_1 = \sum_{n=1}^{m} J'(J''\bar{\xi}\phi_n'' J''\bar{\phi}_n''\xi).$$

Hence the inequality for a finite set $(\phi_1'', \ldots, \phi_m'')$ simply expresses the fact that $M^2(\xi) = M^2(\xi_1) + M^2(\xi_2) \geqq M^2(\xi_1)$, and from this case we at once secure the inequality for an infinite set $(\phi_1'', \ldots, \phi_n'', \ldots)$, as in § 5*f*.

The inequality reduces to that of § 5*f* in case the system Σ' has as range $\mathfrak{P}'$ a singular range (consisting of a single element), while $J'\kappa' \equiv \kappa'$.

The inequality has the second form

$$\sum_n J'(J''\xi\phi_n'' \overline{J''\xi\phi_n''}) \leqq J'J''\xi\bar{\xi} = J\xi\bar{\xi}.$$

(*k*) **Important instances of the generalized Pythagoras inequality.**—Let the two systems Σ', Σ'' of type Σ_5 have identical ranges $\mathfrak{P}'$, $\mathfrak{P}''$, and classes $\mathfrak{M}'$, $\mathfrak{M}''$ and accordingly classes $\mathfrak{K}' \equiv (\mathfrak{M}'\mathfrak{M}')_*$, $\mathfrak{K}'' \equiv (\mathfrak{M}''\mathfrak{M}'')_*$; they differ then at most in functional operations J', J''. We set $\mathfrak{P}' \equiv \mathfrak{P}'' \equiv \mathfrak{P}$; $\mathfrak{M}' \equiv \mathfrak{M}'' \equiv \mathfrak{M}$; $\mathfrak{K}' \equiv \mathfrak{K}'' \equiv \mathfrak{K}$. Then $\mathfrak{K} \equiv (\mathfrak{M}\mathfrak{M})_* \equiv (\mathfrak{M}'\mathfrak{M}'')_*$, that is, the class $\mathfrak{K}$ is the class $(\mathfrak{M}'\mathfrak{M}'')_*$, denoted in § 5*i* by $\mathfrak{M}$, of the $*$-composite system $(\Sigma'\Sigma'')_*$.

For this case, the second form of the generalized Pythagoras inequality is the following:

$$\sum_n J'(J''\kappa\phi_n'' \overline{J''\kappa\phi_n''}) \leqq J'_{(13)} J''_{(24)} \kappa\bar{\kappa},$$

viz.,

$$\sum_n J'_{(s_1t_1)}(J''_{(s_2t_2)} \kappa(s_1, s_2) \phi_n''(t_2) \overline{J''_{(s_2t_2)} \kappa(t_1, s_2) \phi_n''(t_2)}) \leqq J'_{(s_1t_1)} J''_{(s_2t_2)} \kappa(s_1, s_2) \bar{\kappa}(t_1, t_2),$$

where κ is a function of the class $\mathfrak{K}$ and $(\phi_1'', \ldots, \phi_n'', \ldots)$ is a unitary orthogonal system of functions of the class $\mathfrak{M}$ as to the operation J''.

Consider now a single system Σ of type Σ_5. Take a function κ of the class $\mathfrak{K}$ and a unitary orthogonal set $(\phi_1, \ldots, \phi_n, \ldots)$ of functions of the class $\mathfrak{M}$ as to the operation J. We have then two distinct Pythagoras inequalities:

$$\sum_n J(J\kappa\phi_n \overline{J\kappa\phi_n}) \leqq J_{(13)} J_{(24)} \kappa\bar{\kappa};$$

$$\sum_n J(\overline{J\kappa\phi_n} J\kappa\phi_n) \leqq J_{(31)} J_{(24)} \kappa\bar{\kappa} = J_{(13)} J_{(42)} \bar{\kappa}\kappa,$$

derived from the preceding inequality by setting respectively $\Sigma' = \Sigma'' = \Sigma$, $J' = J'' = J$; $\Sigma' = \breve{\Sigma}$, $J' = \breve{J}$, $\Sigma'' = \Sigma$, $J'' = J$. Here $\breve{\Sigma}$ is the transpose system of Σ defined in § 4 e; hence $J'_{(13)} = \breve{J}_{(13)} = J_{(31)}$, while $J'\mu\bar{\mu} = \breve{J}\mu\bar{\mu} = J\bar{\mu}\mu$.

(*l*) **A fundamental convergence theorem.**—*Theorem.* For every function κ of the class $\mathfrak{K}$ of a system Σ of type Σ_5 there exists a function μ_0 of the class $\mathfrak{M}$ of such a nature that for every function μ and infinite unitary orthogonal set $(\phi_1, \ldots, \phi_n, \ldots)$ of functions of the class $\mathfrak{M}$ the infinite series

$$\sum_{n=1}^{\infty} |J\kappa\phi_n J\bar{\phi}_n\mu|$$

converges on the range $\mathfrak{P}$ uniformly as to the scale function μ_0 to a sum $\leqq \mu_0 M(\mu)$.

Written explicitly, the infinite series is

$$\sum_{n=1}^{\infty} |J_{(st)} \kappa(p, s) \phi_n(t) \quad J_{(st)} \bar{\phi}_n(s) \mu(t)|.$$

Proof.—We have the Pythagoras inequalities:

$$\sum_{n=1}^{\infty} \overline{J\bar{\phi}_n\mu} J\bar{\phi}_n\mu \leqq J\bar{\mu}\mu;$$

$$\sum_{n=1}^{\infty} (J\kappa\phi_n \overline{J\kappa\phi_n})_R \leqq (J_{(24)} \kappa\bar{\kappa})_R,$$

where, as indicated by the suffixes R, the four first arguments of the four functions κ are to be set equal. Then, by the Schwarz inequality for the instance III, we have

$$\sum_{n=1}^{\infty} |J\kappa\phi_n J\bar{\phi}_n\mu| \leqq \sqrt{(J_{(24)}\kappa\bar{\kappa})_R J\bar{\mu}\mu},$$

while, by theorem 8 of § 2 h, we see that the infinite series on the left of this inequality converges on the range $\mathfrak{P}$ uniformly as to the scale function $\sqrt{(J_{(24)}\kappa\bar{\kappa})_R}$. Hence the statements of the theorem follow from the fact that the function $J_{(24)}\kappa\bar{\kappa}$ belongs to the class $\mathfrak{K}$ and is dominated by a function $\mu_0\mu_0$, where μ_0 is a real-valued nowhere negative function of the class $\mathfrak{M}$ depending on κ alone, and accordingly the function $\sqrt{(J_{(24)}\kappa\bar{\kappa})_R}$ is dominated by the function μ_0.

This convergence theorem, a generalization of a theorem of E. Schmidt for instance IV, is of importance in the theory of the integral equation G with hermitian kernel κ.

SUR LES RECHERCHES RÉCENTES RELATIVES À LA MEILLEURE APPROXIMATION DES FONCTIONS CONTINUES PAR DES POLYNÔMES

PAR S. BERNSTEIN.

Vous connaissez bien le théorème classique de Weierstrass que toute fonction continue dans un intervalle donné peut être représentée avec une approximation aussi grande qu'on le veut par des polynômes de degré assez élevé.

Depuis que ce théorème a été découvert presque simultanément par Weierstrass et M. Runge, plusieurs mathématiciens en ont donné des démonstrations différentes et ont construit divers polynômes approchés de degré n, $R_n(x)$, tels que le maximum de la différence

$$|f(x) - R_n(x)|$$

tend vers 0, lorsque n croît indéfiniment.

L'approximation fournie par diverses méthodes pour une même fonction et pour une même valeur du degré n n'est pas toujours la même, et il est naturel de rechercher ceux des polynômes approchés P_n pour lesquels le maximum de la différence considérée tend le plus rapidement vers zéro.

Les polynômes $P_n(x)$ jouissant de cette propriété ont reçu le nom de *polynômes d'approximation* et le maximum $E_n[f(x)]$ du module de la différence

$$|f(x) - P_n(x)|$$

le nom *de la meilleure approximation* de la fonction donnée dans l'intervalle considéré.

Je n'ai pas besoin de rappeler que, les polynômes d'approximation avaient été introduits dans la science, par Tchebischeff, encore avant la découverte de Weierstrass; mais ce n'est que ces dernières années qu'on a essayé d'étudier systématiquement la grandeur de la meilleure approximation $E_n[f(x)]$, d'une fonction donnée $f(x)$, pour des valeurs très grandes de n, et de compléter ainsi le théorème de Weierstrass qui exprime que $\lim_{n=\infty} E_n[f(x)] = 0$, quelle que soit la fonction continue $f(x)$. Le fait général qui se dégage de cette étude est l'existence d'une liaison des plus intimes entre les propriétés différentielles de la fonction $f(x)$ et la loi asymptotique de la décroissance des nombres positifs $E_n[f(x)]$. Voici, en effet, les résultats les plus essentiels qui sont obtenus dans cette voie et qui sont résumés dans le Tableau suivant:

Pour qu'une fonction de variable réelle dans un intervalle donné

	il est nécessaire que	Auteur	il est suffisant que	Auteur
1. Soit *analytique*	$\lim_{n=\infty} E_n[f(x)]\rho^n=0$ $(\rho>1)$	S. B.(1)	$\lim_{n=\infty} E_n[f(x)]\rho^n=0$ $(\rho>1)$	S.B.(1)
2. Admette des dérivées de tous les ordres	$\lim_{n=\infty} E_n[f(x)]n^p=0$ (quel que soit p)	S. B.(1)	$\lim_{n=\infty} E_n[f(x)]n^p=0$ (quel que soit p)	S.B.(1)
3. Admette une dérivée d'ordre p satisfaisant à une condition de Lipschitz d'ordre α. (On posera $p=0$ lorsque c'est la fonction elle-même qui satisfait à la condition correspondante de Lipschitz.)	$E_n[f(x)]n^{p+\alpha}<k$ (k étant une constante déterminée)	D. Jackson (2)	$\sum_{n=1}^{n=\infty} E_n[f(x)]n^{p-1+\alpha}$ soit convergente	S.B.(1)
3 bis. Admette une dérivée d'ordre p continue	$E_n[f(x)]n^p<\epsilon_n$ (ϵ_n tendant vers zéro pour n infini)	D. Jackson (2)	$\sum_{n=1}^{n=\infty} E_n[f(x)]n^{p-1}$ soit convergente	S.B.(3)
4. Satisfasse à une condition de Dini-Lipschitz $\left(\lvert f(x+\delta)-f(x)\rvert<\frac{\epsilon}{\lvert\log\delta\rvert}\right.$ où ϵ tend vers zéro avec δ).	$E_n[f(x)]\log n<\epsilon_n$ (où ϵ_n tend vers zéro pour $n=\infty$).	H. Lebesgue (4)	$E_n[f(x)]\log n<\epsilon_n$ (où ϵ_n tend vers zéro pour $n=\infty$).	S.B.(3)

Avant de passer à l'analyse de ce Tableau, je voudrais dire quelques mots sur les méthodes par lesquelles ces résultats ont été obtenus. Pour établir les conditions nécessaires, on se sert de polynômes approchés convenables de la fonction considérée et l'on constate que si la fonction appartient à une des classes indiquées elle est effectivement susceptible de l'approximation correspondante. M. de la Vallée Poussin (5) a le premier suivi cette voie d'une façon systématique et c'est en se servant de ses procédés et en les perfectionnant qu'on a obtenu les résultats énoncés. Pour établir les conditions suffisantes, on procède de la façon suivante: s'il s'agit par exemple de la classe 3 bis, on démontre que lorsque la série $\sum_{n=1}^{n=\infty} E_n[f(x)]n^{p-1}$ est convergente, la fonction considérée peut être développée en une série de polynômes dérivable p fois et admet par conséquent une dérivée continue d'ordre p.

La convergence uniforme des dérivées successives du développement construit résulte essentiellement de certaines propositions d'algèbre concernant les relations entre le module maximum d'un polynôme de degré donné et ceux de ses dérivées successives dans un intervalle déterminé. Ces études algébriques préliminaires forment une suite naturelle de la théorie de Tschebichev des polynômes qui s'écartent le moins de zéro, et en suivant cette voie, je me suis rencontré sur quelques points avec MM. A. et W. Markoff (6) (7), qui, il y a 20 ans, s'étaient occupés de questions semblables.

Passons à présent à l'analyse des résultats. Il suffit d'un coup d'œil superficiel sur le Tableau pour se rendre compte de l'exactitude de ce que je disais tout à l'heure, du lien étroit qui existe entre la meilleure approximation et les propriétés différentielles d'une fonction. Vous voyez d'abord qu'au point de vue réel, les fonctions analytiques laissent se définir comme des fonctions dont la meilleure approximation par des polynômes de degré donné décroît le plus rapidement possible; ce sont les fonctions qui entre toutes les fonctions continues diffèrent le moins possible des polynômes. Une théorie systématique des fonctions de variable réelle admettra donc nécessairement comme premier chapitre la théorie des fonctions analytiques.

Nous arrivons ensuite à la seconde classe de fonctions, celle des fonctions indéfiniment dérivables; cette classe est également déterminée sans ambiguité par la nature de la décroissance de la meilleure approximation. Les résultats relatifs aux classes 3 et 3 bis paraissent, à première vue, moins satisfaisants; ici les conditions nécessaires et suffisantes ne sont plus les mêmes. Il est bien clair, en effet, qu'en posant pour fixer les idées, $p=1$, dans la classe 3 bis, on a d'une part, $E_n n < \epsilon_n$ ou $E_n < \frac{\epsilon_n}{n}$, et d'autre part, ΣE_n convergente, conditions nullement équivalentes. On pourrait croire que de nouvelles études permettraient plus tard d'obtenir également des conditions qui soient à la fois nécessaires et suffisantes. Mais il n'en est rien. J'ai reconnu, en effet, qu'il existe des fonctions à dérivée continue et telles que ΣE_n est divergente; notre condition suffisante n'est donc certainement pas nécessaire; or, d'autre part, si l'on se donne arbitrairement une série divergente à termes positifs Σa_n, il est toujours possible de construire des fonctions $f(x)$ sans dérivées continues, telles que l'on ait $E_n[f(x)] < a_n$. Par conséquent, quelque faible que soit la divergence de la série $\Sigma E_n[f(x)]$ elle est capable de détruire la continuité de la dérivée; il est donc impossible de restreindre la condition suffisante qui, comme nous venons de le voir, n'est cependant pas nécessaire. Ainsi, en général, les fonctions des classes 3 et 3 bis ne peuvent pas être complètement distinguées par leur meilleure approximation; il existe des cas limites, où la nature de la continuité (qui n'est exprimable par aucune condition de Lipschitz) de la dérivée diffère si peu d'une certaine forme de discontinuité qu'il est impossible de décider par la seule considération de la meilleure approximation si cette dérivée est continue ou non. Je pense que l'étude de ces cas critiques, où les fonctions caractérisées par le même ordre de la meilleure approximation jouissent des propriétés différentielles en apparence différentes, pourrait contribuer à éclaircir la notion de la continuité.

Le fait que les conditions nécessaires et suffisantes ne sont pas identiques pour les classes 3 et 3 bis explique la difficulté qu'on a éprouvé à résoudre le problème posé par M. de la Vallée Poussin (8) sur l'ordre de la meilleure approximation de $|x|$. M. de la Vallée Poussin avait démontré que $E_n|x|$ est inférieur à $\frac{k}{n}$, k étant une constante, et cela résulte aussi de notre Tableau, puisque $|x|$ satisfait à une condition de Lipschitz du premier degré et par conséquent appartient à la classe 3 avec $p=0$, $\alpha=1$. Mais le Tableau ne permet aucunement d'affirmer que E_n est nécessairement de l'ordre de $\frac{1}{n}$. La seule chose qu'on puisse affirmer, c'est que $\Sigma E_n |x|$ est divergente, sans quoi $|x|$ aurait une dérivée continue. Ainsi le simple fait de la discontinuité de la dérivée n'exclurait pas, par exemple, la possibilité d'une inégalité de la forme $E_n|x| < \frac{1}{n \log n}$, mais on doit avoir, en général, $E_n|x| > \frac{1}{n \log n^{1+a}}$. Il est clair d'ailleurs, que malgré la divergence de la série $\Sigma E_n|x|$, il pourrait y avoir une infinité de valeurs particulières de n pour lesquelles $E_n|x|$ décroirait même beaucoup plus rapidement. Il est bien facile, en effet, de donner des exemples de fonctions $f(x)$ non dérivables dont la meilleure approximation se comporte pour une infinité de valeurs de n, comme celle des fonctions de la deuxième classe et même de la première classe, qu'on ait par exemple, pour une infinité de valeurs de n convenablement choisies, $E_n[f(x)] < \frac{1}{2^n}$, car on peut imaginer une loi de décroissance de $E_n[f(x)]$ suffisamment irrégulière, pour que la petitesse d'une infinité de termes de la série $\Sigma E_n[f(x)]$ ne l'empêche pas de diverger. La question suivante se pose alors: quelle est donc la nature de ces fonctions sans dérivées, qui pour une infinité des valeurs de n convenablement choisies sont susceptibles de la même approximation que les fonctions indéfiniment dérivables et même les fonctions analytiques? Je n'ai fait que le premier pas de cette étude.

Le théorème que j'ai obtenu me paraît assez curieux pour être signalé ici. Désignons par $\delta_1(\epsilon)$ le maximum de l'oscillation $|f(x+h) - f(x)|$ de la fonction $f(x)$ dans l'intervalle donné AB lorsque $|h|$ est inférieur ou égal à ϵ. La fonction $\delta_1(\epsilon)$ est évidemment non négative et non décroissante. La condition de Lipschitz d'ordre α s'exprime ainsi

$$\delta_1(\epsilon) < k\epsilon^\alpha \quad \text{.................................(I)}$$

quel que soit ϵ. Il est possible que la croissance de la fonction $\delta_1(\epsilon)$ relative à une certaine fonction $f(x)$ soit irrégulière et que l'inégalité (I) sans être vérifiée pour toute valeur de ϵ, soit exacte pour une infinité de valeurs de ϵ aussi petites qu'on le veut. S'il en est ainsi, nous dirons que la fonction satisfait à une condition de Lipschitz généralisée du premier ordre et de degré α.

Considérons également

$$\delta_2(\epsilon) = \text{Max}\,|f(x+2h) - 2f(x+h) + f(x)|,$$

$$\delta_3(\epsilon) = \text{Max}\,|f(x+3h) - 3f(x+2h) + 3f(x+h) - f(x)|,$$

etc. pour $|h| \leqslant \epsilon$.

Si pour une infinité de valeurs de ϵ aussi petites qu'on le veut, on a une inégalité de la forme

$$\delta_1(\epsilon) < k\epsilon^\alpha$$

nous dirons que la fonction $f(x)$ satisfait à une condition généralisée de Lipschitz d'ordre i, de degré α. Cela posé, voici le théorème.

Théorème. S'il existe une infinité de valeurs de n pour lesquelles

$$E_n[f(x)] < \frac{A}{n^p},$$

la fonction $f(x)$ satisfait à des conditions de Lipschitz d'ordre i et de degré $\alpha_i = \frac{ip}{i+p}$ quel que soit i.

Ainsi par exemple, une fonction $f(x)$ qui pourrait même ne pas avoir de dérivée, mais qui jouirait de la propriété que, quel que soit p, il existe une infinité de valeurs de n telles que

$$E_n[f(x)] < \frac{1}{n^p},$$

satisferait à des conditions de Lipschitz généralisées de tout ordre i et de degré α aussi voisin de i qu'on le veut (on aurait donc, pour une infinité de valeurs de ϵ,

$$\delta_1(\epsilon) < k\epsilon^{1-\lambda},$$

$$\delta_2(\epsilon) < k\epsilon^{2-\lambda}, \ldots,$$

où λ est un nombre aussi petit qu'on le veut et k une constante détérminée lorsque λ est fixée).

Ce théorème montre aussi immédiatement que, quel que petit que soit η, on a nécessairement

$$E_n|x| > \frac{1}{n^{1+\eta}} \quad \text{..............................(II)},$$

à partir d'une certaine valeur de n. En effet, si l'inégalité (II) n'était pas vérifiée pour une infinité de valeurs de n, on conclurait, en vertu du théorème précédent, que $|x|$ satisfait à une condition de Lipschitz d'ordre i quelconque et de degré

$$\alpha = \frac{i(1+\eta)}{i+1+\eta}$$

supérieur à 1 pour i suffisamment grand. Or, on reconnait sans peine que la fonction $|x|$ ne satisfait à aucune condition généralisée de Lipschitz de degré supérieur à 1. Notre théorème est cependant insuffisant pour démontrer qu'on a, pour toute valeur de n,

$$E_n|x| > \frac{\epsilon}{n},$$

si ϵ est un nombre fixe suffisamment petit. Ce résultat n'a pu être obtenu jusqu'ici que par des considérations plus spéciales dont je parlerai plus loin, mais je crois qu'on pourra le retrouver également par des considérations générales, en introduisant des propriétés semblables aux conditions généralisées de Lipschitz qui auraient une influence encore plus sensible sur la grandeur de la meilleure approximation.

L'exemple du problème de la meilleure approximation du $|x|$ posé par M. de la Vallée Poussin donne une preuve de plus à l'appui de ce fait qu'une question particulière bien posée donne naissance à des théories d'une portée plus générale.

Avant d'aborder les considérations directes qui conduisirent à la solution complète du problème posé par l'éminent géomètre belge, je veux dire quelques mots d'un théorème fort remarquable dû à M. Lebesgue (4) qui a permis à M. Jackson (2) et à moi (1) de démontrer d'une façon indépendante que

$$E_n|x| > \frac{k}{n\log(n+1)},$$

k étant une constante convenablement choisie. Voici ce théorème :

Si $R_n f(x)$ est l'approximation fournie par une suite de Fourier d'ordre n, la meilleure approximation qu'on puisse obtenir par une suite trigonométrique quelconque du même ordre est supérieure à $\frac{kR_n[f(x)]}{\log(n+1)}$, où k est un facteur déterminé.

Je ne puis m'empêcher de rappeler à cette occasion la belle application de ce théorème faite par M. Lebesgue(4) ; je fais allusion à sa démonstration si simple du fait que toute fonction satisfaisant à une condition de Dini-Lipschitz est développable en une série de Fourier uniformément convergente.

Les mathématiciens qui se sont occupés de la représentation approchée des fonctions ont pu remarquer que les propriétés de l'approximation d'une fonction périodique par des suites trigonométriques limitées sont essentiellement équivalentes à celles de l'approximation des fonctions quelconques par des polynômes. Il est bien facile de voir (3), en particulier, que le théorème de M. Lebesgue s'applique au développement en série de polynômes trigonométriques ($\cos n \operatorname{arc} \cos x$). Ainsi, si $R_n[f(x)]$ est le maximum du reste $\sum_{p=n+1}^{p=\infty} A_p \cos p \operatorname{arc} \cos x$ de la série

$$f(x) = \sum_{p=0}^{p=\infty} A_p \cos p \operatorname{arc} \cos x,$$

on a

$$R_n[f(x)] \geqq E_n[f(x)] > \frac{kR_n[f(x)]}{\log(n+1)} \quad \ldots\ldots\ldots\ldots\ldots \text{(III)},$$

où k est la constante de M. Lebesgue.

Pour $|x|$, on a

$$|x| = \frac{4}{\pi}\left[\frac{1}{2} + \frac{\cos 2 \operatorname{arc} \cos x}{1.3} - \frac{\cos 4 \operatorname{arc} \cos x}{3.5} + \frac{\cos 6 \operatorname{arc} \cos x}{5.7} - \ldots\right],$$

donc

$$R_{2n}|x| = \frac{4}{\pi}\left[\frac{1}{(2n+1)(2n+3)} + \frac{1}{(2n+3)(2n+5)} + \ldots\right] =$$

$$= \frac{\mathbf{2}}{\pi.(2n+1)}.$$

Par conséquent, en vertu de l'inégalité (III), on a

$$\frac{2}{\pi(2n+1)} \geqq E_{2n}|x| > \frac{2k}{\pi(2n+1)\log(2n+1)}.$$

On voit que le théorème de M. Lebesgue fournit dans des cas très étendus l'ordre de la meilleure approximation, au facteur $\frac{1}{\log(n+1)}$ près, mais la question de la valeur exacte de l'ordre de la meilleure approximation reste ouverte.

Nous n'avons pas parlé jusqu'ici de la méthode qui paraît la plus naturelle. Celle qui consisterait à déterminer effectivement les polynômes d'approximation de la fonction donnée et de calculer directement la meilleure approximation E_n ainsi obtenue. Le fait est que, la détermination des polynômes d'approximation qui, en général, ne peut être faite que par approximations successives, devient de plus en plus compliquée à mesure que le degré des polynômes augmente, et il semblait extrêmement difficile de tirer des renseignements plus ou moins précis sur la grandeur de la meilleure approximation en utilisant des considérations de cette nature. C'est M. de la Vallée Poussin (9) qui osa le premier, aborder le problème par cette voie-là. Il serait trop long d'exposer ici sa méthode générale de détermination des polynômes d'approximation. Je me bornerai à signaler le théorème suivant dont l'application conduit immédiatement à des bornes inférieures pour la meilleure approximation. Soit $R_n(x)$ un polynôme de degré n; si le segment AB est divisé en $n+2$ intervalles où la différence

$$f(x) - R_n(x)$$

change successivement de signe, et tels que

$$| f(x) - R_n(x) |$$

dépasse dans chacun d'eux une certaine valeur M, la meilleure approximation $E_n[f(x)]$ de $f(x)$ sur le segment AB par un polynôme de degré n est supérieure à M. Pour appliquer le théorème, il suffit de prendre arbitrairement $n+1$ valeurs de x (appelés *nœuds* par M. de la Vallée Poussin) et de construire le polynôme de degré n, $R_n(x)$, qui en ces $n+1$ points se confond avec $f(x)$. On aura ainsi (sauf des cas exceptionnels) $n+2$ intervalles où la différence $f(x) - R_n(x)$ est de signe contraire et l'on pourra se servir du polynôme $R_n(x)$ pour donner une borne inférieure de la meilleure approximation de $f(x)$. M. de la Vallée Poussin a appliqué son théorème à la recherche d'une borne inférieure de $E_n |x|$; il obtient ainsi (9) $E_n |x| > \frac{k}{n(\log n)^3}$, résultat un peu moins précis que celui de tout à l'heure.

Il est évident que les bornes inférieures qu'on trouvera en appliquant le théorème énoncé dépendent essentiellement du choix des nœuds. Il y aura, en particulier, une position des nœuds pour laquelle le polynôme $R_n(x)$ sera lui-même le polynôme d'approximation et l'on conçoit que plus la position des nœuds sera voisine de celle-là, plus la borne inférieure qu'on trouvera sera voisine de la meilleure approximation. La méthode de M. de la Vallée Poussin ne donne pas d'indications pour diriger ce choix, et c'est sur ce point qu'elle doit être complétée.

Au moment où parut le Mémoire de M. de la Vallée Poussin, je venais de trouver d'une façon un peu moins simple et comme conséquence d'une méthode entièrement différente un théorème équivalent au fond à celui de M. de la Vallée Poussin. L'idée de la méthode (3) dont il s'agit consistait essentiellement à étudier les polynômes d'approximation d'une fonction donnée en utilisant convenablement les polynômes d'approximation connus déjà d'autres fonctions qui diffèrent suffisamment peu de la fonction considérée. L'avantage du point de vue auquel je me plaçais était celui de m'imposer un choix de nœuds peu différents de ceux qui correspondent aux polynômes d'approximation. Dès lors, comme je le disais plus haut, l'application du théorème de M. de la Vallée Poussin devait me conduire à une borne inférieure assez précise de la meilleure approximation. C'est ainsi qu'en revenant à la fonction $|x|$ sur le segment $-1, +1$, j'ai été conduit à prendre comme nœuds du polynôme $R_{2n}(x)$ de degré $2n$ les racines de l'équation

$$x \cos 2n \arccos x = 0$$

et que j'ai obtenu

$$E_{2n} |x| > \frac{2\sqrt{2}}{21(2n+1)}.$$

La question de l'ordre de la meilleure approximation de $|x|$ se trouve donc résolue. Mais en persévérant dans la même voie on peut aller beaucoup plus loin. Je me bornerai à indiquer le résultat le plus essentiel.

Le produit
$$nE_n|x|$$
tend vers une limite λ parfaitement déterminée lorsque n croît indéfiniment. Cette limite λ peut être calculée par approximations successives et l'on a
$$\lambda = 0{\cdot}282 \text{ à } 0{\cdot}004 \text{ près.}$$

Ainsi pour n suffisamment grand,
$$\frac{0{\cdot}286}{n} > E_n|x| > \frac{0{\cdot}278}{n}.$$

La détermination plus précise de λ par la méthode que j'ai suivie ne serait maintenant qu'une question de patience et de calcul.

Il importerait cependant de perfectionner la méthode. Aussi pourrait-on trouver alors des expressions plus ou moins simples pour λ et rattacher peut-être cette constante absolue à d'autres constantes qui se rencontreront certainement dans les recherches relatives à la valeur asymptotique de la meilleure approximation d'autres fonctions que $|x|$. On rencontrera, sans doute, encore bien des difficultés dans ce genre de recherches, mais je ne les crois pas au-dessus des moyens dont dispose l'analyse moderne. Il y a des cas même, où la détermination de la valeur asymptotique de la meilleure approximation d'une fonction se fait beaucoup plus simplement que pour $|x|$. Cela a lieu, en particulier, pour un grand nombre de fonctions entières e^x, $\sin x$, etc. C'est ainsi, par exemple, qu'on trouve immédiatement, par application des théorèmes généraux,
$$\lim \frac{E_n(e^x)\, 2^n (n+1)!}{h^{n+1}} = 1,$$
où la meilleure approximation se rapporte au segment $(-h, +h)$.

J'ai donné une seconde démonstration encore (3), du fait que $E_n|x|$ est de l'ordre de $\frac{1}{n}$. Je me permets seulement d'attirer votre attention sur le théorème qui est à sa base, non pas pour l'appliquer à $|x|$, mais pour en indiquer certaines conséquences qui me paraissent être d'un intérêt général.

Soient $\alpha_1, \alpha_2, \ldots, \alpha_n$ et $\beta_1, \beta_2, \ldots \beta_n$ deux suites de nombres positifs croissants tels que
$$\alpha_1 > \beta_1, \quad \alpha_2 > \beta_2, \ldots, \quad \alpha_n > \beta_n.$$

Si k est inférieur à β_1 et qu'on ait
$$\left| x^k - \sum_{i=1}^{i=n} A_i x^{\alpha_i} \right| \leqslant \epsilon,$$
sur le segment 0, 1 ($A_1, A_2, \ldots$ étant des constantes), il est possible de déterminer des constantes
$$B_1, B_2, \ldots,$$
telles qu'on ait
$$\left| x^k - \sum_{i=1}^{i=n} B_i x^{\beta_i} \right| \leqslant \epsilon$$
sur le segment 0, 1.

Si k est supérieur à α_n on peut conclure inversement qu'une inégalité de la forme

$$\left| x^k - \sum_{i=1}^{i=n} B_i x^{\beta_i} \right| \leqslant \epsilon,$$

entraîne nécessairement la possibilité de l'inégalité

$$\left| x^k - \sum_{i=1}^{i=n} A_i x^{\alpha_i} \right| < \epsilon.$$

On peut appliquer ce théorème à la recherche des conditions pour qu'une suite donnée de puissances positives de x

$$x^{\alpha_1}, \quad x^{\alpha_2}, \ldots, \quad x^{\alpha_n}$$

soit une suite complète, c'est-à-dire une suite telle, qu'il soit possible de rendre la différence

$$f(x) - \sum_{i=1}^{i=n} A_i x^{\alpha_i}$$

aussi petite qu'on le veut dans un intervalle déterminé 0, 1 par un choix convenable des coefficients A_i et en prenant n suffisamment grand, quelle que soit la fonction continue $f(x)$.

Voici les résultats qu'on obtient et qui montrent, comme cela était facile à prévoir, que la nature arithmétique des exposants α_i ne joue aucun rôle, mais que tout dépend de leur croissance.

1°. Si les nombres α_i tendent vers une limite finie, les fonctions x^{α_i} forment toujours une suite complète. Ce résultat un peu paradoxal prouve, par exemple, qu'une fonction continue quelconque peut être indéfiniment approchée au moyen d'une somme de la forme

$$\sum_{i=1}^{i=n} A_i x^{1-\frac{1}{i}}.$$

2°. Si les nombres α_i croissent indéfiniment il suffit pour qu'ils forment une suite complète que

$$\frac{\alpha_n}{n \log n} \text{ tende vers } 0.$$

Au contraire, la suite des puissances x^{α_n} n'est pas complète si l'on peut trouver un nombre ϵ tel que

$$\alpha_n \geqslant n (\log n)^{2+\epsilon}$$

ou bien

$$\alpha_n \geqslant (\log n)^2 (\log \log n)^{1+\epsilon}, \ldots .$$

Ainsi, si la croissance des nombres α_n est trop rapide, le système des fonctions x^{α_n} ne peut être complet, il l'est, au contraire, si cette croissance est suffisamment lente, et il ne reste qu'une incertitude relativement faible au sujet du moment critique où la croissance des exposants α_n devient assez rapide pour que la suite cesse d'être complète. Il serait intéressant de savoir si la condition que la série $\Sigma \frac{1}{\alpha_n}$ diverge ne serait pas une condition nécessaire et suffisante pour que la suite des puissances x^{α_n} soit complète ; il n'est pas certain, d'ailleurs, qu'une condition de cette nature doive nécessairement exister.

Vous voyez que bien des questions des plus diverses se posent dans l'ordre des idées que j'ai taché de vous exposer ici. Les théories dont nous avons parlé sont trop jeunes encore pour qu'il soit temps de juger de la place qu'elles sont appelées à occuper dans la science. Il serait intéressant de savoir si l'on ne pourrait pas en tirer profit pour l'étude des fonctions qui se présentent dans les applications, par exemple, dans les équations différentielles. On ne l'a pas essayé jusqu'ici. Des recherches de cette nature me paraîtraient cependant fort souhaitables, dans le cas de plusieurs variables surtout. On sait, en effet, que l'étude de la nature des fonctions satisfaisant à une équation aux dérivées partielles présente des difficultés considérables. Peut-être, ces difficultés seront-elles levées plus facilement si l'on utilise systématiquement la notion de la meilleure approximation. Il est vrai que les recherches sur l'approximation des fonctions de plusieurs variables sont à peine ébauchées, mais les quelques propositions que j'ai obtenues en appliquant convenablement le Tableau indiqué au début, me paraissent de nature à encourager des recherches du même genre.

Pour terminer, j'indiquerai seulement une de ces propositions: Si une fonction $f(x, y)$ à l'intérieur d'un contour C considérée comme fonction de x seulement, admet une dérivée partielle d'ordre l, $\frac{\partial^l f}{\partial x^l}$, satisfaisant par rapport à x à une condition déterminée de Lipschitz de degré α et si, considérée comme fonction de y seulement elle admet également une dérivée partielle d'ordre l, $\frac{\partial^l f}{\partial y^l}$, satisfaisant par rapport à y à une condition détérminée de Lipschitz de degré α, la fonction $f(x, y)$ admettra, à l'intérieur d'un contour quelconque S intérieur à C, toutes les dérivées partielles d'ordre l $\frac{\partial^l f}{\partial x^i \partial y^{l-i}}$ et ces dernières satisferont, en outre, par rapport aux deux variables, à des conditions de Lipschitz de degré α_1 inférieur à α et aussi voisin de α qu'on le veut.

Travaux contenant les résultats cités (dans l'ordre des citations).

(1) S. Bernstein. Sur l'approximation des fonctions continues par des polynômes. (Comptes rendus de l'Académie des Sciences, t. CLII. 27 février 1911.)

(2) D. Jackson. Über die Genauigkeit der Annäherung stetiger Funktionen durch ganze rationale Funktionen gegebenen Grades und trigonometrische Summen gegebener Ordnung. (Preisschrift und Inaugural-Dissertation, Göttingen, juillet 1911.)

(3) S. Bernstein. Sur la meilleure approximation des fonctions continues par des polynômes de degré donné (en russe). (Communications de la Société mathématique de Charkow, 1912.)

(4) H. Lebesgue. Sur les intégrales singulières. (Annales de la Faculté des Sciences de l'Université de Toulouse, 1909.)

(5) Ch. de la Vallée Poussin. Sur l'approximation des fonctions d'une variable réelle et de leurs dérivées par des polynômes et des suites limitées de Fourier. (Bulletin de l'Académie de Belgique, 1908.)

(6) A. Markow. Sur une question de Mendeleiew (en russe). (Bulletin de l'Académie de Saint Pétersbourg, 1889.)

(7) W. Markow. Sur les fonctions qui s'écartent le moins de zéro (en russe). (Edition de l'Université de Saint Pétersbourg, 1892.)

(8) Ch. de la Vallée Poussin. Sur la convergence des formules d'interpolation entre ordonnées équidistantes. (Bulletin de l'Académie de Belgique, 1908.)

(9) Ch. de la Vallée Poussin. Sur les polynômes d'approximation et la représentation approchée d'un angle. (Bulletin de l'Académie de Belgique, 1910.)

Travaux venant de paraître.

D. Jackson. On the degree of convergence of the development of a continuous function according to Legendre's polynomials. (Transactions of the American Mathematical Society. July 1912.)

S. Bernstein. Sur l'ordre de la meilleure approximation des fonctions continues par des polynômes. (Mémoire publié par l'Académie Royale de Belgique, 1912.)

S. Bernstein. Sur la valeur asymptotique de la meilleure approximation des fonctions analytiques. (Comptes rendus, t. clv. 1912.)

S. Bernstein. Sur la meilleure approximation de $|x|$. (Acta Mathematica, sous presse; paraîtra en 1913.)

ON VECTOR-ANALYSIS AS GENERALISED ALGEBRA

By Alexander Macfarlane.

A vector may be analysed into modulus and space-unit. The quantities of algebra may be viewed as numbers; but they may also be viewed as vectors having a common space-unit. Thus $a+b+c$ may denote a sum of numbers; or it may mean $a\lambda + b\lambda + c\lambda$, where λ denotes the common space-unit (fig. 1). The λ may be left to be understood, because, with the help of $-\lambda$ to denote the opposite unit, it is the same for all the terms. In the same way abc may be viewed as a product of numbers; but it may also be viewed as $(a\lambda)(b\lambda)(c\lambda)$, the product of three vectors having a common space-unit. When we replace the constant by different space-units, we get the sum $a\alpha + b\beta + c\gamma$, and the product $(a\alpha)(b\beta)(c\gamma)$. When the vector as a whole, that is the product of the number and the space-unit, is symbolised by the corresponding italic capital, we get the expressions

$$A+B+C \text{ (fig. 1) and } ABC \text{ (fig. 2).}$$

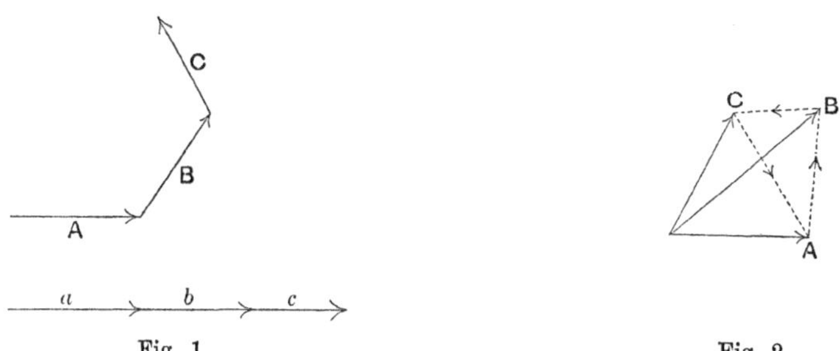

Fig. 1.

Fig. 2.

Our problem then is: When the quantities of common algebra are so generalised as to be in space, and endowed with order whether linear or cyclic, what are the generalisations which necessarily follow in the rules, processes and theorems of algebra? We have to ascend from the particular to a definite general.

For primary space-units we take the total assemblage of rays which emanate from a common point and are cut off by the sphere whose radius is the unit of length. These primary units, as well as the vectors derived from them, have one dimension in length. It is scarcely correct to say that their moduli have one dimension in length; for the algebraic symbols to which physicists attach dimensions are not mere moduli but disguised vectors; for example, v the symbol for velocity.

These units α, β, γ, etc. differ from the Grassmannian e_1, e_2, e_3, etc., in that they are expressly stated to be spherical, in their being infinite in number, and in the circumstance that there are no conditions of orthogonality implied. They differ from the Hamiltonian units in being infinite in number, in having one dimension in length, and in wanting the angular index $\pi/2$, that is the imaginary i as a factor.

But the greatest difference lies in the circumstance that the vectors are here treated as successive; in consequence of which the algebra developed is highly aristocratic. There are successive orders of vectors, and each member of an order has its own place within the order; consequently the problems which arise are largely questions of precedence. When the matter is studied more minutely it is found that there are two kinds of succession which must be carefully distinguished, linear

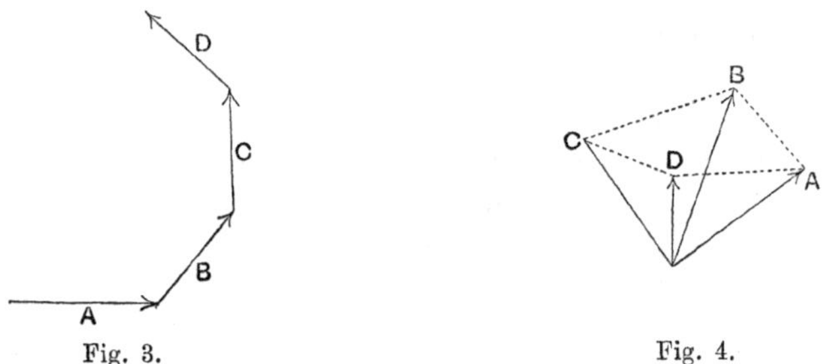

Fig. 3. Fig. 4.

(fig. 3) and cyclic (fig. 4). In common algebra the difference crops out in the two ways of expanding $(a+b+c)^2$; namely

$$a^2+b^2+c^2+2ab+2ac+2bc,$$

and

$$a^2+b^2+c^2+2ab+2bc+2ca,$$

the former being linear, the latter cyclic.

In linear succession $A+B$ denotes the complex vector formed of A followed by B springing from the end of A; whereas in cyclic succession it means A followed by B springing from the same origin as A. In the latter case we have the Round-Table precedence. It follows that $A+B$ is not the same thing as $B+A$ (fig. 5); what is true is that the resultant of $A+B$ is the same as the resultant of $B+A$. Let Σ denote resultant, then

$$\Sigma(A+B)=\Sigma(B+A);$$

but

$$A+B \neq B+A$$

unless A and B have the same unit. In other words the sign + denotes simple addition, not geometric composition. In calculations we cannot substitute the resultant for the complex vector; to do so is to fall into a grievous fallacy which afflicts all the existing forms of space-analysis whether vectorial or quaternionic. It prevented Hamilton from generalising the Exponential Theorem, with the result

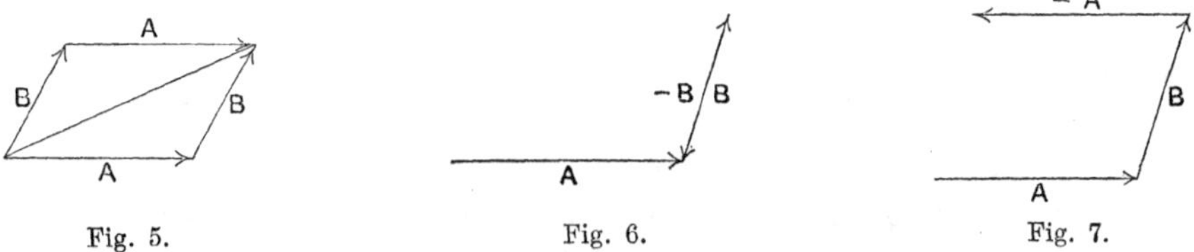

Fig. 5. Fig. 6. Fig. 7.

that the development of the Quaternion Analysis was arrested. In 1892 I succeeded in making the generalisation, and it is the view gained from that eminence which makes possible the present development.

It follows that two adjacent elements of a complex vector cannot be interchanged unless they have the same unit; that is,

$$A + B_1 + B_2 = A + B_2 + B_1$$

if B_1 and B_2 have the same unit β, but not otherwise. It also follows that two elements having opposite units cannot cancel, when they are separated by a different unit. For example

$$A + B - B = A \text{ (fig. 6)},$$

but $$A + B - A \text{ not } = B \text{ (fig. 7)}.$$

Hence two vectors which have different units may be said to be independent, whereas if that condition is not satisfied, they are dependent when adjacent.

Consider next the expression $2(A + B)$. Does it mean

$$2A + 2B \text{ or } A + B + A + B\,?$$

The figure (8) shows the two hypotheses. It means the former, because A is always prior to B; and the latter hypothesis would not enable us to construct $\frac{1}{2}(A + B)$ (fig. 9).

Fig. 8. Fig. 9.

Consider next the square of a vector. By B^2 is meant the product of B by itself; it analyses into $b^2\beta^2$. This unit β^2 has the property that while it refers to the line β it does not have positive or negative direction along the line. Consequently Hamilton and Tait replace β^2 by -1, the minus coming from treating of imaginary units. The substitution of 1 for β^2 may be justified under certain conditions as a reduction, but 1 can never be the full equivalent of β^2; because β having one dimension in length, β^2 must have two; whereas 1 is the purest kind of number. Again, we observe that while B^2 may be represented by an area square as in the second book of Euclid, it does not mean such geometrical quantity; for its factors being identical have necessarily the same direction. It is a scalar square, not an area square, and throughout such squares must be carefully distinguished.

Suppose that in the binary product the second factor is different from the first, as AB. It may be taken to denote the parallelogram formed by A and B; although there are reasons for holding that the simpler and more direct meaning is the triangle AB. On the former interpretation $\alpha\beta$ denotes the rhombus unit; on the latter, the triangular unit. When β is perpendicular to α, the unit $\alpha\beta$ becomes a simple unit of area: in the general case it can be resolved into two components of the second degree, the one scalar, the other areal.

In the ternary product all three factors may be the same, or a pair may be the same, or they may be all different. The cube $B^3 = b^3\beta^3$, having an odd number of line units, has as a result the polar direction of β; and so it is for every odd power of a vector, whereas every even power is like the square in wanting polar direction. Nevertheless the even powers of β cannot be dropped from the unit without violating the principle of dimensions. Similarly B^2C has the polar direction of γ, whereas BC^2 has the polar direction of β. In ABC, which analyses into $abc\alpha\beta\gamma$, the unit $\alpha\beta\gamma$ means the parallelepiped (or rather the tetrahedron) formed by α, β, γ. When α, β, γ form an orthogonal system, $\alpha\beta\gamma$ is simply the unit of volume. It is then a scalar unit of three dimensions.

Does this method of definition fail when we come to the quaternary product? What is the meaning of $ABCD$? It was pronounced "impossible" by Gregory and De Morgan, reasoning from the tri-dimensional character of any resultant vector; but A, B, C, D are not the components of a resultant but the elements of a complex vector. For these there is no limit to the independence. The product of the moduli may be removed; what then is the meaning of $\alpha\beta\gamma\delta$? It does not mean a solid but the solid angle defined by the units $\alpha\beta\gamma\delta$ taken in that order. The quantity $ABCD$ breaks up into a number of components, but each one of them involves the modulus $abcd$. As there is no limit to the number of unit rays which may proceed from a point, so there is no limit to the number of factors in a product of vectors.

The fourth power A^4, being composed of two square factors, has no polar direction; nor has A^2B^2. Again A^3B being divisible into A^2AB has the characteristics of AB.

What is the correct meaning of $(BC)^2$? Is it B^2C^2 or $BCBC$? It is the former, because A has always precedence over B, not only in a sum but in a product. In a product the factors may be associated in any manner so long as the order of the elements is not interfered with.

What is meant by a fractional power of a vector, such as $B^{\frac{1}{2}}$? So far as the modulus is concerned there is no difficulty; it is simply $\sqrt{b}$. But what is $\beta^{\frac{1}{2}}$? It must be such that $(\beta^{\frac{1}{2}})^2 = \beta$. Consequently $\beta^{\frac{1}{2}}$ has the direction of β, but its length dimension is $\frac{1}{2}$. Similarly, $\beta^{\frac{1}{3}}$ has the direction of β, but a length dimension of $\frac{1}{3}$.

In the same manner as $(BC)^2 = B^2C^2$; so $(BC)^{\frac{1}{2}} = B^{\frac{1}{2}}C^{\frac{1}{2}}$.

The reciprocal of B is denoted by $\frac{1}{B}$; it analyses into $\frac{1}{b}\frac{1}{\beta}$, that is to say, not only is its modulus the reciprocal of the modulus of B, but its unit is the reciprocal of the unit of B. In the various books on vector-analysis it is commonly stated that the inversion applies to the modulus only; but a vector-analysis so founded cannot satisfy the principle of dimensions. The systematic notation is B^{-1}, and in general

$$B^{-n} = b^{-n}\beta^{-n}.$$

The quotient $B^{-1}C$ has two different meanings. It primarily means C per B; the English language reads it in the order opposite to the natural order of writing; for B is the antecedent and C the consequent. The modulus is c/b and the unit is $\beta^{-1}\gamma$, which means that the unit β is changed into the unit γ. The other is a reduced meaning got by deriving the amount of the angle, that is, $< \beta^{-1}\gamma$ and the axis of the angle denoted by $[\beta^{-1}\gamma]$; and $\frac{b}{c}[\beta^{-1}\gamma]^{<\beta^{-1}\gamma}$ constitutes the primary quaternion. The former quotient corresponds to the dyad of Gibbs. When B and B^{-1} are adjacent they cancel one another, but not unless.

From the principle of the succession of the elements it follows that

$$(B^{-1}C)^2 = B^{-2}C^2,$$

and

$$(B^{-1}C)^{\frac{1}{2}} = B^{-\frac{1}{2}}C^{\frac{1}{2}}.$$

What is the meaning of $\log B$? We first observe that as log is a kind of power, $\log B = \log(b\beta) = \log b \,.\, \beta^{\circ}$. Its unit has no dimension in length.

Let B and C be two vectors which have the same unit β; then, according to algebra (if we view the algebraic quantity as a vector),

$$(B+C)^2 = B^2 + 2BC + C^2$$
$$= (b^2 + 2bc + c^2)\,\beta^2.$$

Suppose now that C has a different unit γ; does the above formula for the square of the binomial change in form? Here we are at the parting of the ways; quaternionists and vector-analysts take the way of the affirmative; vector-algebra the way of the negative. Hamilton breaks up the square into two factors $(B+C)(B+C)$, applies the distributive rule retaining the order of the binomial factors in the partial products, getting

$$B^2 + BC + CB + C^2;$$

and, as the vector components of BC and CB cancel, the result is

$$B^2 + 2\operatorname{Cos} BC + C^2,$$

where $\operatorname{Cos} BC$ means the product of B and the projection of C on B.

The introduction of the Cos component into the formula instead of the complete product involves a very radical departure from the simple formula of algebra. We note, however, that the expression is the square of $\Sigma(B+C)$; how then, we ask, can it be the square of $B+C$, unless $B+C$ and $\Sigma(B+C)$ mean the same thing? But they are different; hence the above cannot be the true formula.

Grassmann, in his paper *Sur les différents genres de multiplication*, adopts the same rule for expanding the square of a binomial of vectors; but in the second part of the second "Ausdehnungslehre" he deduces the formula $B^2 + 2BC + C^2$ by laying down as the special condition for his algebraic multiplication of vectors that in it $BC = CB$. If that condition is satisfied, then BC and CB involve no order, and the whole subject reduces to common algebra.

The true method of forming the square is after the idea of a symmetrical matrix. Put B^2 and C^2 in the diagonal and BC in each corner; the result is

$$(B+C)^2 = B^2 + 2BC + C^2.$$

As

$$BC = \operatorname{Cos} BC + i \operatorname{Sin} BC,$$

the square of the binomial consists of two components; the scalar or cosine product

$$B^2 + 2\operatorname{Cos} BC + C^2,$$

and the area or Sine product $2i \operatorname{Sin} BC$. The former is the square of the resultant

$\Sigma(B+C)$, and expresses that extension of the Pythagorean Theorem which is demonstrated in the twelfth and thirteenth propositions of the second book of Euclid. The Sine product gives four times the area included between the complex vector and the resultant; with the triangular unit of area it would be twice, which is an indication that the triangular is the natural unit. Here we have an example of a general characteristic of vector-algebra; namely, that for any scalar theorem it supplies a complementary vector theorem, the sum of the two constituting the complete theorem.

Consider next the square of a linear complex of three vectors. It is formed by the matrix

$$\begin{matrix} A^2 & AB & AC \\ AB & B^2 & BC \\ AC & BC & C^2 \end{matrix}$$

or it may be deduced from the preceding principle as follows

$$\begin{aligned}(A+B+C)^2 &= \{A+(B+C)\}^2 \\ &= A^2+2A(B+C)+(B+C)^2 \\ &= A^2+B^2+C^2+2AB+2AC+2BC.\end{aligned}$$

The scalar part

$$A^2+B^2+C^2+2\operatorname{Cos} AB+2\operatorname{Cos} AC+2\operatorname{Cos} BC$$

gives

$$\{\Sigma(A+B+C)\}^2;$$

while the vector part

$$2i\operatorname{Sin} AB+2i\operatorname{Sin} AC+2i\operatorname{Sin} BC$$

gives four times the resultant of the areas enclosed between the complex and the resultant line. In the fig. 10 triangle 1 is $\frac{1}{2}\operatorname{Sin} AB$, and triangle 2 is

$$\tfrac{1}{2}\operatorname{Sin}\Sigma(A+B)C=\tfrac{1}{2}\Sigma\{\operatorname{Sin} AC+\operatorname{Sin} BC\};$$

hence the resultant of the enclosed area is

$$\tfrac{1}{2}\Sigma\{\operatorname{Sin} AB+\operatorname{Sin} AC+\operatorname{Sin} BC\}.$$

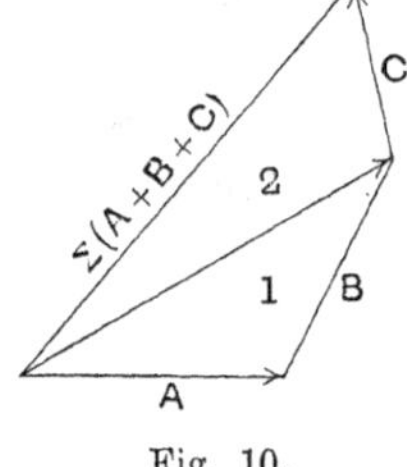

Fig. 10.

If $A+B+C$ is a cyclic complex, its square is derived from the linear square by introducing cyclic symmetry. As the cycle has no ends, there must be no discrimination among the elements; the elements have an order, but no definite point of beginning. Hence in the above case it is

$$A^2+B^2+C^2+2AB+2BC+2CA.$$

We proceed to consider in what manner these principles generalise the fundamental theorems of algebra; and first of the Binomial Theorem.

When n is a positive integer it follows from what has been established that

$$(B+C)^n=B^n+nB^{n-1}C+\frac{n(n-1)}{1.2}B^{n-2}C^2+$$

provided that B is in all the terms placed prior to C. When the number n is even, the scalar part is

$$B^n+\frac{n(n-1)}{1.2}B^{n-2}C^2+\text{etc.}+\operatorname{Cos} BC\left\{nB^{n-2}+\frac{n(n-1)(n-2)}{1.2.3}B^{n-4}C^2+\text{etc.}\right\};$$

while the vector part is an area vector

$$i \operatorname{Sin} BC \left\{ nB^{n-2} + \frac{n(n-1)(n-2)}{1.2.3} B^{n-4} C^2 + \qquad \right\}.$$

When the terms are reduced, these parts become

$$b^n + \frac{n(n-1)}{1.2} b^{n-2} c^2 + \text{etc.} + 2bc \cos \beta\gamma \left\{ nb^{n-2} + \frac{n(n-1)(n-2)}{1.2.3} b^{n-4} c^2 + \qquad \right\}$$

and $$i bc \sin \beta\gamma . [\beta\gamma] \left\{ nb^{n-2} + \frac{n(n-1)(n-2)}{1.2.3} b^{n-4} c^2 + \qquad \right\}.$$

But when n is odd all the terms of the expansion are vectors having a polar axis, and they break up into the two series

$$\beta \left\{ b^n + \frac{n(n-1)}{1.2} b^{n-2} c^2 + \qquad \right\} + \gamma \left\{ nb^{n-1} c + \frac{n(n-1)(n-2)}{1.2.3} b^{n-3} c^3 + \qquad \right\}.$$

Consider next the case where n is fractional; suppose $\frac{1}{2}$. Let B have the greater modulus; then

$$\begin{aligned}(B+C)^{\frac{1}{2}} &= \{B(1+B^{-1}C)\}^{\frac{1}{2}} \\ &= B^{\frac{1}{2}}(1+B^{-1}C)^{\frac{1}{2}} \\ &= B^{\frac{1}{2}} \left\{ 1 + \frac{1}{2} B^{-1} C - \frac{1}{2.4} (B^{-1}C)^2 + \frac{1.3}{2.4.6} (B^{-1}C)^3 - \qquad \right\}.\end{aligned}$$

Now because $(B^{-1}C)^2$ reduces to $\frac{c^2}{b^2}$, the series reduces to two parts

$$\beta^{\frac{1}{2}} \sqrt{b} \left\{ 1 - \frac{1}{2.4} \frac{c^2}{b^2} - \frac{1.3.5}{2.4.6.8} \frac{c^4}{b^4} - \qquad \right\}$$

and $$\beta^{\frac{1}{2}} \beta^{-1} \gamma \sqrt{b} \left\{ \frac{1}{2} \frac{c}{b} + \frac{1.3}{2.4.6} \frac{c^3}{b^3} + \qquad \right\}.$$

What is the meaning of the unit $\beta^{\frac{1}{2}} \beta^{-1} \gamma$? The quotient means that β is changed into γ; hence it means that $\beta^{\frac{1}{2}}$ is changed into $\gamma^{\frac{1}{2}}$. Hence the unit is $\gamma^{\frac{1}{2}}$. The result is a complex having $\beta^{\frac{1}{2}}$ and $\gamma^{\frac{1}{2}}$ for successive units.

The Binomial Theorem also applies when n is negative; for example, b being greater than c,

$$\begin{aligned}(B+C)^{-2} &= B^{-2} - 2B^{-3}C + \frac{2.3}{1.2} B^{-4} C^2 - \\ &= B^{-2} \left\{ 1 - 2B^{-1}C + \frac{2.3}{1.2} B^{-2} C^2 - \qquad \right\}.\end{aligned}$$

When the terms are reduced this becomes

$$= \frac{1}{b^2} \left\{ 1 + \frac{2.3}{1.2} \frac{c^2}{b^2} + \frac{2.3.4.5}{1.2.3.4} \frac{c^4}{b^4} + \qquad \right\} + \frac{1}{b^2} \left\{ 2 \frac{c}{b} + \frac{2.3.4}{1.2.3} \frac{c^3}{b^3} + \qquad \right\} \beta^{-1} \gamma.$$

Consider next the square root of a trinomial. If in $A+B+C$ we have a cyclic succession, the greatest vector can be moved to the front; say B. Then

$$\begin{aligned}\{B+C+A\}^{\frac{1}{2}} &= B^{\frac{1}{2}} \{1 + B^{-1}C + B^{-1}A\}^{\frac{1}{2}} \\ &= B^{\frac{1}{2}} \left\{ 1 + \frac{1}{2}(B^{-1}C + B^{-1}A) - \frac{1}{2.4}(\quad)^2 + \frac{1.3}{2.4.6}(\quad)^3 - \qquad \right\}.\end{aligned}$$

Now the square is

$$(B^{-1}C)^2 + 2B^{-2}CA + (B^{-1}A)^2,$$
$$= \frac{1}{b^2}\{c^2 + 2ca\,\gamma\alpha + a^2\},$$

where $\gamma\alpha$ is of zero dimensions.

And the cube is

$$B^{-3}C^3 + 3B^{-3}C^2A + 3B^{-3}CA^2 + B^{-3}A^3,$$
$$= \frac{1}{b^3}\{c^3\beta^{-1}\gamma + 3c^2a\beta^{-1}\alpha + 3ca^2\beta^{-1}\gamma + a^3\beta^{-1}\alpha\}.$$

Hence the square root of the trinomial is

$$B^{\frac{1}{2}}\left\{1 - \frac{1}{2\,.\,4}\frac{1}{b^2}(c^2 + 2ca\cos\gamma\alpha + a^2) - \qquad\right\} \qquad \ldots\ldots\ldots\ldots(1)$$

$$+ B^{\frac{1}{2}}\left\{\quad - \frac{1}{2\,.\,4}\frac{1}{b^2}2ca\sin\gamma\alpha - \qquad\right\}[\gamma\alpha]^{\pi/2} \ldots\ldots\ldots\ldots(2)$$

$$+ B^{\frac{1}{2}}\left\{\frac{1}{2}\frac{c}{b} + \frac{1\,.\,3}{2\,.\,4\,.\,6}\frac{1}{b^3}(c^3 + 3ca^2) + \qquad\right\}\beta^{-1}\gamma \ldots\ldots\ldots\ldots(3)$$

$$+ B^{\frac{1}{2}}\left\{\frac{1}{2}\frac{a}{b} + \frac{1\,.\,3}{2\,.\,4\,.\,6}\frac{1}{b^3}(a^3 + 3c^2a) + \qquad\right\}\beta^{-1}\alpha \ldots\ldots\ldots\ldots(4).$$

The unit for the series (1) is evidently $\beta^{\frac{1}{2}}$. That for the series (3) is $\beta^{\frac{1}{2}}\beta^{-1}\gamma$, which is $\gamma^{\frac{1}{2}}$. That for the series (4) is $\beta^{\frac{1}{2}}\beta^{-1}\alpha$, which is $\alpha^{\frac{1}{2}}$. Finally the unit for the second series is $\beta^{\frac{1}{2}}[\gamma\alpha]^{\pi/2}$, which means that β is conically rotated round the axis $[\gamma\alpha]$ by a quadrant. Its components are

$$\cos\beta[\gamma\alpha]\,.\,[\gamma\alpha] \quad\text{and}\quad \sin\beta[\gamma\alpha]\,.\,[\beta[\gamma\alpha]].$$

If instead of the cycle of vectors we substitute a cycle of dyads, the same process will hold, and we shall obtain an expression for the square root of a dyadic or linear vector function. The existence of such a quantity was discovered by Tait.

In the case of a product of two factors $(A + B + C)(A' + B' + C')$, where the corresponding elements of the factors have the same units, there is no real order in the factors; the order is in the units, and the expansion is made after the order of the units. If in the above the elements are cyclic, the product is

$$AA' + BB' + CC' + AB' + A'B + BC' + B'C + CA' + C'A,$$

that is

$$aa'\alpha^2 + bb'\beta^2 + cc'\gamma^2 + (ab' + a'b)\alpha\beta + (bc' + b'c)\beta\gamma + (ca' + c'a)\gamma\alpha.$$

But if the two factors are made up of different units, then there is an order of the factors which dominates over the order of the elements in each factor if these have an order. Consider $(A + B)(C + D)$. It means

$$\{\Sigma(A + B)\}\{\Sigma(C + D)\} = \Sigma(AC + AD + BC + BD);$$

in words the product of the resultants is equal to the resultant of the products. This means that the triangle formed by the resultants is the maximum projection of the sum of the several triangles. The Theorem of Moments is simply the vector part of the above theorem; for, from the latter, we also derive that the cosine product of the resultants is equal to the sum of the cosine products of the several triangles.

So long as these two kinds of product are not distinguished, it is impossible to develope vector-algebra. The want of the discrimination is the source of the difficulties in differentiation, and has prevented the development of the transcendental part of vector-algebra.

We are now in a position to attack the difficulties of differentiation. A curve in space is a linear complex of infinitesimal vectors; let it be denoted by S (fig. 11). Let R denote the variable chord from the beginning of the curve, and ΔR an additional chord at its extremity. When ΔR becomes infinitesimal, it coincides with an element of the curve S, and is then denoted by dR. Reciprocally, $R = \int dR$, where $\int$ denotes the resultant of the infinitesimal elements from the origin up to the extremity of R. The difference between S and R is that the former is the sum of the elements, whereas the latter is their resultant.

To find the derivative of R^2, the ordinary process applies with slight modification. The derivative of R^2 is the limit when ΔR is 0 of

$$\{-R^2 + (R + \Delta R)^2\} \frac{1}{\Delta R},$$

that is, of
$$\{2R\Delta R + (\Delta R)^2\} \frac{1}{\Delta R},$$

that is, of
$$2R + \Delta R,$$

therefore is
$$2R.$$

The differential is the derivative multiplied by dR; therefore in this case $2RdR$. It is to be noted that dR must be written after $2R$; also that dR^2/dR, that is, $2RdR/dR$ is a physical quotient, and that it is its reduced value which gives $2R$.

If we take the origin for the vector at a point outside the curve, then instead of the chord we have a radius-vector. Let it for a moment be denoted by R'; then

Fig. 11. Fig. 12.

$R' = \Sigma(A + R)$ (fig. 12) and, as A is constant, $dR' = dR$. Consequently R may be used to denote any radius-vector whether starting from a point in the curve or not.

Since
$$dR = \Sigma\{dr\rho + rd\rho\},$$
$$RdR = r\rho\,\Sigma\{dr\rho + rd\rho\},$$

but this does not break up into

$$r\,dr\,\rho^2 + r^2\rho\,d\rho.$$

It follows from the generalised Binomial Theorem that, for any power of R,

$$dR^n = nR^{n-1}\,dR.$$

Consider next the differential of a product of vectors RS. If R and S are successive factors merely, then

$$\Sigma\,(R + dR)\,\Sigma\,(S + dS) = \Sigma\,(RS + RdS + dR\,.\,S + dRdS)$$

and therefore $$d\,(RS) = \Sigma\,(RdS + dR\,.\,S).$$

Hence, when S is equal to R,

$$d\,(RR) = \Sigma\,(RdR + dR\,.\,R)$$
$$= \text{Cos}\,RdR + \text{Cos}\,dR\,.\,R,$$

of which the modulus is $2r\,dr$.

But suppose that R and S are successive elements; then

$$d\,(RS) = RdS + SdR,$$

for, otherwise, dR^2 would not be $2RdR$.

The result here obtained is sufficiently startling, and shocks the ideas of quaternionists and vector-analysts alike. Nevertheless it is true, and is in harmony with the principle of succession. Any differential is subsequent to each of the original vectors in the complex. The rule of precedence is: Vectors first, then differentials of the first order, then differentials of the second order, and so on; and within these classes the members have the order designated by the order of the alphabet.

Hence, if R, S, T form a linear succession,

$$d\,(RST) = RSdT + RTdS + STdR;$$

but if they form a cyclic succession,

$$d\,(RST) = RSdT + STdR + TRdS.$$

Consider next the differentials of powers of a complex. First of all, it is to be noted that

$$d\,(R + S) = dR + dS$$

in contrast to $$d\Sigma\,(R + S) = \Sigma\,(dR + dS),$$

where the former differential is the sum, and the latter the resultant.

Since $$d\,(R + S)^2 = d\,(R^2 + 2RS + S^2),$$

it follows from the principles already found that

$$d\,(R + S)^2 = 2\,(R + S)\,(dR + dS).$$

In this product the factors are successive, and the elements successive within each factor.

Consider next the differentials of higher order. We have

$$dR^n = nR^{n-1}\,dR;$$

therefore $$d^2R^n = n\,(n-1)\,R^{n-2}\,(dR)^2,$$

d^2R not existing because R is the independent variable. Consequently

$$d^2R^n \frac{1}{(dR)^2} = n(n-1) R^{n-2}.$$

The higher differentials of a product are given by a generalisation of Leibnitz' Theorem. For

$$d(RS) = RdS + SdR,$$
$$d^2(RS) = Rd^2S + 2dRdS + Sd^2R;$$

hence

$$d^n(RS) = Rd^nS + ndRd^{n-1}S + \frac{n(n-1)}{1.2} d^2Rd^{n-2}S + \ldots + nd^2Sd^{n-1}R + dSd^nR.$$

It is to be noted that there is a change at the middle of the series from the R factor being first to the S factor being first. With that understanding we may write

$$d^nRS = (d_S + d_R)^n RS.$$

The same formula applies to component differentials of a single vector when analysed into modulus and unit. Let

$$R = r\rho,$$

then
$$dR = \Sigma\{dr\rho + rd\rho\},$$
$$d^2R = \Sigma\{d^2r\rho + 2dr\,d\rho + rd^2\rho\},$$
and generally
$$d^nR = \Sigma\{d^nr\rho + nd^{n-1}rd\rho + \ldots + rd^n\rho\}$$
$$= \Sigma(d_r + d_\rho)^n r\rho.$$

As r is a modulus, there is no change at the middle.

Consider next the subject of partial differentiation. Let u be a modulus function of the moduli x, y, z and let their respective units be constant and denoted by h, j, k. Then $\frac{\partial u}{\partial x}\Big/ h$, $\frac{\partial u}{\partial y}\Big/ j$, $\frac{\partial u}{\partial z}\Big/ k$ express the gradients along the axes. And their resultant is

$$\Sigma\left\{\frac{\partial u}{\partial x}\Big/ h + \frac{\partial u}{\partial y}\Big/ j + \frac{\partial u}{\partial z}\Big/ k\right\},$$

which is the maximum gradient, and is denoted by ∇u. Let ν denote the unit of this maximum gradient; du and $dn\nu$ corresponding differentials; then

$$\nabla u = \frac{du}{dn}\Big/ \nu.$$

The gradient for any direction σ is

$$\frac{du}{ds}\Big/ \sigma = \frac{du}{dn}\cos\frac{1}{\nu}\sigma \,.\Big/ \sigma.$$

Hence the total differential corresponding to $ds \,.\, \sigma$ is

$$\frac{\partial u}{\partial x}\frac{1}{h}dxh + \frac{\partial u}{\partial y}\frac{1}{j}dyj + \frac{\partial u}{\partial z}\frac{1}{k}dzk,$$

that is,
$$d_su = \frac{\partial u}{\partial x}dx + \frac{\partial u}{\partial y}dy + \frac{\partial u}{\partial z}dz;$$

and
$$\frac{d_su}{ds}\Big/ \sigma = \left\{\frac{\partial u}{\partial x}dx + \frac{\partial u}{\partial y}dy + \frac{\partial u}{\partial z}dz\right\}\Big/ \Sigma\{dxh + dyj + dzk\}.$$

By $\nabla^2 u$ is properly meant the maximum gradient of the second order. As ∇ is cyclic, ∇^2 is the square of a cyclic complex; hence

$$\nabla^2 u = \left\{\frac{\partial^2}{\partial x^2}\Big/ h^2 + \frac{\partial^2}{\partial y^2}\Big/ j^2 + \frac{\partial^2}{\partial z^2}\Big/ k^2 + 2\frac{\partial^2}{\partial y \partial z}\Big/ jk + 2\frac{\partial^2}{\partial z \partial x}\Big/ kh + 2\frac{\partial^2}{\partial x \partial y}\Big/ hj\right\} u.$$

Hence $\nabla^2 u$ consists partly of a scalar gradient, partly of an area-vector gradient.

Hence the second differential of u for the distance $ds . \sigma$ is said to be

$$d_s^2 u = \frac{\partial^2}{\partial x^2} dx^2 + \frac{\partial^2}{\partial y^2} dy^2 + \frac{\partial^2}{\partial z^2} dz^2 + 2\frac{\partial^2}{\partial y \partial z} dy\,dz + 2\frac{\partial^2}{\partial z \partial x} dz\,dx + 2\frac{\partial^2}{\partial x \partial y} dx\,dy.$$

It is however, in reality, twice the differential; for

$$\frac{\partial}{\partial x}\left(\frac{\partial u}{\partial x}\Big/ h\right)\Big/ h = \frac{\partial^2 u}{\partial x^2}\Big/ h\Big/ h,$$

and therefore
$$\frac{1}{1.2}\frac{\partial^2 u}{\partial x^2}\Big/ h^2,$$

on the same principle that g ft. per sec. per sec. gives $\frac{1}{2}g$ ft. per (sec.)^2;

but
$$\frac{\partial}{\partial y}\left(\frac{\partial u}{\partial x}\Big/ h\right)\Big/ j = \frac{\partial^2 u}{\partial y \partial x}\Big/ h\Big/ j,$$

and therefore
$$1\frac{\partial^2 u}{\partial y \partial x}\Big/ (hj).$$

The maximum gradient of the third order $\nabla^3 u$ consists of a vector gradient only.

According to Quaternions $\nabla^2 u$ is

$$-\left(\frac{\partial^2}{\partial x^2} + \frac{\partial^2}{\partial y^2} + \frac{\partial^2}{\partial z^2}\right) u.$$

Leaving out the minus, the expression is derived by reduction from

$$\frac{\partial^2}{\partial x^2}\Big/ h^2 + \frac{\partial^2}{\partial y^2}\Big/ j^2 + \frac{\partial^2}{\partial z^2}\Big/ k^2;$$

the remaining terms of the above square do not appear at all; or if they do appear it is only to be immediately annihilated. I pointed out that the complete nabla of the second order could be obtained by the restoration of the cancelled terms in a paper contributed to the Fourth Congress of Mathematicians, entitled *On the square of Hamilton's delta.*

It is evident from what precedes that Taylor's Theorem retains its simple form, notwithstanding the opposite conclusion reached by Hamilton and Tait. Their theory of expanding a binomial led to a theory of differentiation in which no derivative is possible, and this prevented them from finding the generalisation of Taylor's Theorem. As in common algebra

$$f(B + C) = fB + f'B . C + \frac{1}{1.2}f''B . C^2 + \frac{1}{1.2.3}f'''B . C^3 + \quad .$$

Consider next the subject of integration. Let R denote the radius-vector of a curve in space; then

$$\int_A^B dR = \Sigma(-A + B),$$

that is, the integral is the chord between the limiting vectors A and B, which is evidently the resultant of all the infinitesimal vectors of the curve between A and B. The sign $\int$ denotes taking the resultant of a sum of infinitesimal vectors.

Again
$$\int_A^B RdR = \tfrac{1}{2}(-A^2 + B^2),$$
and
$$\int_0^R RdR = \tfrac{1}{2}R^2.$$

This agrees with the physical formula
$$\int mvdv = m\frac{v^2}{2},$$
for v and dv are vectors having different space-units.

Again
$$\int_A^B R^2 dR = \tfrac{1}{3}\Sigma(-A^3 + B^3),$$
which is a vector.

Hence if n is odd
$$\int_A^B R^n dR = \frac{1}{n+1}\{-A^{n+1} + B^{n+1}\},$$
but if n is even
$$= \frac{1}{n+1}\Sigma\{-A^{n+1} + B^{n+1}\}.$$

Also the formula for integration by parts, namely
$$\int A\,dB = AB - \int BdA.$$

Also
$$\int (A+B)^n (dA + dB) = \frac{1}{n+1}(A+B)^{n+1}.$$

It also follows from the generalised Binomial Theorem that
$$\int R^{1/n} dR = \frac{n}{n+1} R^{n+1/n},$$
where the integral is always a vector; and that
$$\int R^{-n} dR = -\frac{1}{n-1} R^{-(n-1)},$$
and that
$$\int \frac{1}{R} dR = \log R.$$

It is to be noted that
$$\int RdR = \int r\rho\Sigma(dr\rho + rd\rho);$$
and that the Σ makes it different from
$$\int r\rho dr\rho + \int r\rho r d\rho,$$
that is
$$\int rdr\rho^2 + \int r^2 \rho d\rho,$$
the former of which is $\frac{R^2}{2}$, and the latter is twice the resultant of the infinitesimal areas enclosed between the curve and the chord.

Having investigated what may be called the algebraic part of the generalised algebra, we now take up the transcendental complement. Hitherto the elements have been line vectors; now they will be spherical angles. The former part is geometrical in its nature; the complementary part is trigonometrical or quaternionic.

In Euler's notation for a circular angle, namely e^{ib}, b denotes the circular measure; and i which stands for $\sqrt{-1}$ is not in reality a multiplier of b, but the index of an axis which need not be specified so long as only one plane is considered. When angles on the sphere are considered, the axis must be inserted, and also $\pi/2$ for the indefinite $\sqrt{-1}$. Hence for a spherical angle we get the exponential expression $e^{b\beta^{\pi/2}}$. The logarithm of this expression is a quadrantal quaternion, or Hamiltonian vector. Compare this directed logarithm with the line vector; both have a multiplier and both have a direction, but the β of the logarithm is of the second order and zero dimensions, while the β of the line-vector is of the first order and has one dimension in length. But in the logarithm, instead of dimension there is a quadrant of angle; and it enters into the expression in a manner precisely analogous to that in which the dimension enters into the vector. This is the foundation for the profound analogy which was discovered by Hamilton, and which he treated as an identity rather than an analogy. These quadrantal logarithms may be called imaginary vectors, in the sense that they embody the quadrant as an index in the place of a dimension. Hamilton could not pass readily from the one to the other, because he adopted the principle

$$\beta^{\pi/2}\gamma^{\pi/2} = -\cos\beta\gamma + \sin\beta\gamma\,[\beta\gamma]^{\pi/2},$$

instead of having a minus before both terms. For from the true equation

$$\beta^{\pi/2}\gamma^{\pi/2} = -\cos\beta\gamma - \sin\beta\gamma\,[\beta\gamma]^{\pi/2},$$

we derive by dropping both indices and the equivalent minus

$$\beta\gamma = \cos\beta\gamma + \sin\beta\gamma\,[\beta\gamma]^{\pi/2}.$$

By expanding the expression for the angle, we obtain

$$e^{b\beta^{\pi/2}} = 1 + b\beta^{\pi/2} + \frac{1}{2\,!}b^2\beta^{\pi} + \frac{1}{3\,!}b^3\beta^{3\pi/2} +$$

from which we derive by means of $\beta^{\pi} = -1$

$$\begin{aligned} &1 - \frac{1}{2\,!}b^2 + \frac{1}{4\,!}b^4 - \\ &+ \left\{b - \frac{1}{3\,!}b^3 + \frac{1}{5\,!}b^5 - \quad\right\}\beta^{\pi/2} \\ &= \cos b + \sin b\,.\,\beta^{\pi/2}. \end{aligned}$$

In the fig. (13) PQ and QR are two spherical angles of which the logarithms are respectively denoted by $b\beta^{\pi/2}$ and $c\gamma^{\pi/2}$; and PR is the spherical angle which is the spherical resultant of PQ and QR. The question arises: Is $b\beta^{\pi/2} + c\gamma^{\pi/2}$ the log of PR? The binomial is successive; hence the question reduces to the comparison of $e^{b\beta^{\pi/2}}e^{c\gamma^{\pi/2}}$ with $e^{b\beta^{\pi/2}+c\gamma^{\pi/2}}$ when the powers are expanded as those of a successive binomial. The answer is Yes. Hamilton answered No, because he treated the binomial as equivalent to $\{\Sigma(b\beta + c\gamma)\}^{\pi/2}$. To prove that

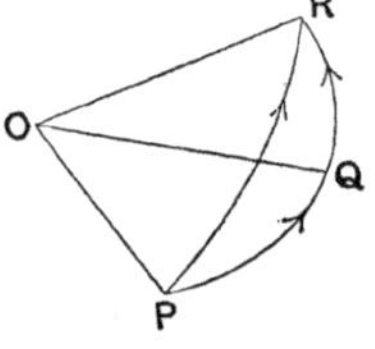

Fig. 13.

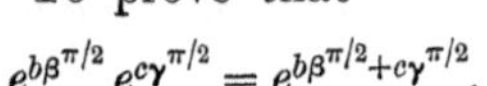

$$e^{b\beta^{\pi/2}}e^{c\gamma^{\pi/2}} = e^{b\beta^{\pi/2}+c\gamma^{\pi/2}}.$$

Since $$e^{b\beta^{\pi/2}} = 1 + b\beta^{\pi/2} + \frac{b^2}{2!}\beta^{\pi} + \frac{b^3}{3!}\beta^{3\pi/2} + \quad ,$$

and $$e^{c\gamma^{\pi/2}} = 1 + c\gamma^{\pi/2} + \frac{c^2}{2!}\gamma^{\pi} + \frac{c^3}{3!}\gamma^{3\pi/2} + \quad ;$$

$$\begin{aligned} e^{b\beta^{\pi/2}}e^{c\gamma^{\pi/2}} = 1 &+ c\gamma^{\pi/2} + \frac{c^2}{2!}\gamma^{\pi} + \frac{c^3}{3!}\gamma^{3\pi/2} + \\ &+ b\beta^{\pi/2} + bc\beta^{\pi/2}\gamma^{\pi/2} + \frac{bc^2}{2!}\beta^{\pi/2}\gamma^{\pi} + \\ &+ \frac{b^2}{2!}\beta^{\pi} + \frac{b^2c}{2!}\beta^{\pi}\gamma^{\pi/2} + \\ &+ \frac{b^3}{3!}\beta^{3\pi/2} + \\ &+ \end{aligned}$$

$$\begin{aligned} &= 1 + b\beta^{\pi/2} + c\gamma^{\pi/2} \\ &+ \frac{1}{2!}\{b^2\beta^{\pi} + 2bc\beta^{\pi/2}\gamma^{\pi/2} + c^2\gamma^{\pi}\} \\ &+ \frac{1}{3!}\{b^3\beta^{3\pi/2} + 3b^2c\beta^{\pi}\gamma^{\pi/2} + 3bc^2\beta^{\pi/2}\gamma^{\pi} + c^3\gamma^{3\pi/2}\} \\ &+ \text{etc.} \end{aligned}$$

The quadratic expression is the square of $b\beta^{\pi/2} + c\gamma^{\pi/2}$ according to the analogy to successive vectors; and the cubic expression is the cube of that same binomial. And so for the other orders of terms. Hence the theorem is proved.

When the terms are reduced by means of the principle

$$\beta^{\pi/2}\gamma^{\pi/2} = -\cos\beta\gamma - \sin\beta\gamma \,.\, [\beta\gamma]^{\pi/2},$$

which includes $\beta^{\pi} = -1$; they fall into the following four components:

$$\begin{aligned} &(1)\quad 1 - \frac{1}{2!}\{b^2 + 2bc\cos\beta\gamma + c^2\} + \frac{1}{4!}\{b^4 + 4b^3c\cos\beta\gamma + 6b^2c^2 \\ &\qquad\qquad + 4bc^3\cos\beta\gamma + c^4\} - \\ &(2)\quad -\left\{bc - \frac{1}{3!}(b^3c + bc^3) + \quad\right\}\sin\beta\gamma \,.\, [\beta\gamma]^{\pi/2} \\ &(3)\quad +\left\{b - \frac{1}{3!}(b^3 + 3bc^2) + \quad\right\}\beta^{\pi/2} \\ &(4)\quad +\left\{c - \frac{1}{3!}(c^3 + 3cb^2) + \quad\right\}\gamma^{\pi/2}. \end{aligned}$$

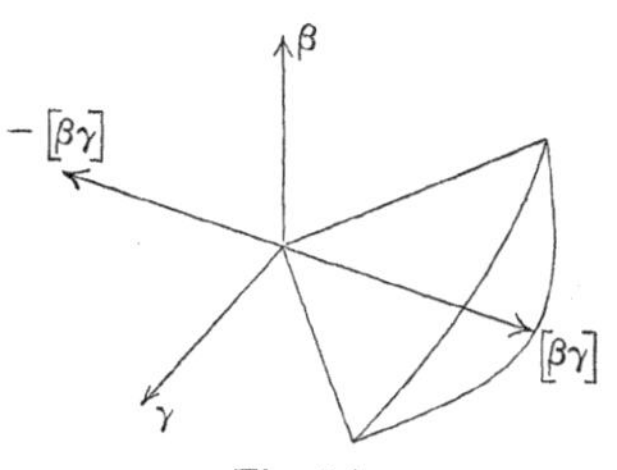

Fig. 14.

The first series gives the cosine of the angle; the second gives the component of the Sine which is along the intersection of the two angles; the third gives the component along the axis of the first angle; and the last gives the component along the axis of the second angle. These axes are represented in fig. 14.

Hence the rules, processes and theorems which apply to successive vectors can all be transferred with slight modification to this transcendental complement which deals with the successive logarithms of angles. For example:

$$e^{b\beta^{\pi/2}+c\gamma^{\pi/2}-c\gamma^{\pi/2}} = e^{b\beta^{\pi/2}};$$

but $$e^{b\beta^{\pi/2}+c\gamma^{\pi/2}-b\beta^{\pi/2}} \text{ not } = e^{c\gamma^{\pi/2}}.$$

Also $$\frac{1}{b\beta^{\pi/2}} = \frac{1}{b}\beta^{-\pi/2},$$

$$(b\beta^{\pi/2})^{\frac{1}{2}} = \sqrt{b}\beta^{\pi/4},$$

$$\log b\beta^{\pi/2} = \log b + \pi/2\,\beta^{\pi/2},$$

$$d\,(b\beta^{\pi/2}) = \{\Sigma\,(db\beta + bd\beta)\}^{\pi/2}.$$

If $b\beta^{\pi/2}$ is an angular velocity, then $\frac{db}{dt}\beta$ is the component changing the speed and $b\frac{d\beta}{dt}$ that changing the axis, the total result being $\left\{\Sigma\left(\frac{db}{dt}\beta + b\frac{d\beta}{dt}\right)\right\}^{\pi/2}$.

I believe that it will be found that the vector-algebra here outlined unifies the methods of Descartes, Grassmann, and Hamilton, and lays the foundations of an analysis which is in perfect logical harmony with algebra.

Note on Notation. In the above paper a capital italic letter is used to denote a vector: the corresponding small italic to denote its modulus, and the corresponding small Greek letter to denote its space-unit. A roman capital denotes a complex of vectors. The functional notations Cos and Sin are used to denote the scalar and vector components, the capital C and the capital S denoting that the appropriate space-unit is involved; whereas cos and sin with small c or s denote the respective moduli. Thus cos is the negative of Hamilton's S; Sin is equivalent to Hamilton's V with the imaginary left out, and sin to his TV with the imaginary left out. The imaginary where needed is denoted explicitly, and for this purpose i or the index $\pi/2$ is used. The triad h, j, k denote a constant orthogonal system of space-units. By $[\beta\gamma]$ is meant the unit which is normal to β and g; Σ denotes resultant of; and the point is used as a separatrix.

THE VALUES THAT CERTAIN ANALYTIC FUNCTIONS CAN TAKE

BY PHILIP E. B. JOURDAIN.

Consider a one-valued function f of a complex variable z and let $z=a$ be any regular point; further let C be a circle round this point such that all the points within it and on its circumference are regular. Let $f(a)=w_0$; we will find under what conditions z can so move—stating the matter picturesquely—so that, corresponding to it, f travels (say, in a straight line) from w_1 to *any* selected value w*.

We have first to prove that, corresponding to C, there is round w_0 an aggregate such that w_0 is isolated from all values which are not taken by f. If not, the points not taken by f condense at w_0; and, if we choose the aggregate contained by C as closed (i.e. including its circumference), the corresponding f-aggregate is also closed†; and consequently each such point is isolated from all the points which are assumed by f. If, then, we describe round every point which is not taken by f the greatest circle within which every point is not taken, we determine what we may call the "border" of all the domains formed by the points not taken by f. This border must correspond to z-points on the circumference of C; for, by the theory of conformal representation, a border-point obviously cannot correspond to a point within C. Thus, if we choose C so that the circumference does not pass through any point z such that $f(z)$ is equal to w_0, we exclude the possibility of the border-points on the w-plane lying arbitrarily close to the point w_0. Thus the points not taken by f cannot condense at the point w_0.

Thus it is possible to determine a path $(a \ldots z_1)$ such that $f(z)$ travels along the straight path $(w_0 \ldots w_1)$ towards w. Since, now, $f(z)$ is again regular about $z=z_1$, we can determine other portions of path $(z_1 \ldots z_2)$ and $(w_1 \ldots w_2)$. We can prove that there is a z-path such that f travels up to w by the considerations used by Lebesgue‡ allied to the following.

In the first place we notice that the z-path can never cut itself; for, suppose z' were a point of contact; then to z' would correspond two values w_1' and w_2' of the function $f(z)$, and this is excluded by the presupposed one-valuedness of $f(z)$.

* In this paper the method which I have followed in my previous paper "On Functions, all of whose Singularities are Non-Essential," *Mess. of Math.* (2), XXXIII. 1904, pp. 166—171, is made quite rigid.

† Cf. Schoenflies, *Die Entwickelung der Lehre von den Punktmannigfaltigkeiten*, Leipzig, 1900, p. 117. The object of making the f-aggregate closed will be quite evident when we consider the method to be used in what follows.

‡ *Leçons sur l'Intégration*, Paris, 1904, p. 105.

In the second place, if the singularities of f are non-essential, the z-path cannot go through any of them, for otherwise $f(z)$ would ultimately increase beyond any finite limit when z approaches one of these singularities *by any path.*

Thus, if $f(z)$ has only non-essential singularities, there can be no obstacle to the attainment of the value w by f. The case is otherwise if there are essential singularities. Then, to attain some particular value for w, it is quite possible that the said path is forced to lead to an essential singularity. Thus the function e^z has its only singularity—an essential one—at infinity; if now we consider any other point $z = a$ and put $w = 0$, we shall never be able to reach, in the manner described above, the point w. Thus it appears that the above method cannot be applied to prove the well-known result of Picard, but is limited to the proof that a function, all of whose singularities are non-essential, can take any assigned value at least once. In this result, of course, the fundamental theorem of algebra is contained.

It should be noticed that Lebesgue's method of proving the Heine-Borel theorem used avoids the use of the multiplicative axiom when the intervals (as here) have no end-points.

ÜBER LIMESBILDUNG UND ALLGEMEINE KÖRPERTHEORIE

Von Josef Kürschák.

Im Folgenden soll in die Körpertheorie oder, in der Terminologie des Herrn Julius König, in die Theorie der orthoiden Bereiche ein neuer Begriff eingeführt werden, der als Verallgemeinerung des absoluten Wertes angesehen werden kann.

Es sei jedem Elemente (jeder Grösse) a eines Körpers K eine reelle Zahl $\|a\|$ so zugeordnet, dass den folgenden Forderungen genügt wird:

(1) *es ist* $\|0\| = 0$; *für jedes von Null verschiedene a ist* $\|a\| > 0$;

(2) *für jedes Element a ist*

$$\|1 + a\| \leqq 1 + \|a\|;$$

(3) *für je zwei Elemente ist*

$$\|ab\| = \|a\| \|b\|;$$

(4) *es gibt in K wenigstens ein solches Element, dass $\|a\|$ von Null und von Eins verschieden ist.*

Jede solche Zuordnung nenne ich eine *Bewertung des Körpers K.* Die Zahl $\|a\|$ nenne ich *die Bewertung von a.*

Die bekanntesten bewerteten Körper erhalten wir, wenn für K ein reeller oder komplexer Zahlkörper gewählt und jeder Zahl ihr *absoluter Wert* zugeordnet wird.

Bedeutet K den Körper der rationalen Zahlen, so gibt es ausser der Bewertung mittels des absoluten Wertes noch unendlich viele andere Bewertungen. Es bedeute nämlich p eine gegebene Primzahl. Dann lässt sich jede von Null verschiedene rationale Zahl in der Gestalt

$$a = \frac{u}{v} p^{\alpha}$$

darstellen, wo u und v zu p teilerfremde ganze Zahlen sind. Der Exponent α kann eine positive oder negative ganze Zahl sein, oder auch Null. Wird nun

$$\|a\| = e^{-\alpha}$$

gesetzt, wo e die Basis der natürlichen Logaritmen bedeutet, und unter $\|0\|$ die Null verstanden: so entspricht diese Festsetzung allen Forderungen, die wir an eine Bewertung gestellt haben. Ich nenne darum $e^{-\alpha}$ *die Bewertung von a in Bezug auf den Äquivalenzmodul p.*

Die wichtigeren Sätze über absolute Werte bleiben für jede Bewertung gültig. Namentlich ist stets $\| 1 \| = 1$, $\| -1 \| = 1$ und

$$\| a - b \| \leqq \| a - c \| + \| c - b \|.$$

Zufolge dieser Ungleichheit ist $\| a - b \|$ ein spezieller Fall des aus der allgemeinen Mengenlehre bekannten Begriffes: *Écart.*

Bemerkt man dies, so liegt es nahe, die bekannten Begriffe des *Limes* und der *Fundamentalreihe* von den reellen und komplexen Zahlen auf beliebige bewertete Grössen zu übertragen. Man braucht nur in den gewöhnlichen Definitionen statt den absoluten Wert überall Bewertung zu setzen. Die auf den Grenzwert bezüglichen, bekannten Sätze behalten zumeist auch nach dieser Verallgemeinerung ihre Gültigkeit.

Z. B.: *Bedeutet A eine beliebige Grösse des bewerteten Körpers K, so gibt es immer eine solche Folge*

$$a_1, a_2, \ldots, a_n, \ldots$$

von Grössen aus K, dass sämtliche Glieder von einander und von A verschieden sind und mit wachsendem n sich dem Limes A nähern.

Trotz der vielen Ähnlichkeiten lassen sich natürlich auch wesentliche Unterschiede zwischen den Körpern finden, wenn wir ihr Verhalten bei der Limesbildung betrachten. Den wichtigsten Unterschied können wir sogar schon bei den gewöhnlichen Grenzübergängen bemerken, wenn wir den Körper der reellen Zahlen mit dem der rationalen Zahlen vergleichen. Im Körper der reellen Zahlen hat jede Fundamentalreihe einen Grenzwert. Hingegen kann im Körper der rationalen Zahlen zu einer Fundamentalreihe aus rationalen Zahlen nur ausnahmsweise ein Limes gefunden werden. Man kann sogar behaupten, dass die irrationalen Zahlen eben dazu eingeführt sind, um die Bestimmung des Limes einer Fundamentalreihe, die im Körper der rationalen Zahlen nur ausnahmsweise möglich ist, im Körper der reellen Zahlen zu einer stets ausführbaren Operation zu machen.

Hat im bewerteten Körper K jede aus ihm entnommene Fundamentalreihe einen Limes, *dann ist dieser Körper identisch mit der Gesamtheit der Limes der in ihm enthaltenen Fundamentalreihen.* Wir nennen darum einen solchen Körper *perfekt.*

Das Ziel meiner Untersuchungen ist zu beweisen, dass jeder bewertete Körper durch die Adjunktion neuer Elemente zu einem solchen *perfekten* Körper erweitert werden kann, der zugleich auch *algebraisch abgeschlossen* ist. Ein Körper K wird algebraisch abgeschlossen genannt, wenn jede rationale ganze Funktion $f(z)$, deren Koeffizienten dem Körper K angehören, in K in *lineare* Faktoren zerfällt.

Die Frage nach der Möglichkeit der gewünschten Erweiterung zerfällt in Teilfragen, die entweder in der Literatur bereits erledigt sind, oder wenigstens durch naheliegende Verallgemeinerungen bekannter Untersuchungen leicht beantwortet werden können.

Vorerst wollen wir die Forderung der algebraischen Abgeschlossenheit unbeachtet lassen und nur die Frage erwägen, ob jeder bewertete Körper zu einem *perfekten* ergänzbar ist. Die Methoden zur Beantwortung dieser Frage stammen von Herrn G. Cantor. Seine Theorie der irrationalen Zahlen ist im Grunde eben die Lösung dieser Frage für den Fall, dass der vorgelegte Körper aus der Gesamtheit der rationalen Zahlen besteht und diese mit ihrem absoluten Werte bewertet sind. Cantors Ideen können in jedem Falle angewandt werden und führen zur kleinsten perfekten Erweiterung des gegebenen Körpers K. Die so erhaltene Erweiterung nennen wir den *derivierten* Körper von K. Ist K ein perfekter Körper, so ist er sein eigener derivierter Körper.

Bei den weiteren Teilfragen wollen wir unser Augenmerk auf die algebraisch abgeschlossenen Erweiterungen richten. Dass jeder Körper zu einem algebraisch abgeschlossenen Körper erweitert werden kann, das hat Herr E. Steinitz* beweisen. Sein Beweis, obzwar derselbe auf den viel bestrittenen Wohlordnungssatz von Herrn E. Zermelo beruht, soll im Folgenden als streng betrachtet werden. Uns wird aber ausser der Existenz der algebraisch abgeschlossenen Erweiterung auch deren Bewertung interessieren. Vom Stammkörper wollen wir dabei voraussetzen, dass er ein perfekter bewerteter Körper sei. Die nächste Frage, die wir zu erledigen haben, gestaltet sich dann in der folgenden Weise: *Kann die kleinste algebraisch abgeschlossene Erweiterung eines perfekten bewerteten Körpers K stets so bewertet werden, dass die Bewertung des Stammkörpers unberührt bleibt?*

Bekanntlich genügt jede Grösse des erweiterten Körpers einer und nur einer solchen algebraischen Gleichung, deren Koeffizienten dem Stammkörper K entnommen sind und die in K irreduzibel ist. Wir nennen diese Gleichung die Definitionsgleichung jener Grösse und fordern, dass solchen Grössen, welche durch dieselbe Definitionsgleichung bestimmt sind, gleiche Bewertung zukomme. Es ist dann evident, dass die der Definitionsgleichung

$$z^n + a_1 z^{n-1} + \ldots + a_n = 0$$

genügende Grösse α keine andre Bewertung erhalten kann, als

$$\| \alpha \| = \| a_n \|^{\frac{1}{n}}.$$

Es fragt sich aber, ob die so zugeordnete Zahl $\| \alpha \|$ in der Tat als Bewertung gelten kann, d. h. ob diese Zuordnung die Forderungen

$$\| \alpha\beta \| = \| \alpha \| \, \| \beta \|$$

$$\| 1 + \alpha \| \leqq 1 + \| \alpha \|$$

befriedigt.

Dass im erweiterten Körper für jedes Produkt in der Tat

$$\| \alpha\beta \| = \| \alpha \| \, \| \beta \|$$

ist, das kann mit Hilfe der Galoisschen Theorie leicht bewiesen werden. Dabei spielt der Umstand, dass der Stammbereich perfekt ist, gar keine Rolle.

Wollen wir aber auch

$$\| 1 + \alpha \| \leqq 1 + \| \alpha \|$$

für den erweiterten Körper beweisen, so gelingt dies nur mit solchen Hilfsmitteln der Analysis, bei denen die Perfektheit des Sammbereiches wesentlich ist. Ich habe diese Hilfsmitteln aus Thèse des Herrn J. Hadamard† entnommen.

* E. Steinitz, "Algebraische Theorie der Körper," *Journal für reine u. ang. Mathematik*, Bd 137 (1909).

† J. Hadamard, "Essai sur l'étude des fonctions données par leur développement de Taylor," *Journal de Math.*, Série 4, T. 8 (1892). Insbesonders Nr. 15—18 (Seite 119—125).

Hadamard hat dort unter Anderem die folgende Frage gelöst: *Was ist die notwendige und hinreichende Bedingung dafür, dass zu einer vorgelegten Potenzreihe* $P(z)$ *eine solche rationale ganze Funktion* $f(z)$ *gefunden werden kann, für welche der Konvergenzradius von* $P(z)f(z)$ *denjenigen von* $P(z)$ *übertrifft?* Desweiteren: *Wenn es solche rationale ganze Funktionen gibt, wie finden wir unter ihnen diejenige vom niedrigsten Grade?*

Bei Hadamard begegnen wir diesen Fragen eigentlich in der folgenden Gestalt: *Wann hat eine Potenzreihe auf ihrem Konvergenzkreise keine anderen Singularitäten als Pole und wie lassen sich diese bestimmen?* In dieser Fassung knüpft sich die Frage an funktionentheoretische Begriffe, die wenigstens an dieser Stelle nicht auf beliebige perfekte Körper übertragen werden können. Wird aber die Frage in unserer obigen Fassung gestellt, so können nicht nur die Resultate, sondern auch die Beweise von Hadamard auf beliebige perfekte bewertete Körper übertragen werden.

Vor Allem kann der Begriff der Potenzreihe auf den Fall übertragen werden, dass die Koeffizienten nicht reelle oder komplexe Zahlen sind, sondern einem perfekten bewerteten Körper entnommen sind, und auch die *Werte*, welche die unabhängige Variable annimmt, Grössen von K sind. Sodann kann der Konvergenzradius für eine solche Potenzreihe

$$c_0 + c_1 z + c_2 z^2 + \dots c_m z^m + \dots$$

einfach in der Weise definiert werden, dass wir darunter den im gewöhnlichen Sinne genommenen Konvergenzradius von

$$\|c_0\| + \|c_1\| r + \|c_2\| r^2 + \dots + \|c_m\| r^m + \dots$$

verstehen. Die Hadamardsche Frage erhält auf diese Weise in jedem perfekten bewerteten Körper einen Sinn, und die Hadamardschen Resultate bleiben samt ihren Ableitungen allgemein gültig.

Übergehen wir zu Potenzreihen, die nach negativen Potenzen fortschreiten, so kann auf Grund der Hadamardschen Resultate leicht der folgende Satz eingesehen werden:

Sind die Koeffizienten von

$$f(z) = z^n + a_1 z^{n-1} + \dots + a_n$$

dem perfekten Körper K *entnommen und ist* $f(z)$ *in* K *irreduzibel, so ist der Konvergenzradius von*

$$\frac{1}{f(z)} = \frac{C_0}{z} + \frac{C_1}{z^2} + \dots$$

gleich $\|a_n\|^{\frac{1}{n}}$.

Nun bedeute α in irgend einer Erweiterung von K eine Wurzel der in K irreduziblen Gleichung $f(z) = 0$, und setzen wir $\|\alpha\| = \|a_n\|^{\frac{1}{n}}$. Es wird dann in der Tat

$$\|1 + \alpha\| \leqq 1 + \|\alpha\|.$$

Es ist dies identisch mit dem Satze, dass zwischen den Konvergenzradien l und l' der Reihen

$$\frac{1}{f(z)} = \frac{C_0}{z} + \frac{C_1}{z^2} + \dots$$

und
$$\frac{1}{f(t-1)} = \frac{C_0'}{t} + \frac{C_1'}{t^2} + \ldots$$
die Ungleichheit
$$l' \leqq 1 + l$$
besteht.

Das Bisherige gestattet uns jeden bewerteten Körper zu einem algebraisch *abgeschlossenen, bewerteten* Körper zu erweitern. Ob dieser Körper auch *perfekt* ist, bleibt noch unbestimmt. Wohl aber wird der derivierte Körper dieses algebraisch abgeschlossenen Körpers, d. h. seine kleinste perfekte Erweiterung, unseren sämtlichen Forderungen genügen. Man kann nämlich beweisen, *dass der derivierte Körper eines algebraisch abgeschlossenen Körpers K stets wieder algebraisch abgeschlossen ist.*

Ist K der Körper der algebraischen Zahlen, sind diese mit ihrem absoluten Werte bewertet, und ist somit der derivierte Körper von K die Gesamtheit der komplexen Zahlen: so erhellt die Richtigkeit meiner Behauptung aus dem fundamentalen Satz der Algebra. Natürlich hat man in der Algebra wenig Grund dafür, dass man diesen Teil des fundamentalen Satzes für sich beweise. Dennoch beginnt Weierstrass* seinen Beweis aus dem Jahre 1891 im Wesentlichen gerade damit. Seine Überlegungen können auch dann wiederholt werden, wenn K ein beliebiger algebraisch abgeschlossener bewerteter Körper ist.

Auf diese Weise können wir mit Benützung der Untersuchungen von Cantor, Steinitz, Hadamard und Weierstrass einen beliebigen bewerteten Körper erst zu einem perfekten, dann zu einem algebraisch abgeschlossenen, endlich zu einem solchen bewerteten Körper erweitern, der sowohl perfekt, als auch algebraisch abgeschlossen ist.

Ich wurde zu diesen Untersuchungen durch Herrn K. Hensels Schöpfung, die Theorie der *p-adischen Zahlen*†, angeregt und zwar insbesonders durch die Theorie der p-adischen rationalen Zahlen. Bewertet man die rationalen Zahlen nicht mit ihrem absoluten Werte, sondern mit ihrer Bewertung *in Bezug auf einen Äquivalenzmodul p*, und bildet dann den derivierten Körper, so bekommt man genau denjenigen Körper, den Hensel mit $K(p)$ bezeichnet und dessen Elemente er *p-adische rationale Zahlen nennt.* Die Grössen dieses Körpers sind (mit Ausname der rationalen Zahlen) weder Zahlen, noch Funktionen im gewöhnlichen Sinne, sondern neue Symbole, die eben so beschaffen sind, dass sie den Körper der in Bezug auf den Äquivalenzmodul p bewerteten rationalen Zahlen zu einem perfekten Körper erweitern.

* K. Weierstrass, "Neuer Beweis des Satzes, dass jede ganze rationale Funktion dargestellt werden kann als ein Produkt aus linearen Funktionen derselben Veränderlichen," *Sitzungsberichte d. Akad. Berlin* 1891, Seite 1085—1101, Werke Bd 3 (1903), Seite 251—168.

† Siehe insbesonders: K. Hensel, *Theorie der algebraischen Zahlen*, Bd 1, Leipzig (1908).

SOME APPLICATIONS OF QUATERNIONS

By L. Silberstein.

This communication will be published in the *Philosophical Magazine* in the course of 1913.

SOME EQUATIONS OF MIXED DIFFERENCES OCCURRING IN THE THEORY OF PROBABILITY AND THE RELATED EXPANSIONS IN SERIES OF BESSEL'S FUNCTIONS

By H. Bateman.

Equations of mixed differences appear to have been first obtained in the study of vibrating systems. When John Bernoulli* discussed the problem of a vibrating string he used a difference equation which is practically equivalent to

$$\frac{d^2u_n}{dt^2} = k^2 [u_{n-1} + u_{n+1} - 2u_n] \quad \text{.......................(1)},$$

and this difference equation was used by d'Alembert to obtain the well-known partial differential equation

$$\frac{\partial^2 u}{\partial t^2} = a^2 \frac{\partial^2 u}{\partial x^2}.$$

Equation (1) plays a prominent part in Euler's investigation of the propagation of sound in air† and has been discussed more recently by other writers‡.

The equation is usually solved by forming periodic solutions of the type

$$u_n = f(n)\, e^{ipt},$$

but I find that for some purposes the particular solution

$$u_n(t) = J_{2(n-m)}(2kt)$$

can be used more advantageously. For instance the solution which satisfies the conditions

$$u_n(t) = f(n), \qquad \frac{du_n}{dt} = 0 \qquad (n \text{ an integer})$$

when $t = 0$, is given generally by the formula§

$$u_n(t) = \sum_{m=-\infty}^{\infty} J_{2(n-m)}(2kt) f(m).$$

* *Petrop. Comm.* 3 (1728) [32], p. 13. *Collected Works*, Vol. III. p. 198. The problem of the string of beads was discussed in detail by Lagrange, *Mécanique Analytique*, t. I. p. 390.

† *Nova theoria lucis et colorum*, § 28, Berol. (1746), p. 184. An equation slightly different from the above has been obtained by Airy in a theory of the aether.

‡ See for instance M. Born und Th. v. Kármán, "Über Schwingungen in Raumgittern," *Phys. Zeitschr.* April 15 (1912), p. 297.

§ An expansion of this kind usually converges within a circle $|t| = \rho$ but with suitable values of the coefficients $f(m)$ it may converge for all values of t.

Many interesting expansions may be obtained from this result by using particular solutions $u_n(t)$, for instance we find that

$$n^2 + k^2t^2 \sum_{m=-\infty}^{\infty} m^2 J_{2(n-m)}(2kt):$$

this equation holds for integral values of n and for all values of t.

A solution satisfying the conditions

$$u_n(t) = 0, \qquad \frac{du_n}{dt} = \kappa [g(n) - g(n+1)] \qquad (n \text{ an integer})$$

when $t = 0$ is given generally by the formula

$$u_n(t) = \sum_{m=-\infty}^{\infty} g(m) J_{2(n-m)+1}(2kt),$$

and this result may also be used to obtain interesting expansions.

Equations of mixed differences have been solved hitherto* either by forming periodic solutions of the form $f_p(n) e^{ipt}$ or by using symbolical methods†; in the case of the numerous equations‡ which occur in the theory of vibrations the method of periodic solutions is used almost invariably. It will be of interest then to develop the present method and mention a few problems of chance for which the solutions obtained are naturally adapted.

The equation $$\frac{dF_n}{dx} = \tfrac{1}{2}[F_{n-1}(x) + F_{n+1}(x) - 2F_n(x)] \quad \text{.................. (2)},$$

which is analogous to the equation of the conduction of heat, possesses the particular solution $e^{-x} I_n(x)$ where $I_n(x)$ denotes the Bessel's function

$$\frac{1}{i^n} J_n(ix).$$

This particular solution may be generalised so as to provide us with a formula

$$F_n(x) = e^{-(x-a)} \sum_{m=-\infty}^{\infty} I_{n-m}(x-a) F_m(a) \quad \text{.................. (3)},$$

which gives the value of $F_n(x)$ when the value of $F_m(a)$ is known§.

* See however the remark on p. 387 of Nielsen's *Cylinderfunktionen* (1904).

† See for instance J. F. W. Herschel, *Calculus of Finite Differences*, Cambridge (1820), pp. 37—43; D. F. Gregory, *Mathematical Writings* (1865), pp. 38—41; Boole's *Finite Differences* (1860), pp. 193—207; G. Oltramare, *Assoc. Franc.* Bordeaux (1895), pp. 175—186; *Calcul de généralisation*, Paris (1899).

‡ Equations of mixed differences sometimes occur in the theory of radiation, see for instance Lord Rayleigh's paper "On the propagation of waves along connected systems of similar bodies," *Phil. Mag.* XLIV. pp. 356—362 (1897). *Scientific Papers*, Vol. IV.

§ The solution of the equation

$$\frac{du_n}{dx} = \tfrac{1}{2}(u_{n-1} - u_{n+1})$$

which satisfies the conditions $u_n(0) = f(n)$ is given by

$$u_n(x) = \sum_{m=-\infty}^{\infty} J_{n-m}(x) f(m).$$

This formula is due to Sonin, *Math. Ann.* Bd. XVI. p. 4 (1880).

Thus the particular solution $n^2 - 2nx + x^2$ gives rise to the expansion

$$n^2 - 2nx + x^2 = \sum_{m=-\infty}^{\infty} m^2 J_{n-m}(x).$$

By using particular solutions we may obtain the following expansions, most of which are well known:

$$I_n(x) = \sum_{m=-\infty}^{\infty} I_{n-m}(x-a)\, I_m(a), \qquad F_n(x) = e^{-x} I_n(x),$$

$$(n^2+x)\, e^{x-a} = \sum_{m=-\infty}^{\infty} (m^2+a)\, I_{n-m}(x-a), \qquad F_n(x) = n^2 + x,$$

$$e^{x\cos 2\alpha} \cos 2n\alpha = \sum_{m=-\infty}^{\infty} e^{a\cos 2\alpha} \cos 2m\alpha\, I_{n-m}(x-a), \qquad F_n(x) = e^{-2x\sin^2\alpha} \cos 2n\alpha.$$

The expansion (3) gives the solution of the following problem.

Consider a large number of boxes and a continually increasing large number of objects each of which is marked with either $+1$ or -1. The objects are placed in the boxes, there being no restriction as to the number in each, and when the average number of objects in a box is a the sum of the numbers in each box is supposed to be known. The problem is to find the chance that a box chosen at random contains numbers which add up to n when the average number of objects in each box has increased to x. It is supposed that an additional object is just as likely to go into one box as another.

Denoting by $F_n(x)$ the chance that a box contains numbers adding up to n and treating κ the number of boxes as very large we obtain the difference equation

$$F_n\left(x + \frac{1}{\kappa}\right) = \frac{1}{2\kappa}[F_{n-1}(x) + F_{n+1}(x)] + \left(1 - \frac{1}{\kappa}\right) F_n(x)$$

which reduces to (2) when squares of $\frac{1}{\kappa}$ are neglected*. Since

$$\sum_{r=-\infty}^{\infty} I_r(x-a) = e^{x-a}$$

it follows that when $F_n(x)$ is given by (3) we generally have

$$\sum_{n=-\infty}^{\infty} F_n(x) = \sum_{n=-\infty}^{\infty} F_n(a).$$

Consequently if $\sum_{n=-\infty}^{\infty} F_n(a) = 1$, as should be the case, we also have

$$\sum_{n=-\infty}^{\infty} F_n(x) = 1.$$

The function $e^{-(x-a)} I_{n-m}(x-a)$ is a type of Green's function for the equation (2), for a solution of the equation

$$\frac{du_n}{dx} - \tfrac{1}{2}(u_{n+1} + u_{n-1} - 2u_n) = f(n, x)$$

is given by

$$u_n = \sum_{m=-\infty}^{\infty} \int_{-\infty}^{x} e^{-(x-a)}\, I_{n-m}(x-a) f(m, a)\, da,$$

and this solution generally satisfies the conditions

$$u_n = 0 \text{ for } x = -\infty, \text{ or for } n = \pm\infty.$$

* It is interesting to compare the present problem with those considered by Lord Rayleigh, *Theory of Sound*, Vol. I. p. 37. "Dynamical problems in illustration of the kinetic theory of gases," *Phil. Mag.* Vol. XXXII. 1891. *Scientific Papers*, Vol. III.

If in our probability problem objects with *positive* signs only are added after the initial distribution has been examined the difference equation (2) must be replaced by the simpler one

$$\frac{dF_n}{dx} = F_{n-1} - F_n \quad \text{.....................(4).}$$

Poisson's formula*

$$F_n(x) = \frac{x^n}{\lfloor n} e^{-x}$$

gives the required probability in the case when there is no initial distribution of objects; from this particular solution we may derive the more general solution

$$F_n(x) = \sum_{p=0}^{\infty} \frac{(x-a)^p}{\lfloor p} \phi(n-p) e^{-(x-a)} \quad \text{.....................(5)}$$

which satisfies the condition $F_n(a) = \phi(n)$. If $\phi(n) = 0$ when n is negative the summation extends from $p = 0$ to $p = n$ and we obtain the solution of the problem for the case in which all the objects have the positive sign.

Interesting expansions may be derived from (5) by substituting particular solutions for $F_n(x)$, but the simplest ones are so well known that it will not be necessary to write them down here.

It is easily seen that in general

$$\sum_{n=-\infty}^{\infty} F_n(x) = \sum_{n=-\infty}^{\infty} F_n(a) = 1.$$

It should be mentioned that the problems considered here can be solved accurately as well as approximately by using partial difference equations in place of the equations of mixed differences. The equation which takes the place of (4) is

$$\kappa\,[F_n(m+1) - F_n(m)] = F_{n-1}(m) - F_n(m) \quad \text{..............(6)}$$

and has already been considered by Lagrange†. A particular solution is given by

$$F_n(m) = \frac{m(m-1)\ldots(m-n+1)}{n!} \frac{1}{\kappa^n}\left(1 - \frac{1}{\kappa}\right)^{m-n} \quad \text{...........(7),}$$

the well-known formula for the chance that a box contains a number n when we have no previous knowledge of the distribution of objects among the boxes‡. A more general solution is

$$F_n(m) = \sum_{s=0}^{m-r} \frac{(m-r)_s}{s!} \kappa^{-s}\left(1 - \frac{1}{\kappa}\right)^{m-r-s} F_{n-s}(r).$$

If there are no negative signs on the objects we must sum from 0 to n. Using the particular solution (7) the formula gives Vandermonde's theorem.

* *Recherches sur la probabilité des jugements*, Paris (1837), pp. 205—207. See also L. von Bortkewitsch, *Das Gesetz der kleinen Zahlen*, Leipzig (Teubner) (1898), this memoir contains tables of the function in the formula; L. Seidel, *Münch. Ber.* (1876), pp. 44—50. Smoluchowski, *Boltzmann Festschrift* (1904). Bateman, *Phil. Mag.* (1910) and (1911).

† *Miscellanea Taurinensia*, t. I. p. 33.

‡ And the signs are supposed to be all positive.

FONCTIONS IMPLICITES OSCILLANTES

Par Michel Petrovitch.

1. Les propositions de Sturm sur les intégrales de l'équation différentielle linéaire du second ordre

$$y'' + f(x) y = 0$$

expriment les relations entre les particularités (signe, limites de variation etc.) de la *valeur relative de la dérivée seconde* de y, c'est-à-dire de la valeur du rapport

$$\Delta(y) = \frac{y''}{y},$$

dans un intervalle considéré (a, b) de la variable indépendante x et l'*allure* de l'intégrale y lorsque celle-ci est finie et continue dans cet intervalle.

Je voudrais faire remarquer que les considérations élémentaires, conduisant à ces propositions, s'étendent à des classes plus générales d'équations de tous les ordres et conduisent à des particularités d'allure des intégrales réelles, finies et continues (ainsi que leurs dérivées) dans l'intervalle considéré de x.

2. Soit donnée une fonction, algébrique ou transcendante,

$$\phi(x, \xi_o, \xi_1, \ldots \xi_n)$$

de $n+2$ variables, à variations arbitraires ou assujetties à des conditions (C) s'exprimant par des égalités ou inégalités. Nous dirons:

1° qu'elle est, dans un intervalle (a, b) de la variable x, *supérieurement limitée* par une fonction donnée $\mu(x)$ si, pendant que x varie de a à b, la valeur de la fonction ϕ reste inférieure à $\mu(x)$ pour tout système de valeurs $\xi_o, \xi_1 \ldots \xi_n$ compatible avec les conditions (C);

2° qu'elle est, dans l'intervalle (a, b), *inférieurement limitée* par une fonction donnée $\lambda(x)$ si, pendant que x varie de a à b, la valeur de ϕ reste supérieure à $\lambda(x)$ pour tout système de valeurs $\xi_o, \xi_1, \ldots \xi_n$ compatible avec les conditions (C).

Un polynôme P ne contenant que des puissances *paires* des ξ_i à coefficients tous positifs pour x compris entre a et b est une fonction inférieurement limitée par son terme $\phi(x)$ indépendant des ξ_i; il est supérieurement limitée par ce terme si tous les coefficients sont négatifs.

La fonction $\frac{1}{P}$ est inférieurement ou supérieurement limitée par $\frac{1}{\phi}$ suivant que ses coefficients, supposés tous du même signe, sont négatifs ou positifs.

La fonction
$$\phi = \frac{f_1 + \phi_1 \xi_\kappa^2}{f_2 + \phi_2 \xi_\kappa^2}$$

où f_1, f_2, ϕ_1, ϕ_2 sont des fonctions finies, telles que f_1 et ϕ_1 ont un même signe ϵ et f_2, ϕ_2 un même signe ϵ' pour x compris entre a et b, est inférieurement et supérieurement limitée par les fonctions

$$\frac{f_1}{f_2} \text{ et } \frac{\phi_1}{\phi_2},$$

ou inversement, suivant les signes ϵ et ϵ'.

Un cas analogue se présente avec la fonction

$$\phi = f + \phi e^P,$$

où P est un des polynômes précédents à coefficients fonctions toutes négatives pour x compris entre a et b, f et ϕ étant du même signe dans cet intervalle ; ou bien avec les fonctions

$$\phi = \Sigma f_\kappa \xi_\kappa^2$$

où les ξ_κ sont assujetties à verifier une condition de la forme

$$\psi + \Sigma \phi_\kappa \xi_\kappa^2 \geqslant 0,$$

les f_κ, ϕ_κ, ψ étant des fonctions de x positives dans l'intervalle (a, b) etc.

3. Étant donné un système d'équations différentielles (E) d'un ordre quelconque, définissant n variables

$$y_1, \; y_2, \ldots y_n$$

comme fonction d'une variable indépendante x, il arrive qu'une expression ϕ, dépendant de x, des y_κ et de leurs dérivées successives, égale, en vertu des équations (E) mêmes, à la valeur relative $\Delta(y_\kappa)$ de la dérivée seconde d'une des variables y_κ, soit une fonction inférieurement ou supérieurement limitée pour x compris entre a et b, par des fonctions de x ne changeant pas de signe dans cet intervalle.

Ceci arrive, par exemple, dans le cas de l'équation du premier ordre

$$y' = f(x, y)$$

lorsque l'expression
$$\frac{1}{y}\left(\frac{\partial f}{\partial x} + f\frac{\partial f}{\partial y}\right)$$

est inférieurement ou supérieurement limitée par une telle fonction. Pour qu'il en soit ainsi de l'équation linéaire

$$y' = My + N$$

où M et N sont fonctions de x, il faut et il suffit qu'on ait

$$M = -\frac{N'}{N}$$

et que la fonction
$$M' + M^2,$$

égale alors à $\Delta(y)$, ne change pas de signe dans l'intervalle (a, b). Pour qu'il en soit ainsi de l'équation

$$y' = My + \sqrt{P + Ny^2}$$

il faut et il suffit qu'on ait à la fois

$$N' + 4MN = 0,$$
$$P' + 2MP = 0,$$

et que la fonction $$M' + M^2 + N,$$

à laquelle se réduit alors l'expression $\Delta(y)$, ne change pas de signe dans l'intervalle (a, b).

Les équations, d'un ordre quelconque, de la forme

$$y'' + y\phi(x, y, y' \ldots y^{(n)}) = 0,$$

où ϕ est l'une des expressions précédentes, et dont l'équation linéaire

$$y'' + f(x)\, y = 0$$

n'est qu'un cas particulier, appartiennent au type d'équations considérées. Tel est, par exemple, le cas de l'équation

$$y'' + my^3 + ny = 0$$

(où m, n sont des constantes du même signe) s'intégrant par les fonctions elliptiques.

Le système
$$\frac{dy_1}{dx} = f_1(x, y_1 \ldots y_n)$$
$$\ldots\ldots\ldots\ldots\ldots\ldots$$
$$\ldots\ldots\ldots\ldots\ldots\ldots$$
$$\frac{dy_n}{dx} = f_n(x, y_1 \ldots y_n)$$

forme un système (E) toutes les fois qu'au moins l'une des expressions

$$\frac{1}{y_\kappa}\left(\frac{\partial f_\kappa}{\partial x} + \Sigma f_i \frac{\partial f_\kappa}{\partial y_i}\right)$$

est une fonction ϕ inférieurement ou supérieurement limitée.

Ceci arrive, par exemple, dans le cas du système

$$\frac{dy_1}{dx} = My_1 + Ny_2$$
$$\frac{dy_2}{dx} = Py_1 + Qy_2,$$

où M, N, P, Q sont fonctions de x liées par la relation

$$N' + N(M + Q) = 0$$

et lorsque la fonction $$M' + M^2 + NP$$

à laquelle se reduit alors l'expression $\Delta(y_1)$, ne change pas de signe pour x compris entre a et b.

Dans le cas du système
$$\frac{dy_1}{dx} = my_2y_3,$$
$$\frac{dy_2}{dx} = ny_1y_3,$$
$$\frac{dy_3}{dx} = py_1y_2,$$

qu'on rencontre dans le problème du mouvement d'un corps solide et qui s'intègre par des fonctions elliptiques, on aura

$$\Delta(y_1) = m(py_2^2 + ny_3^2),$$

de sorte que le système appartient au type considéré toutes les fois que mp et mn sont du même signe et que y_1, y_2, y_3 (ainsi que leurs dérivées) sont réelles, finies et continues dans l'intervalle (a, b); en effet, $\Delta(y_1)$ ne saurait s'annuler pour une valeur $x = \alpha$ que si pour cette valeur de x on avait à la fois

$$y_2 = 0, \quad y_3 = 0,$$

dans quel cas en vertu des équations du système et celles qu'on obtient en différentiant celles-ci toutes les dérivées successives de y_1 seraient nulles pour $x = \alpha$.

Dans le cas de problèmes de la Dynamique, il arrive que l'équation des forces vives fournit des conditions d'inégalités, en vertu desquelles le système d'équations du second ordre, auxquelles se ramène le problème, fournit pour une ou plusieurs coordonnées q l'expression $\Delta(q)$ inférieurement ou supérieurement limitée.

4. Une fonction y de x sera dite *régulière* dans l'intervalle de $x = a$ à $x = b$ si, dans cet intervalle, elle et toutes ses dérivées sont réelles, finies et continues.

Lorsque le système donné (E) d'équations définit la valeur relative $\Delta(y)$ de la dérivée seconde d'une intégrale y comme fonction ϕ inférieurement ou supérieurement limitée, il y a des rapports entre les particularités (signe, mode de croissance, limites de variation etc.) de la fonction dans un intervalle (a, b) et la repartition des zéros simples de toute l'intégrale y régulière dans cet intervalle. Ces rapports sont résumés dans les faits suivants:

(a) Supposons que la fonction ϕ soit *inférieurement limitée* par une fonction $\lambda(x)$ finie et continue dans l'intervalle (a, b); soit u une intégrale quelconque de l'équation

$$u'' - \lambda(x)\, u = 0.$$

Deux zéros simples consécutifs de u compris dans (a, b) comprennent au plus un zéro de y; deux zéros simples de y comprennent au moins un zéro de u. Lorsque y et u ont un zéro commun $x = \alpha$ la variable x, en croissant à partir de α, atteindra d'abord un zéro de u et ensuite un zéro de y.

(b) Supposons que ϕ soit supérieurement limitée par une fonction $\mu(x)$ finie et continue dans (a, b); soit v une intégrale quelconque de l'équation

$$v'' - \mu(x)\, v = 0.$$

Deux zéros simples de v compris dans (a, b) comprennent au moins un zéro de y; deux zéros simples consécutifs de y comprennent au plus un zéro de v. Lorsque y et v ont un zéro commun $x = \alpha$, la variable x, en croissant à partir de α, atteindra d'abord un zéro de y et ensuite un zéro de v.

Ces propositions, qui ne sont autres que celles de Sturm pour les équations linéaires du second ordre, se démontrent de la manière bien élémentaire connue ne supposant que les inégalités

$$\lambda(x) \leqslant \phi \leqslant \mu(x)$$

et la régularité de l'intégrale y dans (a, b).

En prenant pour fonctions de comparaison u et v diverses fonctions dont on connaît, d'une part la manière dont varient les valeurs relatives de leurs dérivées secondes, et d'autre part la repartition des zéros dans un intervalle donné (a, b), on aura les propositions suivantes mettant en évidence les rapports entre les particularités de la fonction ϕ correspondante à une intégrale y d'un système (E), régulière dans (a, b), et la repartition des valeurs de x, comprises dans cet intervalle, pour lesquelles y change de signe en s'annulant.

1°. La fonction de comparaison

$$u = e^{rx}$$

ayant comme $\Delta(u)$ la valeur r^2, conduit à la règle suivante: *si ϕ est inférieurement limitée par une fonction constamment positive dans (a, b), l'intégrale y ne change de signe plus d'une fois dans cet intervalle.*

2°. La fonction de comparaison

$$v = \sin x\sqrt{N} \qquad (N = \text{const.})$$

ayant pour $\Delta(v)$ la valeur $-N$, conduit à la règle suivante: *si ϕ est supérieurement limitée par une fonction constamment négative dans (a, b), où elle a $-N$ comme une limite supérieure, l'intégrale y change de signe, dans cet intervalle, au moins autant de fois qu'il y a d'unités entières dans*

$$\frac{(b-a)\sqrt{N}}{\pi}.$$

La fonction de comparaison

$$u = \sin x\sqrt{M} \qquad (M = \text{const.})$$

ayant pour $\Delta(u)$ la valeur $-M$, conduit à la règle suivante: *si ϕ est inférieurement limitée par une fonction constamment négative dans (a, b), où elle a $-M$ comme une limite inférieure, l'intégrale y change de signe, dans cet intervalle, au plus autant de fois qu'il y a d'unités entières dans*

$$\frac{(b-a)\sqrt{M}}{\pi} + 2.$$

Ce sont les règles connues de Sturm pour le cas spécial d'une équation linéaire du second ordre, applicables, également, à d'autres types d'équations.

3°. Les fonctions de comparaison de la forme

$$x^{1-2m} \sin px$$

(m et p étant des constantes positives), ayant pour valeur relative de la dérivée seconde l'expression

$$\frac{C - Dx^{2m}}{x^2},$$

avec

$$C = \frac{m^2-1}{4}, \qquad D = p^2 m^2,$$

conduisent à la règle suivante: *les constantes positives C, D étant choisies de manière que dans (a, b) ϕ soit supérieurement limitée par la fonction*

$$\frac{C - Dx^{2m}}{x^2}$$

négative dans (a, b), y change de signe, dans cet intervalle, au moins autant de fois qu'il y a d'unités entières dans

$$\frac{b^m - a^m}{2\pi} p,$$

où

$$m = \sqrt{1+4C}, \qquad p = \sqrt{\frac{D}{1+4C}}.$$

De même: *les constantes positives C', D' étant choisies de manière que dans (a, b) ϕ soit inférieurement limitée par la fonction*

$$\frac{C' - D'x^{2m'}}{x^2}$$

négative dans (a, b), y change de signe, dans cet intervalle, au plus autant de fois qu'il y a d'unités entières dans

$$\frac{b^{m'} - a^{m'}}{2\pi} p' + 2,$$

où

$$m' = \sqrt{1+4C'}, \qquad p' = \sqrt{\frac{D'}{1+4C'}}.$$

Dans le cas, par exemple, de l'équation linéaire

$$y'' = \left(\frac{3}{4x^2} + 4x^2\right) y$$

la règle 2° fournit pour l'intervalle $(1, 4)$ et en prenant $N = 63, 9$ le nombre 9 comme limite supérieure du nombre de changement de signe de y; la règle 3° en prenant

$$C' = \tfrac{3}{4}, \qquad D' = \tfrac{25}{4},$$

ce qui donne

$$m' = 2, \qquad p' = \tfrac{5}{4},$$

en fournit 7 comme une limite supérieure. De même, pour une limite inférieure de ce nombre la règle 2° fournit 1 et la règle 3° le nombre 2.

4°. Les fonctions de comparaison de la forme

$$e^{-kx} \sin (pe^{2kx})$$

(où k et p sont des constantes positives), ayant pour valeur relative de la dérivée seconde l'expression

$$A - Be^{4kx},$$

avec

$$A = k^2, \qquad B = 4p^2k^2,$$

conduisent à la règle suivante: *les constantes positives A et B étant choisies de manière que, dans (a, b), ϕ soit supérieurement limitée par la fonction*

$$A - Be^{4kx},$$

négative dans (a, b), y change de signe, dans cet intervalle, au moins autant de fois qu'il y a d'unités entières dans

$$\frac{1}{2\pi}\left(e^{2b\sqrt{A}} - e^{2a\sqrt{A}}\right)\sqrt{\frac{B}{A}}.$$

De même: *les constantes positives A', B' étant choisies de manière que, dans (a, b), ϕ soit inférieurement limitée par la fonction*

$$A' - B' e^{4k'x},$$

négative dans (a, b), *y change de signe au plus autant de fois qu'il y a d'unités dans*

$$2 + \frac{1}{2\pi}\left(e^{2b\sqrt{A'}} - e^{2a\sqrt{A'}}\right)\sqrt{\frac{B'}{A'}}.$$

Dans le cas, par exemple, de l'équation linéaire

$$y'' = (1 - 4e^{4x})\, y,$$

en prenant
$$A = 1, \qquad B = 1,$$
$$A' = 1, \qquad B' = 9,$$

on trouve que l'intervalle $(0, 3)$ y doit changer de signe au moins 62 et au plus 188 fois, tandis que la règle 2° en fournit les limites 1 et 740.

5°. Les fonctions de comparaison de la forme $\sqrt{x}\sin(p\log x)$ (p étant une constante positive), ayant pour valeur relative de la dérivée seconde l'expression

$$-\frac{A}{x^2},$$

avec
$$A = p^2 + \tfrac{1}{4},$$

conduisent à la règle suivante : *les constantes positives A et A′ étant choisies de manière que dans* (a, b) *on ait constamment*

$$-\frac{A'}{x^2} \leqslant \phi \leqslant -\frac{A}{x^2},$$

le nombre de changement de signe de y dans (a, b) *est compris entre les deux valeurs*

$$\frac{\sqrt{4A - 1}}{2\pi}\log\frac{b}{a},$$

et
$$2 + \frac{\sqrt{4A' - 1}}{2\pi}\log\frac{b}{a}.$$

Dans le cas, par exemple, de l'équation linéaire

$$y'' = -\frac{65}{4x^2}\, y,$$

en prenant
$$A = \tfrac{25}{2}, \qquad A' = \tfrac{41}{2},$$

on trouve que dans l'intervalle $(1, 100)$ y doit changer de signe au moins 2 et au plus 4 fois ; la règle 2° fournit 1 et 128 comme limites de ce nombre.

5. Les règles précédentes s'appliquent, par exemple, aux équations du premier ordre

$$y' = f(x, y),$$

pour lesquelles l'expression

$$\phi = \frac{1}{y}\left(\frac{\partial f}{\partial x} + f\frac{\partial f}{\partial y}\right)$$

est limitée par une des fonctions de comparaison λ ou μ. On les verifie directement, par exemple, sur les équations de la forme

$$y' = -\frac{1}{2}\frac{P'}{P}\, y + \sqrt{P - k^2P^2y^2}$$

(où P est fonction de x ou constante, k une constante) ayant comme intégrale générale

$$y = \frac{1}{k\sqrt{P}} \sin\left(k\int P dx + C\right),$$

et comme ϕ l'expression

$$\frac{3}{4}\left(\frac{P'}{P}\right)^2 - k^2P^2 - \frac{1}{2}\frac{P''}{P}.$$

Appliquées à l'équation du second ordre

$$y'' + fy + \phi y^3 = 0,$$

où f et ϕ sont des fonctions positives de x, elles mettent en évidence le caractère oscillant des intégrales régulières et fournissent des limites inférieures du nombre d'oscillations de ces intégrales dans un intervalle de x donné; l'expression ϕ correspondante est supérieurement limitée par la fonction constamment négative $-f(x)$. On vérifie directement les règles sur le cas particulier de l'équation

$$y'' + Ay + By^3 = 0$$

(où A et B sont des constantes positives), s'intégrant par les fonctions elliptiques.

Appliquées au système

$$\frac{dy_i}{dx} = f_i(x, y_1, \ldots y_n), \quad (i = 1, 2 \ldots n)$$

ces règles mettent en évidence le caractère oscillant de toute intégrale régulière y_k pour laquelle l'expression

$$\phi = \frac{1}{y_k}\left(\frac{\partial f_k}{\partial x} + \Sigma f_i \frac{\partial f_k}{\partial y_i}\right)$$

est supérieurement limitée par une fonction négative; la connaissance d'une fonction de comparaison μ conduit à la connaissance d'une limite inférieure du nombre d'oscillations comprises entre les limites données de x. Les règles se vérifient directement sur le système de la forme

$$\frac{dy_1}{dx} = My_1 + Ny_2,$$

$$\frac{dy_2}{dx} = Py_1 - \left(M + \frac{N'}{N}\right) y_2,$$

où M, N, P sont des fonctions de x telles que l'expression

$$M' + M^2 + NP,$$

à laquelle se réduit l'expression ϕ relative à l'intégrale y_1, soit constamment négative, et plus particulièrement sur le cas où M, N, P sont des constantes.

SUR LA SÉRIE DE STIRLING

PAR J. HADAMARD.

La série de Stirling,

$$\log\sqrt{2\pi} - x + (x+\tfrac{1}{2})\log x + \frac{B_1}{1\,.\,2}\frac{1}{x} - \frac{B_2}{3\,.\,4}\frac{1}{x^3} + \ldots + (-1)^{n-1}\frac{B_n}{(2n-1)\,2n}\frac{1}{x^{2n-1}} + \ldots \qquad \ldots\ldots(1),$$

correspondant à l'expression asymptotique de $\log\Gamma(x+1)$ pour x très grand, est, on le sait, divergente. Elle renseigne sur l'allure de la fonction, mais n'en fournit pas la valeur numérique.

M. Borel s'est, le premier, posé le problème de remplacer une pareille série par une quantité ayant, pour chaque valeur de la variable, un sens bien défini. Il l'a résolu (pour une classe très étendue de développements, à laquelle appartient celui que nous considérons en ce moment) en substituant aux séries considérées des intégrales définies.

J'ai montré depuis* qu'on peut arriver au même résultat en suivant la méthode classique de M. Mittag Leffler, c'est-à-dire en ajoutant à chaque terme de la série divergente un terme de *décroissance plus rapide* lequel ne change pas, par conséquent, l'allure du terme primitif, mais qui produit la convergence lorsqu'on le considère pour $n = 1, 2, \ldots, \infty$.

On peut, comme j'ai eu l'occasion de le constater précédemment, indiquer *effectivement* une forme de série de cette espèce pour la fonction de Bessel. Je me propose aujourd'hui de traiter la même question pour la série de Stirling.

La série que nous allons former ne représentera pas $\log\Gamma(x)$; mais en désignant sa somme par $S(x)$, on aura

$$\log\Gamma(x) = S(x) + \psi(x),$$

$\psi(x)$ décroissant lui aussi, plus vite que n'importe quelle puissance de x.

Pour arriver à ce résultat, nous appliquerons la méthode qui jusqu'ici semble la plus naturelle pour l'évaluation asymptotique de $\log\Gamma(x)$, celle qui est fondée sur l'intégrale de Raabe

$$\int_x^{x+1}\log\Gamma(x)\,dx = x\log x - x + \log\sqrt{2\pi},$$

* *Transactions of the American Mathematical Society*, **1902.**

et, pour déduire de là la valeur de $\log \Gamma(x)$ lui-même, nous prendrons comme point de départ les mémoires* dans lesquels M. Lindelöf étudie les relations qui existent entre une intégrale définie et la somme des valeurs de la fonction à intégrer correspondant à des valeurs en progression arithmétique données à la variable.

Nous emprunterons aux recherches de M. Lindelöf la formule†

$$\tfrac{1}{2}[F(x)+F(x+1)]=\int_x^{x+1} F(x)\,dx+2\int_0^\infty \frac{Q(x+1,t)-Q(x,t)}{e^{2\pi t}-1}\,dt,$$

$$Q(x,t)=\frac{F(x+it)-F(x-it)}{2i}.$$

Pour $F(x)=\log\Gamma(x)$ (et, par conséquent, $F(x+1)-F(x)=\log x$) ceci devient

$$\tfrac{1}{2}[\log\Gamma(x)+\log\Gamma(x+1)]=\log\Gamma(x)+\tfrac{1}{2}\log x$$

$$=\int_x^{x+1}\log\Gamma(x)\,dx+\int_0^\infty \frac{\log(x+it)-\log(x-it)}{i(e^{2\pi t}-1)}\,dt \quad \ldots\ldots(2).$$

La première intégrale du dernier membre est celle de Raabe. C'est le second terme que nous nous proposons de développer.

Si la quantité sous le signe $\int$ était développable (dans tout l'intervalle), suivant les puissances de x, ce second terme s'écrirait:

$$2\Sigma\frac{(-1)^{n-1}}{(2n-1)\,x^{2n-1}}\int_0^\infty \frac{t^{2n-1}}{e^{2\pi t}-1}\,dt,$$

c'est-à-dire l'expression (3), puisque‡

$$\int_0^\infty \frac{t^{2n-1}}{e^{2\pi t}-1}\,dt=\frac{B_n}{4n}.$$

Pour appliquer ce même développement dans des conditions où il soit légitime, nous limiterons la deuxième intégrale de la formule (2) à l'intervalle $(0, x)$.

Le reste $$\int_x^\infty \frac{\log(x+it)-\log(x-it)}{i(e^{2\pi t}-1)}\,dt$$

satisfait à la condition de décroissance imposée ci-dessus et peut, dès lors, être laissé de côté.

Dans l'intégrale $\int_0^x$, notre développement est applicable: son terme général est de la forme

$$\frac{2(-1)^{n-1}}{(2n-1)\,x^{2n-1}}\int_0^x \frac{t^{2n-1}}{e^{2\pi t}-1}\,dt=\frac{2.(-1)^{n-1}}{(2n-1)\,x^{2n-1}}\left(\frac{B_n}{4n}-\int_x^\infty \frac{t^{2n-1}}{e^{2\pi t}-1}\,dt\right).$$

* Voir en particulier "Quelques applications d'une formule sommatoire générale," *Acta Soc. Fennicae*, t. XXXI. (1902); voir aussi *Acta Math.* t. XXVII., et *Leçons sur le calcul des résidus*, Paris, Gauthier Villars, 1905, Ch. III.

† C'est la formule (35) du No. 9 du Mémoire des *Acta Soc. Fennicae* (note précédente) dans laquelle on fait $f(n)=F(x+n)$, $m=0$, $n=1$, ou l'une des formules du No. 30 (page 62) des *Leçons sur le calcul des résidus*.

‡ Voir, par exemple, *Leçons sur le calcul des résidus*, p. 71.

Dans l'intégrale qui constitue le terme soustractif, nous remplacerons la fraction $\frac{1}{e^{2\pi t}-1}$ par son développement fini :

$$\frac{1}{e^{2\pi t}-1}=e^{-2\pi t}+e^{-4\pi t}+\ldots+e^{-2\lambda\pi t}+\frac{e^{-2\lambda\pi t}}{e^{2\pi t}-1}.$$

En prenant $\lambda=n$ (ou plus généralement, $\lambda=kn$, k étant un entier fixe) *les restes,* $\frac{e^{-2\lambda\pi t}}{e^{2\pi t}-1}$, *reportés dans l'intégrale, donnent une série convergente,* dont la somme satisfait d'ailleurs à la condition de décroissance.

Il en résulte qu'on rend la série de Stirling convergente en ajoutant au $n^{\text{ème}}$ terme la quantité

$$\frac{2(-1)^{n-1}}{(2n-1)x^{2n-1}}\int_x^\infty t^{2n-1}\left(e^{-2\pi t}+e^{-4\pi t}+\ldots+e^{-2n\pi t}\right)dt=\Phi_n\left(\frac{1}{x},\, e^{-2\pi x}\right),$$

étant un polynôme entier de degré $2n-1$ par rapport à l'une des variables, n par rapport à l'autre.

ÜBER EINE AUFGABE VON HERMITE AUS DER THEORIE DER MODULFUNKTION

VON L. SCHLESINGER.

In einem Briefe an Stieltjes* vom 23 Februar 1891 schreibt Hermite das folgende: "Permettez moi...de vous demander s'il y aurait lieu de suivre cette idée qui m'a traversé l'esprit: Je considère la relation entre les modules pour la transformation de second ordre

$$l = \frac{2\sqrt{k}}{1+k}.$$

"Si l'on pose $\omega = \frac{iK'}{K}$ et $k = \phi^4(\omega)$

elle devient

$$\phi^4\left(\frac{\omega}{2}\right) = \frac{2\phi^2(\omega)}{1+\phi^4(\omega)} \qquad \text{.............................(1).}$$

"Or il m'a semblé voir qu'une telle relation ne peut être satisfaite, si l'on suppose que $\phi(\omega)$ soit une fonction uniforme dans tout le plan. D'où cette conséquence non inutile à remarquer que la présence d'une coupure rend possible des relations d'une certaine forme, d'une nature et d'un caractère de grande importance."

Ich möchte heute einen ersten Teil meiner Untersuchungen vorlegen, die sich auf die von Hermite aufgeworfene und daran anschliessende weitere Fragen beziehen, indem ich mir die Aufgabe stelle, stufenweise fortschreitend die allgemeinsten monogenen Funktionen aufzustellen, die der Hermiteschen Funktionalgleichung genügen. Es wird dabei eine Bemerkung von Gauss zu besprechen sein, die sich vor kurzem im handschriftlichen Nachlasse vorgefunden hat, und wir werden im Anschlusse hieran zu einem einfachen System von Bedingungen gelangen, durch das die Modulfunktion charakterisiert werden kann.

1. Wir gehen der Bequemlichkeit wegen statt von $\phi(\omega)$ von der vierten Potenz dieser Funktion aus und schreiben die Hermitesche Funktionalgleichung (1) in der Form

$$f\left(\frac{t}{2}\right) = \frac{2\sqrt{f(t)}}{1+f(t)} \qquad \text{................................(2)}$$

oder

$$f^2\left(\frac{t}{2}\right) = \frac{4f(t)}{[1+f(t)]^2} \qquad \text{.............................(2A).}$$

* *Correspondance Hermite-Stieltjes*, II. (1905), S. 148.

Es sollen die monogenen Funktionen der komplexen Variabeln t charakterisiert werden, die der Gleichung (2) Genüge leisten. Dabei soll die Aussage, dass eine mehrdeutige Funktion der Funktionalgleichung genügt, bedeuten, dass diese Gleichung durch jeden eindeutig definierten Zweig jener Funktion befriedigt wird. In diesem Sinne sind auch alle anderen im folgenden zu betrachtenden Funktionalgleichungen zu verstehen.

Es sei $x=f(t)$, und $t=F(x)$ die inverse Funktion von $f(t)$. Dann folgt aus (2) die Gleichung

$$2F\left(\frac{2\sqrt{x}}{1+x}\right)=F(x) \qquad \text{.............................(3)}$$

die wir, um die Quadratwurzel zu vermeiden, durch die Substitution

$$\sqrt{x}=\xi$$

in die Gestalt

$$2F\left(\frac{2\xi}{1+\xi^2}\right)=F(\xi^2) \qquad \text{.............................(4)}$$

transformieren wollen.

Wir nehmen nun zuvörderst an, dass die Funktion $F(x)$ in der Umgebung von $x=1$ holomorph sei.

Dann folgt aus (4) durch Differentiation nach ξ

$$2\,\frac{1-\xi^2}{(1+\xi^2)^2}\,\frac{1}{\xi}\,F'\left(\frac{2\xi}{1+\xi^2}\right)=F'(\xi^2),$$

also für $\xi=1$

$$F'(1)=0.$$

Durch fortgesetzte Differentiation ergibt sich ebenso, dass alle Derivierten von $F(x)$ für $x=1$ verschwinden, sodass also $F(x)$ eine Konstante sein muss. Wie man sieht, bleibt dieser Schluss bestehen, wenn man an Stelle von (4) die allgemeinere Funktionalgleichung

$$p\,.\,F\left(\frac{2\xi}{1+\xi^2}\right)=F(\xi^2) \qquad \text{.............................(5)}$$

zu Grunde legt, wo p irgend eine endliche Konstante bedeutet, und es folgt sogar, dass für $p\neq 1$ die Konstante $F(x)$ notwendig den Wert Null haben muss. Wir haben also den Satz:

I. *Eine Funktion, die der Gleichung* (5) *Genüge leistet und in der Umgebung von* $x=1$ *holomorph ist, ist notwendig eine Konstante und für* $p\neq 1$, *also insbesondere für die Funktionalgleichung* (4), *identisch gleich Null.*

2. Wir stellen zunächst eine partikuläre Lösung der Funktionalgleichung (4) her, die nicht identisch verschwindet.

Es werde

$$x=\frac{c}{a},\quad b=\sqrt{a^2-c^2} \qquad \text{.............................(6)}$$

gesetzt, wo a, c beliebige komplexe Zahlen bedeuten, die nur der Bedingung unterworfen sind, dass a^2-c^2 von 0 und 1 verschieden ist.

Wir bilden aus a, b einerseits, aus a, c andererseits den Algorithmus des arithmetisch-geometrischen Mittels (agM.)

$$\left.\begin{array}{lll} a_1 = \dfrac{a+b}{2}, & b_1 = \sqrt{ab}, & c_1 = \sqrt{a^2 - b^2} \\ a_2 = \dfrac{a_1+b_1}{2}, & b_2 = \sqrt{a_1 b_1}, & c_2 = \sqrt{a_1^2 - b_1^2} \\ \ldots\ldots & & \end{array}\right\} \quad \ldots\ldots (7)$$

$$\left.\begin{array}{lll} \bar{a}_1 = \dfrac{a+c}{2}, & \bar{c}_1 = \sqrt{ac}, & \bar{b}_1 = \sqrt{\bar{a}_1^2 - \bar{c}_1^2} \\ \bar{a}_2 = \dfrac{\bar{a}_1+\bar{c}_1}{2}, & \bar{c}_2 = \sqrt{\bar{a}_1 \bar{c}_1}, & \bar{b}_2 = \sqrt{\bar{a}_2^2 - \bar{c}_2^2} \\ \ldots\ldots & & \end{array}\right\} \quad \ldots\ldots (8)$$

Dann ist bekanntlich*

$$\lim_n a_n = \lim_n b_n = M(a, b), \quad \lim_n c_n = 0 \quad \ldots\ldots (9)$$

$$\lim_n \bar{a}_n = \lim_n \bar{c}_n = M(a, c), \quad \lim_n \bar{b}_n = 0 \ldots\ldots (10)$$

und es bestehen die Beziehungen

$$M(\rho a, \rho b) = \rho M(a, b) \quad \ldots\ldots (11)$$

für ein beliebiges von Null verschiedenes ρ,

$$M(a, b) = M(a_n, b_n) = 2^n M(\bar{a}_n, \bar{b}_n) \ldots\ldots (12)$$

$$M(a, c) = M(\bar{a}_n, \bar{c}_n) = 2^n M(a_n, c_n) \quad \ldots\ldots (13).$$

Da

$$x_1 = \frac{2\sqrt{x}}{1+x} = \frac{\bar{c}_1}{\bar{a}_1} \quad \ldots\ldots (14)$$

ist, so genügt die durch die Gleichung

$$\omega = i \frac{M(a, b)}{M(a, c)} = i \frac{M(1, \sqrt{1-x^2})}{M(1, x)} \quad \ldots\ldots (15)$$

definierte Funktion von x, zufolge der Gleichungen (11), (12), (13) der Funktionalgleichung (4). Da bekanntlich

$$\left.\begin{array}{l} \lim\limits_{x=1} M(1, x) = 1 \\ \lim\limits_{x=1} M(1, \sqrt{1-x^2}) \log \dfrac{4}{\sqrt{1-x^2}} = \dfrac{\pi}{2} \end{array}\right\} \quad \ldots\ldots (16)$$

ist, so besitzt die Funktion ω von x für $x = 1$ eine Singularität, wie es nach unserem Satze I. sein muss. Die inverse dieser Funktion, $x = \phi(\omega)$, ist nichts anderes als die Modulfunktion†.

* Vergl. für die hier anzugebenden Sätze über das agM. etwa Schlesinger, *Handbuch der Theorie der linearen Differentialgleichungen*, Bd. II. 2 (1898), S. 7 ff.

† In der Bezeichnung von Hermite, $x = \phi^4(\omega)$.

3. Bedeutet $F(x)$ irgend eine nicht identisch verschwindende Lösung der Funktionalgleichung (4), so genügt offenbar

$$G(x) = F(x) \cdot \frac{M(1, x)}{iM(1, \sqrt{1-x^2})} \quad \text{.......................(17)}$$

der Funktionalgleichung
$$G\left(\frac{2\xi}{1+\xi^2}\right) = G(\xi^2) \quad \text{.............................(18)}$$

die aus der allgemeinen Gleichung (5) hervorgeht, wenn $p=1$ gesetzt wird. Wir wollen diese Funktionalgleichung die Gauss'sche nennen, die Berechtigung hierfür wird sich sogleich ergeben. Es erscheint somit durch Zuhilfenahme der Funktion (15) die allgemeine Lösung der Funktionalgleichung (4) auf die der Gauss'schen zurückgeführt. Wir beschäftigen uns also mit der letzteren, von der bereits feststeht (Satz I), dass eine Lösung derselben, die für $x=1$ holomorph ist, notwendig konstant sein muss.

Setzt man
$$\left.\begin{aligned} x_1 &= \frac{2\sqrt{x}}{1+x} = \frac{\bar{c}_1}{\bar{a}_1} \\ x_2 &= \frac{2\sqrt{x_1}}{1+x_1} = \frac{\bar{c}_2}{\bar{a}_2} \\ &\cdots\cdots\cdots\cdots \\ x_n &= \frac{\bar{c}_n}{\bar{a}_n} \end{aligned}\right\} \quad \text{.............................(19)}$$

so ist für eine Funktion $G(x)$ die der Gauss'schen Gleichung (18) genügt

$$G(x) = G(x_1) = G(x_2) = \ldots = G(x_n) \quad \text{.................(20).}$$

Da nach den Gleichungen (10) und (19)

$$\lim_n x_n = 1$$

ist, so folgt, dass die Lösung der Gauss'schen Gleichung (18) auch dann eine Konstante sein muss, wenn für sie nur feststeht, dass sie im Punkte $x=1$ eine Singularität besitzt, an der sie bestimmt ist*. In diesem Falle existiert nämlich $\lim_{x=1} G(x)$ als endlicher oder unendlich grosser Wert, der unabhängig ist von dem Wege, auf dem die Variable x in den Punkt $x=1$ einrückt. Es ist also

$$\lim_n G(x_n) = \lim_{x=1} G(x).$$

Bedeutet nun x einen Wert, für den die Funktion $G(x)$ holomorph ist, so haben wir nach (20)

$$G(x) = \lim_n G(x_n),$$

* Im Sinne der von mir (zuerst *Handbuch*, I. 1895, S. 16) angegebenen Modifikation der von Fuchs (1886, *Werke*, II. S. 349) aufgestellten Definition, wird von einer Funktion $F(x)$ gesagt, dass sie an der Stelle $x=a$ bestimmt (nicht unbestimmt) ist, wenn (1) $F(x)$ an allen Stellen einer gewissen Umgebung von $x=a$ holomorph ist und (2) für jede gegen die Null konvergierende Wertenfolge $\delta_1, \delta_2, \ldots$, die so beschaffen ist, dass sich $F(x)$ in den Punkten $a+\delta_k$ holomorph verhält, die Folge, $F(a+\delta_1)$ $F(a+\delta_2), \ldots$ einem bestimmten, von der Wahl der besonderen Folge $\delta_1, \delta_2, \ldots$ unabhängigen endlichen oder unendlich grossen Grenzwerte $\lim_{x=a} F(x)$ zustrebt, vorausgesetzt, dass zur Berechnung der Funktionswerte $F(a+\delta_k)$ ein beliebig gewählter, aber dann eindeutig festzuhaltender Zweig der Funktion $F(x)$ benutzt wird.

der Wert $\lim_{x=1} G(x) = C$ ist also ein endlicher, und es ist $G(x) = C$. Wir haben also den Satz:

II. *Eine Lösung der Funktionalgleichung* (18), *die für $x=1$ nicht unbestimmt wird, ist eine Konstante.*

In dem handschriftlichen Nachlass von Gauss findet sich eine Aufzeichnung, die demnächst an anderer Stelle veröffentlicht werden soll*, in der Gauss sagt, es sei leicht zu sehen, dass eine Funktion, die der Funktionalgleichung

$$F(t^2) = F\left(\frac{1+t^2}{2t}\right), \quad \ldots\ldots\ldots\ldots\ldots\ldots\ldots\ldots\ldots (21)$$

genügt, notwendig eine Konstante ist. Setzt man

$$\xi = \frac{1}{t}, \quad G\left(\frac{1}{t^2}\right) = F(t^2),$$

so verwandelt sich die Funktionalgleichung (18) in die Gauss'sche Gleichung (21). Gauss hatte bei dieser Aussage offenbar den Fall vor Augen, wo die Funktion $F(t)$ als Funktion der reellen Variabeln t für $t=1$ stetig ist. In der Tat werden wir in der Nr. 5 die allgemeinste Lösung der Gauss'schen Funktionalgleichung (18) aufstellen und an einem einfachen Beispiele erkennen, dass diese Lösung sich nicht notwendig auf eine Konstante reduziert, sodass also die in unserem Satze II. enthaltene Einschränkung für die Funktion $G(x)$ nicht überflüssig ist.

4. Die Gleichungen (16) lehren, dass die Funktion (15)

$$\omega = i\frac{M(1, \sqrt{1-x^2})}{M(1, x)}$$

für $x=1$ singulär, aber bestimmt ist. Bedeutet $F(x)$ eine Lösung der Funktionalgleichung (4), die in $x=1$ eine Singularität besitzt, von der Beschaffenheit, dass das Produkt

$$F(x) \cdot \frac{1}{\omega}$$

für $x=1$ bestimmt ist†, so ist dieses Produkt eine Konstante, es unterscheidet sich also $F(x)$ von ω nur durch einen konstanten Faktor. Für die inverse Funktion $x=f(t)$ von $t=F(x)$, die der Funktionalgleichung (2) genügt, erhalten wir also die Darstellung

$$f(t) = \phi(C \cdot t),$$

wo ϕ die Charakteristik der Modulfunktion und C eine Konstante bedeutet. Zufolge der Gleichungen (16) kann die Bedingung für $F(x)$ so formuliert werden, dass

$$F(x) \cdot \log\frac{4}{\sqrt{1-x^2}}$$

für $x=1$ bestimmt sein soll. Wir haben also den Satz:

* *Materialien für eine wissenschaftliche Biographie von Gauss, gesammelt von Klein und Brendel* (Leipzig, Teubner), Heft III. Anhang S. 120.

† Es genügt nicht etwa zu sagen, dass $F(x)$ selbst in $x=1$ bestimmt sein soll, denn das Produkt zweier Funktionen, die in $x=1$ bestimmt sind, kann sehr wohl in diesem Punkte unbestimmt sein. Z. B. sind $(x-1)^{-\frac{1}{2}+i}$ und $(x-1)^{\frac{1}{2}}$ beide in $x=1$ bestimmt, aber ihr Produkt $(x-1)^i$ ist (vergl. Brodén, *Arkiv för Mathem., Astron. och Fysik*, I. 1903–4, S. 446) in $x=1$ unbestimmt.

III. *Eine monogene Lösung $x=f(t)$ der Hermite'schen Funktionalgleichung* (2), *deren inverse Funktion in $x=1$ singulär, aber mit $\log\frac{4}{\sqrt{1-x^2}}$ multipliziert für $x=1$ bestimmt ist, geht aus der Modulfunktion $\phi(\omega)$ durch die Substitution $Ct=\omega$ hervor, wo C eine von Null verschiedene Konstante bedeutet.*

Für die lemniskatische Funktion ist $\omega=i$, $x=i$, wir können also für die Modulfunktion $x=\phi(\omega)$ die folgenden, diese Funktion eindeutig charakterisierenden Bedingungen angeben:

(1) Sie genügt der Funktionalgleichung

$$\phi^2\left(\frac{\omega}{2}\right)=\frac{4\,\phi(\omega)}{(1+\phi(\omega))^2},$$

(2) ihre inverse Funktion ist mit $\log\frac{4}{\sqrt{1-x^2}}$ muliplziert in $x=1$ bestimmt,

(3) sie nimmt für $\omega=i$ den Wert i an.

5. Die Betrachtungen der vorigen Nummer zeigen, dass die allgemeine Lösung der Funktionalgleichung (4) aus der allgemeinen Lösung der Gauss'schen Funktionalgleichung (18) durch Multiplikation mit einer partikulären Lösung von (4), also etwa mit ω hervorgeht. Wir wollen daher jetzt die Gauss'sche Funktionalgleichung (18) ganz allgemein zu lösen suchen.

Wir setzen zu diesem Zweck in $G(x)$ der Reihe nach $x=f(t)$, $t=e^u$ ein, wo $x=f(t)$ irgend eine Lösung der Hermite'schen Funktionalgleichung (2) bedeutet. Dann ist nach (18)

$$G(f(t))=G\left(f\left(\frac{t}{2}\right)\right),$$

und

$$G(f(e^u))=G(f(e^{u-\log 2})).$$

Bezeichnen wir also mit $P(u)$ irgend eine monogene Funktion der komplexen Variabeln u mit der Periode $\log 2$, so ist $P(\log t)$ die allgemeine Lösung der Gauss'schen Funktionalgleichung und folglich $t\,.\,P(\log t)$ die allgemeine Lösung der Funktionalgleichung (4). Wählen wir z. B. für die Lösung $x=f(t)$ von (2) die Modulfunktion $x=\phi(\omega)$, so stellt uns

$$G(x)=P(\log\omega)=P\left(\log i\,\frac{M(1,\sqrt{1-x^2})}{M(1,x)}\right) \qquad \text{..........(22)}$$

die allgemeine Lösung der Gauss'schen Funktionalgleichung, und folglich

$$t=F(x)=P\left(\log i\,\frac{M(1,\sqrt{1-x^2})}{M(1,x)}\right)i\,\frac{M(1,\sqrt{1-x^2})}{M(1,x)} \qquad \text{.........(23)}$$

die allgemeine Lösung der Funktionalgleichung (4) dar. Die Umkehrung der Funktion (23) liefert demnach die allgemeine Lösung der Hermite'schen Funktionalgleichung (2). Die Wahl $P(u)=e^{\frac{2\pi i u}{\log 2}}$ würde z. B. die Lösung

$$G(x)=\left\{\frac{iM(1,\sqrt{1-x^2})}{M(1,x)}\right\}^{\frac{2\pi i}{\log 2}}$$

der Gauss'schen Funktionalgleichung liefern, die offenbar keine Konstante ist, aber in $x=1$ unbestimmt wird. Da die periodische Funktion $P(u)$ in $u=\infty$ stets unbestimmt ist, so ist überhaupt, in Uebereinstimmung mit unserem Satze II., die durch (22) dargestellte Lösung der Gauss'schen Funktionalgleichung im Punkte $x=1$ stets unbestimmt, wenn sie sich nicht auf eine Konstante reduziert.

DIRECT DERIVATION OF THE COMPLEMENTARY THEOREM FROM ELEMENTARY PROPERTIES OF THE RATIONAL FUNCTIONS

BY J. C. FIELDS.

Let $$f(z, u) = u^n + f_{n-1}u^{n-1} + \dots + f_0 = 0 \quad \dots\dots\dots\dots\dots\dots (1)$$

be an equation in which the coefficients $f_{n-1}, \dots, f_0$ are rational functions of z. The equation may be reducible or irreducible. We shall assume however that it does not involve a repeated factor. Any rational function of (z, u) can be written in the reduced form

$$h_{n-1}u^{n-1} + h_{n-2}u^{n-2} + \dots + h_0 \dots\dots\dots\dots\dots\dots (2),$$

where the coefficients $h_{n-1}, \dots, h_0$ are rational functions of z. The term involving u^{n-1} we call the *principal term*, the coefficient h_{n-1} the *principal coefficient* of the function. By the *principal residue* of a rational function relative to a given value of the variable z we shall mean the residue of the principal coefficient of the function for this value of z.

In the neighbourhood of a value $z = a$ (or $z = \infty$) the equation (1) can be represented in the form

$$(u - P_1) \dots (u - P_n) = 0 \quad \dots\dots\dots\dots\dots\dots (3),$$

where $P_1, \dots, P_n$ are power-series in $z - a$ $\left(\text{or } \frac{1}{z}\right)$ involving, it may be, fractional exponents and possibly also a finite number of negative exponents. They will group themselves into a number r of cycles of orders $\nu_1, \dots, \nu_r$, the exponents in the branches belonging to the several cycles being integral multiples of the numbers $\frac{1}{\nu_1}, \dots, \frac{1}{\nu_r}$ respectively. Only for a finite number of values of the variable z will any of the corresponding power-series involve negative exponents, and for all but a finite number of values of the variable we shall have $\nu_1 = \dots = \nu_r = 1$, and therefore $r = n$.

A rational function of (z, u) will have certain *orders of coincidence* with the branches of the fundamental equation corresponding to a given value of the variable z. These orders of coincidence will be the same for the branches of the same cycle. The orders of coincidence of the function $f_u'(z, u)$ with the branches of the r_κ cycles corresponding to a finite value $z = a_\kappa$ we designate by the notation $\mu_1^{(\kappa)}, \dots, \mu_{r_\kappa}^{(\kappa)}$, its orders of coincidence with the branches of the r_∞ cycles corresponding to the value

$z = \infty$ by the notation $\mu_1^{(\infty)}, \ldots, \mu_{r_\infty}^{(\infty)}$. If the orders of coincidence of a rational function for the value $z = a_\kappa$ do not fall short of the numbers

$$\mu_1^{(\kappa)} - 1 + \frac{1}{\nu_1^{(\kappa)}}, \ldots, \mu_{r_\kappa}^{(\kappa)} - 1 + \frac{1}{\nu_{r_\kappa}^{(\kappa)}} \quad \ldots\ldots\ldots\ldots\ldots\ldots\ldots(4)$$

respectively, we say that the orders of coincidence of the function are adjoint* for the value of the variable in question, and we also say that the function itself is adjoint for this value of the variable.

It is readily seen that we can construct a rational function which actually possesses an arbitrarily assigned set of orders of coincidence corresponding to any value of z.—Here it is to be understood of course that the orders of coincidence in question are integral multiples of the corresponding numbers $\frac{1}{\nu_1}, \ldots, \frac{1}{\nu_r}$ respectively.—Assigning a system of sets of orders of coincidence $\tau_1^{(\kappa)}, \ldots, \tau_{r_\kappa}^{(\kappa)}$ for all values of the variable z, the value $z = \infty$ included, we shall designate this system by the notation (τ) and we shall say of a rational function that it is built on the basis (τ) when its orders of coincidence with the several branches corresponding to any value of the variable z in no case fall short of the orders of coincidence given by the basis. In assigning the orders of coincidence of a basis we shall assume that for all but a finite number of these orders of coincidence we have taken the value 0. We shall find it convenient to employ the notation $(\tau)'$ to designate that part of the basis (τ) which corresponds to finite values of the variable z and to designate by $(\tau)^{(\infty)}$ the partial basis corresponding to the value $z = \infty$.

The general rational function conditioned by the set of orders of coincidence $\tau_1^{(\kappa)}, \ldots, \tau_{r_\kappa}^{(\kappa)}$ for the value $z = a_\kappa$ can be written in the form

$$\frac{\phi^{(i_\kappa)}(z, u)}{(z - a_\kappa)^{i_\kappa}} + ((z - a_\kappa, u)) \quad \ldots\ldots\ldots\ldots\ldots\ldots\ldots\ldots(5),$$

where $\phi^{(i_\kappa)}(z, u)$ is a polynomial in (z, u) of degree $< i_\kappa$ in z, and where the notation $((z - a_\kappa, u))$ signifies a polynomial in u with coefficients which, expanded in powers

* It might be well in this connection to draw attention to the divergence which exists between the meaning which the writer has found convenient to attach to the term *adjoint* and the sense in which the word *adjungirt* has been defined by Brill and Nöther in their classical paper "Ueber die algebraischen Funktionen und ihre Anwendung in der Geometrie," *Math. Annalen*, Bd. VII. Brill and Nöther speak of "adjungirte Curven" meaning thereby curves which fulfil certain conditions with regard to the singular points of the fundamental curve. The writer however finds that it fits in better with his methods to define the term "*adjoint*" with reference to any individual value of z and to say of a function of (z, u) that it is *adjoint* for one or more values of the variable z—it may be for all values of the variable, the value $z = \infty$ included. If the fundamental equation (1) is an integral algebraic equation it represents an algebraic curve as it stands and if $g(z, u) = 0$ is an *adjungirte Curve* the function $g(z, u)$ is, as a matter of fact, *adjoint* for all finite values of the variable z when we regard u as a function of z satisfying equation (1). For the value $z = \infty$ however a function $g(z, u)$ will not in general be *adjoint* when $g(z, u) = 0$ is an *adjungirte Curve* of a given degree. The property of adjointness relative to the value $z = \infty$ in fact limits the degree of a rational function. If equation (1) is integral and if the function $g(z, u)$ is *adjoint* for all finite values of z it must, in its reduced form, be integral in (z, u) and the curve $g(z, u) = 0$ is an *adjungirte Curve* of a certain degree. If, however, the equation (1) is not integral, a function $g(z, u)$ may be *adjoint* for all finite values of z without being integral.

of $z-a_\kappa$, involve no negative exponents. The general rational function conditioned by the partial basis $(\tau)'$ will have the form

$$H(z, u)=\frac{N(z, u)}{Q(z)}+P(z, u) \quad \ldots\ldots\ldots\ldots\ldots\ldots\ldots(6),$$

where $Q(z)$ is a polynomial in z and where $N(z, u)$ and $P(z, u)$ are polynomials in (z, u), the former having a degree in z less than that of $Q(z)$. Under the fractional element in (6) is evidently included the fractional part of the expression (5)—that is, on imposing further conditions on the arbitrary constants involved in the fractional part of (6) we could reduce it to the fractional part of (5).

It is readily shewn that we can construct a rational function which, for finite values of the variable z, actually possesses the orders of coincidence assigned by the partial basis $(\tau)'$. Also we can construct a rational function which actually possesses for the value $z=\infty$ an arbitrarily assigned set of orders of coincidence $\tau_1^{(\infty)}, \ldots, \tau_{r_\infty}^{(\infty)}$. We may or may not however be able to construct a rational function which is simultaneously conditioned by the partial bases $(\tau)'$ and $(\tau)^{(\infty)}$, that is, we may or may not be able to build a function on the basis (τ).

We say that a partial basis $(t)'$ is *lower* than a partial basis $(\tau)'$ if the orders of coincidence indicated in the former partial basis nowhere exceed those designated by the latter basis and if, at the same time, one or more of the orders of coincidence given by the partial basis $(t)'$ fall short of the corresponding orders of coincidence given by the partial basis $(\tau)'$. In like manner with reference to the value $z=\infty$ we speak of a partial basis $(t)^{(\infty)}$ which is lower than the partial basis $(\tau)^{(\infty)}$. It is evident that we impose on the general rational function conditioned by the partial basis $(t)'$ a number of conditions given by the sum

$$\sum_\kappa{}' \sum_{s=1}^{r_\kappa} (\tau_s^{(\kappa)} - t_s^{(\kappa)})\, \nu_s^{(\kappa)} \quad \ldots\ldots\ldots\ldots\ldots\ldots\ldots\ldots\ldots(7)$$

in order to obtain the general rational function conditioned by the partial basis $(\tau)'$. Here the accent over the sign of summation signifies that the summation is extended only to terms having reference to finite values of z. Also we impose on the general rational function conditioned by the partial basis $(t)^{(\infty)}$ a number of conditions given by the sum

$$\sum_{s=1}^{r_\infty} (\tau_s^{(\infty)} - t_s^{(\infty)})\, \nu_s^{(\infty)} \quad \ldots\ldots\ldots\ldots\ldots\ldots\ldots\ldots\ldots(8)$$

in order to obtain the general rational function conditioned by the partial basis $(\tau)^{(\infty)}$.

If two sets of orders of coincidence $\tau_1^{(\kappa)}, \ldots, \tau_{r_\kappa}^{(\kappa)}$ and $\bar{\tau}_1^{(\kappa)}, \ldots, \bar{\tau}_{r_\kappa}^{(\kappa)}$, corresponding to the same value $z=a_\kappa$, satisfy the inequalities

$$\tau_s^{(\kappa)} + \bar{\tau}_s^{(\kappa)} \gtreqless \mu_s^{(\kappa)} - 1 + \frac{1}{\nu_s^{(\kappa)}}; \quad s=1, 2, \ldots, r_\kappa \quad \ldots\ldots\ldots\ldots(9),$$

we say that they are complementary adjoint. If they satisfy the inequalities

$$\tau_s^{(\kappa)} + \bar{\tau}_s^{(\kappa)} \gtreqless i + \mu_s^{(\kappa)} - 1 + \frac{1}{\nu_s^{(\kappa)}}; \quad s=1, 2, \ldots, r_\kappa \ldots\ldots\ldots(10),$$

we say that they are complementary adjoint to the order i. Two bases (τ) and $(\bar{\tau})$ we say are complementary bases if the sets of orders of coincidence which they furnish for finite values of the variable z are complementary adjoint, while the sets of orders of coincidence furnished by them for the value $z=\infty$ are complementary adjoint to the order 2. Where then (τ) and $(\bar{\tau})$ are complementary bases, the orders of coincidence furnished by them for finite values of the variable $z=a_\kappa$ satisfy the inequalities (9), while for the value $z=\infty$ they furnish orders of coincidence which satisfy the inequalities

$$\tau_s^{(\infty)}+\bar{\tau}_s^{(\infty)} \gtreqless \mu_s^{(\infty)}+1+\frac{1}{\nu_s^{(\infty)}}; \quad s=1,2,\ldots,r_\infty \ldots\ldots\ldots\ldots(11).$$

From now on we shall find it convenient to assume that the equation (1) is an integral algebraic equation. In this case the writer has shewn* that a rational function which is adjoint for a finite value $z=a_\kappa$ must be integral with regard to the element $z-a_\kappa$. He has also shewn† that the necessary and sufficient conditions in order that a rational function $\psi(z,u)$ may have for the value $z=a_\kappa$ orders of coincidence which are complementary adjoint to the orders of coincidence $\tau_1^{(\kappa)}, \ldots, \tau_{r_\kappa}^{(\kappa)}$ are obtained on equating to 0 the principal residue relative to the value $z=a_\kappa$ in the product of $\psi(z,u)$ and the general function (5). Supposing $\psi(z,u)$ to be an *integral* rational function of (z,u) the necessary and sufficient conditions here in question are evidently also obtained on equating to 0 the principal residue relative to the value $z=a_\kappa$ in the product

$$\frac{\phi^{(i_\kappa)}(z,u)}{(z-a_\kappa)^{i_\kappa}}\cdot\psi(z,u)\ldots\ldots\ldots\ldots\ldots\ldots\ldots\ldots(12).$$

In this product no residues can possibly present themselves other than those which correspond to the values $z=a_\kappa$ and $z=\infty$. Since however the sum of the residues of a rational function of z must be 0, it follows that the necessary and sufficient conditions in order that an integral rational function $\psi(z,u)$ should have for the value $z=a_\kappa$ orders of coincidence which are complementary adjoint to the orders of coincidence $\tau_1^{(\kappa)}, \ldots, \tau_{r_\kappa}^{(\kappa)}$ are obtained on equating to 0 the principal residue relative to the value $z=\infty$ in the product (12).

Now the function (6) in the neighbourhood of the value $z=a_\kappa$ has the form given in (5). It is therefore necessary that the principal residue relative to the value $z=a_\kappa$ in the product of $\psi(z,u)$ and the function (6) should have the value 0 in order that the former function should have for the value $z=a_\kappa$ orders of coincidence which are complementary adjoint to the orders of coincidence $\tau_1^{(\kappa)}, \ldots, \tau_{r_\kappa}^{(\kappa)}$. In order then that the integral rational function $\psi(z,u)$ should have, for the value $z=a_\kappa$, orders of coincidence which are complementary adjoint to the orders of coincidence $\tau_1^{(\kappa)}, \ldots, \tau_{r_\kappa}^{(\kappa)}$ it is necessary that the principal residue relative to the value $z=a_\kappa$ in the product

$$\frac{N(z,u)}{Q(z)}\cdot\psi(z,u)\ \ldots\ldots\ldots\ldots\ldots\ldots\ldots\ldots(13)$$

* "On the foundations of the theory of algebraic functions of one variable," *Phil. Trans. Roy. Soc.* 1912.

† *loc. cit.*

should have the value 0. In order therefore that the integral rational function $\psi(z, u)$ should have for all finite values of the variable z orders of coincidence which are complementary adjoint to the corresponding orders of coincidence furnished by the partial basis $(\tau)'$ it is necessary that the principal residues of the product (13) relative to the several values $z = a_\kappa$ should all have the value 0. The sum of these residues will in this case be 0 and the principal residue of the product (13) relative to the value $z = \infty$ must consequently have the value 0 when the orders of coincidence of the function $\psi(z, u)$, for all finite values of the variable z, are complementary adjoint to the corresponding orders of coincidence furnished by the partial basis $(\tau)'$.

Conversely, if the principal residue relative to the value $z = \infty$ in the product (13) is 0 the integral rational function $\psi(z, u)$ must, for all finite values of the variable z, have orders of coincidence which are complementary adjoint to the corresponding orders of coincidence furnished by the partial basis $(\tau)'$. For if the principal residue, relative to the value $z = \infty$, in the product (13) is 0, so also is 0 the value of the principal residue relative to the value $z = \infty$, in the product (12), since the first factor in (12) may be obtained from the first factor in (13) by imposing on it certain conditions. If then the principal residue relative to the value $z = \infty$ in the product (13) has the value 0, the integral rational function $\psi(z, u)$ must, for the value $z = a_\kappa$, have orders of coincidence which are complementary adjoint to the corresponding orders of coincidence furnished by the partial basis $(\tau)'$. This holds for all the finite values of the variable z in question. The necessary and sufficient conditions then in order that an integral rational function $\psi(z, u)$ may, for all finite values of the variable z, have orders of coincidence which are complementary adjoint to the corresponding orders of coincidence furnished by the partial basis $(\tau)'$, are obtained on equating to 0 the principal residue relative to the value $z = \infty$ in the product (13).

Expanding the coefficients of the powers of u in the first factor of the product (13) in powers of $\frac{1}{z}$, the principal residue of the product relative to the value $z = \infty$ will evidently depend on a finite number of terms only in the expanded first factor. If, for example, $\psi(z, u)$ is to be of degree in z not greater than a definite number M, we could name an integer i such that the principal residue of the product (13) relative to the value $z = \infty$ would be independent of any term in the expanded first factor involving $\frac{1}{z}$ to a power as high as $\left(\frac{1}{z}\right)^i$. We do not for our purpose need to select the smallest integer i for which this holds. In the sequel M will designate the greatest degree in z of a rational function of (z, u) which is compatible with the possession by the function of the orders of coincidence $\bar{\tau}_1^{(\infty)}, \ldots, \bar{\tau}_{r'_\infty}^{(\infty)}$ for the value $z = \infty$.

In the general rational function of (z, u) conditioned by the partial basis $(\tau)^{(\infty)}$ we shall suppose the coefficients of the powers of u to be expanded in powers of the element $\frac{1}{z}$ and the general function so conditioned we shall represent in the form

$$R\left(\frac{1}{z}, u\right) = R^{(i)}\left(\frac{1}{z}, u\right) + z^{-i}\left(\left(\frac{1}{z}, u\right)\right) \quad \ldots\ldots\ldots\ldots\ldots\ldots(14),$$

where $R^{(i)}\left(\frac{1}{z}, u\right)$ is a polynomial in u in whose coefficients $\frac{1}{z}$ does not appear to a power higher than $\left(\frac{1}{z}\right)^{i-1}$, and where the notation $\left(\left(\frac{1}{z}, u\right)\right)$ signifies a polynomial in u whose coefficients expanded in powers of $\frac{1}{z}$ involve no negative exponents. The integer i we shall choose sufficiently large to suit our purpose. In the memoir already cited we have shewn that the necessary and sufficient conditions in order that a rational function $\psi(z, u)$ may have, for the value $z=\infty$, orders of coincidence $\bar{\tau}_1^{(\infty)}, \ldots, \bar{\tau}_{r_\infty}^{(\infty)}$ which are complementary adjoint to the order 2 to the orders of coincidence $\tau_1^{(\infty)}, \ldots, \tau_{r_\infty}^{(\infty)}$, are obtained on equating to 0 the principal residue relative to the value $z=\infty$ in the product

$$R\left(\frac{1}{z}, u\right) . \psi(z, u) \qquad \text{.............................(15).}$$

Supposing $\psi(z, u)$ to be a rational function of (z, u) of limited degree in z and choosing i sufficiently large, the principal coefficient in the product

$$z^{-i}\left(\left(\frac{1}{z}, u\right)\right) . \psi(z, u) \qquad \text{..........................(16)}$$

will present no residue relative to the value $z=\infty$ whatever values the constant coefficients in the function $\left(\left(\frac{1}{z}, u\right)\right)$ may have. Choosing i sufficiently large then, the necessary and sufficient conditions, in order that a rational function $\psi(z, u)$ of specified degree in z may have, for the value $z=\infty$, orders of coincidence $\bar{\tau}_1^{(\infty)}, \ldots, \bar{\tau}_{r_\infty}^{(\infty)}$ which are complementary adjoint to the order 2 to the orders of coincidence furnished by the partial basis $(\tau)^{(\infty)}$, are obtained on equating to 0 the principal residue relative to the value $z=\infty$ in the product

$$R^{(i)}\left(\frac{1}{z}, u\right) . \psi(z, u) \qquad \text{.............................(17).}$$

We shall now suppose $\psi(z, u)$ to be an integral rational function of (z, u) and furthermore we shall assume that it has the form

$$\psi(z, u) = \sum_{t=1}^{n} \sum_{r=1}^{i-1} a_{r-1, t-1} z^{r-1} u^{t-1} \qquad \text{.....................(18).}$$

In this case also we shall prove that the vanishing of the principal residue in the product (17) furnishes the necessary and sufficient conditions in order that the function $\psi(z, u)$ may have the orders of coincidence $\bar{\tau}_1^{(\infty)}, \ldots, \bar{\tau}_{r_\infty}^{(\infty)}$ for the value $z=\infty$—if only the integer i be chosen large enough. Among the conditions here in question then will be found those which reduce the degree in z of the function $\psi(z, u)$ to M.

Let us consider the product

$$z^{-\rho} u^{n-\tau} \sum_{t=1}^{n} \sum_{r=1}^{i-1} a_{r-1, t-1} z^{r-1} u^{t-1}; \quad \rho \geqq 1,\ 1 \leqq \tau \leqq n \qquad \text{............(19).}$$

Bearing in mind the fact that the fundamental equation (1) is integral, we readily see that in the reduced form of the product (19), the principal residue is the sum of the coefficient $\alpha_{\rho-1,\tau-1}$ and a linear expression in terms of coefficients $\alpha_{r-1,t-1}$ in which $r \gtreqless \rho$, $t > \tau$. If we equate this residue to 0 we obtain for $\alpha_{\rho-1,\tau-1}$ an expression which is linear in terms of coefficients $\alpha_{r-1,t-1}$ in which $r \gtreqless \rho$, $t > \tau$. In particular when $\tau = n$ we obtain $\alpha_{\rho-1,n-1} = 0$.

Writing $$R\left(\frac{1}{z}, u\right) = \sum_{t=1}^{n} \sum_{r} \beta_{-r,n-t} z^{-r} u^{n-t} \qquad (20),$$

let us consider the product

$$\sum_{t=1}^{n} \sum_{r=1}^{i-1} \beta_{-r,n-t} z^{-r} u^{n-t} \,.\, \sum_{t=1}^{n} \sum_{r=1}^{i-1} \alpha_{r-1,t-1} z^{r-1} u^{t-1} \qquad (21).$$

By the letter j we shall indicate an integer so large that the coefficients $\beta_{-r,n-t}$ are all arbitrary for which $r \geqq j$ and at the same time so large that $j-2 \geqq M$—the greatest degree in z which a rational function of (z, u) can have and yet possess the orders of coincidence $\bar{\tau}_1^{(\infty)}, \ldots, \bar{\tau}_{r_\infty}^{(\infty)}$ for the value $z = \infty$. Furthermore we shall take i so large that in the product

$$z^{-\rho} u^{n-\tau} \sum_{t=1}^{n} \sum_{r=1}^{j-1} \alpha_{r-1,t-1} z^{r-1} u^{t-1} \qquad (22)$$

the principal residue is identically 0 for arbitrary values of the $n(j-1)$ coefficients $\alpha_{r-1,t-1}$ here in question for $\rho \geqq i$.

Among the $n(j-1)$ coefficients $\beta_{-r,n-t}$ for which $r \gtreqless j-1$ we shall suppose that there are just $n(j-1)-\lambda$ which are arbitrary. The remaining λ coefficients are then linearly expressible in terms of these $n(j-1)-\lambda$ coefficients. On equating to 0 the principal residue in the product

$$\sum_{t=1}^{n} \sum_{r=1}^{j-1} \beta_{-r,n-t} z^{-r} u^{n-t} \,.\, \sum_{t=1}^{n} \sum_{r=1}^{j-1} \alpha_{r-1,t-1} z^{r-1} u^{t-1} \qquad (23)$$

it is readily seen that we impose on the coefficients $\alpha_{r-1,t-1}$ in the second factor just $n(j-1)-\lambda$ independent conditions. To see this we note, in the first place, that we impose a condition on the coefficients $\alpha_{r-1,t-1}$ of the second factor in the product (23), on taking for the first factor a specific function and equating to 0 the principal residue in the product. For if in the specific function ρ is the greatest value of r in a coefficient $\beta_{-r,n-t}$ which is different from 0 and if τ is the least value of t in a coefficient $\beta_{-\rho,n-t}$ which is different from 0 we obtain $\alpha_{\rho-1,\tau-1}$ in terms of coefficients $\alpha_{r-1,t-1}$ in which* $r < \rho$, or $r = \rho$, $t > \tau$.

If in the product (23) we take turn about for the first factor a number of specific linearly independent functions, it is evident that the principal residues in the several products, regarded as linear expressions in the arbitrary constants $\alpha_{r-1,t-1}$, are linearly independent of one another. For if they were connected by a linear relation, the linear expression in the specific first factors just referred to, constructed with the like multipliers, would furnish us with a specific first factor for the product (23) such that

* The truth of the statement here made follows from the fact that the equation (1) is integral and that therefore no negative powers of z are introduced in reducing a power of u higher than u^{n-1} by the aid of this equation.

the principal residue in the product would be 0 independently of the values of the constants $\alpha_{r-1,t-1}$. This however we have seen to be impossible. It follows that on substituting turn about for the first factor in (23) a number of specific linearly independent functions and equating to 0 the principal residues in the several products we impose on the coefficients $\alpha_{r-1,t-1}$ in the second factor as many linearly independent conditions. On equating to 0 then the principal residue in the product (23) where $n(j-1)-\lambda$ is the number of the coefficients $\beta_{-r,n-t}$ which are arbitrary we impose on the otherwise arbitrary coefficients $\alpha_{r-1,t-1}$ in the second factor of the product just $n(j-1)-\lambda$ independent conditions. There remain therefore among the coefficients $\alpha_{r-1,t-1}$ of the second factor of the product (23) just λ which are arbitrary. In like manner, on equating to 0 the principal residue in the product (21) we impose on the coefficients $\alpha_{r-1,t-1}$ in the second factor $n(i-1)-\lambda$ conditions, so that just λ of these $n(i-1)$ coefficients remain arbitrary.

Let us now consider more in detail the conditions imposed on the coefficients $\alpha_{r-1,t-1}$ in the second factor of the product (21) when we equate to 0 the principal residue in this product. These conditions are plainly made up of the conditions obtained on equating to 0 the principal residue in the product (23) together with the conditions obtained on equating to 0 the principal residues in the $n(i-j)$ products

$$z^{-\rho}u^{n-\tau}\sum_{t=1}^{n}\sum_{r=1}^{i-1}\alpha_{r-1,t-1}z^{r-1}u^{t-1};\quad \rho=j, j+1, \ldots, i-1;\ \tau=1, 2, \ldots, n \ldots(24).$$

The $n(i-j)$ conditions so obtained give us the $n(i-j)$ coefficients $\alpha_{\rho-1,\tau-1}$, severally expressed linearly in terms of coefficients $\alpha_{r-1,t-1}$ in which $r \gtreqless \rho$, $t>\tau$. Among others, these conditions evidently include the $i-j$ conditions $\alpha_{\rho-1,n-1}=0$. In any case we can plainly combine the $n(i-j)$ conditions so as to express ultimately each of the $n(i-j)$ coefficients $\alpha_{\rho-1,\tau-1}$ here in question linearly in terms of the coefficients $\alpha_{r-1,t-1}$ in which $r<j$. From this fact follows that the $n(i-j)$ conditions here in question are linearly independent. The effect of equating to 0 the principal residue in the product (21) then is to express each of the $n(i-j)$ coefficients $\alpha_{r-1,t-1}$ in which $r \geqq j$ linearly in terms of the $n(j-1)$ coefficients in which $r<j$; these latter $n(j-1)$ coefficients being at the same time expressed linearly in terms of the λ ones among them which remain arbitrary after equating to 0 the principal residue in the product (23).

Now, as we have already seen, the necessary and sufficient conditions in order that a rational function $\psi(z, u)$ may have, for the value $z=\infty$, orders of coincidence $\bar{\tau}_1^{(\infty)}, \ldots, \bar{\tau}_{r_\infty}^{(\infty)}$ which are complementary adjoint to the order 2 to the orders of coincidence $\tau_1^{(\infty)}, \ldots, \tau_{r_\infty}^{(\infty)}$ are obtained on equating to 0 the principal residue relative to the value $z=\infty$ in the product (15). Supposing $\psi(z, u)$ as before to be the integral rational function represented in (18), the necessary and sufficient conditions in order that it may have the orders of coincidence $\bar{\tau}_1^{(\infty)}, \ldots, \bar{\tau}_{r_\infty}^{(\infty)}$ for the value $z=\infty$ are then evidently obtained on equating to 0 the principal residue in the product

$$\sum_{t=1}^{n}\sum_{r=1}^{\infty}\beta_{-r,n-t}z^{-r}u^{n-t}\cdot\sum_{t=1}^{n}\sum_{r=1}^{i-1}\alpha_{r-1,t-1}z^{r-1}u^{t-1}\ \ldots\ldots\ldots\ldots\ldots(25),$$

where the first factor is derived from the function $R\left(\frac{1}{z}, u\right)$ in (20) by omitting all those terms which do not involve a negative power of z. Among the conditions so arrived at are included the conditions obtained on equating to 0 the principal residue in the product (21)—and only these conditions are included, as we may readily shew. For any further condition would have to be obtained on equating to 0 the principal residue in a product of the form

$$z^{-\rho}u^{n-\tau}\sum_{t=1}^{n}\sum_{r=1}^{i-1}\alpha_{r-1,t-1}z^{r-1}u^{t-1} \quad \text{.......................(26)},$$

where $\rho \geqq i$. We have however chosen i so large that the principal residue in the product (22) is 0 for arbitrary values of the $n(j-1)$ coefficients $\alpha_{r-1,t-1}$ in which $r<j$. On equating to 0 the principal residue in the product (26) then we impose no restriction on the values of the coefficients $\alpha_{r-1,t-1}$ in which $r<j$. The only condition so obtained would have to be a linear relation connecting the $n(i-j)$ coefficients $\alpha_{r-1,t-1}$ in which r has one of the values $j, j+1, \ldots, i-1$. Such condition however could be nothing new, for on substituting in it for each of the $n(i-j)$ coefficients $\alpha_{r-1,t-1}$ in question its expression in terms of the coefficients $\alpha_{r-1,t-1}$ in which $r<j$, obtained on equating to 0 the principal residue in the product

$$\sum_{t=1}^{n}\sum_{r=1}^{i-1}\beta_{-r,n-t}z^{-r}u^{n-t}\,.\sum_{t=1}^{n}\sum_{r=1}^{i-1}\alpha_{r-1,t-1}z^{r-1}u^{t-1} \quad \text{..............(27)}$$

we arrive at an expression which must be identically 0, since otherwise we should have a linear relation connecting the coefficients $\alpha_{r-1,t-1}$ in which $r<j$, as a result of equating to 0 the principal residue in the product (26).

On equating to 0 the principal residue in the product (25) we then impose no further conditions on the coefficients $\alpha_{r-1,t-1}$ than those already obtained on equating to 0 the principal residue in the product (21). The necessary and sufficient conditions in order that the integral rational function $\psi(z, u)$ represented in (18) may have the orders of coincidence $\bar{\tau}_1^{(\infty)}, \ldots, \bar{\tau}_{r_\infty}^{(\infty)}$ are therefore obtained on equating to 0 the principal residue in the product (21).

To obtain the general rational function built on the basis (τ), that is to obtain the general rational function conditioned simultaneously by the partial bases $(\tau)'$ and $(\tau)^{(\infty)}$, we subject the function $H(z, u)$ in (6) and the function $R\left(\frac{1}{z}, u\right)$ in (14) to the conditions implied in identifying them with each other. The polynomial $N(z, u)$ in (6) involves a number l of arbitrary constants and the function $R^{(i)}\left(\frac{1}{z}, u\right)$ in (14) involves a number l_∞ of arbitrary constants. These numbers we do not here determine, and in the sequel we shall find that we do not need to know their actual values in order to prove the complementary theorem. The $l+l_\infty$ arbitrary constants in question we shall refer to as the arbitrary constants δ.

Identifying the function $H(z, u)$ in (6) with the function $R\left(\frac{1}{z}, u\right)$ in (14), we may write our identity in the form

$$\frac{N(z, u)}{Q(z)} - R^{(i)}\left(\frac{1}{z}, u\right) = -P(z, u) + z^{-i}\left(\left(\frac{1}{z}, u\right)\right) \quad \text{...........(28)}.$$

This identity evidently determines the integral rational function $P(z, u)$ as that part of $R^{(i)}\left(\frac{1}{z}, u\right)$ which is integral in (z, u). It also determines each of the otherwise arbitrary constant coefficients in $z^{-i}\left(\left(\frac{1}{z}, u\right)\right)$ in terms of the constants δ which appear in the function which stands on the left-hand side of the identity. The conditions to which these $l + l_\infty$ constants δ are subjected by virtue of the identity are obtained on expanding in powers of $\frac{1}{z}$ the coefficients of the powers of u on the left-hand side of the identity, and equating to 0 the aggregate constant coefficient of each term which involves a power $\left(\frac{1}{z}\right)^r$, where r has one of the values 1, 2, ..., $i-1$. Writing

$$\frac{N(z, u)}{Q(z)} - R^{(i)}\left(\frac{1}{z}, u\right) = \sum_{t=1}^{n} \sum_{r} c_{-r, n-t} z^{-r} u^{n-t} \quad \text{..............(29)}$$

the coefficients $c_{-r,n-t}$ are linear in the constants δ and the conditions to which these constants are subjected by the identity (28) are embodied in the identity

$$\sum_{t=1}^{n} \sum_{r=1}^{i-1} c_{-r, n-t} z^{-r} u^{n-t} = 0 \quad \text{..........................(30).}$$

The constants δ then are subjected to the $n(i-1)$ conditions

$$c_{-r, n-t} = 0\,; \quad r = 1, \ldots, i-1\,; \; t = 1, \ldots, n \quad \text{..............(31)}$$

which may or may not all be linearly independent.

Now we have seen that the necessary and sufficient conditions in order that the function $\psi(z, u)$ may, for finite values of the variable z, have orders of coincidence which are complementary adjoint to the orders of coincidence furnished by the partial basis $(\tau)'$ are obtained on equating to 0 the principal residue relative to the value $z = \infty$ in the product (13) for arbitrary values of the constants δ involved in $N(z, u)$. Where the function $\psi(z, u)$ is of limited degree M in z we have also seen that the principal residue relative to the value $z = \infty$ in the product (13) is independent of any term in the expanded first factor involving $\frac{1}{z}$ to a power as high as $\left(\frac{1}{z}\right)^i$—when i is chosen large enough. Furthermore, supposing $\psi(z, u)$ to be the integral rational function represented in (18), we have shewn that, on equating to 0 the principal residue in the product (21) we obtain the necessary and sufficient conditions in order that the function $\psi(z, u)$ may have, for the value $z = \infty$, the orders of coincidence $\bar{\tau}_1^{(\infty)}, \ldots, \bar{\tau}_{r_\infty}^{(\infty)}$—the integer i being taken sufficiently large. It evidently follows that the necessary and sufficient conditions in order that $\psi(z, u)$ may be built on the basis $(\bar{\tau})$ complementary to the basis (τ) are obtained on equating to 0, independently of the values of the arbitrary constants δ, the principal residue relative to the value $z = \infty$ in the product

$$\sum_{t=1}^{n} \sum_{r=1}^{i-1} c_{-r, n-t} z^{-r} u^{n-t} \,.\, \psi(z, u) \quad \text{......................(32),}$$

where $\psi(z, u)$ is supposed to have the form given in (18).

The conditions obtained on equating to 0 the principal residue in the product (32) include the conditions obtained on equating to 0 the principal residue in the product (21) and therefore involve the reduction of the degree of $\psi(z, u)$ in z to M. The vanishing of the principal residue in the product (32), with the degree of $\psi(z, u)$ in z reduced to M, then further involves the vanishing of the residue relative to the value $z=\infty$ in the product (13)—the integer i being of course assumed to have been chosen sufficiently large in the first place.

Among the $n(i-1)$ coefficients $c_{-r,n-t}$ in (32), regarded as expressions linear in the arbitrary constants δ, we shall suppose that d are linearly independent and d such coefficients we shall for the moment represent by the notation $c_1, c_2, \ldots, c_d$. Expressing the remaining coefficients in the first factor of the product (32) linearly in terms of the d coefficients here in question we can evidently represent the first factor in the form

$$\sum_{s=1}^{d} c_s \phi_s\left(\frac{1}{z}, u\right) \qquad \text{(33)},$$

where the d functions $\phi_s\left(\frac{1}{z}, u\right)$ are specific linearly independent functions. Since the d coefficients c_s are linearly independent expressions in the constants δ they may be given any arbitrarily assigned values by properly choosing the values of the constants δ. In order then that the principal residue in the product

$$\sum_{s=1}^{d} c_s \phi_s\left(\frac{1}{z}, u\right) . \psi(z, u) \qquad \text{(34)}$$

may be 0 for arbitrary values of the constants δ, it must be 0 for arbitrary values of the d coefficients c_s. That is, the principal residue must be 0 in each of the products

$$\phi_s\left(\frac{1}{z}, u\right) . \psi(z, u); \quad s=1, 2, \ldots, d \qquad \text{(35)}.$$

Conversely, if the principal residue in each of the d products (35) is 0 it is evident that the principal residue in the product (34)—that is, in the product (32)—is 0 for arbitrary values of the constants δ. Because of the linear independence of the functions $\phi_s\left(\frac{1}{z}, u\right)$ we see, on recalling the reasoning employed in connection with the product (23), that the d conditions on the coefficients of the function $\psi(z, u)$, obtained on equating to 0 the principal residues in the products (35), are linearly independent of one another. The number of the conditions imposed on the coefficients of the integral rational function $\psi(z, u)$ on equating to 0 the principal residue in the product (32) independently of the values of the constants δ is then d, the number of the $n(i-1)$ coefficients $c_{-r,n-t}$ in the first factor of the product which are linearly independent of one another, when regarded as expressions in the arbitrary constants δ. This is therefore also the number of the conditions to which we subject the $n(i-1)$ coefficients of the integral rational function $\psi(z, u)$ in order that it may be built on the basis $(\bar{\tau})$ which is complementary to the basis (τ). It follows that the number of the arbitrary constants involved in the general *integral* rational function $\psi(z, u)$ built on the basis $(\bar{\tau})$ is just equal to the number $n(i-1)-d$ of the $n(i-1)$ coefficients $c_{-r,n-t}$ in the first factor of (32) which are linearly dependent on the remaining coefficients, when we regard the coefficients as linear expressions in the arbitrary constants δ.

We shall assume for the moment that the orders of coincidence furnished by the basis (τ) for *finite* values of the variable z are none of them positive. Writing

$$\tau_s^{(\kappa)} = -\sigma_s^{(\kappa)}, \quad s = 1, 2, \ldots, r_\kappa,$$

in the case of every finite value $z = a_\kappa$ the numbers $\sigma_s^{(\kappa)}$ are 0 or positive. For the orders of coincidence furnished by the basis $(\bar{\tau})$ for finite values of the variable z we then have from (9)

$$\bar{\tau}_s^{(\kappa)} = \sigma_s^{(\kappa)} + \mu_s^{(\kappa)} - 1 + \frac{1}{\nu_s^{(\kappa)}}; \quad s = 1, 2, \ldots, r_\kappa.$$

A rational function of (z, u) built on the basis $(\bar{\tau})$ then must be adjoint for all finite values of the variable z and must therefore be integral. It follows that the number of the coefficients $c_{-r, n-t}$ in the first factor of (32) which are linearly dependent on the remaining coefficients is equal to the number of the arbitrary constants involved in the expression of the general rational function built on the basis $(\bar{\tau})$—for we have shewn that it is equal to the number of the arbitrary constants involved in the expression of the general *integral* rational function built on this basis.

Employing the notation $N_{\bar{\tau}}$ to designate the number of the arbitrary constants involved in the expression of the general rational function built on the basis $(\bar{\tau})$ and going back to the $n(i-1)$ conditions (31) imposed on the constants δ in identifying the function $H(z, u)$ in (6) with the function $R\left(\frac{1}{z}, u\right)$ in (14), we see that just $n(i-1) - N_{\bar{\tau}}$ of the equations of condition here in question are linearly independent, for we have proved that just $N_{\bar{\tau}}$ of the $n(i-1)$ expressions $c_{-r, n-t}$ here involved are linearly dependent on the remaining ones. By virtue of the identity (28) then the $l + l_\infty$ constants δ involved in the functions $N(z, u)$ and $R^{(i)}\left(\frac{1}{z}, u\right)$ are subjected to just $n(i-1) - N_{\bar{\tau}}$ independent conditions.—We of course impose no condition on the constants δ by equating the polynomial $P(z, u)$ to the integral part of the function $R^{(i)}\left(\frac{1}{z}, u\right)$.—The number of the constants δ which remain arbitrary we readily see* gives the number N_τ of the arbitrary constants involved in the expression of the general rational function $H(z, u)$ built on the basis (τ) and we therefore have

$$N_\tau = l + l_\infty - n(i-1) + N_{\bar{\tau}} \quad \text{.......................(36)}.$$

This formula we have obtained on the assumption that the orders of coincidence furnished by the basis (τ) for finite values of the variable z are none of them positive.

Where $H(z, u)$ and $\bar{H}(z, u)$ are the general rational functions built on the bases (τ) and $(\bar{\tau})$ respectively, we can evidently so choose a polynomial $S(z)$ that the functions $S(z) . H(z, u)$ and $\frac{\bar{H}(z, u)}{S(z)}$ are the general rational functions built on complementary bases (ρ) and $(\bar{\rho})$ respectively, the orders of coincidence furnished by the basis $(\bar{\rho})$ being at the same time all 0 or negative. Remembering that the

* For more detail consult the writer's memoir already cited in this paper.

orders of coincidence for finite values of the variable z furnished by the partial basis $(\bar{\tau})'$ require a rational function to be integral, we see that the general rational function conditioned by the partial basis $(\bar{\rho})'$ may be written in the form

$$\frac{\bar{N}(z, u)}{S(z)} + \bar{P}(z, u) \qquad \text{.............................(37)},$$

where $\bar{P}(z, u)$ is an arbitrary polynomial in (z, u), and where $\bar{N}(z, u)$ is a polynomial in (z, u) whose degree in z is less than that of $S(z)$. It may be too that $\bar{N}(z, u)$ has a factor in common with $S(z)$. The general rational function conditioned by the partial basis $(\bar{\rho})^{(\infty)}$ can be written in the form

$$\bar{R}\left(\frac{1}{z}, u\right) = \bar{R}^{(i)}\left(\frac{1}{z}, u\right) + z^{-i}\left(\left(\frac{1}{z}, u\right)\right) \qquad \text{..................(38)}.$$

The polynomial $\bar{N}(z, u)$ will involve a number $\bar{l}$ of arbitrary constants, while a number $\bar{l}_\infty$ of arbitrary constants will present themselves in the function $\bar{R}^{(i)}\left(\frac{1}{z}, u\right)$. In analogy with (36) we evidently have the formula

$$N_{\bar{\rho}} = \bar{l} + \bar{l}_\infty - n(i-1) + N_\rho \qquad \text{.......................(39)}.$$

It is understood throughout that we choose i sufficiently large and we are plainly at liberty to give it the same value in the two formulae (36) and (39). The numbers l_∞ and $\bar{l}_\infty$ depend for their values on i. Evidently we have $N_\rho = N_\tau$, $N_{\bar{\rho}} = N_{\bar{\tau}}$. From (39) we therefore derive

$$N_{\bar{\tau}} = \bar{l} + \bar{l}_\infty - n(i-1) + N_\tau \qquad \text{.......................(40)}.$$

Let us now take any partial basis $(t)'$ which is lower than the partial bases $(\tau)'$ and $(\bar{\rho})'$. On starting out from the general rational function $h(z, u)$ conditioned by the partial basis $(t)'$, we arrive at the general rational function (6) conditioned by the partial basis $(\tau)'$ after imposing on the coefficients of the function $h(z, u)$ just

$$\Sigma'_{\kappa} \sum_{s=1}^{r_\kappa} (\tau_s^{(\kappa)} - t_s^{(\kappa)})\, \nu_s^{(\kappa)}$$

conditions. These conditions evidently affect only the arbitrary constants involved in the essentially fractional part of $h(z, u)$ and reduce their number to l, the number of the arbitrary constants involved in the numerator $N(z, u)$ in (6). On starting out from the general rational function $h(z, u)$ conditioned by the partial basis $(t)'$, we arrive at the general rational function (37) conditioned by the partial basis $(\bar{\rho})'$ after imposing on the coefficients of the function $h(z, u)$ just

$$\Sigma'_{\kappa} \sum_{s=1}^{r_\kappa} (\bar{\rho}_s^{(\kappa)} - t_s^{(\kappa)})\, \nu_s^{(\kappa)}$$

conditions. These conditions affect only the arbitrary constants involved in the essentially fractional part of $h(z, u)$ and reduce their number to $\bar{l}$, the number of the arbitrary constants involved in the numerator $\bar{N}(z, u)$ in (37). From the preceding, for the number of the arbitrary constants involved in the essentially fractional part of $h(z, u)$ we derive

$$l + \Sigma'_{\kappa} \sum_{s=1}^{r_\kappa} (\tau_s^{(\kappa)} - t_s^{(\kappa)})\, \nu_s^{(\kappa)} = \bar{l} + \Sigma'_{\kappa} \sum_{s=1}^{r_\kappa} (\bar{\rho}_s^{(\kappa)} - t_s^{(\kappa)})\, \nu_s^{(\kappa)},$$

whence

$$l + \Sigma'_{\kappa} \sum_{s=1}^{r_\kappa} \tau_s^{(\kappa)}\, \nu_s^{(\kappa)} = \bar{l} + \Sigma_{\kappa} \sum_{s=1}^{r_\kappa} \bar{\rho}_s^{(\kappa)}\, \nu_s^{(\kappa)} \qquad \text{.....................(41)}$$

Again let us take a partial basis $(t)^{(\infty)}$ which is lower than the partial bases $(\tau)^{(\infty)}$ and $(\bar{\rho})^{(\infty)}$. On starting out from the general rational function $h_\infty(z, u)$ conditioned by the partial basis $(t)^{(\infty)}$, we arrive at the general rational function (14) conditioned by the partial basis $(\tau)^{(\infty)}$ after imposing on the coefficients of the function $h_\infty(z, u)$ just $\sum_{s=1}^{r_\infty}(\tau_s^{(\infty)} - t_s^{(\infty)})\,\nu_s^{(\infty)}$ conditions. These conditions leave us l_∞ arbitrary constants involved in the function $R^{(i)}\left(\frac{1}{z}, u\right)$ in (14). On starting out from the general rational function $h_\infty(z, u)$ conditioned by the partial basis $(t)^{(\infty)}$, we arrive at the general rational function (38) conditioned by the partial basis $(\bar{\rho})^{(\infty)}$ after imposing on the coefficients of the function $h_\infty(z, u)$ just $\sum_{s=1}^{r_\infty}(\bar{\rho}_s^{(\infty)} - t_s^{(\infty)})\,\nu_s^{(\infty)}$ conditions. These conditions leave us $\bar{l}_\infty$ arbitrary constants involved in the function $\bar{R}^{(i)}\left(\frac{1}{z}, u\right)$ in (38). We derive

$$l_\infty + \sum_{s=1}^{r_\infty}(\tau_s^{(\infty)} - t_s^{(\infty)})\,\nu_s^{(\infty)} = \bar{l}_\infty + \sum_{s=1}^{r_\infty}(\bar{\rho}_s^{(\infty)} - t_s^{(\infty)})\,\nu_s^{(\infty)},$$

whence

$$l_\infty + \sum_{s=1}^{r_\infty}\tau_s^{(\infty)}\,\nu_s^{(\infty)} = \bar{l}_\infty + \sum_{s=1}^{r_\infty}\bar{\rho}_s^{(\infty)}\,\nu_s^{(\infty)} \quad\ldots\ldots\ldots\ldots(42).$$

From (41) and (42) by addition we obtain

$$l + l_\infty + \sum_\kappa \sum_{s=1}^{r_\kappa}\tau_s^{(\kappa)}\,\nu_s^{(\kappa)} = \bar{l} + \bar{l}_\infty + \sum_\kappa \sum_{s=1}^{r_\kappa}\bar{\rho}_s^{(\kappa)}\,\nu_s^{(\kappa)} \quad\ldots\ldots\ldots\ldots(43).$$

Since however $\bar{H}(z, u)$ and $\frac{\bar{H}(z, u)}{S(z)}$ are the general rational functions built on the bases $(\bar{\tau})$ and $(\bar{\rho})$ respectively and since the aggregate sum of the orders of coincidence of the rational function $S(z)$ with the branches of all the cycles is 0, it follows that we have

$$\sum_\kappa \sum_{s=1}^{r_\kappa}\bar{\tau}_s^{(\kappa)}\,\nu_s^{(\kappa)} = \sum_\kappa \sum_{s=1}^{r_\kappa}\bar{\rho}_s^{(\kappa)}\,\nu_s^{(\kappa)}.$$

From (43) we then derive

$$l + l_\infty + \sum_\kappa \sum_{s=1}^{r_\kappa}\tau_s^{(\kappa)}\,\nu_s^{(\kappa)} = \bar{l} + \bar{l}_\infty + \sum_\kappa \sum_{s=1}^{r_\kappa}\bar{\tau}_s^{(\kappa)}\,\nu_s^{(\kappa)} \quad\ldots\ldots\ldots\ldots(44).$$

From (36) and (40) we have

$$l + l_\infty - 2N_\tau = \bar{l} + \bar{l}_\infty - 2N_{\bar{\tau}} \quad\ldots\ldots\ldots\ldots(45).$$

Combining (44) and (45) we deduce

$$N_\tau + \tfrac{1}{2}\sum_\kappa \sum_{s=1}^{r_\kappa}\tau_s^{(\kappa)}\,\nu_s^{(\kappa)} = N_{\bar{\tau}} + \tfrac{1}{2}\sum_\kappa \sum_{s=1}^{r_\kappa}\bar{\tau}_s^{(\kappa)}\,\nu_s^{(\kappa)} \quad\ldots\ldots\ldots\ldots(46).$$

This is the complementary theorem.

In deducing the complementary theorem we have here assumed that no positive orders of coincidence are furnished for finite values of the variable z by the basis (τ). We shall now suppose (τ) to be any arbitrary basis and $(\bar{\tau})$ the complementary basis.

The general rational functions built on these bases we shall designate by $H(z, u)$ and $\bar{H}(z, u)$ respectively. It is evident that we can select a polynomial $S(z)$ so that $\frac{H(z, u)}{S(z)}$ and $S(z)\bar{H}(z, u)$ are the general rational functions built on a pair of complementary bases (t) and $(\bar{t})$, of which the former furnishes no positive orders of coincidence for finite values of the variable z. The complementary theorem, as we have proved it, then holds good for the bases (t) and $(\bar{t})$, and we therefore have

$$N_t + \tfrac{1}{2}\sum_{\kappa}\sum_{s=1}^{r_\kappa} t_s^{(\kappa)}\nu_s^{(\kappa)} = N_{\bar{t}} + \tfrac{1}{2}\sum_{\kappa}\sum_{s=1}^{r_\kappa} \bar{t}_s^{(\kappa)}\nu_s^{(\kappa)}.$$

We plainly have, however,

$$N_t = N_\tau, \quad N_{\bar{t}} = N_{\bar{\tau}}, \quad \sum_{\kappa}\sum_{s=1}^{r_\kappa} t_s^{(\kappa)}\nu_s^{(\kappa)} = \sum_{\kappa}\sum_{s=1}^{r_\kappa} \tau_s^{(\kappa)}\nu_s^{(\kappa)}, \quad \sum_{\kappa}\sum_{s=1}^{r_\kappa} \bar{t}_s^{(\kappa)}\nu_s^{(\kappa)} = \sum_{\kappa}\sum_{s=1}^{r_\kappa} \bar{\tau}_s^{(\kappa)}\nu_s^{(\kappa)},$$

and the formula just obtained goes over into the formula (46). The complementary theorem therefore holds for any pair of complementary bases (τ) and $(\bar{\tau})$. The theorem has been derived on the assumption that the fundamental equation is integral. This restriction, however, is readily removed and we can at the same time obtain the theorem in a somewhat more general form given in Chapter XII of the writer's book* on the algebraic functions.

* *Theory of the algebraic functions of a complex variable*, Mayer and Müller, Berlin, 1906.

AXIOMS OF ORDINAL MAGNITUDES

By A. B. Frizell.

The word "axiom" is not used here as a substitute for either of the modern terms "assumption" or "postulate"; it is rather to be taken in the same sense as in Euclid.

1. *Axioms of Magnitude.*

M. 1. Between every two magnitudes of the same class one of the relations $A = B$ (A is equal to B) or $A \neq B$ (A is not equal to B) must hold.

2. The relations $A = B$ and $A \neq B$ cannot both hold for the same A and B.

3. The simultaneous relations $A = B$ and $B = C$ shall always involve the relation $A = C$.

4. The assertions $B = A$ and $A = B$ are to be identical in meaning.

5. For every magnitude, $A = A$.

A class of magnitudes is any set of symbols (e.g. those denoting the points in an arbitrary space) satisfying Axioms M.

There is a temptation to treat M. 5 as a consequence of 4 and 3, viz. $A = B$ and $B = A$, hence $A = A$. This, however, tacitly assumes that the class in question contains, corresponding to every member A, a distinct element $B = A$; and for the purposes of the present paper this possibility is to be expressly excluded by the

2. *Axioms of Arrangement.*

A. 1. Between every two magnitudes of an arranged set one of the relations $A < B$ (A precedes B) or $A > B$ (A follows B) must hold.

2. The relations $A < B$ and $A > B$ cannot both hold for the same A and B.

3. From the simultaneous relations $A < B$ and $B < C$ the relation $A < C$ shall always follow.

4. The statements $B > A$ and $A < B$ are to be identical in meaning.

5. No two magnitudes in an arranged set can be equal.

For example, the definition of a line postulates an arranged set of points.

Definitions. Of two symbols in an arranged set, that which precedes will be called lower, and that which follows higher, than the other.

A set of magnitudes will be said to be ordered according to a rule denoted by the sign $\bigcirc$ if for all members of the set $A < A \bigcirc B$, $A < B \bigcirc A$ and Axioms A. 1—4 all hold.

The only indispensable restriction to be imposed on the sign $\bigcirc$ (or any equivalent sign) is contained in the

3. *Group Axioms.*

G. 1. There shall be a class of magnitudes and a rule denoted by the sign $\bigcirc$ whereby equals with equals give equals for all members of the class for which this combination shall have been defined.

2. There shall be a set comprised in the class postulated in G. 1 such that if A and B both belong to the set so does $A \bigcirc B$ also.

3. The rule denoted by the sign $\bigcirc$ shall be associative for all members of the set postulated by G. 2.

4. Equals combined with unequals by the rule $\bigcirc$ shall give unequals.

5. There shall be a set which satisfies G. 1—4 and contains corresponding to every pair A, B of its members an element X and an element Y such that

$$B \bigcirc X = A = Y \bigcirc B.$$

A rule obeying Axiom G. 1 will be called a group rule or g-rule.

A set of magnitudes satisfying Axioms G. 1 and 2 is said to possess the fundamental group property.

A set that satisfies G. 1—4 will be called a *semigroup*. The purpose of Axioms G. is heuristic rather than subsumptive. If it were desired merely to define a group, part of the set would be redundant.

4. *Axioms of Numbering.*

N. 1. There shall be a g-rule denoted by the sign $\bigcirc$.

2. There shall be a symbol π and a symbol $\lambda \leqq \pi$.

3. There shall be a set $[\pi\lambda]$ composed of π and other magnitudes to be postulated in N. 4.

4. If κ belongs to $[\pi\lambda]$ so does $\lambda \bigcirc \kappa$ when $\lambda = \pi$ and so does $\kappa \bigcirc \lambda$ if $\lambda < \pi$.

5. The set $[\pi\lambda]$ shall be an arranged set.

6. For all members of $[\pi\lambda]$, $\kappa < \lambda \bigcirc \kappa$ and $\kappa < \kappa \bigcirc \lambda$.

Every set of magnitudes satisfying Axioms N. will be called a numbered set. To illustrate, π and λ may denote rectilinear segments and $\bigcirc$ the operation of a *Streckenüberträger*.

Note. It is easy to show inductively that a numbered set is ordered according to its defining g-rule.

5. The magnitudes to be postulated by the following sets of axioms will be called ordinal magnitudes.

Axioms of Infinity.

I. 1. There shall be an arranged set of magnitudes $[M]_A$.

2. The set $[M]_A$ shall have a first element.

3. Every member of the set $[M]_A$ shall have an immediate successor in the set.

4. Every member of the set $[M]_A$ except the first shall have an immediate predecessor in the set.

Every set satisfying Axioms I. will be called an infinite set.

It will be shown later (Theorem IV) that every numbered set is an infinite set.

6. *Transfinite Axioms.*

T. 1. There shall be an infinite set of g-rules $\boxed{\nu}$.

2. There shall be a symbol ω; we will call it the unit of transfinite magnitudes.

3. To every g-rule shall belong a numbered set of *primitive* elements having $\pi = \omega = \lambda$.

4. There shall be a set of *monomial* elements consisting of the primitives and others to be postulated in T. 5.

5. To every monomial element M shall belong a numbered set of monomials having $\pi = M = \lambda$ for every g-rule < the lowest used in postulating M.

6. To every monomial M shall belong a numbered set of *binomial* elements having $\pi = M$ for every monomial $\lambda < p$ (where $p > \omega$ is the highest primitive used in postulating M) for every g-rule $\leqq$ the lowest used in postulating M, the sign $=$ being excluded when M is itself a primitive element.

7. There shall be a set of *polynomial* elements composed of the binomials and others to be postulated by T. 8.

8. To every polynomial P shall belong a numbered set of polynomials having $\pi = P$ for every monomial $\lambda < p$, where $p > \omega$ is the highest primitive in the lowest term of P, and for every g-rule $\leqq$ the lowest used in postulating P.

7. The following notation naturally suggests itself as a means of making clear the content of Axioms T.

Put $\omega \boxed{\nu} \omega = \omega^{(1)}_{(\nu)}$, $\omega \boxed{\nu} \omega^{(1)}_{(\nu)} = \omega^{(2)}_{(\nu)}$, &c., so that the formula for a primitive element is $p = \omega^{(\iota)}_{(\nu)}$, where ν, ι are merely arbitrary symbols of an infinite set.

Let $$\omega^{(\iota)}_{(\nu_1)} \boxed{\nu_2} \omega^{(\iota)}_{(\nu_1)} = \omega^{(\iota,\ 1)}_{(\nu_1,\ \nu_2)},\quad \omega^{(\iota)}_{(\nu_1)} \boxed{\nu_2} \omega^{(\iota,\ 1)}_{(\nu_1,\ \nu_2)} = \omega^{(\iota,\ 2)}_{(\nu_1,\ \nu_2)},\ \text{\&c.},$$

in general $M = \omega^{(\iota_1,\ \iota_2)}_{(\nu_1,\ \nu_2)}$ for this type of monomial.

Writing $$\omega^{(\iota_1,\,\iota_2)}_{(\nu_1,\,\nu_2)} \boxed{\nu_3}\, \omega^{(\iota_1,\,\iota_2)}_{(\nu_1,\,\nu_2)} = \omega^{(\iota_1,\,\iota_2,\,1)}_{(\nu_1,\,\nu_2,\,\nu_3)}$$

and so on, we have as the most general formula for a monomial

$$M = \omega^{(\iota_1,\,\iota_2,\,\ldots\iota_n)}_{(\nu_1,\,\nu_2,\,\ldots\nu_n)}.$$

Thus if we denote the two lowest rules by $\boxed{\mu}$ and $\boxed{\nu}$ respectively the primitives $\omega^{(\iota)}_{(\mu)}$ and $\omega^{(\kappa)}_{(\nu)}$ may be written ω^{ι} and $\kappa\omega$ respectively, the most general monomial is $\omega^{(\iota,\,\kappa)}_{(\mu,\,\nu)} = \kappa\omega^{\iota}$, there are binomials $\kappa_1\omega^{\iota_1} \boxed{\nu}\, \kappa_2\omega^{\iota_2}$ and polynomials $\Sigma\kappa\omega^{\iota}$ where Σ denotes merely repeated application of the sign $\boxed{\nu}$.

Familiar notation has been used purposely, detached from its connotation, to emphasize the fact that the concepts in question justify their existence on the basis of postulating pure and simple. They have a foundation which is in so far simpler than that of the natural numbers that it requires fewer axioms and none which need proof to show that the set is self-consistent. There can be no contradiction between the different g-rules posited in Axioms T. so long as we postulate no relations connecting them.

8. The preceding sets of axioms are connected with each other and with those that are to follow by certain propositions.

Theorem I. No arranged set can form a group with regard to a g-rule according to which it is ordered.

For every group contains a modulus μ for its g-rule, so that we should have

$$\alpha = \alpha \bigcirc \mu$$

contrary to definition and A. 5.

Corollary. No arranged set can constitute a semigroup with modulus for a g-rule according to which it is ordered.

Theorem II. An arranged set ordered according to a g-rule with respect to which it possesses the fundamental group property contains at least one numbered set.

For if κ belongs to the set so does $\kappa' = \kappa \bigcirc \kappa$ by G. 2. Therefore also $\kappa'' = \kappa \bigcirc \kappa'$, $\kappa''' = \kappa \bigcirc \kappa''$, and so on.

But this set obeys Axioms N., having $\pi = \kappa = \lambda$.

Corollary. An arranged set ordered according to a g-rule with respect to which it constitutes a semigroup contains at least one numbered set.

Scholium. A necessary condition of an arranged set ordered according to a g-rule constituting a semigroup on that rule is the existence in it of a subset within which

$$\lambda \bigcirc \alpha \bigcirc \beta = \lambda \bigcirc (\alpha \bigcirc \beta),$$

where λ is a fixed element of the subset and α, β arbitrary members of it.

Lemma. A numbered set in which $\lambda = \pi$ will constitute a semigroup on its g-rule if combinations of its members by this rule are defined inductively in accordance with the formula

$$\lambda \bigcirc \alpha \bigcirc \beta = \lambda \bigcirc (\alpha \bigcirc \beta).$$

Proof. If $\alpha \bigcirc \beta$ belongs to $[\lambda\lambda]$ so does $\lambda \bigcirc (\alpha \bigcirc \beta)$ by N. 4. Therefore so does $(\lambda \bigcirc \alpha) \bigcirc \beta$ by definition. Hence if $\lambda \bigcirc \beta$ belongs to $[\lambda\lambda]$ so do $\lambda' \bigcirc \beta, \lambda'' \bigcirc \beta, \ldots$ where $\lambda' = \lambda \bigcirc \lambda$, $\lambda'' = \lambda \bigcirc \lambda'$, &c.

But $\lambda \bigcirc \beta$ belongs to $[\lambda\lambda]$ if β does (N. 4), and since every member of $[\lambda\lambda]$ is comprehended in the series $\lambda, \lambda', \lambda'', \ldots$ the fundamental group property is proved.

Suppose now that the associative relation

$$\alpha \bigcirc (\beta \bigcirc \gamma) = \alpha \bigcirc \beta \bigcirc \gamma$$

has been established for every member of $[\lambda\lambda]$ up to and including a certain element α and for every β, γ. Then

$$(\lambda \bigcirc \alpha) \bigcirc (\beta \bigcirc \gamma) = \lambda \bigcirc \{\alpha \bigcirc (\beta \bigcirc \gamma)\} = \lambda \bigcirc (\alpha \bigcirc \beta \bigcirc \gamma)$$
$$= \lambda \bigcirc (\alpha \bigcirc \beta) \bigcirc \gamma = \lambda \bigcirc \alpha \bigcirc \beta \bigcirc \gamma.$$

But $\lambda \bigcirc \beta \bigcirc \gamma = \lambda \bigcirc (\beta \bigcirc \gamma)$ by definition (hypothesis). Therefore

$$\lambda' \bigcirc \beta \bigcirc \gamma = \lambda' \bigcirc (\beta \bigcirc \gamma),$$

and so on. Hence, by strict induction, the rule denoted by $\bigcirc$ is associative.

By hypothesis no two members of $[\lambda\lambda]$ are equal (N. 5 and A. 5). Therefore by the associative property Axiom G. 4 is verified.

Thus Axioms G. 1—4 all hold and we have a semigroup. Q. E. D.

THEOREM III. Sufficient conditions of an arranged set constituting a semigroup are

(1) that it be a numbered set; (2) $\lambda = \pi$; (3) inductive definitions according to the formula

$$(\lambda \bigcirc \alpha) \bigcirc \beta = \lambda \bigcirc (\alpha \bigcirc \beta).$$

THEOREM IV. Every numbered set is an infinite set. For it is arranged (N. 5), has a first element (N. 3 and 6), contains an immediate successor to every one of its members (N. 4 and 6), and the immediate predecessor of every one except the first (N. 3 and 6).

THEOREM V. An infinite set ordered according to a g-rule with respect to which it constitutes a semigroup is a numbered set.

Proof. The given set K is arranged (I. 1) and ordered according to the given rule (hypothesis). Whence K contains at least one numbered set $[\lambda\lambda]$ (Th. II, Cor.).

The set $[\lambda\lambda]$ is infinite (Th. IV) and associative (G. 3). Therefore it constitutes a semigroup on the rule $\bigcirc$ (Lemma).

If now K is included in $[\lambda\lambda]$ the theorem is proved.

If K contains an element κ that is not in $[\lambda\lambda]$, then κ cannot follow $[\lambda\lambda]$ since thus κ would have no immediate predecessor in K, contrary to I. 4. But by G. 2

K contains $\kappa' = \kappa \bigcirc \kappa$, $\kappa'' = \kappa \bigcirc \kappa'$, ..., which is likewise an infinite set and cannot precede λ, since thus λ would have no immediate predecessor in K. Hence λ is some member of $[\kappa\kappa]$, say $\lambda = \kappa^{(\nu)}$. Therefore every $\lambda^{(\iota)}$ is a member of $[\kappa\kappa]$ and we have only a single numbered set. Like reasoning holds if K contains an element κ_1 that is not in $[\kappa\kappa]$, and so on for other such elements $\kappa_2, \kappa_3, \ldots$. But there must be a last κ_ν, since otherwise K would have no first element. Therefore K is only a single numbered set. Q. E. D.

THEOREM VI. No set of symbols can form a group with regard to each of two g-rules whereof one is distributive over the other.

For it would contain a modulus ν for the latter rule so that

$$\alpha \bigcirc \nu = \alpha.$$

Whence by the distributive law (hypothesis) and G. 1

$$\alpha\beta = (\alpha \bigcirc \nu)\, \beta = \alpha\beta \bigcirc \nu\beta.$$

Therefore we should have

$$\nu\beta' = \nu\beta,$$

when $\beta' \neq \beta$, which is contrary to G. 4.

Corollary. No set of symbols can constitute a semigroup on both rules with modulus for that over which the other is distributed.

Scholium. A necessary condition of an infinite set ordered according to each of two g-rules constituting a semigroup on both is the existence of a formula connecting the two rules.

For such a set is a numbered set (Th. V) with reference to each rule. Taking the lower rule the given set may therefore be written λ, $\lambda' = \lambda \bigcirc \lambda$, $\lambda'' = \lambda \bigcirc \lambda'$, ... Then indicating combination by the higher rule by juxtaposition

$$\lambda\lambda = \lambda^{(\nu)} \quad \text{(hypothesis and G. 2).}$$

THEOREM VII. Sufficient conditions of an infinite set that constitutes a semigroup on a g-rule $\bigcirc$ according to which it is ordered constituting also a semigroup on a rule $\square$ are that the rule $\square$ be distributive over the rule $\bigcirc$ and that

$$\lambda \square \lambda = \lambda^{(\nu)}.$$

Proof. If $\lambda^{(\nu)} = \lambda$, then $\lambda \square (\lambda \square \lambda) = \lambda \square \lambda = \lambda$, and so on. If $\lambda^{(\nu)} > \lambda$, let $\lambda^{(\nu)}$ denote the immediate predecessor of $\lambda^{(\nu)}$. Then

$$\lambda \square (\lambda \square \lambda) = \lambda \square (\lambda \bigcirc \lambda^{(\nu)}) = (\lambda \square \lambda) \bigcirc (\lambda \square \lambda^{(\nu)}),$$

and hence belongs to $[\lambda\lambda]$ if $\lambda \square \lambda^{(\nu)}$ does.

But if $\lambda \square \lambda^{(\mu)}$ belongs to $[\lambda\lambda]$, so does $\lambda \square \lambda^{(\bar{\mu})}$, where $\lambda^{(\bar{\mu})} = \lambda \bigcirc \lambda^{(\mu)}$, for

$$\lambda \square \lambda^{(\bar{\mu})} = \lambda \square (\lambda \bigcirc \lambda^{(\mu)}) = (\lambda \square \lambda) \bigcirc (\lambda \square \lambda^{(\mu)}) = \lambda^{(\nu)} \bigcirc (\lambda \square \lambda^{(\mu)}),$$

and

$$\lambda \square \lambda' = \lambda \square (\lambda \bigcirc \lambda) = (\lambda \square \lambda) \bigcirc (\lambda \square \lambda) = \lambda^{(\nu)} \bigcirc \lambda^{(\nu)},$$

whence the fundamental group property by strict induction.

The associative property follows readily from the distributive by the reasoning which Weierstrass used for complex numbers. In like manner we get the relation of equals with unequals.

Corollary. The set is also ordered according to the rule $\square$ provided

$$\lambda \square \lambda > \lambda.$$

Scholium. All sets postulated in accordance with Th. VII are holoedrically isomorphic with one another. Consequently we naturally take as representative of this set of sets the simplest of them, viz. that in which $\lambda \square \lambda = \lambda$.

THEOREM VIII. Given two g-rules, the higher distributive over the lower, and an infinite semigroup on the lower rule, ordered according to it, if further

$$\lambda \square \lambda = \lambda,$$

then λ is modulus for the higher rule denoted by the sign $\square$. For

$$\lambda \square \lambda' = \lambda \square (\lambda \bigcirc \lambda) = (\lambda \square \lambda) \bigcirc (\lambda \square \lambda) = \lambda \bigcirc \lambda = \lambda',$$

and $$\lambda \square (\lambda \bigcirc \lambda^{(\iota)}) = (\lambda \square \lambda) \bigcirc (\lambda \square \lambda^{(\iota)}) = \lambda \bigcirc \lambda^{(\iota)},$$

whenever it has already been proved that $\lambda \bigcirc \lambda^{(\iota)} = \lambda^{(\iota)}$.

Whence the theorem follows readily by strict induction.

9. The above is perhaps a sufficiently exhaustive account of the logical status of the system of natural numbers. They cannot logically be postulated by a single g-rule. Addition alone does not distinguish them from any other numbered set. Multiplication alone cannot yield the complete set but only successive subsets—the powers of 2, 3, ... Multiplication and addition together, *with no further restriction*, yield the transfinite set $\Sigma\iota\omega^{\kappa}$. This is an arranged set constituting a semigroup on each rule, ordered according to both. We cannot have an arranged set forming a group on either rule, nor, if we postulate the distributive relation, a semigroup on both with modulus for the lower rule. If we demand a set ordered according to both rules, this excludes the modulus of the higher rule, and we get only a subset. Consequently we postulate according to the enunciation of Th. VIII, or, in other words, we formulate the

Axioms of the Finite Unit.

U. 1. There shall be g-rules denoted by the signs $\boxed{\mu}$ and $\boxed{\nu}$.

2. The rule denoted by $\boxed{\mu}$ shall be distributive over the rule $\boxed{\nu}$.

3. There shall be a symbol μ; we will call it the finite unit.

4. There shall be an infinite set of symbols

$$\mu, \quad \mu' = \mu \bigcirc \mu, \quad \mu'' = \mu \bigcirc \mu', \ldots$$

forming a semigroup on the rule $\boxed{\nu}$, ordered according to it.

5. $$\mu \boxed{\mu} \mu = \mu.$$

It follows by Th. VIII that μ is modulus for the rule $\boxed{\mu}$, and by Th. VII that the set postulated in U. 4 constitutes a semigroup with respect to the rule $\boxed{\mu}$, but is not ordered according to it.

Thus the system of natural numbers, however simple it may be in epistemology, is connected with the equally simple concept of a rule for combining symbols by relations which exhibit a certain degree of complexity. And it is these relationships which are of interest in mathematics. Even if we take the naive point of view with which Cantor contents himself in *Annalen* XLVI., and posit the natural numbers by a twofold abstraction, the logical structure is there and cannot be ignored. The validity of Axioms U., on the basis of any such external evidence, supplies *eo ipso* a consistency proof for Axioms T. in the special case of two g-rules only. But if we can postulate two independent g-rules we can also postulate an infinite set of them without involving any contradictions.

10. The necessity of distinguishing different sets of symbols resulting from sets of two, three, ... g-rules suggests supplementing Axioms T. by a set of Axioms T_N. which differs from the former only in replacing Axiom T. 1 by *Axiom* T_N. 1. There shall be for each value of $N=2, 3 \ldots$ a finite set of g-rules denoted by signs $\boxed{1}$, $\boxed{2}$, ... $\boxed{N}$.

We now proceed in the direction of Cantor's "Fortsetzung der natürlichen Zahlenreihe über das Unendliche hinaus," and first we will postulate his second ordinal class by the

Axioms for a Denumerable Set.

D. 1. There shall be a set of symbols N postulated by Axioms U.

2. There shall be a symbol $\omega > N$ for all values of N.

3. There shall be g-rules denoted by signs $\boxed{\lambda}$, $\boxed{\mu}$ and $\boxed{\nu}$.

4. There shall be a set of symbols postulated by Axioms T_3.

5. Between every two symbols α, $\alpha \boxed{\nu}\, \omega$ of those postulated in 4, we interpolate a numbered set $\alpha \boxed{\nu}\, N$, $N = \mu, \mu', \mu'', \ldots$.

6. The whole set postulated in Axioms D. 1—5 shall be ordered in accordance with the

11. *Axioms of Ordering.*

O. 1. Of two polynomials, $P > Q$ if $M > N$, where M, N are the highest terms in P, Q respectively.

2. Of two monomials, $M > N$ if $p > q$, where p, q are the highest primitives in M, N respectively.

3. Of two primitive elements, $p = \omega_{(\mu)}^{(\iota)}$ and $q = \omega_{(\nu)}^{(\kappa)}$, $p > q$ if $\mu > \nu$ or if $\mu = \nu$ and $\iota > \kappa$.

4. $\alpha + N_1 > \alpha + N_2$ if $N_1 > N_2$, where N_1, N_2 are natural numbers and α infinite or transfinite.

Using Axioms O. we also postulate non-enumerable sets by the

12. *Axioms of Transfinite Numbers.*

${}_N$T. 1. There shall be a set of symbols postulated by Axioms U.

2. There shall be a symbol $\omega > N$ (every value of N).

3. There shall be a set of g-rules $\boxed{\nu_1}$, $\boxed{\nu_2}$, ... $\boxed{\nu_N}$.

4. There shall be a set of symbols postulated by Axioms T_N.

5. Between every two symbols α, $\alpha \boxed{\nu_N} \omega$ of the set postulated in 4, shall be a numbered set $\alpha \boxed{\nu_N} N$. N.B. The suffix N is constant; the other N is variable.

6. The whole set of symbols postulated in Axioms ${}_N$T. 1—5 shall be ordered in accordance with Axioms O.

Thus Axioms ${}_3$T. are identical with Axioms D., while Axioms ${}_2$T. furnish a series of type ω.

13. We next formulate *Axioms* T^1.

This set differs from the set denoted by ${}_N$T. only in replacing ${}_N$T. 3 and ${}_N$T. 4 respectively by

Axiom T^1. 3. There shall be an infinite set of g-rules.

Axiom T^1. 4. There shall be a set of symbols postulated by the totality of Axioms T_N. ($N = 2, 3, \ldots$).

And now we prepare to take the last steps by laying down the

Axioms of Postulating.

ω. 1. There shall be an arranged set of sets of Axioms T^τ. consisting of the set T^1. and other sets to be postulated in Axioms T^τ.

2. Every set of sets of axioms composed of all sets that precede each set T^τ. (where τ is to have the meaning stated in ω. 3) shall postulate a set of symbols of ordinal type υ.

3. The symbols τ shall be taken in order from among those postulated by Axioms T^τ. ($\tau = 2, 3, \ldots \omega, \ldots$).

4. The symbol T^o shall have the same meaning as U, so that to $\tau = 1$ shall belong the value $\upsilon = f(1) = \omega$.

14. *The Last Axioms.*

T^τ. 1. There shall be a set of g-rules whose ordinal type is υ.

2. To every g-rule shall belong an υ-set of *primitive* elements having $\pi = \upsilon = \lambda$.

3. There shall be a set of *monomial* elements composed of the primitives and others to be postulated in T^τ. 4.

4. To every monomial M shall belong an υ-set of monomials having $\pi = M = \lambda$ for every g-rule $<$ lowest rule used in postulating M.

5. To every monomial M shall belong an υ-set of *binomial* elements having $\pi = M$ for every monomial $\lambda < p$, where $p > \upsilon$ is the highest primitive used in postulating M, for every g-rule $\leqq$ lowest used in postulating M (the sign $=$ excluded when $M = p$).

6. There shall be a set of *polynomial* elements consisting of the binomials and others to be postulated in T^τ. 7.

7. To every polynomial P shall belong an υ-set of new polynomials having $\pi = P$ for every monomial $\lambda < p$, where $p > \upsilon$ is the highest primitive in the lowest term of P, for every g-rule $\leqq$ lowest used in postulating P.

8. Between every two symbols β and $\beta \boxed{1} \upsilon$ of the set postulated in T^{τ}. 2—7 shall be an υ-set of new symbols.

9. The whole set just postulated shall be ordered by Axioms O. extended by the substitution of υ for ω in O. 3 and of an υ-set for the infinite set used in O. 4.

15. By virtue of Axiom ω. 4 the natural numbers are comprehended in the set of symbols postulated by Axioms T^{τ}., and the next value of υ, which may be denoted by $\upsilon = f(2)$, is brought into the general scheme by Axiom ω. 2. Then ω. 3 directs us to take, corresponding to $\tau = 3$, the set of Axioms T^2. which uses a set of g-rules with ordinal type $\upsilon = f(2)$ and yields a set of symbols having the ordinal type $\upsilon = f(3)$. The transfinite induction from $f(N)$ to $f(\omega)$ is provided for by Axiom ω. 2 and this is also enough for all cases of transfinite induction. Thus the Axioms ω. and T^{τ}. constitute a recurring procedure which carries forward the postulating process in a manner that *ipso facto* excludes all idea of limitation.

There will always be found between the set of values already used for τ and that of the symbols which they have made it possible to postulate a gap from which new values of τ may be drawn. And the word "always" is not here liable to the same ambiguity as in the classic example of Achilles and the tortoise. The present gap does not admit of closure by the introduction of a new symbol. That possibility depended essentially upon the external conditions imposed by the statement of the problem in question. But in our case there are no external conditions.

It remains to state explicitly what is involved in the above, viz.

THEOREM IX. Every set of symbols postulated by Axioms T^{τ}. is well ordered. For all these sets are built up out of infinite sets of symbols ordered by the sequence of the sets in infinite sets of g-rules.

16. The theory may now be completed by establishing

THEOREM X. The set of symbols postulated by Axioms T^{τ}. may be made to include any *given* well-ordered type of order. For otherwise there would be, just as in functions of a real variable, a type constituting a limit to the T^{τ}. set and therefore likewise a limit to the set of values for τ. But by Axioms ω. every limiting value to a sequence of the symbols τ is taken up into the series.

Scholium. The expression "the totality of the symbols τ" has no meaning and we have the

Corollary. It is not possible to postulate all well-ordered sets.

There is thus a second kind of infinity that sets a natural boundary to the series of postulates. Perhaps this is the infinity of the continuum. If so the continuum concept would turn out to be arithmetically a "*Ding an sich,*" the existence of which may be surmised but never proved.

UNE QUESTION DE MAXIMUM OU DE MINIMUM

PAR ALESSANDRO PADOA.

1. Soit
$$z = y_1 y_2 \dots y_n \quad \dots\dots (1),$$
où n est un nombre entier plus grand que 1 et où, pour toute valeur entière de r depuis 1 jusqu'à n,
$$y_r = a_{r1}x_1 + a_{r2}x_2 + \dots a_{r,n-1}x_{n-1} + a_{rn} \quad \dots\dots (2).$$

Remarquons tout de suite que, si l'on pose
$$n\Delta = \begin{vmatrix} a_{11} & a_{12} \dots a_{1n} \\ a_{21} & a_{22} \dots a_{2n} \\ \dots & \dots \\ a_{n1} & a_{n2} \dots a_{nn} \end{vmatrix} \quad \dots\dots (3),$$
on peut *toujours* supposer
$$\Delta \geqq 0 \quad \dots\dots (4)$$
parce que, s'il avait été $\Delta < 0$, il aurait suffit d'échanger entre eux préalablement deux facteurs quelconques de z, pour changer le signe de Δ.

Mais, pour une raison que j'expliquerai tout à l'heure, ici je suppose que le *mineur* d'un élément quelconque de la dernière colonne soit différent de zéro; donc, en désignant par A_{rs} le mineur (signé) qui correspond à l'élément a_{rs}, je suppose que
$$A = A_{1n}A_{2n} \dots A_{nn} \neq 0 \quad \dots\dots (5).$$

L'Algèbre élémentaire nous apprend que*, si
$$y_r \geqq 0 \text{ et si } \sum_{r=1}^{r=n} y_r = ns,$$
où s est un nombre positif *donné*, et s'il peut se faire que
$$y_1 = y_2 = \dots = y_n,$$
alors ces valeurs des y_r font atteindre à z son *maximum*, par rapport aux dites conditions, de manière que
$$\max z = s^n.$$

Mais ici, en assujettissant les variables à la *seule* condition que y_r ait toujours le même signe que A_{rn}, c'est-à-dire que
$$A_{rn}y_r \geqq 0 \quad \dots\dots (6),$$
je vais démontrer d'une façon élémentaire que z atteint *toujours* un *maximum* ou un *minimum* selon que A est *positif* ou *négatif*, que la valeur de ce *maximum* ou de ce *minimum* est déterminée, dans les deux cas, par la formule
$$z' = \Delta^n : A \quad \dots\dots (7)$$

* La première des conditions suivantes est nécessaire seulement lorsque $n > 2$.

et que cette valeur z' de z correspond toujours à *un seul* système de valeurs (déterminées et finies) des variables $x_1, x_2, \dots x_{n-1}$.

En effet, la valeur de notre déterminant (3) ne change pas si à la dernière colonne on ajoute les produits des autres par $x_1, x_2, \dots x_{n-1}$, c'est-à-dire que, par la formule (2),

$$n\Delta = \begin{vmatrix} a_{11} & a_{12} \dots y_1 \\ a_{21} & a_{22} \dots y_2 \\ \dots & \dots \\ a_{n1} & a_{n2} \dots y_n \end{vmatrix}$$

d'où, en développant selon la dernière colonne,

$$A_{1n}y_1 + A_{2n}y_2 + \dots + A_{nn}y_n = n\Delta \quad \dots\dots\dots (8).$$

Mais, en multipliant membre à membre les égalités (1) et (5), on obtient

$$Az = (A_{1n}y_1)(A_{2n}y_2)\dots(A_{nn}y_n);$$

de manière que, par rapport à Az, les formules (4), (6) et (8) nous ramènent immédiatement à la question d'Algèbre élémentaire que je viens de vous rappeler*.

Donc, *pourvu* qu'on puisse déterminer un système de valeurs des variables tel que

$$A_{1n}y_1 = A_{2n}y_2 = \dots = A_{nn}y_n \dots\dots\dots (9),$$

ces valeurs feront atteindre à Az son *maximum*, de manière que

$$\max(Az) = \Delta^n \dots\dots\dots (10);$$

et par suite, A étant différent de zéro (5), la formule (7) nous donnera le *maximum* ou le *minimum* de z selon que A est *positif* ou *négatif*.

A cause de la formule (8), la condition (9) est verifiée seulement si, pour toute valeur de r,

$$A_{rn}y_r = \Delta \dots\dots\dots (11).$$

En considérant comme *inconnues* $x_1, x_2, \dots x_{n-1}$, la formule (11) désigne un système de n équations linéaires; mais, bien que le nombre des équations y soit plus grand que celui des inconnues, ce système admet *toujours* une solution.

La solution dont je parle est représentée par la formule

$$x_s' = \frac{1}{n}\left(\frac{A_{1s}}{A_{1n}} + \frac{A_{2s}}{A_{2n}} + \dots + \frac{A_{ns}}{A_{nn}}\right) \dots\dots\dots (12)$$

où, pour toute valeur entière de s depuis 1 jusqu'à $n-1$, x_s' désigne la valeur qu'il faut donner à l'inconnue x_s; valeur qui est *toujours déterminée et finie* en conséquence de l'hypothèse (5), dont maintenant on voit l'importance.

En effet, si nous désignons par y_r' la valeur qui prend y_r lorsque, dans la formule (2) préalablement multipliée par n, on remplace les variables

$$x_1, x_2, \dots x_{n-1}$$

par les nombres

$$x_1', x_2', \dots x_{n-1}',$$

* En conséquence de la note précédente, même dans la question dont je m'occupe la condition (6) est nécessaire seulement lorsque $n > 2$.

nous obtenons

$$\begin{aligned} ny_r' = a_{r1}\left(\frac{A_{11}}{A_{1n}} + \frac{A_{21}}{A_{2n}} + \ldots + \frac{A_{n1}}{A_{nn}}\right) \\ + a_{r2}\left(\frac{A_{12}}{A_{1n}} + \frac{A_{22}}{A_{2n}} + \ldots + \frac{A_{n2}}{A_{nn}}\right) \\ + \ldots\ldots\ldots\ldots\ldots\ldots\ldots\ldots\ldots \\ + a_{r,n-1}\left(\frac{A_{1,n-1}}{A_{1n}} + \frac{A_{2,n-1}}{A_{2n}} + \ldots + \frac{A_{n,n-1}}{A_{nn}}\right) \\ + a_{rn}\left(\frac{A_{1n}}{A_{1n}} + \frac{A_{2n}}{A_{2n}} + \ldots + \frac{A_{nn}}{A_{nn}}\right) \end{aligned}$$

où le dernier terme n'est qu'une transformation de na_{rn}.

En faisant les multiplications indiquées et puis en faisant l'addition par colonnes, par une propriété bien connue des déterminants chacune de ces sommes est zéro, sauf celle de place r qui, à cause de la formule (3), vaut $\frac{n\Delta}{A_{rn}}$; on obtient donc $ny_r' = \frac{n\Delta}{A_{rn}}$, d'où

$$A_{rn}y_r' = \Delta \ldots\ldots\ldots\ldots\ldots\ldots\ldots\ldots\ldots\ldots\ldots (13)$$

et par conséquent le système (11) est verifié.

D'ailleurs, ce système ne pourrait avoir d'autres solutions; parce que, en laissant de côté une quelconque des n équations, on obtient un système de $n-1$ équations à $n-1$ inconnues, dont le déterminant des coefficients, étant le mineur d'un élément de la dernière colonne du déterminant $n\Delta$, est toujours différent de zéro, par l'hypothèse (5); par suite ce système admet toujours une seule solution.

Donc la formule (12) nous donne *le seul* système de valeurs des variables auquel correspond le *maximum* ou le *minimum* de z, que nous savons déjà être déterminé par la formule (7).

2. Comme *cas particulier*, si tous les facteurs de z peuvent devenir simultanément *zéro*, de la formule (8) on déduit $\Delta = 0$. Réciproquement: si $\Delta = 0$, alors la formule (13) devient $A_{rn}y_r' = 0$, d'où, en conséquence de l'hypothèse (5), $y_r' = 0$; c'est-à-dire que tous les facteurs de z deviennent simultanément *zéro*, pour les valeurs (12) des variables.

D'ailleurs, en ce cas, ces valeurs sont les seules compatibles avec la condition (6), ce qu'on déduit de $\Delta = 0$ et des formules (6), (8), (5); par conséquent, *zéro* étant en ce cas *la seule* valeur de z que nous avons à considérer, il est en même temps le *maximum* et le *minimum*, soit A positif ou négatif.

3. Voici un exemple d'*application géométrique* du théorème que je viens de démontrer.

Soient $$y_1 = 0, \quad y_2 = 0, \quad y_3 = 0, \quad y_4 = 0$$

les équations normales, en coordonnées cartésiennes, de 4 plans quelconques de l'espace ordinaire.

L'hypothèse (5) $A \neq 0$ a le rôle d'exclure que 3 de ces plans soient parallèles à une même droite. Remarquons qu'on peut même toujours supposer que

$$A > 0 \ldots\ldots\ldots\ldots\ldots\ldots\ldots\ldots\ldots\ldots\ldots (14)$$

parce que, s'il avait été $A < 0$, il aurait suffit de changer préalablement tous les signes d'un des polinomes donnés, par ex. de y_4; alors A_{14}, A_{24}, A_{34} auraient changés de signe, tandis que A_{44} l'aurait conservé, et ainsi A aurait changé de signe.

Du cas particulier que nous venons d'examiner il résulte (ainsi que nous l'apprend la Géométrie analytique) que $\Delta = 0$ seulement lorsque le point de coordonnées x_1', x_2', x_3' appartient aux 4 plans donnés (qui, par conséquent, appartiennent à une même gerbe, dont ce point est le centre).

Maintenant, soit $\Delta \neq 0$, c'est-à-dire (4)

$$\Delta > 0 \qquad \qquad (15);$$

en ce cas, les 4 plans donnés sont les faces d'un vrai tétraèdre.

Si nous désignons par y_1'' ce que devient y_1 lorsqu'on remplace les variables x_1, x_2, x_3 par les coordonnées du point d'intersection des plans $y_2 = 0$, $y_3 = 0$, $y_4 = 0$, la formule (8) nous donne

$$A_{14} y_1'' = 4\Delta,$$

et en général

$$A_{r4} y_r'' = 4\Delta \qquad \qquad (16),$$

où y_r'' sont les *hauteurs* du tétraèdre donné, que la formule (16) nous apprend à calculer aisément.

Des formules (15), (16) on déduit $A_{r4} y_r'' > 0$; par conséquent $A_{r4} y_r > 0$ pour tout point qui, par rapport au plan $y_r = 0$, se trouve du même côté de l'intersection des autres 3 plans. On voit ainsi que la condition (6) porte à s'occuper *seulement* des points qui appartiennent au tétraèdre dans l'acception élémentaire du mot. Alors, en se souvenant de la formule (14), on arrive au résultat suivant: parmi les points internes au tétraèdre donné, il y en a *un seul* dont *le produit des distances des faces* du tétraèdre est un *maximum*, ses *coordonnées* sont x_1', x_2', x_3' et la *valeur* de ce maximum est $\Delta^4 : A$.

4. D'ailleurs, par la même méthode mais d'une façon géométrique élémentaire, on peut traiter directement la question particulière dont je viens de parler.

Si v est le volume du tétraèdre considéré et si (pour toute valeur entière de r depuis 1 jusqu'à 4) l'on désigne par b_r un tiers de l'aire d'une de ses faces et par y_r la distance de cette face d'un point variable interne au tétraèdre, alors le produit

$$z = y_1 y_2 y_3 y_4$$

devient *maximum* ensemble au produit

$$(b_1 y_1)(b_2 y_2)(b_3 y_3)(b_4 y_4)$$

où

$$b_r y_r > 0 \quad \text{et} \quad \sum_1^4 b_r y_r = v,$$

c'est-à-dire lorsque

$$b_r y_r = \frac{v}{4},$$

c'est-à-dire lorsque les distances du point des faces sont la quatrième partie des hauteurs correspondantes.

On connaît ainsi 4 planes (parallèles aux faces du tétraèdre) qui passent par le point cherché et dont 3 suffisent pour le déterminer.

NEUE EMPIRISCHE DATEN ÜBER DIE ZAHLENTHEORETISCHE FUNKTION $\sigma(n)$

Von R. v. Sterneck.

Die von F. Mertens mit $\sigma(n)$ bezeichnete zahlentheoretische Funktion ist durch die Gleichung $\sigma(n) = \sum_{\lambda=1}^{n} \mu(\lambda)$ definiert, wobei $\mu(\lambda)$ den Wert ± 1 hat, je nachdem λ das Produkt einer geraden oder ungeraden Anzahl verschiedener Primzahlen ist und den Wert 0 für alle durch ein Quadrat, grösser als 1, teilbaren Argumente λ.

Mertens hat das Gesetz $|\sigma(n)| \leqq \sqrt{n}$ bis zur Grenze $n = 10{,}000$ nachgewiesen und zugleich gezeigt, dass mit dem allgemeinen Nachweis dieses Gesetzes auch die Richtigkeit der Riemann'schen Vermutung bewiesen wäre, dass die komplexen Nullstellen der Funktion $\zeta(s)$ sämtlich den reellen Bestandteil $\frac{1}{2}$ haben.

In den Jahren 1897 und 1901 habe ich durch Herstellung einer bis 500,000 reichenden Tabelle der Werte $\sigma(n)$ den Nachweis geführt, dass bis zu dieser Grenze, wenn man von einigen ganz im Anfange gelegenen Stellen in der Umgebung von $n = 200$ absieht, sogar die Relation $|\sigma(n)| < \frac{1}{2}\sqrt{n}$ strenge erfüllt ist, indem der Quotient $\frac{\sigma(n)}{\sqrt{n}}$ in diesem Intervalle stets zwischen den Grenzen $\pm 0{\cdot}46$ hin und herschwankt.

Den Anlass, mich neuerdings der empirischen Prüfung des erwähnten Gesetzes zuzuwenden, fand ich in einer Arbeit von E. Landau, die sich mit einer Abhandlung von Franel aus dem Jahre 1896 beschäftigt und in der gezeigt wird, dass eine von Franel dortselbst ohne Beweis gegebene Relation mit der Riemann'schen Vermutung im Widerspruch steht. Wenn nun auch für die Richtigkeit dieser von Franel angegebenen Beziehung nicht der geringste Wahrscheinlichkeitsgrund geltend gemacht werden kann, so schien es mir doch angesichts der Sachlage wünschenswert, die empirische Grundlage des Gesetzes $|\sigma(n)| < \frac{1}{2}\sqrt{n}$ so ausgedehnt als möglich zu gestalten. Dabei dachte ich nicht an eine Fortsetzung meiner Tabelle sondern nur an einzelne Stichproben für möglichst grosse Argumente n. Unterstützt durch die kaiserliche Akademie der Wissenschaften in Wien, welche mir eine Subvention zur Entlohnung von Rechnern gewährte, konnte ich diesem Plane entsprechend im Laufe des vergangenen Winters 16 neue Funktionswerte $\sigma(n)$ ermitteln, deren Argumente bis zur Grenze 5,000,000 hinaufreichen.

Die Formel, nach der die Berechnungen durchgeführt wurden, ist die folgende:

$$\sum_{\lambda' \leqq g} \sigma\left(\frac{n}{\lambda'}\right) + \sum_{\lambda=1}^{g} \mu(\lambda)\, \omega_i\left(\frac{n}{\lambda}\right) - \omega_i(g)\, \sigma(g) = 0.$$

In derselben bedeutet g die grösste in $\sqrt{n}$ enthaltene ganze Zahl, $\omega_i(m)$ die Anzahl der die Zahl m nicht übersteigenden, durch keine der i ersten Primzahlen teilbaren Zahlen; λ' hat alle derartigen Zahlen, welche g nicht übertreffen, zu durchlaufen. Die Formel gilt für alle jene n, welche grösser oder gleich dem Produkte der i ersten Primzahlen sind.

Setzt man in derselben z. B. $i = 4$, so erhält man eine Formel, in welcher in der ersten Summe auf $\sigma(n)$ sogleich $\sigma\left(\frac{n}{11}\right)$ folgt, die sich somit zur Berechnung von $\sigma(n)$ bis zum 11fachen äussersten Argumente der vorhandenen Tabelle eignet.

Auf Grund dieser Formeln wurden folgende Funktionswerte ermittelt:

$\sigma(600{,}000) = -230$, $\sigma(1{,}800{,}000) = +406$,
$\sigma(700{,}000) = +226$, $\sigma(2{,}000{,}000) = -247$,
$\sigma(800{,}000) = -20$, $\sigma(2{,}500{,}000) = +364$,
$\sigma(900{,}000) = -225$, $\sigma(3{,}000{,}000) = +109$,
$\sigma(1{,}000{,}000) = +214$, $\sigma(3{,}500{,}000) = -136$,
$\sigma(1{,}200{,}000) = -153$, $\sigma(4{,}000{,}000) = +194$,
$\sigma(1{,}400{,}000) = -247$, $\sigma(4{,}500{,}000) = +177$,
$\sigma(1{,}600{,}000) = +168$, $\sigma(5{,}000{,}000) = -705$.

Bildet man für jeden derselben den Quotienten $\frac{|\sigma(n)|}{\sqrt{n}}$, so erhält man, auf 3 Dezimalstellen abgerundet, der Reihe nach folgende Werte: 0·297, 0·270, 0·022, 0·237, 0·214, 0·140, 0·209, 0·133, 0·302, 0·175, 0·230, 0·063, 0·073, 0·097, 0·083, 0·315.

Die neuberechnenden Funktionswerte erfüllen also, wie man sieht, sämtlich das Gesetz $|\sigma(n)| < \frac{1}{2}\sqrt{n}$. Da dies aber nur von diesen einzelnen Stellen feststeht, ist damit natürlich nicht etwa der vollständige Beweis erbracht, dass dieses Gesetz bis zur Grenze 5,000,000 gültig bleibt; doch hat dies auf Grund obiger Zahlen, wie man sich leicht klar machen kann, einen sehr hohen Grad von Wahrscheinlichkeit.

Die eben erhaltenen Verhältniszahlen geben im Mittel 0·179. Vergleichen wir diesen Wert mit dem Mittelwerte des Quotienten $\frac{|\sigma(n)|}{\sqrt{n}}$ für das Intervall von 0 bis 500,000, für welchen man durch Mittelbildung aus den Werten dieses Quotienten nach je 10,000 Argumenten n den Näherungswert 0·140 erhält, so können wir eine ganz befriedigende Uebereinstimmung konstatieren, da man ja bei einer so geringen Zahl von nur 16 zufällig herausgegriffenen Funktionswerten nicht erwarten kann, durch einfache Mittelbildung dem wahren Mittelwerte dieses Quotienten für das ganze Intervall besonders nahe zu kommen. Alle Anzeichen sprechen daher dafür, dass das Gesetz $|\sigma(n)| < \frac{1}{2}\sqrt{n}$ auch bis zur Grenze 5,000,000 richtig bleibt und dass der Quotient $\frac{\sigma(n)}{\sqrt{n}}$ ungefähr zwischen denselben Grenzen hin und herschwankt wie in dem früher untersuchten zehnmal kleineren Intervall; es waren dies die Grenzen $\pm 0{\cdot}46$.

Als Ergebnis der Untersuchung ist somit festzustellen, dass sich die Funktion $\frac{\sigma(n)}{\sqrt{n}}$ im Intervall von 0 bis 5,000,000 genau ebenso zu verhalten scheint, wie in dem zehnmal kleineren Intervall bis 500,000, dass somit die Relation

$$|\sigma(n)| < \tfrac{1}{2}\sqrt{n}$$

ein zwar unbewiesenes, aber ausserordentlich wahrscheinliches zahlentheoretisches Gesetz darstellt, und somit auch die Riemann'sche Vermutung mit einem hohen Grad von Wahrscheinlichkeit als richtig angesehen werden kann.

SOME USES IN THE THEORY OF FORMS OF THE FUNDAMENTAL PARTIAL FRACTION IDENTITY

By E. B. Elliott.

1. If $$F(\phi) \equiv (\phi - a_1)(\phi - a_2) \dots (\phi - a_n) \quad \text{.....................(1)},$$
where $a_1, a_2, \dots a_n$ are different constants, the identity is
$$1 = \sum_{s=1}^{s=n} \left(\frac{1}{F'(a_s)} \frac{F(\phi)}{\phi - a_s} \right) \quad \text{.........................(2)}.$$
Here $F(\phi)/(\phi - a_s)$ is the *integral* function
$$(\phi - a_1)(\phi - a_2) \dots (\phi - a_{s-1}) \cdot (\phi - a_{s+1}) \dots (\phi - a_n) \text{...........(3)}.$$

The ϕ, which disappears from the right of (2), need not be a quantity. But, if we take it to be a symbol of operation, it must not operate on $a_1, a_2, \dots a_n$, as the factors in (1) have to be commutative.

Let it be a symbol of *direct* differential operation $\phi\left(x, y, z, \dots \frac{\partial}{\partial x}, \frac{\partial}{\partial y}, \frac{\partial}{\partial z}, \dots\right)$ on functions of any number of letters $x, y, z, \dots$. Then, if u is any function of $x, y, z, \dots$,
$$u = \sum_{s=1}^{s=n} \left(\frac{1}{F'(a_s)} \frac{F(\phi)}{\phi - a_s} u \right) \quad \text{.........................(4)}.$$
We look upon this as a formula for the separation of u into n parts by direct operation.

2. Now let u denote any solution of the differential equation
$$F(\phi) u = f(x, y, z, \dots) \quad \text{.........................(5)}.$$
By direct operation, as in (4), we have
$$u = u_1 + u_2 + \dots + u_n \quad \text{.............................(6)},$$
where, for $s = 1, 2, \dots n$, u_s satisfies
$$(\phi - a_s) u_s = \frac{1}{F'(a_s)} f(x, y, z, \dots) \quad \text{....................(7)}.$$

We further remark that any u_s satisfying (7) satisfies
$$F(\phi) u_s = \frac{1}{F'(a_s)} \frac{F(\phi)}{\phi - a_s} f(x, y, z, \dots) \quad \text{.................(8)},$$
so that, by (4), any sum (6) of solutions, each of one of the n equations (7), for $s = 1, 2, \dots n$, satisfies (5).

It follows that the general solution u of (5) is the sum of the general solutions of (7) for all the separate values $s = 1, 2, \dots n$.

In particular we have a compact proof *ab initio* that the general solution of the linear equation, with constant coefficients and no repeated operating factors on the left,

$$F\left(\frac{d}{dx}\right) y = f(x) \qquad \text{.............................(9)}$$

is the sum of the general solutions of the several equations

$$\left(\frac{d}{dx} - a_s\right) y = \frac{1}{F'(a_s)} f(x), \quad (s = 1, 2, \dots n) \qquad \text{..............(10)},$$

i.e. is

$$y = \sum_1^n \frac{1}{F'(a_s)} e^{a_s x} \{\textstyle\int e^{-a_s x} f(x)\, dx + C_s\} \qquad \text{..................(11)};$$

so that it involves n arbitrary constants.

3. The integration of equations like (5) is not, however, our present concern. We are going to deal with *known* solutions u of such equations; separate them into parts (6) by direct operation; examine the parts; and draw conclusions from the separability.

In all the examples chosen the right-hand side of (5) will be zero; and we shall deal with solutions of particular types (for instance with rational integral solutions), under circumstances when it is clear or demonstrable that the separation is into parts of the same types.

Separation of gradients into seminvariant and other parts.

4. For the notation and nomenclature reference is made to my *Algebra of Quantics*, Chap. VII.

If $C_r x^{\varpi - r} y^r$ is any term in a covariant of order ϖ of $(a_0, a_1, \dots a_p)(x, y)^p$, C_r being of degree i and weight w in $a_0, a_1, \dots a_p$, so that $ip - 2w = \varpi - 2r = \eta$, say, then

$$\{\Omega O - (r+1)(\varpi - r)\}\, C_r = 0,$$

i.e.

$$\{\Omega O - (r+1)(\eta + r)\}\, C_r = 0 \qquad \text{......................(12)}.$$

Conversely, if this is satisfied then C_r is the coefficient of $x^{\eta + r} y^r$ in a covariant of order $\eta + 2r$, provided $\eta \nless -1$, as we for the present assume.

Now consider the most general gradient (rational integral homogeneous isobaric function) G of type w, i, p, where $ip - 2w = \eta$. Gradients included in G, by proper limitation of the arbitrary numerical multipliers, figure as:

(1) seminvariant coefficients of x^η in covariants of order η, when $\eta \nless 0$,

(2) coefficients of $x^{\eta+1} y$ in covariants of order $\eta + 2$,

(3) .... $x^{\eta+2} y^2$ .... $\eta + 4$,

etc., etc.

$(w+1)$ the coefficient of $x^{\eta+w} y^w$ in the one covariant of order $\eta + 2w = ip$, i.e. in the i-th power of $(a_0, a_1, \dots a_p)(x, y)^p$, arbitrarily multiplied.

By the above these included gradients satisfy respectively, and are in their generality determined by, the corresponding differential equations:

$$\left.\begin{array}{rl}(1) & (\Omega O - 1\,.\,\eta)\,G_1 = 0 \\ (2) & (\Omega O - 2\,.\,\eta + 1)\,G_2 = 0 \\ (3) & (\Omega O - 3\,.\,\eta + 2)\,G_3 = 0 \\ & \text{etc., etc.} \\ (w+1) & (\Omega O - w + 1\,.\,\eta + w)\,G_{w+1} = 0\end{array}\right\} \quad \ldots\ldots\ldots\ldots\ldots(13).$$

Consequently every one of them satisfies

$$(\Omega O - 1\,.\,\eta)(\Omega O - 2\,.\,\eta + 1)\ldots(\Omega O - w + 1\,.\,\eta + w)\,G = 0 \quad \ldots\ldots(14),$$

i.e.

$$O^{w+1}\Omega^{w+1}G = 0 \quad \ldots\ldots\ldots\ldots\ldots\ldots(14')$$

(cf. *Algebra of Quantics*, § 125, Ex. 3).

Now the most general G of type w, i, p satisfies this equation; for it satisfies $\Omega^{w+1}G = 0$, since operation with Ω lowers weight by 1, and there is no gradient of negative weight. Hence, by § 1, we have an identity

$$G = \sum_{1}^{w+1} A_s F_s \quad \ldots\ldots\ldots\ldots\ldots\ldots(15),$$

where, for $s = 1, 2, \ldots w+1$,

$$F_s \equiv (\Omega O - 1\,.\,\eta)\ldots(\Omega O - s - 1\,.\,\eta + s - 2)\,.\,(\Omega O - s + 1\,.\,\eta + s)\ldots(\Omega O - w + 1\,.\,\eta + w)\,G$$

$$\equiv O^{s-1}\Omega^{s-1}\,.\,(\Omega O - s + 1\,.\,\eta + s)\ldots(\Omega O - w + 1\,.\,\eta + w)\,G \quad \ldots\ldots\ldots(16),$$

and

$$(\Omega O - s\,.\,\eta + s - 1)\,F_s = 0 \quad \ldots\ldots\ldots\ldots(17).$$

Also every A_s has a definitely determined numerical value.

Thus we have expressed the general G, of its type, with η zero or positive, as a sum of seminvariants A_1F_1 and other coefficients in covariants. (With $\eta = -1$ the seminvariant part is of course absent.)

In the general $G = \Sigma A_s F_s = \Sigma G_s$, every G_s is general for purposes of (1) to $(w+1)$ above. For if, instead of the general G on the left, we take only the general G_s which is annihilated by $\Omega O - s\,.\,\eta + s - 1$, everything on the right vanishes except the corresponding $A_s F_s$, which must therefore be G_s.

Accordingly the general gradient G of type w, i, p, with $\eta \nless -1$, has been expressed by direct operation as the sum of (1) the general seminvariant G_1 of the type (absent if $\eta = -1$), (2) the general second coefficient G_2 in a covariant led by a seminvariant $\left(\frac{1}{\eta}\Omega G_2\right)$ of type $w-1, i, p$, (3) the general third coefficient G_3 in a covariant led by a seminvariant of type $w-2, i, p$, and so on, and lastly $(w+1)$ a numerical multiple of the $(w+1)$-th coefficient in the i-th power of $(a_0, a_1, \ldots a_p)(x, y)^p$.

A variation of the above reasoning (here omitted) succeeds in expressing the most general G for which $ip - 2w = -\eta'$ is negative as a sum of coefficients in covariants of orders $\eta', \eta' + 2, \ldots ip$.

Separation of a homogeneous function into orthogonal invariants.

5. It has been proved, somewhat circuitously*, that every rational integral function H of degree i in the coefficients of $(a_0, a_1, \dots a_p)(x, y)^p$ satisfies the differential equation

$$\prod_{m=0}^{m=ip} \{O - \Omega - \iota(ip - 2m)\} H = 0 \quad \dots\dots\dots\dots(18),$$

where ι denotes $\sqrt{-1}$. The equation may be written

$$\{(O - \Omega)^2 + (ip)^2\}\{(O - \Omega)^2 + (ip - 2)^2\} \dots \{(O - \Omega)^2 + 2^2\}(O - \Omega) H = 0 \dots (18'),$$

or

$$\{(O - \Omega)^2 + (ip)^2\}\{(O - \Omega)^2 + (ip - 2)^2\} \dots \{(O - \Omega)^2 + 1^2\} H = 0 \dots (18''),$$

according as ip is even or odd.

It follows, by § 1, that every homogeneous H of degree i can by direct operation be expressed as a sum of parts

$$H = H_0 + H_1 + H_2 + \dots + H_{ip} \quad \dots\dots\dots\dots(19),$$

where, for $m = 0, 1, 2, \dots ip$,

$$\{O - \Omega - \iota(ip - 2m)\} H_m = 0 \quad \dots\dots\dots\dots(20).$$

Now this last equation is the condition, necessary and sufficient, for H_m to be an orthogonal invariant of factor $e^{\iota(ip-2m)\theta}$ in the expression of invariancy for the transformation $x = X\cos\theta - Y\sin\theta$, $y = X\sin\theta + Y\cos\theta$, as we learn by employing the infinitesimal transformation (cf. *Proc. Lond. Math. Soc.* loc. cit.). Accordingly every rational integral homogeneous function of degree i can by direct operation be expressed as a sum of orthogonal invariants of the various possible factors for degree i. If H is general, so is every H_m; for if we take the general H_m for H on the left in (19), we obtain H_m alone and complete on the right.

For one of the parts to be an *absolute* orthogonal invariant (for direct turning) it is necessary that ip be even. This being so, the most general absolute orthogonal invariant of degree i is extracted from the most general H in the form

$$H_{\frac{1}{2}ip} = \frac{1}{2^{ip}\{(\frac{1}{2}ip)!\}^2} \cdot \frac{\prod_0^{ip}\{O - \Omega - \iota(ip - 2m)\}}{O - \Omega} H$$
$$= \{(O - \Omega)^2 + (ip)^2\}\{(O - \Omega)^2 + (ip - 2)^2\} \dots \{(O - \Omega)^2 + 2^2\} H \dots (21).$$

Universally, for all the $ip + 1$ values of m, the general H_m which is a part of H is extracted from H as a definite numerical multiple of the result of operating on H with the product of all the operating factors in (18) except the corresponding $O - \Omega - \iota(ip - 2m)$.

* Mr Berry showed (*Proc. Camb. Phil. Soc.* Vol. XIII. pt. II. p. 55) that the equation is satisfied by the leading coefficient in any absolute orthogonal covariant (for direct turning) of order ip, and I subsequently remarked (*Quarterly Journal*, Vol. XXXVII. p. 93) that every rational integral function of degree i in the coefficients is such a leading coefficient. Probably a better order of ideas would be (1) that orthogonal invariants (non-absolute and absolute) of degree i are just as numerous as products of coefficients of that degree, and (2) that every such invariant is annihilated by some one of the operating factors in (18) (cf. *Proc. Lond. Math. Soc.* Vol. XXXIII. pp. 226, etc.).

Sources of ternary covariants and mixed concomitants.

6. In a recent paper (*Proc. Lond. Math. Soc.* Ser. 2, Vol. XI. p. 269) I have shown how by differential operations to exhibit all ternary covariant sources in a double system of coefficients

$$\begin{array}{c} c_{00}, \\ c_{10},\ c_{01}, \\ c_{20},\ c_{11},\ c_{02}, \\ c_{30},\ c_{21},\ c_{12},\ c_{03}, \\ \dots\dots\dots\dots\dots, \end{array}$$

i.e. all rational integral functions of letters contained in this double system, regarded as unending, which are coefficients of highest powers z^{ϖ} of z in covariants of a ternary quantic

$$\sum_{r=0,\ s=0}^{r+s=p} \frac{p\,!}{r\,!\,s\,!\,(p-r-s)\,!} c_{rs} x^r y^s z^{p-r-s} \quad \dots\dots\dots\dots\dots\dots(22)$$

of sufficiently high order p. In connexion with any particular source, p may be any number not less than the greatest value of $r+s$ for any c_{rs} present in that source.

The sources are those homogeneous rational integral functions of the letters c_{rs} which are of constant equal weights throughout in first and second suffixes, and which have the four annihilators

$$\Omega_{yx} \equiv \sum_{r=1}^{r \to \infty} \left\{ c_{r0} \frac{\partial}{\partial c_{r-1,1}} + 2c_{r-1,1} \frac{\partial}{\partial c_{r-2,2}} + \dots + rc_{1,r-1} \frac{\partial}{\partial c_{0r}} \right\} \quad \dots\dots\dots(23),$$

$$\Omega_{xy} \equiv \sum_{r=1}^{r \to \infty} \left\{ rc_{r-1,1} \frac{\partial}{\partial c_{r0}} + (r-1)\, c_{r-2,2} \frac{\partial}{\partial c_{r-1,1}} + \dots + c_{0r} \frac{\partial}{\partial c_{1,r-1}} \right\} \dots(24),$$

$$\Omega_{xz} \equiv \sum_{r=0,\ s=0}^{r+s \to \infty} \left\{ (r+1)\, c_{rs} \frac{\partial}{\partial c_{r+1,s}} \right\} \quad \dots\dots\dots\dots\dots\dots\dots\dots\dots\dots\dots(25),$$

$$\Omega_{yz} \equiv \sum_{r=0,\ s=0}^{r+s \to \infty} \left\{ (s+1)\, c_{rs} \frac{\partial}{\partial c_{r,s+1}} \right\} \quad \dots\dots\dots\dots\dots\dots\dots\dots\dots\dots\dots(26).$$

Let us denote the degree of one of them by i, and its equal weights by q, q.

The fact proved (loc. cit.) was that if $G_{i,q,q}$ is the most general rational integral homogeneous doubly isobaric function of degree i and weights q, q with arbitrary numerical coefficients—say the most general gradient of type i, q, q—then

$$\phi\psi \frac{\partial}{\partial c_{00}} G_{i,q,q} \quad \dots\dots\dots\dots\dots\dots\dots\dots\dots\dots(27)$$

is the most general covariant source of the type, where

$$\psi \equiv c_{00} - (c_{10}\Omega_{xz} + c_{01}\Omega_{yz}) + \frac{1}{1\,.\,2}(c_{20}\Omega^2{}_{xz} + 2c_{11}\Omega_{xz}\Omega_{yz} + c_{02}\Omega^2{}_{yz}) - \dots \ (28),$$

and $$\phi \equiv 1 - \frac{1}{1\,.\,2}\Omega_{xy}\Omega_{yx} + \frac{1}{1\,.\,2^2\,.\,3}\Omega^2{}_{xy}\Omega^2{}_{yx} - \frac{1}{1\,.\,2^2\,.\,3^2\,.\,4}\Omega^3{}_{xy}\Omega^3{}_{yx} + \dots \ (29).$$

The general $\frac{\partial}{\partial c_{00}} G_{i,q,q}$ is of course the general $G_{i-1,q,q}$. $\psi G_{i-1,q,q}$ is the most general gradient of type i, q, q which is annihilated by Ω_{xz} and by Ω_{yz}, and $\phi\psi G_{i-1,q,q}$ is the most general gradient included in this which has Ω_{yx} and Ω_{xy} as well as Ω_{xz} and Ω_{yz} for annihilators. The method of the present paper will now be used to express the whole $\psi G_{i-1,q,q}$ as a sum of parts of which $\phi\psi G_{i-1,q,q}$ is one, while the rest, as well as this, can be interpreted.

The general $\psi G_{i-1,q,q}$, and in fact the general gradient of second weight q, is annihilated by Ω_{yx}^{q+1}; for operation with Ω_{yx} diminishes second weight by 1. We accordingly have

$$\Omega_{xy}^{q+1}\,\Omega_{yx}^{q+1}\,\psi G_{i-1,q,q}=0,$$

i.e. (see § 4),

$$\Omega_{yx}\Omega_{xy}(\Omega_{yx}\Omega_{xy}-1\,.\,2)(\Omega_{yx}\Omega_{xy}-2\,.\,3)\ldots(\Omega_{yx}\Omega_{xy}-q\,.\,q+1)\,\psi G_{i-1,q,q}=0 \quad\ldots(30).$$

Consequently $\psi G_{i-1,q,q}$ is separable, by direct operation as in § 1, into a sum of $q+1$ parts

$$G^{(0)}+G^{(1)}+G^{(2)}+\ldots+G^{(q)},$$

where

$$\Omega_{yx}\Omega_{xy}G^{(0)}=0 \quad\ldots\ldots\ldots\ldots(31),$$

and, for $r=1, 2, \ldots q$,

$$(\Omega_{yx}\Omega_{xy}-r\,.\,r+1)\,G^{(r)}=0 \quad\ldots\ldots\ldots\ldots(32).$$

We have, in fact,

$$G^{(0)}=(-1)^q\frac{1}{1\,.\,2^2\,.\,3^2\ldots q^2\,.\,q+1}(\Omega_{yx}\Omega_{xy}-1\,.\,2)(\Omega_{yx}\Omega_{xy}-2\,.\,3)\ldots$$
$$\ldots(\Omega_{yx}\Omega_{xy}-q\,.\,q+1)\,\psi G_{i-1,q,q} \quad\ldots\ldots(33),$$

and, for $r=1, 2, \ldots q$,

$$G^{(r)}=A_r\Omega_{yx}\Omega_{xy}(\Omega_{yx}\Omega_{xy}-1\,.\,2)\ldots(\Omega_{yx}\Omega_{xy}-\overline{r-1}\,.\,r)\,.\,(\Omega_{yx}\Omega_{xy}-\overline{r+1}\,.\,\overline{r+2})\ldots$$
$$\ldots(\Omega_{yx}\Omega_{xy}-q\,.\,q+1)\,\psi G_{i-1,q,q}$$
$$=A_r\Omega_{xy}^r\Omega_{yx}^r(\Omega_{yx}\Omega_{xy}-\overline{r+1}\,.\,\overline{r+2})(\Omega_{yx}\Omega_{xy}-\overline{r+2}\,.\,\overline{r+3})\ldots$$
$$\ldots(\Omega_{yx}\Omega_{xy}-q\,.\,q+1)\,\psi G_{i-1,q,q} \quad\ldots\ldots(34),$$

where

$$(2r+1)\,A_r=(-1)^{q-r}(q-r)\,!\,(q+r+1)\,!\,.$$

7. The first part $G^{(0)}$ is the most general covariant source of type i, q, q. It is in fact $\phi\psi G_{i-1,q,q}$, otherwise written.

The other q parts $G^{(1)}, G^{(2)}, \ldots G^{(q)}$ specify mixed concomitants. It will be seen that they are, respectively, the most general gradients of type i, q, q which occur as coefficients of $\xi\eta\zeta^0, \xi^2\eta^2\zeta^0, \ldots \xi^q\eta^q\zeta^0$ in covariant sources of the universal concomitant $\xi x+\eta y+\zeta z$ and a ternary quantic (22) jointly, i.e. which occur as coefficients of $\xi\eta, \xi^2\eta^2, \ldots \xi^q\eta^q$ in quantics $(\xi, \eta, \zeta)^2, (\xi, \eta, \zeta)^4, \ldots (\xi, \eta, \zeta)^{2q}$ which present themselves as co-factors with highest powers z^{ϖ} of z in mixed concomitants.

We have, in fact, as in § 4, that $G^{(r)}$ is the middle coefficient in a $2r$-ic covariant of the binary system

$$\left.\begin{array}{c} c_{00} \\ c_{10}x+c_{01}y \\ c_{20}x^2+2c_{11}xy+c_{02}y^2 \\ \ldots\ldots\ldots\ldots \end{array}\right\} \quad\ldots\ldots\ldots\ldots(35).$$

Moreover all the coefficients in this binary covariant have Ω_{xz} and Ω_{yz} for annihilators, by my earlier paper. If we put ξ for y and η for $-x$ in it, we obtain a $2r$-ic contravariant of the binary system, annihilated as such by $\Omega_{yx}+\xi\dfrac{\partial}{\partial\eta}$ and $\Omega_{xy}+\eta\dfrac{\partial}{\partial\xi}$, and also annihilated by Ω_{xz} and Ω_{yz}. This (we will see) is the part free from ζ in a $2r$-ic $(\xi, \eta, \zeta)^{2r}$ obeying differential equations which tell us that it is the ξ, η, ζ co-factor of the highest power z^{ϖ} of z in a mixed concomitant, which can be determined from it by differential operation in the usual manner.

The following order of construction of the $(\xi, \eta, \zeta)^{2r}$ will make the truth of the statements clear. Operate on $\frac{\partial}{\partial c_{00}} G^{(r)} \xi^r \eta^r$ with

$$c_{00} - \left\{c_{10}\left(\Omega_{xz} + \zeta\frac{\partial}{\partial\xi}\right) + c_{01}\left(\Omega_{yz} + \zeta\frac{\partial}{\partial\eta}\right)\right\}$$
$$+ \frac{1}{1.2}\left\{c_{20}\left(\Omega_{xz} + \zeta\frac{\partial}{\partial\xi}\right)^2 + 2c_{11}\left(\Omega_{xz} + \zeta\frac{\partial}{\partial\xi}\right)\left(\Omega_{yz} + \zeta\frac{\partial}{\partial\eta}\right) + c_{02}\left(\Omega_{yz} + \zeta\frac{\partial}{\partial\eta}\right)^2\right\} - \dots \quad \dots\dots\dots(36).$$

The result is a $2r$-ic in ξ, η, ζ. It has the annihilators

$$\Omega_{xz} + \zeta\frac{\partial}{\partial\xi}, \qquad \Omega_{yz} + \zeta\frac{\partial}{\partial\eta} \quad \dots\dots\dots\dots\dots\dots\dots\dots(37),$$

as direct operation at once verifies. It has also $iG^{(r)}\xi^r\eta^r$ for its only term free from ζ; because $\psi\frac{\partial}{\partial c_{00}}G^{(r)} = iG^{(r)}$, in virtue of the annihilation of $G^{(r)}$ by Ω_{xz} and Ω_{yz}. Now operate again on the result with

$$1 - \frac{1}{1.2}\left(\Omega_{xy} + \eta\frac{\partial}{\partial\xi}\right)\left(\Omega_{yx} + \xi\frac{\partial}{\partial\eta}\right) + \frac{1}{1.2^2.3}\left(\Omega_{xy} + \eta\frac{\partial}{\partial\xi}\right)^2\left(\Omega_{yx} + \xi\frac{\partial}{\partial\eta}\right)^2 - \dots \quad \dots\dots\dots(38).$$

The new result, which is a $(\xi, \eta, \zeta)^{2r}$, has the two former annihilators (37), and also the two

$$\Omega_{yx} + \xi\frac{\partial}{\partial\eta}, \qquad \Omega_{xy} + \eta\frac{\partial}{\partial\xi} \quad \dots\dots\dots\dots\dots\dots\dots\dots(39).$$

It has, then, the four annihilators which exactly suffice to express that it is the co-factor of the highest power of z in a mixed concomitant. Also the terms free from ζ in the $(\xi, \eta, \zeta)^{2r}$, being obtained by operation with (38) on $iG^{(r)}\xi^r\eta^r$, compose as stated that contravariant of the binary system (35) which contains the term $G^{(r)}\xi^r\eta^r$, numerically multiplied.

On examining the effect of (36) in its operation on $\frac{\partial}{\partial c_{00}}G^{(r)}$, and remembering from my former paper that, for every m and n,

$$\frac{1}{m!\,n!}\Omega_{xz}^m\,\Omega_{yz}^n\frac{\partial}{\partial c_{00}}G^{(r)} = (-1)^{m+n}\frac{\partial}{\partial c_{mn}}G^{(r)},$$

because $\Omega_{xz}G^{(r)} = 0$ and $\Omega_{yz}G^{(r)} = 0$, we find that the coefficient of the highest power ζ^{2r} of ζ in the result of operation is

$$\left\{c_{rr}\frac{\partial}{\partial c_{00}} + \left(c_{r+1,r}\frac{\partial}{\partial c_{10}} + c_{r,r+1}\frac{\partial}{\partial c_{01}}\right) + \left(c_{r+2,r}\frac{\partial}{\partial c_{20}} + c_{r+1,r+1}\frac{\partial}{\partial c_{11}} + c_{r,r+2}\frac{\partial}{\partial c_{02}}\right) + \dots\right\}G^{(r)} = D_{rr}G^{(r)}, \text{ say}\dots\dots\dots(40).$$

Consequently the coefficient of $\zeta^{2r}z^{\varpi}$ in that ternary mixed concomitant which is provided by $G^{(r)}$ is $\phi D_{rr}G^{(r)}$. It is a gradient of type $i, q+r, q+r$, and, from its formation by means of ϕ, is an invariant of the binary system (35). Call it a mixed concomitant source of type i, $q+r$, $q+r$.

Notice that the source $\phi D_{rr} G^{(r)}$, and the $(\xi, \eta, \zeta)^{2r}$ in which it is the coefficient of ζ^{2r}, are independent of p, the order of our ternary quantic (22), provided this be great enough for every c_{mn} introduced in the source to occur as a coefficient in the p-ic. The lower limit for p is apparently $2q+2r$, but I shall not be surprised if this turns out to be excessive. The order ϖ in x, y, z of the mixed concomitant finally obtained depends, of course, on p, and is known to be $ip-3q-r$.

8. By reasoning as in § 4, every $G^{(r)}$ is general of its type and with its properties when the $G_{i-1,q,q}$ from which we start is general. It is the sum of arbitrary multiples of all the particular $G^{(r)}$s of its type which yield mixed concomitants as above. Remembering this we are provided with the following theorem of enumeration:

The whole number of the linearly independent gradients of type i, q, q that are annihilated by Ω_{xz} and by Ω_{yz} is equal to the sum of (1) *the number of linearly independent covariant sources of type i, q, q, and* (2) *the aggregate of the numbers of the linearly independent sources of the q possible types i, $q+r$, $q+r$ $(r=1, 2, \ldots q)$ of those mixed concomitants of the q corresponding classes (dimensions in ξ, η, ζ) $2r$ which contain terms free from ζ in their co-factors with highest powers z^{ϖ} of z.* [Mixed concomitants without such terms are results of multiplying other concomitants by powers of the universal concomitant $\xi x+\eta y+\zeta z$.]

For all the sources derived from the general $G_{i-1,q,q}$, and referred to in the above enumeration, to be relevant to a p-ic, p has to be not less than a certain number, which may be as great as $4q$.

There is no upper limit to p, and we may take it as infinite. If we do, every $\varpi\,(=ip-3q-r)$ is also infinite. We may look upon our results as affording by direct operations all the finite sources of infinitely continued covariants and mixed concomitants in the perpetuant theory of the infinitely continued

$$c_{00}+(c_{10}x+c_{01}y)+\frac{1}{1\,.\,2}(c_{20}x^2+2c_{11}xy+c_{02}y^2)+\ldots \quad \ldots\ldots\ldots(41).$$

Completeness will have been given to the theory of the enumeration of all linearly independent concomitant sources, when it has been found possible to enumerate gradients of given type with Ω_{xz} and Ω_{yz} for annihilators, i.e. to ascertain the number of arbitraries in $\psi G_{i-1,q,q}$, with $G_{i-1,q,q}$ general of its type. The problem of the enumeration of *irreducible* ternary perpetuant sources appears to be a subsequent one.

ON REGULAR AND IRREGULAR SOLUTIONS OF SOME INFINITE SYSTEMS OF LINEAR EQUATIONS

By Helge von Koch.

1. If the infinite determinant

$$\Delta = \begin{vmatrix} A_{11}, & A_{12}, & \dots \\ A_{21}, & A_{22}, & \dots \\ \dots & \dots & \dots \end{vmatrix}$$

is supposed absolutely convergent*, the system of equations

$$\left.\begin{array}{l} A_{11}x_1 + A_{12}x_2 + \dots = 0 \\ A_{21}x_1 + A_{22}x_2 + \dots = 0 \\ \dots\dots\dots\dots\dots\dots\dots \end{array}\right\} \quad \dots\dots(1)$$

will have properties quite similar to those of a finite system of equations, *provided that the unknown quantities*

$$x_1, \ x_2, \ \dots \quad \dots\dots(2)$$

are subjected to certain additional conditions depending on the nature of the coefficients A_{ik}.

For instance, if Δ is a normal determinant† having a value $\neq 0$, and if the x_k are subjected to the condition

$$|x_k| < \text{Const.} \quad (k = 1, 2, \dots),$$

the only solution of the system will be

$$x_1 = 0, \ x_2 = 0, \ \dots \quad \dots\dots(3);$$

and the corresponding non-homogeneous system

$$\left.\begin{array}{l} A_{11}x_1 + A_{12}x_2 + \dots = C_1 \\ A_{21}x_1 + A_{22}x_2 + \dots = C_2 \\ \dots\dots\dots\dots\dots\dots\dots \end{array}\right\} \quad \dots\dots(4)$$

* At the Scandinavian Congress of Mathematicians in Stockholm 1909, a report was given on the theory of infinite determinants and the resolution of infinite systems of linear equations. See the *Compte rendu* of the Congress (Leipzig, Teubner, 1910).

† This signifies that the product $\Pi_i A_{ii}$ and the sum $\Sigma_i \Sigma_k A_{ik}$ (extended to all i, k verifying the condition $i \neq k$) are absolutely convergent.

will, if the C_k are supposed inferior, in absolute value, to a given constant, have an unique solution

$$x_k = \frac{1}{\Delta}(\alpha_{1k} C_1 + \alpha_{2k} C_2 + \dots) \qquad \text{..........(5)},$$

where α_{ik} denotes the minor of Δ corresponding to the element A_{ik}.

If, on the contrary, no additional conditions for the x_k are introduced, the analogy with finite systems wholly disappears and there will, in general, appear solutions of quite a new kind.

This I propose to illustrate by certain systems of the special type*

$$\left.\begin{array}{r} x_1 + a_{12}x_2 + a_{13}x_3 + \dots = c_1 \\ x_2 + a_{23}x_3 + \dots = c_2 \\ x_3 + \dots = c_3 \\ \dots\dots\dots\dots \end{array}\right\} \qquad \text{..........(6)}.$$

Whatever may be the values of the a_{ik}, the infinite determinant

$$\Delta = \left| \begin{array}{cccc} 1, & a_{12}, & a_{13}, & \dots \\ & 1, & a_{23}, & \dots \\ & & 1, & \dots \\ & & & \dots \end{array} \right| \qquad \text{..........(7)},$$

and all the minors of Δ of any finite order are certainly absolutely convergent. We have $\Delta = 1$ and

$$\alpha_{11} = 1, \ \alpha_{21} = -a_{12}, \ \dots, \ \alpha_{\nu+1.1} = (-1)^\nu \left| \begin{array}{cccc} a_{12}, & a_{13}, & \dots & a_{1\,\nu+1} \\ 1, & a_{23}, & \dots & a_{2\,\nu+1} \\ & 1, & \dots & a_{3\,\nu+1} \\ & \dots & \dots & \dots \\ & & 1, & a_{\nu\,\nu+1} \end{array} \right| \qquad \text{......(8)}$$

and similar formulas for the other minors†.

Supposing now the a_{ik} such as to make the series

$$1 + |a_{i.\,i+1}|^2 + |a_{i.\,i+2}|^2 + |a_{i.\,i+3}|^2 + \dots \qquad \text{..........(9)}$$

convergent and not superior, for any value of i, to a certain positive quantity K^2, it is easily seen, by the application of a well-known theorem of Hadamard, that

$$|\alpha_{\nu+1.\,1}| \leqq K^\nu$$

and, generally,

$$|\alpha_{\nu+k.\,k}| \leqq K^\nu \quad (\nu = 0, 1, 2, \dots; \ k = 1, 2, \dots),$$

and hence it is not difficult to prove the following theorem.

* It may be remarked that a very general class of systems of the type (4) can be, by suitable transformation, reduced to this type.

† These minors will evidently satisfy the relations $\alpha_{\nu\nu} = 1$ and

$$\alpha_{\nu.\,\nu}\, a_{\nu.\,\nu+p} + \alpha_{\nu+1.\,\nu}\, a_{\nu+1.\,\nu+p} + \dots + \alpha_{\nu+p.\,\nu} = 0 \quad (p = 1, 2, \dots),$$

$$\alpha_{\nu+p.\,\nu} + \alpha_{\nu+p.\,\nu+1}\, a_{\nu.\,\nu+1} + \dots + \alpha_{\nu+p.\,\nu+p}\, a_{\nu.\,\nu+p} = 0 \quad (p = 1, 2, \dots).$$

THEOREM 1. *If in* (6) *the* c_k *as well as the unknown quantities are subjected to the conditions**

$$\left.\begin{aligned}\overline{\lim}\sqrt[n]{|c_n|} < \frac{1}{K}\\ \overline{\lim}\sqrt[n]{|x_n|} < \frac{1}{K}\end{aligned}\right\} \qquad \text{..............................(10)},$$

the system will have for unique solution

$$x_k = c_k + \alpha_{k+1.k}\, c_{k+1} + \alpha_{k+2.k}\, c_{k+2} + \ldots\dagger \quad (k = 1, 2, \ldots) \text{.........(11)}.$$

* According to the notation proposed by Prof. Pringsheim, $\overline{\lim}\, c$ denotes the greatest limiting value of c.

† Demonstration: Supposing x_k a solution with the property (10), we can choose a positive number $\epsilon < 1$ and an integer n_1 such that

$$|x_n| < \left(\frac{\epsilon}{K}\right)^n \text{ for } n \geqq n_1.$$

Hence

$$|x_n| + |a_{n.n+1}|\,|x_{n+1}| + |a_{n.n+2}|\,|x_{n+2}| + \ldots$$
$$< \left(\frac{\epsilon}{K}\right)^n \left(1 + |a_{n.n+1}|\frac{\epsilon}{K} + |a_{n.n+2}|\left(\frac{\epsilon}{K}\right)^2 + \ldots\right)$$
$$< K_1 . \left(\frac{\epsilon}{K}\right)^n \text{ for } n \geqq n_1,$$

K_1 being a certain positive constant.

Hence the series

$$\sum_{\nu=n}^{\infty} |\alpha_{\nu.n}|\,(|x_\nu| + |a_{\nu.\nu+1}|\,|x_{\nu+1}| + \ldots)$$

is convergent and we are justified in writing

$$\sum_{\nu=n}^{\infty} \alpha_{\nu n}\,(x_\nu + a_{\nu.\nu+1}\, x_{\nu+1} + \ldots)$$
$$= \alpha_{nn}\, x_n$$
$$+ (\alpha_{nn} a_{n.n+1} + \alpha_{n+1.n})\, x_{n+1}$$
$$+ (\alpha_{nn} a_{n.n+2} + \alpha_{n+1.n} a_{n+1.n+2} + \alpha_{n+2.n})\, x_{n+2}$$
$$+ \ldots$$
$$= x_n \text{ (cf. note *, p. 352)}.$$
$$\therefore\ x_n = \sum_{\nu=n}^{\infty} \alpha_{\nu.n}\, c_\nu,$$

which proves (11) to be a *necessary* form for x_k.

Supposing next x_k defined by (11); we easily find a positive $\epsilon < 1$ and an integer n_1 such that (for $n \geqq n_1$)

$$\lim \sqrt[n]{\bar{x}_n} < \frac{\epsilon}{K},$$

where $\bar{x}_k$ denotes the expression obtained by taking in (11) every term with its absolute value. Hence

$$\bar{x}_\nu + |a_{\nu.\nu+1}|\,\bar{x}_{\nu+1} + |a_{\nu.\nu+2}|\,\bar{x}_{\nu+2} + \ldots$$

is convergent and

$$x_\nu + a_{\nu.\nu+1} x_{\nu+1} + a_{\nu.\nu+2} x_{\nu+2} + \ldots$$

can be written in the form

$$c_\nu + c_{\nu+1}\,(\alpha_{\nu+1.\nu} + \alpha_{\nu+1.\nu+1} a_{\nu.\nu+1})$$
$$+ c_{\nu+2}\,(\alpha_{\nu+2.\nu} + \alpha_{\nu+2.\nu+1} a_{\nu.\nu+1} + \alpha_{\nu+2.\nu+2} a_{\nu.\nu+2})$$
$$+ \ldots$$

which is $= c_\nu$ (cf. note *, p. 352). Thus the theorem is proved.

As an immediate consequence of this we have:

THEOREM 2. *The only solution of the homogeneous system*

$$\left.\begin{array}{r} x_1 + a_{12}x_2 + a_{13}x_3 + \dots = 0 \\ x_2 + a_{23}x_3 + \dots = 0 \\ x_3 + \dots = 0 \\ \dots\dots\dots\dots \end{array}\right\} \quad \dots\dots\dots\dots(12)$$

which verifies the additional condition

$$\overline{\lim}\sqrt[n]{|x_n|} < \frac{1}{K} \quad \dots\dots\dots\dots(12')$$

is

$$x_1 = 0, \; x_2 = 0, \; \dots \quad \dots\dots\dots\dots(13).$$

As the solutions thus found correspond to those of a similar finite system, I will in the following call them *regular* solutions. Thus the regular solution of (12) is (13) and that of (6) is (11). All other solutions of the same systems shall be called *irregular* solutions.

By a suitable change of the constant $\frac{1}{K}$, the above results will remain true if instead of the conditions imposed on the series (9) we only suppose that a_{ik} satisfies the condition

$$|a_{ik}| \leqq T^{k-i} \quad (i, k = 1, 2, \dots) \quad \dots\dots\dots\dots(9'),$$

T being a given positive constant. For, taking a constant $S < \frac{1}{T}$ and writing (6) under the form

$$\bar{x}_i + \bar{a}_{i.\,i+1}\,\bar{x}_{i+1} + \bar{a}_{i.\,i+2}\,\bar{x}_{i+2} + \dots = \bar{c}_i \quad (i = 1, 2, \dots),$$

where

$$\bar{a}_{ik} = a_{ik}S^{k-i}, \quad \bar{c}_i = \frac{c_i}{S^i}, \quad \bar{x}_i = \frac{x_i}{S^i},$$

the new coefficients $\bar{a}_{ik}$ will evidently satisfy the conditions above imposed on (9).

Theorems 1 and 2 also remain true if in (10) and (12′) we substitute for $\frac{1}{K}$ a value $C < \frac{1}{K}$, and in many cases they may remain true even for values $C > \frac{1}{K}$. For any given system of a_{ik} which satisfies (9′) there exists, according to a fundamental theorem of Weierstrass, an upper limit $\bar{C} > 0$ of all values C such that

(1) each series $x_i + a_{i.\,i+1}x_{i+1} + \dots$ converges if

$$\overline{\lim}\sqrt[n]{|x_n|} < C;$$

(2) Theorem 2 remains true if $\frac{1}{K}$ is replaced by C.

From the definition of $\bar{C}$ we hence derive the following results.

If $\bar{C}$ is finite, any solution of (12) which satisfies the condition

$$\overline{\lim}\sqrt[n]{|x_n|} < \bar{C} \quad \dots\dots\dots\dots(12'')$$

must be identical with the regular solution

$$x_1 = 0, \; x_2 = 0, \; \dots.$$

At the same time, if ϵ is an arbitrarily small positive quantity, there exists at least one system of values

$$x_1, x_2, \ldots$$

satisfying the condition $$\bar{C} \leqq \overline{\lim} \sqrt[n]{|x_n|} < \bar{C} + \epsilon,$$

and such as either to satisfy the equations (12) or to render one or several of the series in the left members divergent.

If $\bar{C} = \infty$, any solution of (12) for which $\overline{\lim} \sqrt[n]{|x_n|}$ is finite must be identical with the solution $x_1 = 0, x_2 = 0, \ldots$; and if (12) admits another solution x_k we must have

$$\overline{\lim} \sqrt[n]{|x_n|} = \infty .$$

Corresponding results may be obtained for the non-homogeneous system (6).

An important simplification will be obtained if in addition to (9′) we suppose, for instance, that the power-series

$$1 + a_{i.\,i+1} t + a_{i.\,i+2} t^2 + \ldots \quad (i = 1, 2, \ldots)$$

converges for any finite value of t. For then the series

$$x_i + a_{i.\,i+1} x_{i+1} + a_{i.\,i+2} x_{i+2} + \ldots$$

will certainly converge for any system of values x_k such that $\overline{\lim} \sqrt[n]{|x_n|}$ is finite and we conclude:

THEOREM 3. *There exists a positive limit $\bar{C}$ (finite or infinite) which divides the solutions of system* (12) *in the following manner:*

If $\bar{C}$ is finite, any solution of (12) *satisfying* (12″) *must be identical with the regular solution $x_1 = 0, x_2 = 0, \ldots$.*

At the same time, if ϵ is an arbitrarily small positive quantity, the system (12) *will admit of at least one irregular solution satisfying the condition*

$$\bar{C} \leqq \overline{\lim} \sqrt[n]{|x_n|} < \bar{C} + \epsilon.$$

If $\bar{C}$ is infinite, any solution of (12) *which is not identical with the solution $x_1 = 0, x_2 = 0, \ldots$, must satisfy the condition*

$$\overline{\lim} \sqrt[n]{|x_n|} = \infty .$$

For the study of the above systems, the determination of the limit $\bar{C}$ evidently is of great importance. In general this determination seems very complicated but I will point out some interesting cases where the value of $\bar{C}$ is easily obtained.

2. Consider the system

$$\left.\begin{aligned} x_1 + c_1 x_2 + c_2 x_3 + c_3 x_4 + \ldots &= 0 \\ x_2 + c_1 x_3 + c_2 x_4 + \ldots &= 0 \\ x_3 + c_1 x_4 + \ldots &= 0 \\ \ldots\ldots\ldots\ldots\ldots\ldots & \end{aligned}\right\} \quad \ldots\ldots\ldots\ldots\ldots\ldots\ldots(14),$$

the coefficients c_k being supposed such that the Taylor-series

$$F(t) = 1 + c_1 t + c_2 t^2 + \ldots \quad \ldots\ldots\ldots\ldots\ldots\ldots\ldots\ldots(15)$$

is convergent in a certain circle with radius $R > 0$* and that $F(t)$ has one or several zeros in this circle. I shall denote by ξ the least possible absolute value of these zeros. Then we have

$$0 < \xi < R.$$

It is now easy to prove the following

THEOREM. *Every solution x_k of (14) such that*

$$\overline{\lim} \sqrt[n]{|x_n|} < \xi \quad \text{.............................(16)}$$

must be identical with the regular solution

$$x_1 = 0, \ x_2 = 0, \ \ldots;$$

at the same time, there are irregular solutions satisfying the condition

$$\overline{\lim} \sqrt[n]{|x_n|} = \xi \quad \text{.............................(16').}$$

Thus the exact value of the above-mentioned number $\overline{C}$ is in this case $= \xi$.

Demonstration. Suppose x_k to be a solution of (14) satisfying the condition (16). Then the series

$$G\left(\frac{1}{t}\right) = \frac{x_1}{t} + \frac{x_2}{t^2} + \frac{x_3}{t^3} + \ldots$$

will certainly converge for $|t| > R_1$ and thus the function

$$F(t)\, G\left(\frac{1}{t}\right) \quad \text{..............................(17)}$$

will be regular and uniform in the region

$$R_1 < |t| < R,$$

where

$$R_1 = \overline{\lim} \sqrt[n]{|x_n|} < \xi.$$

According to a theorem of Weierstrass (17) can then be developed in a Laurent-series, proceeding after positive and negative powers of t. The coefficients of the negative powers will all vanish because of equations (14) and thus $F(t)\, G\left(\frac{1}{t}\right)$ will be a regular function in the neighbourhood of $t = 0$. This is, $F(t)$ being $\neq 0$ for $|t| \leqq R_1$, impossible unless $G\left(\frac{1}{t}\right)$ vanishes identically:

$$x_1 = 0, \ x_2 = 0, \ \ldots$$

which proves the first part of the theorem.

To prove the second part we denote by ρ the root (or one of the roots) satisfying the condition

$$|\rho| = \xi.$$

Then putting in (14)

$$x_1 = 1, \quad x_2 = \rho, \quad x_3 = \rho^2, \ldots \quad \text{......................(18),}$$

the left-hand members will all converge absolutely and all be $= 0$. Thus (18) is an irregular solution of (14) satisfying the condition (16').

* Should $F(t)$ converge for every value of t we put $R = \infty$.

Denoting by $1, \quad \alpha_1, \quad \alpha_2, \quad \alpha_3, \ldots$

those minors of the determinant

$$\begin{vmatrix} 1, & c_1, & c_2, & c_3, \ldots \\ & 1, & c_1, & c_2, \ldots \\ & & 1, & c_1, \ldots \\ & & & \ldots \end{vmatrix}$$

which correspond to the first column so that

$$\alpha_k = (-1)^k \begin{vmatrix} c_1, & c_2, & \ldots & c_{k-1} \\ 1, & c_1, & \ldots & c_{k-2} \\ & 1, & \ldots & c_{k-3} \\ & & 1, & c_1 \end{vmatrix} \quad \ldots\ldots\ldots\ldots\ldots\ldots\ldots\ldots(19),$$

it appears immediately that these quantities satisfy the equations

$$\left.\begin{aligned} &c_1 + \alpha_1 = 0 \\ &c_2 + c_1\alpha_1 + \alpha_2 = 0 \\ &\ldots\ldots\ldots\ldots\ldots\ldots \\ &c_k + c_{k-1}\alpha_1 + \ldots + \alpha_k = 0 \\ &\ldots\ldots\ldots\ldots\ldots\ldots\ldots\ldots \end{aligned}\right\} \quad \ldots\ldots\ldots\ldots\ldots\ldots\ldots\ldots(20),$$

and the condition $\overline{\lim} \sqrt[n]{|\alpha_n|} = \dfrac{1}{\xi}$ ....(21).

For $1, \alpha_1, \alpha_2, \ldots$ will be the coefficients in the development of $\dfrac{1}{F(t)}$ accordi to ascending powers of t and the radius of convergence of this development is since this is the absolute value of the zero of $F(t)$ which is nearest origin.

By this remark it is easy to obtain the following theorem.

Let $a_0, a_1, a_2, \ldots$ be given quantities such as to satisfy the condition

$$\overline{\lim} \sqrt[n]{|a_n|} < \xi,$$

ξ being defined as before by means of the given quantities $1, c_1, c_2, \ldots$. Then if x_1, x_2 is a solution of the system

$$\left.\begin{aligned} x_1 + c_1 x_2 + c_2 x_3 + \ldots &= a_0 \\ x_2 + c_1 x_3 + \ldots &= a_1 \\ x_3 + \ldots &= a_2 \\ \ldots\ldots \end{aligned}\right\} \quad \ldots\ldots\ldots\ldots\ldots\ldots\ldots\ldots(22)$$

satisfying the additional condition

$$\overline{\lim} \sqrt[n]{|x_n|} < \xi,$$

this solution is identical with the regular solution

$$\left.\begin{aligned} x_1 &= a_0 + \alpha_1 a_1 + \alpha_2 a_2 + \ldots \\ x_2 &= \phantom{a_0 + {}} a_1 + \alpha_1 a_2 + \ldots \\ x_3 &= \phantom{a_0 + \alpha_1 a_1 + {}} a_2 + \ldots \\ &\ldots\ldots\ldots\ldots\ldots\ldots\ldots\ldots \end{aligned}\right\} \quad \ldots\ldots\ldots\ldots\ldots\ldots\ldots(23);$$

at the same time there exists a singular solution $y_1, y_2, \ldots$ *such that*

$$\overline{\lim} \sqrt[n]{|y_n|} = \xi.$$

Thus even in this case ξ will be the barrier between the regular and the irregular solutions of the system*.

3. As an application of these results we take

$$F(x) = \frac{e^{hx}-1}{hx} = 1 + \frac{hx}{\underline{|2}} + \frac{h^2x^2}{\underline{|3}} + \ldots,$$

$$c_1 = \frac{h}{\underline{|2}}, \qquad c_2 = \frac{h^2}{\underline{|3}}, \ldots.$$

$$\therefore\ \xi = \frac{2\pi}{H} \qquad (H = |h|),$$

so that system (14) takes the form

$$\left.\begin{array}{r} x_1 + \dfrac{h}{\underline{|2}} x_2 + \dfrac{h^2}{\underline{|3}} x_3 + \dfrac{h^3}{\underline{|4}} x_4 + \ldots = 0 \\ x_2 + \dfrac{h}{\underline{|2}} x_3 + \dfrac{h^2}{\underline{|3}} x_4 + \ldots = 0 \\ x_3 + \dfrac{h}{\underline{|2}} x_4 + \ldots = 0 \\ \ldots\ldots\ldots \end{array}\right\} \quad \ldots\ldots\ldots\ldots\ldots\ldots(24).$$

This system, as well known, is nearly connected with Euler-Maclaurin's sum-formula.

It evidently possesses the irregular solution

$$x_1 = 1, \qquad x_2 = \frac{2\pi i}{h}, \qquad x_3 = \left(\frac{2\pi i}{h}\right)^2, \ldots \qquad \ldots\ldots\ldots\ldots\ldots(25),$$

* The systems (14) and (22) will be of quite a different nature if the function (15) has no zero-points within the circle of convergence $|t| < R$. In this case it is easy to prove that $\overline{C} = R$. Thus if $R = \infty$, any solution of (14) satisfying the condition

$$\overline{\lim} \sqrt[n]{|x_n|} < \text{finite constant}$$

must be identical with the regular solution

$$x_1 = 0, \ x_2 = 0, \ \ldots.$$

Taking, for example,

$$F(x) = e^x,$$

$$c_1 = 1, \ c_2 = \frac{1}{\underline{|2}}, \ c_3 = \frac{1}{\underline{|3}}, \ \ldots,$$

(22) will be identical with a system considered by E. Borel (*Annales de l'Ecole Normale*, Sér. III, t. 12, p. 3; t. 13, p. 79; see also *Comptes Rendus*, t. 124, p. 673 (Paris, 1897)), which is of great importance in his researches concerning the representation of functions of real variables. The method given by Borel furnishes an infinity of solutions of the considered system for any values of the quantities a_k (these solutions are, according to our terminology, *irregular*) and applies to a general class of systems which contains (22) as a special case.

and according to the results obtained (No. 2) we can state that *a solution* $x_k\ (k=1,2,\ldots)$ *of* (24) *cannot have the property*

$$\overline{\lim}\sqrt[n]{|x_n|}<\frac{2\pi}{H},$$

unless

$$x_1=x_2=\ldots=0.$$

From this result it is easy to deduce the following remark concerning periodic functions:

If the series

$$k_1x+\frac{k_2}{\underline{|2}}x^2+\frac{k_3}{\underline{|3}}x^3+\ldots \qquad (26)$$

represents a periodic function $\phi(x)$ *with the period* h, $|h|=H$ *being supposed inferior to the radius of convergence of the series, there are only two cases possible, viz.*:

$$(1)\quad \overline{\lim}\sqrt[n]{|k_n|}\geqq\frac{2\pi}{H}.$$

$$(2)\quad k_1=k_2=\ldots=0.$$

To prove this, suppose

$$\overline{\lim}\sqrt[n]{|k_n|}<\frac{2\pi}{H} \qquad (27),$$

and

$$\phi(x+h)=\phi(x).$$

Hence we have

$$\phi(h)=\phi(0),\quad \phi'(h)=\phi'(0),\quad \phi''(h)=\phi''(0)\text{ etc.}$$

or

$$\frac{h}{\underline{|1}}\phi'(0)+\frac{h^2}{\underline{|2}}\phi''(0)+\frac{h^3}{\underline{|3}}\phi'''(0)+\ldots=0,$$

$$\frac{h}{\underline{|1}}\phi''(0)+\frac{h^2}{\underline{|2}}\phi'''(0)+\ldots=0,$$

$$\frac{h}{\underline{|1}}\phi'''(0)+\ldots=0,$$

$$\ldots\ldots\ldots,$$

so that

$$x_1=\phi'(0)=k_1,\quad x_2=\phi''(0)=k_2,\ldots$$

is a solution of system (24). Hence because of (27) and of the preceding result:

$$k_1=k_2=\ldots=0,$$

which proves the statement.

Evidently the limit $\frac{2\pi}{H}$ given by this remark is a real minimum, for by the function

$$e^{\frac{2\pi ix}{h}}-1=k_1x+\frac{k_2}{\underline{|2}}x^2+\frac{k_3}{\underline{|3}}x^3+\ldots$$

this limit is attained:

$$\overline{\lim}\sqrt[n]{|k_n|}=\frac{2\pi}{H}.$$

From this point of view, the exponential function appears as a limiting-case of all periodic functions with the same period.

It has already been remarked that (25) represents an irregular solution of (24).

It is now easy to form an infinity of such solutions and a general formula embracing them*.

Put
$$u = e^{\frac{2\pi i x}{h}},$$
$$\phi(x) = \sum_{\lambda=-\infty}^{+\infty} K_\lambda u^\lambda,$$
with the conditions
$$\overline{\lim_{\lambda=+\infty}} \sqrt[\lambda]{|K_\lambda|} < e^{-2\pi}, \qquad \overline{\lim_{\lambda=+\infty}} \sqrt[\lambda]{|K_{-\lambda}|} < e^{-2\pi} \quad \text{.........(28).}$$

Then $\phi(x)$ will be a periodic function (period $= h$) which is regular within a circle of radius $> |h|$ and with $x = 0$ as centre. Thus
$$X_\nu = \left(\frac{2\pi i}{h}\right)^\nu \sum_{\lambda=-\infty}^{+\infty} K_\lambda \lambda^\nu \qquad (\nu = 1, 2, \ldots) \quad \text{..............(29)}$$
will be a solution of the system (24). One at least of the K_λ being supposed $\neq 0$, it follows that the function
$$X_1 x + X_2 x^2 + \ldots$$
can not vanish identically. Hence, by our general result we must have
$$\overline{\lim} \sqrt[n]{|X_n|} \geqq \frac{2\pi}{H} \quad \text{..............................(30),}$$
and our X_ν form an irregular solution of (24). This solution contains an infinite number of arbitrary constants, viz.
$$K_\lambda \quad (\lambda = -\infty \ldots +\infty).$$

Remark. The formula (30) may be written thus:
$$\overline{\lim_{\nu=+\infty}} \left| \sum_{\lambda=-\infty}^{+\infty} K_\lambda \lambda^\nu \right|^{\frac{1}{\nu}} \geqq 1,$$
the K_λ being supposed to satisfy (28). As a very particular consequence of this we infer that the infinite system
$$\sum_{\lambda=-\infty}^{+\infty} K_\lambda \lambda^\nu = 0 \qquad (\nu = 0, 1, 2, \ldots)$$
can not admit any solution with the property (28) other than the evident solution
$$K_\lambda = 0 \quad (\lambda = -\infty \ldots +\infty).$$

4. Considering instead of (24) the non-homogeneous system
$$\left.\begin{aligned} x_1 + \frac{h}{\lfloor 2} x_2 + \frac{h^2}{\lfloor 3} x_3 + \frac{h^3}{\lfloor 4} x_4 + \ldots &= a_0 \\ x_2 + \frac{h}{\lfloor 2} x_3 + \frac{h^2}{\lfloor 3} x_4 + \ldots &= a_1 \\ x_3 + \frac{h}{\lfloor 2} x_4 + \ldots &= a_2 \\ \ldots\ldots\ldots & \end{aligned}\right\} \quad \text{.................(31),}$$

* Cf. a research of P. Stäckel published in *Festschrift*, Heinrich Weber gewidnet (Leipzig, 1912), and concerning a system identical with (24). (This research was unknown to the author of the present paper until about two months after it had been sent to the General Secretary.)

and supposing that $$\overline{\lim}\sqrt[n]{|a_n|} < \frac{2\pi}{H},$$

we arrive at the following result.

The formula

$$x_k = a_{k-1} + \alpha_1 a_k + \alpha_2 a_{k+1} + \dots \qquad (k = 1, 2, \dots) \qquad \dots\dots\dots\dots(32)$$

gives a solution of (31) *with the property*

$$\overline{\lim}\sqrt{|x_n|} < \frac{2\pi}{H};$$

and any solution having this property must be identical with (32).

Here $$\alpha_1 = -\frac{h}{\underline{|2}}, \qquad \alpha_2 = \frac{h^2}{\underline{|2}}, \dots,$$

$$\alpha_k = (-1)^k h^k \begin{vmatrix} \frac{1}{\underline{|2}}, & \frac{1}{\underline{|3}}, & \frac{1}{\underline{|4}} \cdots & \frac{1}{\underline{|k}} \\ 1, & \frac{1}{\underline{|2}}, & \frac{1}{\underline{|3}} \cdots & \frac{1}{\underline{|k-1}} \\ & 1, & \frac{1}{\underline{|2}} \cdots & \frac{1}{\underline{|k-2}} \\ & & \dots\dots & \dots \\ & & 1, & \frac{1}{\underline{|2}} \end{vmatrix},$$

or

$$\left.\begin{aligned} &\alpha_1 = -\frac{h}{\underline{|2}}, \quad \alpha_{2k+1} = 0 \quad (k = 1, 2, \dots) \\ &\alpha_{2k} = (-1)^{k-1} \frac{B_k}{\underline{|2k}} h^{2k} \quad (k = 1, 2, \dots) \end{aligned}\right\} \qquad \dots\dots\dots\dots(33),$$

where $B_1, B_2, B_3, \dots$ is the suite of Bernoulli's numbers:

$$B_1 = \frac{1}{6}, \qquad B_2 = \frac{1}{30}, \qquad B_3 = \frac{1}{42}, \dots,$$

so that the first of formulas (32) may be written thus:

$$x_1 = a_0 - \frac{h}{2} a_1 + \frac{B_1}{\underline{|2}} h^2 a_2 - \frac{B_2}{\underline{|4}} h^4 a_4 + \frac{B_3}{\underline{|6}} h^6 a_6 - \dots \qquad \dots\dots\dots\dots(34).$$

Applications. Suppose

$$f(x) = f(0) + \frac{x}{\underline{|1}} f'(0) + \frac{x^2}{\underline{|2}} f''(0) + \dots \qquad \dots\dots\dots\dots(35)$$

to be an *integral* function such that

$$\overline{\lim}\sqrt[n]{|f^{(n)}(0)|} < \frac{2\pi}{P} \qquad \dots\dots\dots\dots(36),$$

P being a given positive quantity. Then Euler-Maclaurin's series

$$\Delta f(x) - \frac{h}{2}\Delta f'(x) + \frac{B_1}{\underline{|2}} h^2 \Delta f''(x) + \dots \qquad \dots\dots\dots\dots(37),$$

where
$$\Delta f(x) = f(x+h) - f(x),$$
$$\Delta f'(x) = f'(x+h) - f'(x),$$
$$\dots\dots\dots\dots\dots\dots$$

will converge for
$$|h| \leqq P \quad \dots\dots\dots\dots(38)$$
and for any values of x.

For we have*
$$\overline{\lim} \sqrt[2n]{\frac{B_n}{\lfloor 2n}} = \frac{1}{2\pi} \quad \dots\dots\dots\dots(39),$$

and according to (36) we can choose a positive number $\epsilon < 1$ and an integer n_1 such that
$$|f^{(n)}(0)| < \left(\frac{2\pi\epsilon}{P}\right)^n \text{ for } n \geqq n_1,$$

and this gives—x being supposed less than a given constant R in absolute value—
$$|f^{(n)}(x)| < \left(\frac{2\pi\epsilon}{P}\right)^n \left(1 + \frac{2\pi\epsilon R}{P} + \frac{1}{\lfloor 2}\left(\frac{2\pi\epsilon R}{P}\right)^2 + \dots\right) \text{ for } n \geqq n_1,$$

and thus
$$|\Delta f^{(n)}(x)| < \left(\frac{2\pi\epsilon}{P}\right)^{n+1} . K,$$
where K is a certain constant.

These formulas also prove that, under the condition (38), the series (37) will converge absolutely and uniformly in any finite portion of the plane of x.

It is easy to see that *the series* (37) *is equal to* $h . f'(x)$ *for any value of* x.

For we have
$$f'(x) + \frac{h}{\lfloor 2} f''(x) + \frac{h^2}{\lfloor 3} f'''(x) + \dots = \frac{\Delta f(x)}{h},$$
$$f''(x) + \frac{h}{\lfloor 2} f'''(x) + \dots = \frac{\Delta f'(x)}{h},$$
$$\dots\dots\dots\dots\dots\dots,$$

and as in this system the right-hand members and the $f^{(n)}(x)$ have the property
$$\overline{\lim} \sqrt[n]{\frac{\Delta f^{(n)}(x)}{h}} < \frac{2\pi}{P} \leqq \frac{2\pi}{|h|},$$
$$\overline{\lim} \sqrt[n]{|f^{(n)}(x)|} < \frac{2\pi}{P} \leqq \frac{2\pi}{|h|},$$
we must have by (34)
$$hf'(x) = \Delta f(x) - \frac{h}{2}\Delta f'(x) + \frac{B_1 h^2}{\lfloor 2}\Delta f''(x) - \dots + (-1)^{n-1}\frac{B_n h^{2n}}{\lfloor 2n}\Delta f^{(2n)}(x) + \dots \quad \dots(40).$$

Next consider the functional equation
$$\phi(x+h) = \phi(x) + C(x) \quad \dots\dots\dots\dots(41),$$

* By the well-known formula:
$$B_n = 2 . \frac{\lfloor 2n}{(2\pi)^{2n}} . \left(1 + \frac{1}{2^{2n}} + \frac{1}{3^{2n}} + \dots\right)$$
or by (21).

$\phi(x)$ denoting the unknown function, h a given constant and $C(x)$ a given integral function with the property*

$$\overline{\lim} \sqrt[n]{|C^{(n)}(0)|} < \frac{2\pi}{H} \qquad (H = |h|) \quad \text{.................(42).}$$

Then if in (31) we put

$$a_0 = C(x), \qquad a_1 = C'(x), \qquad a_2 = C''(x), \ldots,$$
$$x_1 = h\phi'(x), \quad x_2 = h\phi''(x), \quad x_3 = h\phi'''(x), \ldots,$$

this system will be satisfied if $\phi(x)$ is a solution of (41). Thus if we impose on $\phi(x)$ the condition

$$\overline{\lim} \sqrt[n]{|\phi^{(n)}(0)|} < \frac{2\pi}{H} \quad \text{..........................(43),}$$

from which it follows that

$$\overline{\lim} \sqrt[n]{|\phi^{(n)}(x)|} < \frac{2\pi}{H} \quad \text{..........................(44)}$$

for every finite value of x, the general result obtained (formula 34) proves that

$$h\phi'(x) = C(x) - \frac{h}{2}C'(x) + \frac{B_1 h^2}{\lfloor 2}C''(x) - \ldots + (-1)^{n-1}\frac{B_n h^{2n}}{\lfloor 2n}C^{(2n)}(x) + \ldots \quad \text{...(45).}$$

As the right-hand member converges uniformly in any finite region of the x-plane we thus obtain $\phi(x)$ by integrating term by term this expression and we arrive at the conclusion that the solution defined by (45) is *the only solution of* (41) *satisfying the additional condition* (43)†.

5. In a similar manner we may study the functional equation,

$$\phi(x+h) - \phi(x) - hf'(x) = C(x)$$

or, more generally,

$$\phi(x+h) - \phi(x) - h\phi'(x) - \frac{h^2}{\lfloor 2}\phi''(x) - \ldots - \frac{h^m}{\lfloor m}\phi^{(m)}(x) = C(x).$$

This problem will lead to a system of equations of the form

$$x_1 + \frac{h}{m+2}x_2 + \frac{h^2}{(m+2)(m+3)}x_3 + \frac{h^3}{(m+2)(m+3)(m+4)}x_4 + \ldots = a_0,$$
$$x_2 + \frac{h}{m+2}x_3 + \frac{h^2}{(m+2)(m+3)}x_4 + \ldots = a_1,$$
$$x_3 + \frac{h}{m+2}x_4 + \ldots = a_2,$$
$$\ldots\ldots\ldots\ldots,$$

* As remarked above, it follows from this property that

$$\overline{\lim} \sqrt[n]{|C^{(n)}(x)|} < \frac{2\pi}{H}$$

for any given value of x.

† C. Guichard has proved (*Annales de l'Ecole Normale*, Sér. III, t. 4, p. 285, 1887. Cf. Hurwitz, *Acta Mathematica*, t. 20, p. 985) that if $C(x)$ is an arbitrary integral function, equation (41) will always be satisfied by an integral function $\phi(x)$. Thus the values

$$h\phi'(0), \quad h\phi''(0), \quad h\phi'''(0),$$

will form an *irregular* solution of the corresponding system (31) if in (42) the sign $<$ is replaced by $>$.

and if we put

$$\frac{h^{m+1}x^{m+1}}{\lfloor m+1} \frac{1}{e^{hx}-1-\frac{hx}{\lfloor 1}-\dots-\frac{h^m x^m}{\lfloor m}} = 1 + g_1 x + g_2 x^2 + \dots,$$

it will follow that $1, \quad g_1, \quad g_2, \dots$

are the minors belonging to the first column of the determinant

$$\begin{vmatrix} 1, & \frac{h}{m+2}, & \frac{h^2}{(m+2)(m+3)}, & \dots \\ & 1, & \frac{h}{m+2}, & \dots \\ & & 1, & \dots \\ & & & \dots \end{vmatrix}.$$

In consequence, in order to have the division between the regular and the irregular solutions, we must introduce instead of 2π the absolute value ξ of the root of the equation

$$e^x - 1 - \frac{x}{\lfloor 1} - \dots - \frac{x^m}{\lfloor m} = 0,$$

which is nearest $x = 0$.

ON THE FUNCTIONS ASSOCIATED WITH THE ELLIPTIC CYLINDER IN HARMONIC ANALYSIS

By E. T. Whittaker.

1. *Introduction.*

It is well-known that the solution of the wave-equation

$$\frac{\partial^2 V}{\partial x^2} + \frac{\partial^2 V}{\partial y^2} = k^2 \frac{\partial^2 V}{\partial t^2}$$

for circular bodies, or of the potential-equation

$$\frac{\partial^2 V}{\partial x^2} + \frac{\partial^2 V}{\partial y^2} + \frac{\partial^2 V}{\partial z^2} = 0$$

for circular-cylindrical distributions, leads to the functions of Bessel: in the same way, the solution of these equations for elliptic bodies or elliptic-cylindrical distributions leads to the "elliptic-cylinder functions," which are defined by the differential equation

$$\frac{d^2y}{dz^2} + (a + k^2 \cos^2 z)\, y = 0 \quad \text{.........................(1)},$$

where a and k denote constants*.

One reason for the importance of these elliptic-cylinder functions lies in the fact that they are not, like the functions of Legendre and Bessel, mere particular or degenerate cases of the hypergeometric function: the differential equation (1) is, indeed, the equation which most naturally presents itself for study, in the theory of linear differential equations, when the hypergeometric equation has been disposed of. It has a further practical interest in connexion with Hill's theory of the motion of the moon's perigee.

The solutions of the above differential equation are not, in general, periodic functions of z: but there are an infinite number of solutions which are periodic functions of z, of period 2π. (This is analogous to the fact that the solutions of Legendre's differential equation are not in general polynomials in z, although there are an infinite number of them—namely $P_n(z)$ when n is any integer—which are

* Mathieu, *Liouville's Journal*, (2), XIII. (1868), p. 137. Recent papers by Butts, *Amer. Journ. Math.* XXX. (1908), p. 129 and Marshall, *ibid.* XXXI. (1909), p. 311: recent inaugural dissertations at Zurich by Dannacher (1906) and Wiesmann (1909).

polynomials in z.) It is these periodic solutions of the differential equation which are required in mathematical physics, and it is with them that the present paper is concerned.

2. *The elliptic-cylinder functions are the solutions of a certain integral-equation.*

The starting-point of the present investigation is a result previously published by the author*, that the general solution of the potential-equation is

$$V = \int_0^{2\pi} f(x \cos \theta + y \sin \theta + iz, \theta)\, d\theta,$$

where f denotes an arbitrary function of its two arguments. In order to obtain the solution in elliptic-cylinder functions, we replace x and y by the variables which are appropriate to the elliptic cylinder, namely ξ and η, where

$$x = h \cos \xi \cosh \eta,$$
$$y = h \sin \xi \sinh \eta.$$

The potential-equation in terms of these variables becomes

$$\frac{\partial^2 V}{\partial \xi^2} + \frac{\partial^2 V}{\partial \eta^2} + h^2 (\cosh^2 \eta - \cos^2 \xi) \frac{\partial^2 V}{\partial z^2} = 0,$$

and the general solution of the equation becomes

$$V = \int_0^{2\pi} f(h \cos \xi \cosh \eta \cos \theta + h \sin \xi \sinh \eta \sin \theta + iz, \theta)\, d\theta.$$

The solutions required in the harmonic analysis appropriate to the elliptic cylinder are of the form

$$V = e^{imz} X(\xi)\, Y(\eta),$$

where $X(\xi)$ denotes a function of ξ alone and $Y(\eta)$ denotes a function of η alone: substituting, we have for the determination of X and Y the equations

$$\frac{d^2 X}{d\xi^2} + (A + m^2 h^2 \cos^2 \xi)\, X = 0,$$
$$\frac{d^2 Y}{d\eta^2} + (-A - m^2 h^2 \cosh^2 \eta)\, Y = 0,$$

each of which is equivalent to the equation (1). The corresponding solution of the potential-equation evidently takes the form

$$V = \int_0^{2\pi} e^{m (h \cos \xi \cosh \eta \cos \theta + h \sin \xi \sinh \eta \sin \theta + iz)}\, \phi(\theta)\, d\theta,$$

where $\phi(\theta)$ is a function as yet undetermined. From this it follows that the differential equation (1) must be satisfied by an integral of the form

$$y(z) = \int_0^{2\pi} e^{k (\cos z \cosh \eta \cos \theta + \sin z \sinh \eta \sin \theta)}\, \phi(\theta)\, d\theta,$$

where η is arbitrary. Taking η to be zero, we have the result that the differential equation (1) must be satisfied by an integral of the form

$$y(z) = \int_0^{2\pi} e^{k \cos z \cos \theta}\, \phi(\theta)\, d\theta.$$

* *Mathematische Annalen*, LVII. (1903), p. 333.

In order to determine the function $\phi(\theta)$, we substitute this integral in the differential equation, which gives

$$\int_0^{2\pi} e^{k\cos z\cos\theta}(k^2\cos^2\theta + k^2\cos^2 z\sin^2\theta - k\cos z\cos\theta + a)\,\phi(\theta)\,d\theta = 0,$$

or, integrating by parts, and supposing that $\phi(\theta)$ is periodic,

$$\int_0^{2\pi} e^{k\cos z\cos\theta}\{\phi''(\theta) + a\phi(\theta) + k^2\cos^2\theta\,\phi(\theta)\}\,d\theta = 0,$$

which is evidently satisfied provided

$$\frac{d^2\phi}{d\theta^2} + (a + k^2\cos^2\theta)\,\phi = 0.$$

Thus we see that $\phi(\theta)$ must be a periodic elliptic-cylinder function of θ, formed with the same constants a and k as $y(z)$ itself; but there does not exist more than one distinct periodic solution of the equation (1) with the same constants a and k: so that $\phi(\theta)$ must be (save for a multiplicative constant) the same function of θ as $y(z)$ is of z. Thus finally we have the result that *the periodic solutions of the equation* (1) *satisfy the homogeneous integral-equation*

$$y(z) + \lambda\int_0^{2\pi} e^{k\cos z\cos\theta}\,y(\theta)\,d\theta = 0 \quad\ldots\ldots\ldots\ldots\ldots\ldots\ldots(2).$$

It is known from the general theory of integral-equations that this equation (2) does not possess a solution except when λ has one of a certain set of values λ_0, λ_1, λ_2, λ_3, ...: and when λ has one of these values, say λ_r, there exists a corresponding solution $y_r(z)$. This set of solutions $y_0(z)$, $y_1(z)$, $y_2(z)$, ... are the periodic solutions of the differential equation (1); that is to say, they are the elliptic-cylinder functions required in mathematical physics*.

3. *Determination of the elliptic-cylinder function of zero order, $ce_0(z)$.*

We shall now make use of the result just obtained in order to derive the periodic solutions of the differential equation (1).

When k is zero, the solutions of the equation (1) with period 2π are obtained by taking a to be the square of an integer: the solutions are then

$$1,\quad \cos z,\quad \sin z,\quad \cos 2z,\quad \sin 2z,\quad \cos 3z,\quad \sin 3z,\quad \ldots.$$

As we shall see, the periodic solutions of the equation (1) when k is different from zero correspond respectively to these, and reduce to them when k tends to zero. We shall call these solutions

$$ce_0(z),\quad ce_1(z),\quad se_1(z),\quad ce_2(z),\quad se_2(z),\quad ce_3(z),\quad se_3(z),\quad \ldots,$$

$ce_r(z)$ being the solution which reduces to $\cos rz$ when k is zero, and $se_r(z)$ being the solution which reduces to $\sin rz$.

* This integral-equation was mentioned, and ascribed to myself, by Mr H. Bateman in *Trans. Camb. Phil. Soc.* XXI. p. 193 (1909): but I have not previously published anything on the subject, the theorem having merely been communicated to Mr Bateman in conversation.

The elliptic-cylinder functions $ce_r(z)$ and $se_r(z)$ both reduce to the circular-cylinder or Bessel function $J_r(ik\cos z)$ when the eccentricity of the elliptic cylinder reduces to zero.

We shall first determine the elliptic-cylinder function of zero order, $ce_0(z)$.

Since $ce_0(z)$ reduces to unity when k reduces to zero, we see at once from the integral-equation (2) that λ must reduce to $-1/2\pi$ when k reduces to zero. So write

$$-\frac{1}{2\pi\lambda} = 1 + a_1k + a_2k^2 + a_3k^3 + \dots,$$

$$ce_0(z) = 1 + kb_1(z) + k^2b_2(z) + k^3b_3(z) + \dots,$$

where $b_1(z)$, $b_2(z)$, $b_3(z)$, ... are periodic functions of z, of period 2π, having no constant term.

Substituting in the integral-equation (2), we have

$$\{1 + a_1k + a_2k^2 + \dots\}\{1 + kb_1(z) + k^2b_2(z) + \dots\}$$
$$= \frac{1}{2\pi}\int_0^{2\pi}\left\{1 + k\cos z\cos\theta + \frac{k^2}{2!}\cos^2 z\cos^2\theta + \dots\right\}\{1 + kb_1(\theta) + k^2b_2(\theta) + \dots\}\,d\theta.$$

Equating the coefficients of k on both sides of this equation, we have

$$b_1(z) + a_1 = \frac{1}{2\pi}\int_0^{2\pi}\{\cos z\cos\theta + b_1(\theta)\}\,d\theta.$$

Since the integral on the right-hand side vanishes, we have

$$b_1(z) = 0, \quad a_1 = 0.$$

Next equating the coefficients of k^2 on both sides of the equation, we have

$$b_2(z) + a_2 = \frac{1}{2\pi}\int_0^{2\pi}\{b_2(\theta) + \tfrac{1}{2}\cos^2 z\cos^2\theta\}\,d\theta$$
$$= \tfrac{1}{4}\cos^2 z.$$

Since $b_2(z)$ is to contain no constant term, we must have

$$b_2(z) = \tfrac{1}{8}\cos 2z, \qquad a_2 = \tfrac{1}{8}.$$

Similarly by equating the coefficients of k^3 and k^4, we find

$$b_3(z) = 0, \qquad a_3 = 0,$$

$$b_4(z) = \frac{1}{2^9}\cos 4z, \quad a_4 = \frac{7}{2^9}.$$

The first terms of the lowest-order solution of the integral-equation

$$ce(z) = \lambda\int_0^{2\pi} e^{k\cos z\cos\theta}\,ce(\theta)\,d\theta$$

are therefore given by the equation

$$ce_0(z) = 1 + \frac{k^2}{8}\cos 2z + \frac{k^4}{2^9}\cos 4z + \dots;$$

the corresponding value of λ is at once obtained by writing $z=\frac{1}{2}\pi$ in the integral-equation, which becomes

$$ce_0(\tfrac{1}{2}\pi)=\lambda\int_0^{2\pi} ce_0(\theta)\,d\theta=2\pi\lambda,$$

so we have

$$ce_0(z)=\frac{1}{2\pi}\,ce_0(\tfrac{1}{2}\pi)\int_0^{2\pi} e^{k\cos z\cos\theta}\,ce_0(\theta)\,d\theta.$$

By continuing the above procedure we could obtain as many terms as are required of the expansions, but we should not obtain a formula for the general term. In order to obtain this, we write

$$ce_0(z)=1+\sum_{r=1}^{\infty} A_{2r}(k)\cos 2rz\,;$$

then since it is known that

$$e^{m\cos z}=J_0(im)-2iJ_1(im)\cos z-2J_2(im)\cos 2z+\ldots,$$

we can equate coefficients of $\cos 2rz$ in the integral-equation, and thus obtain

$$A_{2r}(k)=\frac{1}{2\pi}\,ce_0(\tfrac{1}{2}\pi)\int_0^{2\pi}(-1)^r.\,2J_{2r}(ik\cos\theta)\,ce_0(\theta)\,d\theta,\qquad (r=1,\,2,\,\ldots\,\infty).$$

When on the right-hand side we substitute the expansions, and use the formula

$$\frac{1}{2\pi}\int_0^{2\pi}\cos^{2q}\theta\cos 2p\theta\,d\theta=\frac{2q\,!}{2^{2q}\,\overline{q+p}\,!\,\overline{q-p}\,!},$$

the last equation becomes

$$A_{2r}(k)=\frac{k^{2r}}{2^{4r-1}r\,!\,r\,!}-\frac{r(3r+4)\,k^{2r+4}}{2^{4r+7}\,\overline{r+1}\,!\,\overline{r+1}\,!}+\ldots.$$

We thus find for the elliptic-cylinder function of zero order the expansion

$$ce_0(z)=1+\sum_{r=1}^{\infty}\left\{\frac{k^{2r}}{2^{4r-1}r\,!\,r\,!}-\frac{k^{2r+4}r(3r+4)}{2^{4r+7}.\,\overline{r+1}\,!\,\overline{r+1}\,!}+\ldots\right\}\cos 2rz.$$

4. *The other integral-equation satisfied by the elliptic-cylinder functions.*

In the same way we can show that the elliptic-cylinder functions are the solutions of the integral-equation

$$y(z)+\lambda\int_0^{2\pi} e^{ik\sin z\sin\theta}\,y(\theta)\,d\theta=0,$$

and in particular that

$$ce_0(z)=\frac{1}{2\pi}\,ce_0(0)\int_0^{2\pi}\cos(k\sin z\sin\theta)\,ce_0(\theta)\,d\theta.$$

5. *The elliptic-cylinder functions of order unity, $ce_1(z)$ and $se_1(z)$.*

The two elliptic-cylinder functions of order unity, which we denote by $ce_1(z)$ and $se_1(z)$, may be obtained in the same way: the integral-equation for $ce_1(z)$ is found to be

$$ce_1(z)=-\frac{1}{k\pi}\,ce_1'(\tfrac{1}{2}\pi)\int_0^{2\pi} e^{k\cos z\cos\theta}\,ce_1(\theta)\,d\theta,$$

and from it we derive the expansion

$$ce_1(z) = \cos z + \sum_{r=1}^{\infty} \left\{ \frac{k^{2r}}{2^{4r} \,.\, \overline{r+1}\,!\, r\,!} - \frac{r k^{2r+2}}{2^{4r+4} \,.\, \overline{r+1}\,!\, \overline{r+1}\,!} + \frac{k^{2r+4}}{2^{4r+10} \,.\, \overline{r-1}\,!\, \overline{r+2}\,!} + \dots \right\} \cos (2r+1) z.$$

Similarly

$$se_1(z) = \sin z + \sum_{r=1}^{\infty} \left\{ \frac{k^{2r}}{2^{4r} \,.\, \overline{r+1}\,!\, r\,!} + \frac{r k^{2r+2}}{2^{4r+4} \,.\, \overline{r+1}\,!\, \overline{r+1}\,!} + \frac{k^{2r+4}}{2^{4r+10} \,.\, \overline{r-1}\,!\, \overline{r+2}\,!} + \dots \right\} \sin (2r+1) z.$$

The elliptic-cylinder functions of higher order can be determined by the same method directly from the integral-equation: the method may indeed be applied to solve a large class of integral-equations.

SUR LES SINGULARITÉS DES ÉQUATIONS DIFFÉRENTIELLES

PAR GEORGES RÉMOUNDOS.

1. Dans un travail antérieur (*Bul. Soc. Mathématique de France*, t. XXXVI. 1908, "Contribution à la théorie des singularités des équations différentielles du premier ordre") nous avons étudié le problème de l'existence d'une intégrale de l'équation différentielle:

$$x^2 \frac{dy}{dx} = \alpha y + f(x, y), \quad (\alpha \neq 0) \quad \ldots\ldots\ldots\ldots\ldots\ldots(1)$$

s'annulant pour $x = 0$ et holomorphe dans le voisinage de ce point; la fonction $f(x, y)$ est supposée holomorphe dans le voisinage de $x = 0$ et $y = 0$ s'annulant pour ces valeurs et ne contenant pas de terme de la forme αy. Briot et Bouquet ont étudié* le même problème dans le cas particulier:

$$x^2 \frac{dy}{dx} = \alpha y + x\phi(x), \quad \ldots\ldots\ldots\ldots\ldots\ldots(2),$$

$\phi(x)$ désignant une fonction holomorphe dans le voisinage de $x = 0$, et ils ont établi un théorème intéressant cité dans mon travail ci-dessus indiqué. Ce théorème de Briot et Bouquet et mes recherches conduisent à la conclusion qu'il n'existe pas, en général, une intégrale holomorphe dans le voisinage de $x = 0$ et s'annulant pour $x = 0$; il n'en existe pas *sûrement* dans le cas où les coefficients de la fonction $f(x, y)$ (développée en série de Maclaurin) sont tous réels et négatifs et le nombre α est positif.

L'équation (1) permet de calculer de proche en proche les coefficients d'une série Taylorienne qui satisfait *formellement* à l'équation; mais dans ce calcul fait par dérivations successives ce n'est pas le terme x^2y' qui joue le rôle prépondérant, comme il arrive ordinairement, mais le terme αy qui ne contient pas la dérivée y'; *c'est là le fait essentiel qui caractérise la singularité qui fait l'objet de cette communication:* cette propriété morphologique est la cause essentielle pour laquelle la série entière est, en général, divergente. On s'en rend bien compte en étudiant mon travail plus haut cité, où l'on voit la présence de quelques facteurs qui entraînent la divergence et que j'ai appelés *facteurs de divergence.*

Je me propose d'indiquer une généralisation systématique de cette théorie en étudiant une singularité analogue qui se présente aux équations différentielles d'ordre quelconque.

* Recherches sur les proprietés des fonctions définies par des équations différentielles (*Journal de l'École Polytechnique*, Cahier XXXVI, 1856, p. 161).

2. Nous donnerons, d'abord, une définition: Considérons une expression

$$\sigma(x, y, y', y'', \dots y^{(m)}),$$

dont chaque terme est un monôme entier par rapport à x, y et aux dérivées

$$y', y'', \dots y^{(m)}$$

et faisons sur cette expression n dérivations successives; si, après la n^{me} dérivation, nous remplaçons x et y par zéro, il peut se faire que l'expression (E) ainsi obtenue ne contienne pas la dérivée $y^{(m+n)}$; si nous désignons par V l'ordre le plus élevé des dérivées qui figurent dans l'expression (E), la différence $V-n$ sera appelée *force** de l'expression donnée.

Cette définition entraîne les conclusions suivantes:

(*a*) Toute expression de la forme x^n aura sa force égale à $-n$.

(*b*) La force de toute expression de la forme $x^p y^q$ est égale à $-p-q+1$, les nombres p et q étant supposés entiers.

(*c*) La force de toute expression P ne contenant ni x ni y est égale à l'ordre le plus élevé des dérivées qui y figurent. Si nous désignons par f la force de l'expression P, la force du produit $x^q . P$ est égale à $f-q$.

(*d*) La force d'un terme de la forme ay^q est égale à $1-q$.

(*e*) La force du produit d'une expression P par une puissance y^q est égale à $f-q$, où le nombre f désigne la force de l'expression P. Il est, enfin, évident que la force d'une somme est égale à la plus grande des forces des termes.

Cela posé, considérons une équation différentielle de la forme:

$$y = \phi(x, y, y', y'', \dots y^{(m)}), \quad \dots\dots\dots\dots\dots\dots\dots\dots (3)$$

dont le premier membre est de force zéro. Si la force du second membre est plus petite que celle du premier, c'est-à-dire négative, l'équation différentielle a deux propriétés importantes: 1°. *Elle permet toujours de calculer par des dérivations successives les coefficients d'une série entière satisfaisant formellement à l'équation différentielle.* 2°. *Le terme* y *du premier membre joue le rôle prépondérant dans le calcul des coefficients de la série et non pas les termes qui contiennent la dérivée de l'ordre le plus élevé, comme il arrive dans les cas réguliers.*

Nous supposons, bien entendu, que l'expression $\phi(x, y, y', y'', \dots y^{(m)})$ s'annulle pour $x=0$ et $y=0$, quelques que soient les $y', y'', \dots y^{(m)}$; dans le cas contraire, les conditions initiales $(x=0,\ y=0)$ ne seraient pas singulières pour l'équation différentielle.

3. Nous ferons des applications utiles de nos considérations en prenant comme point de départ l'équation différentielle:

$$x^2 y' = \alpha y + bx, \quad (\alpha \neq 0, b \neq 0)$$

qui appartient à la catégorie des équations (3), parce que le terme αy est de force zéro, tandis que la force des autres termes est négative. Il est facile de voir que cette équation n'admet pas d'intégrale holomorphe s'annulant pour $x=0$, quelques soient les nombres α et b (voir le *Traité d'Analyse* de M. Picard, tome III. p. 39).

* Il serait, peut-être, mieux de remplacer le mot *force* par le mot *poids*, d'après l'opinion de M. Borel, qu'il a bien voulu me faire connaître à la fin de ma communication.

Considérons maintenant l'équation différentielle:

$$y = ax^2y' + bx + F(x, y, y', y'', \dots y^{(m)}) \quad \dots\dots\dots\dots\dots\dots(4),$$

où F désigne un polynôme entier par rapport aux dérivées $y', y'', \dots y^{(m)}$, dont les coefficients sont des fonctions quelconques des x et y holomorphes dans le voisinage de $x=0$ et $y=0$; tous les coefficients des termes monômes du second membre sont supposés positifs. *Si la force de l'expression* F *est négative, nous pouvons démontrer que l'equation n'admet pas d'intégrale holomorphe s'annulant pour* x = 0.

En effet, on constate aisément que la série entière, qui satisfait formellement (il en existe, parce que la force du second membre est négative) à l'équation (4), a ses coefficients respectivement plus grands que ceux de la série divergente donnée par l'équation différentielle: $y = ax^2y' + bx$. Nous en concluons que la série entière, qui satisfait formellement à l'équation (4), est aussi *divergente* et, par conséquent, elle n'admet pas d'intégrale holomorphe s'annulant pour $x = 0$.

Les coefficients des deux séries, que nous venons de comparer, sont tous positifs. Nous avons ainsi obtenu un résultat très général concernant des équations différentielles d'ordre quelconque.

Il n'est pas sans intérêt de signaler particulièrement les équations de la forme:

$$y = bx + f(x, y) + a_1x^2y' + a_2x^{\mu_2}y'' + a_3x^{\mu_3}y''' + \dots + a_mx^{\mu_m}y^{(m)} \dots\dots(5),$$

où $f(x, y)$ désigne une fonction holomorphe dans le voisinage des valeurs $x=0$ et $y=0$ à coefficients réels et positifs et les nombres $b, a_1, a_2, a_3, \dots a_m$ sont aussi supposés réels et positifs; il faut faire aussi les hypothèses

$$\mu_2 \geqslant 3, \ \mu_3 \geqslant 4, \dots \mu_m \geqslant m+1,$$

pour que la force du second membre soit négative. L'équation (5) n'admet pas d'intégrale holomorphe s'annulant pour $x=0$*; nous pouvons même remplacer le terme a_1x^2y' par $a_1x^{\mu_1}y'$, où $\mu_1 \geqslant 2$.

On peut faire un certain rapprochement du résultat concernant les équations différentielles (5) avec le théorème bien connu de Fuchs sur l'existence d'intégrales régulières des équations linéaires.

4. Je tiens ici à résumer les propriétés morphologiques des équations différentielles, qui présentent la singularité sur laquelle je veux appeler l'attention des mathématiciens:

1°. L'équation différentielle est satisfaite par les valeurs $x=0$ et $y=0$, quelles que soient les valeurs des dérivées qui y figurent.

2°. L'équation différentielle nous permet de calculer une série entière unique qui satisfait formellement à l'équation et qui répond aux conditions initiales

$$(x=0, \ y=0).$$

3°. Ce n'est pas les termes contenant la dérivée du plus grand ordre qui jouent le rôle prépondérant dans le calcul des coefficients de la série par des dérivations successives; les termes de la plus grande force jouent ce rôle. *Aucun des termes contenant la dérivée du plus grand ordre n'a la force maximum.*

* Il est, d'ailleurs, facile de prouver qu'il n'existe même pas d'intégrale algébroïde dans le voisinage du point $x=0$ et s'annulant pour $x=0$, parce que la force du second membre de l'équation (5) ne saurait jamais croître par la substitution $x=t^{\rho}$, l'exposant ρ étant entier.

ON THE CONTINUATION OF THE HYPER-GEOMETRIC SERIES

By M. J. M. Hill.

ABSTRACT.

The object of this paper is to call attention to certain difficulties which arise in the attempt to apply *the method of ordinary algebraic expansion*, which has been successfully applied to series having one or two points of singularity to a series with three such points, viz.:—the hypergeometric series.

The equation to be proved is

$$F(\alpha, \beta, \alpha+\beta-\gamma+1, 1-x) = \frac{\Pi(\alpha+\beta-\gamma)\,\Pi(-\gamma)}{\Pi(\alpha-\gamma)\,\Pi(\beta-\gamma)} F(\alpha, \beta, \gamma, x)$$

$$+\frac{\Pi(\gamma-2)\,\Pi(\alpha+\beta-\gamma)}{\Pi(\alpha-1)\,\Pi(\beta-1)} x^{1-\gamma} F(\alpha-\gamma+1, \beta-\gamma+1, 2-\gamma, x).$$

The series $\quad F(\alpha, \beta, \alpha+\beta-\gamma+1, 1-x)$

and $\quad x^{1-\gamma} F(\alpha-\gamma+1, \beta-\gamma+1, 2-\gamma, x)$

cannot be expressed in a series of integral powers of x.

Hence in each series concerned only $n+1$ terms are taken, and an attempt is made to prove that the difference between the two sides tends to zero as n tends to ∞, it being known that $|x|$ and $|1-x|$ are each less than unity.

The series $x^{1-\gamma}$ {the first $(n+1)$ terms of $F(\alpha-\gamma+1, \beta-\gamma+1, 2-\gamma, x)$} is treated thus:—

$x^{1-\gamma}$ is expanded in powers of $1-x$ by the binomial theorem and $(n+1)$ terms are retained. The terms retained are multiplied by the first $(n+1)$ terms of

$$F(\alpha-\gamma+1, \beta-\gamma+1, 2-\gamma, x)$$

and it is shown that the terms of more than n dimensions in x and $1-x$ can be neglected.

The terms in the product, which are retained, are arranged according to powers of x.

Then the first $(n+1)$ terms of $F(\alpha, \beta, \alpha+\beta-\gamma+1, 1-x)$ are also arranged according to powers of x, and the coefficient of x^r is transformed into two parts; one of which is the coefficient of x^r in

$$\frac{\Pi(\alpha+\beta-\gamma)\,\Pi(-\gamma)}{\Pi(\alpha-\gamma)\,\Pi(\beta-\gamma)} F(\alpha, \beta, \gamma, x),$$

The *ratio* of the other part to the coefficient of x^r in

$$\frac{\Pi(\gamma-2)\,\Pi(\alpha+\beta-\gamma)}{\Pi(\alpha-1)\,\Pi(\beta-1)}\,x^{1-\gamma}\,F(\alpha-\gamma+1,\,\beta-\gamma+1,\,2-\gamma,\,x)$$

tends to unity as n tends to ∞, but this does not prove that the *difference* between the other part of the coefficient of x^r in $F(\alpha,\,\beta,\,\alpha+\beta-\gamma+1,\,1-x)$ and the coefficient of x^r in

$$\frac{\Pi(\gamma-2)\,\Pi(\alpha+\beta-\gamma)}{\Pi(\alpha-1)\,\Pi(\beta-1)}\,x^{1-\gamma}\,F(\alpha-\gamma+1,\,\beta-\gamma+1,\,2-\gamma,\,x)$$

tends to zero as n tends to ∞.

I have in the paper expressed the difference in the form

$$k_0(1-x)^n+k_1x(1-x)^{n-1}+k_2x^2(1-x)^{n-2}+\ldots+k_nx^n.$$

From the forms of the coefficients k it is clear that, when v is small compared with n, the term $k_vx^v(1-x)^{n-v}$ tends to zero as n tends to ∞, taking account of the fact that $|x|$ and $|1-x|$ are each less than unity. But when v is comparable with n, a further investigation is necessary to make the investigation complete.

1. In the year 1902 I obtained *by ordinary algebraic expansion* the continuations of two series which have one singular point, viz.:—the binomial and logarithmic series, and of two series which have two singular points, viz.:—the series for arc tan x and arc sin x. (*Proceedings of the London Mathematical Society*, Vol. XXXV. pp. 388—416.)

The object of the present paper is to call attention to the difficulties involved in attempting to apply the same processes to the case next in order of simplicity, viz. the hypergeometric series with three singular points.

Although I have not surmounted all the difficulties, yet the attempt has given two interesting results, which I will just mention. Employing the notation a_r to denote the product $a(a+1)\ldots(a+r-1)$, the first result (published in the *Proceedings of the London Mathematical Society*, pp. 335—341, 1907) is this:—

$$1+\frac{\alpha\beta}{\gamma}+\frac{\alpha_2\beta_2}{2!\,\gamma_2}+\ldots+\frac{\alpha_s\beta_s}{s!\,\gamma_s}$$
$$=\frac{(\gamma-\alpha)_{t+1}(\gamma-\beta)_{t+1}}{\gamma_{t+1}(\gamma-\alpha-\beta)_{t+1}}\left(1+\frac{\alpha\beta}{\gamma+t+1}+\ldots+\frac{\alpha_s\beta_s}{s!\,(\gamma+t+1)_s}\right)$$
$$-\frac{\alpha_{s+1}\beta_{s+1}}{(\gamma-\alpha-\beta)\,s!\,\gamma_{s+1}}\left[1+\frac{(\gamma-\alpha)(\gamma-\beta)}{(\gamma-\alpha-\beta+1)(\gamma+s+1)}+\ldots+\frac{(\gamma-\alpha)_t(\gamma-\beta)_t}{(\gamma-\alpha-\beta+1)_t(\gamma+s+1)_t}\right]\quad\ldots\ldots\ldots\text{(I)}$$

And this when $t=\infty$ becomes

$$1+\frac{\alpha\beta}{\gamma}+\ldots+\frac{\alpha_s\beta_s}{s!\,\gamma_s}=\frac{\Pi(\gamma-1)\,\Pi(\gamma-\alpha-\beta-1)}{\Pi(\gamma-\alpha-1)\,\Pi(\gamma-\beta-1)}$$
$$-\frac{\Pi(s+1,\,\gamma-1)(s+1)^{\alpha+\beta-\gamma}}{(\gamma-\alpha-\beta)\,\Pi(s+1,\,\alpha-1)\,\Pi(s+1,\,\beta-1)}\left\{1+\frac{(\gamma-\alpha)(\gamma-\beta)}{(\gamma-\alpha-\beta+1)(\gamma+s+1)}+\ldots\right.$$
$$\left.+\frac{(\gamma-\alpha)_t(\gamma-\beta)_t}{(\gamma-\alpha-\beta+1)_t(\gamma+s+1)_t}+\ldots\text{ to }\infty\right\}\quad\ldots\ldots\ldots\text{(II)},$$

which when the real part of $\gamma-\alpha-\beta$ is positive gives Gauss's Formula for the infinite series

$$1+\frac{\alpha\beta}{\gamma}+\ldots+\frac{\alpha_s\beta_s}{s!\,\gamma_s}+\ldots,$$

viz.
$$\frac{\Pi(\gamma-1)\,\Pi(\gamma-\alpha-\beta-1)}{\Pi(\gamma-\alpha-1)\,\Pi(\gamma-\beta-1)}.$$

The second result can be expressed thus:—

Let
$$u_r=\frac{\alpha_{r-1}\beta_{r-1}}{(\gamma-1)_r(\delta-1)_r},$$

$$H_r\begin{pmatrix}\alpha, & \beta\\ \gamma, & \delta\end{pmatrix}=\sum_1^r u_r.$$

Then
$$H_r\begin{pmatrix}\alpha & \beta\\ \gamma & \delta\end{pmatrix}-\frac{(\gamma-\alpha)_s(\gamma-\beta)_s}{(\gamma-1)_s(\gamma+\delta-\alpha-\beta-1)_s}H_r\begin{pmatrix}\alpha & \beta\\ \gamma+s & \delta\end{pmatrix}$$
$$=H_s\begin{pmatrix}\gamma-\alpha & \gamma-\beta\\ \gamma & \gamma+\delta-\alpha-\beta\end{pmatrix}-\frac{\alpha_r\beta_r}{(\gamma-1)_r(\delta-1)_r}H_s\begin{pmatrix}\gamma-\alpha & \gamma-\beta\\ \gamma+r & \gamma+\delta-\alpha-\beta\end{pmatrix}\quad\text{.........(III)},$$

from which (I) may be deduced by multiplying through by $(\delta-1)$ and then putting $\delta=1$.

When both r, s are ∞ and the real parts of $\gamma+\delta-\alpha-\beta$ and δ are >1, this gives

$$\frac{1}{(\gamma-1)(\delta-1)}\left(1+\frac{\alpha\beta}{\gamma\delta}+\frac{\alpha_2\beta_2}{\gamma_2\delta_2}+\ldots\text{ to }\infty\right)$$
$$=\frac{1}{(\gamma-1)(\gamma+\delta-\alpha-\beta-1)}\left(1+\frac{(\gamma-\alpha)(\gamma-\beta)}{\gamma(\gamma+\delta-\alpha-\beta)}+\frac{(\gamma-\alpha)_2(\gamma-\beta)_2}{\gamma_2(\gamma+\delta-\alpha-\beta)_2}+\ldots\text{ to }\infty\right)\quad\text{.........(IV)}.$$

The equation just written was first obtained by Dr Barnes. The more complete form from which it is deduced was published by Mr Whipple and myself in the *Quarterly Journal of Pure and Applied Mathematics*, No. 162, 1910.

2. I proceed now to the problem which I have been endeavouring to solve:—

It is to obtain a verification *by ordinary algebraic processes* of the equation

$$F(\alpha,\beta,\alpha+\beta-\gamma+1,1-x)=\frac{\Pi(\alpha+\beta-\gamma)\,\Pi(-\gamma)}{\Pi(\alpha-\gamma)\,\Pi(\beta-\gamma)}F(\alpha,\beta,\gamma,x)$$
$$+\frac{\Pi(\gamma-2)\,\Pi(\alpha+\beta-\gamma)}{\Pi(\alpha-1)\,\Pi(\beta-1)}x^{1-\gamma}F(\alpha-\gamma+1,\beta-\gamma+1,2-\gamma,x)\ldots\ldots(\text{V}).$$

If we take the series
$$F(\alpha,\beta,\alpha+\beta-\gamma+1,1-x),$$

and endeavour to expand it in powers, not of $(1-x)$, but of x, the coefficient of x^r is

$$(-1)^r\frac{\alpha_r\beta_r}{r!(\alpha+\beta-\gamma+1)_r}\left[1+\frac{(\alpha+r)(\beta+r)}{1!(\alpha+\beta-\gamma+1+r)}+\frac{(\alpha+r)_2(\beta+r)_2}{2!(\alpha+\beta-\gamma+1+r)_2}+\ldots\text{ to }\infty\right],$$

but the term in brackets is ∞ if the real part of $(\alpha+\beta-\gamma+1+r-(\alpha+r)-(\beta+r))$ is positive, i.e. if the real part of $(1-\gamma-r)$ is positive, which is the case for all the terms for which r exceeds the real part of $1-\gamma$.

Hence it is necessary to limit our consideration to a finite number of terms, say $(n+1)$, of the three hypergeometric series in the equation, and then to prove that the difference between the left- and right-hand sides of the equation will tend to zero as n tends to infinity.

3. Taking then only $(n+1)$ terms of

$$F(\alpha, \beta, \alpha+\beta-\gamma+1, 1-x),$$

and expanding these in powers of x, the coefficient of x^r is

$$(-1)^r \frac{\alpha_r \beta_r}{r!(\alpha+\beta-\gamma+1)_r}\left[1+\frac{(\alpha+r)(\beta+r)}{1!(\alpha+\beta-\gamma+1+r)}+\ldots+\frac{(\alpha+r)_{n-r}(\beta+r)_{n-r}}{(n-r)!(\alpha+\beta-\gamma+1+r)_{n-r}}\right] \quad \text{.........(VI).}$$

I transform the square bracket in (VI) by the aid of formula (I).

The result consists of two parts, viz.:—

the first part is

$$(-1)^r \frac{\alpha_r \beta_r}{r!(\alpha+\beta-\gamma+1)_r} \frac{(\alpha-\gamma+1)_{t+1}(\beta-\gamma+1)_{t+1}}{(\alpha+\beta-\gamma+1+r)_{t+1}(1-\gamma-r)_{t+1}} \ldots \text{(VII)},$$

multiplied by the series

$$1+\frac{(\alpha+r)(\beta+r)}{1!(\alpha+\beta-\gamma+2+r+t)}+\ldots+\frac{(\alpha+r)_{n-r}(\beta+r)_{n-r}}{(n-r)!(\alpha+\beta-\gamma+2+r+t)_{n-r}} \quad \text{.........(VIII);}$$

the second part is

$$(-1)^{r+1} \frac{\alpha_r \beta_r}{r!(\alpha+\beta-\gamma+1)_r} \frac{(\alpha+r)_{n-r+1}(\beta+r)_{n-r+1}}{(1-\gamma-r)(n-r)!(\alpha+\beta-\gamma+1+r)_{n-r+1}} \ldots \text{(IX)},$$

multiplied by

$$1+\frac{(\alpha-\gamma+1)(\beta-\gamma+1)}{(2-\gamma-r)(\alpha+\beta-\gamma+n+2)}+\ldots+\frac{(\alpha-\gamma+1)_t(\beta-\gamma+1)_t}{(2-\gamma-r)_t(\alpha+\beta-\gamma+n+2)_t} \quad \text{.........(X).}$$

4. Consider the expression marked (VII). It is

$$\frac{\alpha_r \beta_r}{r!\gamma_r}(-1)^r \gamma_r \frac{(\alpha-\gamma+1)_{t+1}(\beta-\gamma+1)_{t+1}}{(\alpha+\beta-\gamma+1)_{r+t+1}(1-\gamma-r)_{t+1}}$$

$$= \frac{\alpha_r \beta_r}{r!\gamma_r} \frac{(\alpha-\gamma+1)_{t+1}(\beta-\gamma+1)_{t+1}}{(\alpha+\beta-\gamma+1)_{t+1}(1-\gamma)_{t+1}} \frac{(-1)^r \gamma_r (1-\gamma+t+1-r)_r}{(\alpha+\beta-\gamma+t+2)_r(1-\gamma-r)_r}.$$

Now $$(-1)^r \gamma_r = (1-\gamma-r)_r.$$

Hence the expression (VII) is

$$\frac{\alpha_r \beta_r}{r!\gamma_r} \frac{(\alpha-\gamma+1)_{t+1}(\beta-\gamma+1)_{t+1}}{(\alpha+\beta-\gamma+1)_{t+1}(1-\gamma)_{t+1}} \frac{(2-\gamma-r+t)_r}{(\alpha+\beta-\gamma+2+t)_r}.$$

Now suppose t to become ∞ in comparison with r.

Then $$\frac{(2-\gamma-r+t)_r}{(\alpha+\beta-\gamma+2+t)_r} \text{ becomes } 1,$$

and $$\frac{(\alpha-\gamma+1)_{t+1}(\beta-\gamma+1)_{t+1}}{(\alpha+\beta-\gamma+1)_{t+1}(1-\gamma)_{t+1}} \text{ becomes } \frac{\Pi(\alpha+\beta-\gamma)\,\Pi(-\gamma)}{\Pi(\alpha-\gamma)\,\Pi(\beta-\gamma)}.$$

Also the expression (VIII) tends to 1, as t tends to ∞ in comparison with r.

Hence the product of the expressions (VII) and (VIII) becomes, when $t=\infty$,

$$\frac{\Pi(\alpha+\beta-\gamma)\,\Pi(-\gamma)}{\Pi(\alpha-\gamma)\,\Pi(\beta-\gamma)}\,\frac{\alpha_r\beta_r}{r!\,\gamma_r},$$

which is the coefficient of x^r in the first term on the right-hand side of (V).

5. Consider next the expression (IX). It can be written

$$+(-1)^{r+1}\frac{\alpha_{n+1}\beta_{n+1}}{(1-\gamma-r)\,r!\,(n-r)!\,(\alpha+\beta-\gamma+1)_{n+1}}$$

$$=\frac{\alpha_{n+1}\beta_{n+1}}{(\alpha+\beta-\gamma+1)_{n+1}\,(\gamma-1)_{n+1}}\,\frac{(-1)^{r+1}(\gamma-1)_{n+1}}{r!\,(n-r)!\,(1-\gamma-r)}.$$

It should be noted that as n tends to ∞

$$\frac{\alpha_{n+1}\beta_{n+1}}{(\alpha+\beta-\gamma+1)_{n+1}\,(\gamma-1)_{n+1}}$$

tends to

$$\frac{\Pi(\alpha+\beta-\gamma)\,\Pi(\gamma-2)}{\Pi(\alpha-1)\,\Pi(\beta-1)},$$

which is the coefficient in the second term on the right-hand side of (V).

6. It remains therefore to see whether the difference between the first $(n+1)$ terms in

$$x^{1-\gamma}F(\alpha-\gamma+1,\,\beta-\gamma+1,\,2-\gamma,\,x)$$

and the series

$$\sum_{r=0}^{n}\frac{(-1)^{r+1}(\gamma-1)_{n+1}x^r}{r!\,(n-r)!\,(1-\gamma-r)}\left\{1+\frac{(\alpha-\gamma+1)(\beta-\gamma+1)}{(2-\gamma-r)(\alpha+\beta-\gamma+n+2)}+\dots+\frac{(\alpha-\gamma+1)_t(\beta-\gamma+1)_t}{(2-\gamma-r)_t(\alpha+\beta-\gamma+n+2)_t}\right\}\quad\text{.........(XI)}$$

will tend to zero as n tends to infinity, where t is ∞ compared with r, and as r may range up to n, we must regard t as ∞ compared with n.

Let us write (XI) thus

$$\sum_{r=0}^{n}\frac{(-1)^{r+1}(\gamma-1)_{n+1}x^r}{r!\,(n-r)!\,(1-\gamma-r)}\{1+\epsilon_r\}\quad\text{..................(XII).}$$

I proceed to show how to replace $x^{1-\gamma}F(\alpha-\gamma+1,\beta-\gamma+1,2-\gamma,x)$ by a polynomial of degree n which differs from it by a quantity which tends to zero as n tends to ∞, when $|x|<1$ and $|(1-x)|<1$.

Now

$$x^{1-\gamma}=(1-(1-x))^{1-\gamma}$$

$$=1-(1-\gamma)(1-x)+\frac{(1-\gamma)(-\gamma)}{1.2}(1-x)^2+\dots+(-1)^n\frac{(1-\gamma)\dots(1-\gamma-n+1)}{n!}(1-x)^n+\dots,$$

the remainder can be neglected if n is large and $|1-x|<1$.

$$\therefore\ x^{1-\gamma}=1+(\gamma-1)(1-x)+\frac{(\gamma-1)_2}{2!}(1-x)^2+\dots+\frac{(\gamma-1)_n}{n!}(1-x)^n\quad\text{approximately}$$

$$=\tau_0+\tau_1(1-x)+\tau_2(1-x)^2+\dots+\tau_n(1-x)^n$$

where

$$\tau_0=1,\quad \tau_n=\frac{(\gamma-1)_n}{n!}.$$

Let $\quad F(\alpha-\gamma+1, \beta-\gamma+1, 2-\gamma, x) = \sigma_0 + \sigma_1 x + \ldots + \sigma_n x^n,$

the remainder being negligible if $|x| < 1$, and n large enough.

$$x^{1-\gamma} F(\alpha-\gamma+1, \beta-\gamma+1, 2-\gamma, x)$$
$$= [\tau_0 + \tau_1(1-x) + \ldots + \tau_n(1-x)^n][\sigma_0 + \sigma_1 x + \ldots + \sigma_n x^n] \text{ approximately.}$$

The terms in the product whose dimensions in x and $1-x$ are greater than n are

$$\begin{aligned} &\tau_1(1-x)\sigma_n x^n \\ &+ \tau_2(1-x)^2(\sigma_{n-1}x^{n-1} + \sigma_n x^n) \\ &+ \ldots\ldots\ldots \\ &+ \tau_n(1-x)^n(\sigma_1 x + \ldots + \sigma_n x^n) \quad \ldots\ldots\ldots\ldots\ldots\ldots\text{(XIII)}. \end{aligned}$$

Now let τ be the numerical value of the greatest of the coefficients $\tau_1, \tau_2, \ldots \tau_n$.

Let σ be the numerical value of the greatest of the coefficients $\sigma_1, \sigma_2, \ldots \sigma_n$.

Let ρ be the greater of the numbers $|x|$, $|1-x|$. Then the modulus of the sum of the terms in (XIII) is less than

$$\sigma\tau\rho^{n+1}\left\{\begin{array}{l} 1 \\ +(1+\rho) \\ \ldots\ldots\ldots\ldots\ldots\ldots\ldots\ldots \\ +(1+\rho+\ldots+\rho^{n-1}) \end{array}\right\} = \frac{\sigma\tau\rho^{n+1}}{1-\rho}\left\{\begin{array}{l} 1-\rho \\ +1-\rho^2 \\ \ldots\ldots\ldots \\ +1-\rho^n \end{array}\right\} = \frac{\sigma\tau\rho^{n+1}}{1-\rho}\left(n - \frac{\rho(1-\rho^n)}{1-\rho}\right).$$

This is a finite multiple of $n\sigma\tau\rho^{n+1}$ which tends to zero as n tends to ∞ since $\rho < 1$.

The terms of less than $(n+1)$ dimensions in x and $(1-x)$ are

$$\begin{aligned} &\sigma_0(\tau_0 + \tau_1(1-x) + \ldots + \tau_n(1-x)^n) \\ &+ \sigma_1 x(\tau_0 + \tau_1(1-x) + \ldots + \tau_{n-1}(1-x)^{n-1}) \\ &+ \sigma_2 x^2(\tau_0 + \tau_1(1-x) + \ldots + \tau_{n-2}(1-x)^{n-2}) \\ &+ \ldots\ldots\ldots \\ &+ \sigma_{n-1}x^{n-1}(\tau_0 + \tau_1(1-x)) \\ &+ \sigma_n x^n(\tau_0). \end{aligned}$$

The coefficient of x^r is

$$\begin{aligned} &\sigma_0(-1)^r[\tau_r + \tau_{r+1}({}_{r+1}C_r) + \ldots + \tau_n({}_nC_r)] \\ &+ \sigma_1(-1)^{r-1}[\tau_{r-1} + \tau_r({}_rC_{r-1}) + \ldots + \tau_{n-1}({}_{n-1}C_{r-1})] \\ &+ \ldots\ldots\ldots \\ &+ \sigma_p(-1)^{r-p}[\tau_{r-p} + \tau_{r-p+1}({}_{r-p+1}C_{r-p}) + \ldots + \tau_{n-p}({}_{n-p}C_{r-p})] \\ &+ \ldots\ldots\ldots \\ &+ \sigma_{r-1}(-1)[\tau_1 + \tau_2({}_2C_1) + \ldots + \tau_{n-r+1}({}_{n-r+1}C_1)] \\ &+ \sigma_r[\tau_0 + \tau_1 + \ldots + \tau_{n-r}]. \end{aligned}$$

Now

$$\tau_{r-p}+\tau_{r-p+1}({}_{r-p+1}C_{r-p})+\ldots+\tau_{n-p}({}_{n-p}C_{r-p})$$

$$=\frac{(\gamma-1)_{r-p}}{(r-p)!}\left[1+\frac{\gamma-1+r-p}{1!}+\frac{(\gamma-1+r-p)_2}{2!}+\ldots+\frac{(\gamma-1+r-p)_{n-r}}{(n-r)!}\right]^*$$

$$=\frac{(\gamma-1)_{r-p}}{(r-p)!}\,\frac{(\gamma+r-p)_{n-r}}{(n-r)!}=\frac{(\gamma-1)_{n-p+1}}{(r-p)!(n-r)!(\gamma-1+r-p)}$$

$$=\frac{{}_{n-p}C_{n-r}}{(n-p)!}\,\frac{(\gamma-1)_{n-p+1}}{\gamma-1+r-p}.$$

Hence the coefficient of x^r is

$$\sigma_0(-1)^r\frac{{}_nC_{n-r}}{n!}\,\frac{(\gamma-1)_{n+1}}{\gamma-1+r}+\sigma_1(-1)^{r-1}\frac{{}_{n-1}C_{n-r}}{(n-1)!}\,\frac{(\gamma-1)_n}{\gamma-2+r}$$

$$+\ldots+\sigma_p(-1)^{r-p}\frac{{}_{n-p}C_{n-r}}{(n-p)!}\,\frac{(\gamma-1)_{n-p+1}}{\gamma-1+r-p}+\ldots$$

$$+\sigma_{r-1}(-1)\frac{{}_{n-r+1}C_{n-r}}{(n-r+1)!}\,\frac{(\gamma-1)_{n-r+2}}{\gamma}+\sigma_r\frac{{}_{n-r}C_{n-r}}{(n-r)!}\,\frac{(\gamma-1)_{n-r+1}}{\gamma-1}.$$

Putting in the values of the σ's this becomes

$$\frac{(-1)^r}{r!(n-r)!}\,\frac{(\gamma-1)_{n+1}}{\gamma-1+r}+\frac{(\alpha-\gamma+1)(\beta-\gamma+1)}{(2-\gamma)}(-1)^{r-1}\frac{1}{(r-1)!(n-r)!}\,\frac{(\gamma-1)_n}{\gamma-2+r}$$

$$+\frac{(\alpha-\gamma+1)_2(\beta-\gamma+1)_2}{2!(2-\gamma)_2}(-1)^{r-2}\frac{1}{(n-r)!(r-2)!}\,\frac{(\gamma-1)_{n-1}}{\gamma-3+r}$$

$$+\ldots\ldots\ldots$$

$$+\frac{(\alpha-\gamma+1)_p(\beta-\gamma+1)_p}{p!(2-\gamma)_p}(-1)^{r-p}\frac{1}{(n-r)!(r-p)!}\,\frac{(\gamma-1)_{n-p+1}}{\gamma-1+r-p}+\ldots$$

$$+\frac{(\alpha-\gamma+1)_{r-1}(\beta-\gamma+1)_{r-1}}{(r-1)!(2-\gamma)_{r-1}}(-1)\frac{1}{(n-r)!1!}\,\frac{(\gamma-1)_{n-r+2}}{\gamma}$$

$$+\frac{(\alpha-\gamma+1)_r(\beta-\gamma+1)_r}{r!(2-\gamma)_r}\,\frac{(\gamma-1)_{n-r+1}}{\gamma-1}\,\frac{1}{(n-r)!}$$

$$=\frac{(-1)^{r+1}(\gamma-1)_{n+1}}{(n-r)!r!(1-\gamma-r)}\quad\text{multiplied by}$$

$$\left[1-{}_rC_1\frac{(\alpha-\gamma+1)(\beta-\gamma+1)}{2-\gamma}\,\frac{\gamma-1+r}{\gamma-2+r}\,\frac{1}{\gamma-1+n}\right.$$

$$+{}_rC_2\frac{(\alpha-\gamma+1)_2(\beta-\gamma+1)_2}{(2-\gamma)_2}\,\frac{\gamma-1+r}{\gamma-3+r}\cdot\frac{1}{(\gamma+n-2)(\gamma+n-1)}$$

$$-\ldots\ldots\ldots$$

$$+(-1)^p\,{}_rC_p\frac{(\alpha-\gamma+1)_p(\beta-\gamma+1)_p}{(2-\gamma)_p}\,\frac{\gamma-1+r}{\gamma-1+r-p}\cdot\frac{1}{(\gamma+n-p)\ldots(\gamma+n-1)}$$

$$+\ldots\ldots\ldots$$

$$+(-1)^{r-1}\,{}_rC_{r-1}\frac{(\alpha-\gamma+1)_{r-1}(\beta-\gamma+1)_{r-1}}{(2-\gamma)_{r-1}}\,\frac{\gamma-1+r}{\gamma}\,\frac{1}{(\gamma+n-r+1)\ldots(\gamma+n-1)}$$

$$\left.+(-1)^r\,{}_rC_r\frac{(\alpha-\gamma+1)_r(\beta-\gamma+1)_r}{(2-\gamma)_r}\,\frac{\gamma-1+r}{\gamma-1}\,\frac{1}{(\gamma+n-r)\ldots(\gamma+n-1)}\right].$$

* $1+\frac{a}{1}+\frac{a_2}{2!}+\ldots+\frac{a_n}{n!}=\frac{(a+1)_n}{n!}.$

Let us write this

$$\frac{(-1)^{r+1}(\gamma-1)_{n+1}}{(n-r)!\,r!\,(1-\gamma-r)}(1+\eta_r).$$

Then the polynomial which replaces the first $(n+1)$ terms of

$$x^{1-\gamma}F(\alpha-\gamma+1,\ \beta-\gamma+1,\ 2-\gamma,\ x)$$

is

$$\sum_{r=0}^{n}\frac{(-1)^{r+1}(\gamma-1)_{n+1}x^r}{(n-r)!\,r!\,(1-\gamma-r)}(1+\eta_r)\ \ldots\ldots\ldots\ldots\ldots\ldots\text{(XIV)}.$$

7. It will be seen that the *ratio* of the coefficient of x^r in (XIV) to that of x^r in (XII) is

$$\frac{1+\eta_r}{1+\epsilon_r},$$

and this tends to unity as n tends to ∞. But this is not enough to identify the two polynomials.

It is necessary to show that their difference, viz.:

$$\sum_{r=0}^{n}\frac{(-1)^{r+1}(\gamma-1)_{n+1}x^r}{(n-r)!\,r!\,(1-\gamma-r)}(\eta_r-\epsilon_r)\ \ldots\ldots\ldots\ldots\ldots\ldots\text{(XV)},$$

tends to zero as n tends to ∞, $|x|$ and $|1-x|$ being each less than unity.

I think that this can be done by expressing it in the form

$$k_0(1-x)^n+k_1x(1-x)^{n-1}+k_2x^2(1-x)^{n-2}+\ldots+k_nx^n\ \ldots\ldots\text{(XVI)},$$

and examining the forms of the k's.

I proceed to show how the values of the k's can be obtained.

8. Let us now express the polynomial

$$\delta_0+\delta_1x+\ldots+\delta_nx^n$$

in the form

$$k_0(1-x)^n+k_1x(1-x)^{n-1}+\ldots+k_rx^r(1-x)^{n-r}+\ldots+k_nx^n.$$

Equating coefficients of powers of x we get

$$\delta_0=k_0$$
$$\delta_1=k_1-{}_nC_1k_0$$
$$\delta_2=k_2-{}_{n-1}C_1k_1+{}_nC_2k_0$$
$$\delta_3=k_3-{}_{n-2}C_1k_2+{}_{n-1}C_2k_1-{}_nC_3k_0,$$

and generally

$$\delta_t=k_t-{}_{n-t+1}C_1k_{t-1}+{}_{n-t+2}C_2k_{t-2}-\ldots+(-1)^t{}_nC_tk_0.$$

Solving successively for the k's we get

$$k_0=\delta_0$$
$$k_1={}_nC_1\delta_0+\delta_1$$
$$k_2={}_nC_2\delta_0+{}_{n-1}C_1\delta_1+\delta_2$$
$$k_3={}_nC_3\delta_0+{}_{n-1}C_2\delta_1+{}_{n-2}C_1\delta_2+\delta_3,$$

and generally

$$k_t={}_nC_t\delta_0+{}_{n-1}C_{t-1}\delta_1+{}_{n-2}C_{t-2}\delta_2+\ldots+{}_{n-t+1}C_1\delta_{t-1}+\delta_t.$$

9. If
$$\sum_{r=0}^{n} \frac{(-1)^{r+1}(\gamma-1)_{n+1}x^r}{(n-r)!\,r!\,(1-\gamma-r)}\eta_r$$
be expressed in the form (XVI) then I find that
$$k_v=\frac{(\gamma-1)_{n+1}}{(n-v)!}\left[\frac{\sigma_1}{(\gamma+n-1)(\gamma-1)_v}+\frac{\sigma_2}{(\gamma+n-2)_2(\gamma-1)_{v-1}}+\ldots+\frac{\sigma_v}{(\gamma+n-v)_v(\gamma-1)}\right]$$
where the σ's are as defined in Art. 6.

If
$$\sum_{r=0}^{n} \frac{(-1)^{r+1}(\gamma-1)_{n+1}x^r}{(n-r)!\,r!\,(1-\gamma-r)}\epsilon_r$$
be expressed in the same form, then I find that
$$k_v=\frac{(\gamma-1)_{n+1}}{(n-v)!\,v!\,(\gamma-1)_{v+1}}\left[\frac{\sigma_1(v+1)!}{\alpha+\beta-\gamma+n+2}+\frac{\sigma_2(v+2)!}{(\alpha+\beta-\gamma+n+2)_2}+\ldots\text{ to }\infty\right].$$
The condition for the convergence of this last series is satisfied for $v \lesseqgtr n$, which is the case.

It is clear from the forms for k_v that when v is small compared with n, then
$$k_v x^v (1-x)^{n-v}$$
tends to zero as n tends to ∞.

But when v is not small compared with n a further investigation of the forms of k_v is necessary before the demonstration can be completed.

ON MERSENNE'S NUMBERS

By Allan Cunningham.

These are numbers of form $M_q = (2^q - 1)$, with q a prime. They have the peculiarity of having no algebraic divisors, and of being for the most part composite numbers.

In 1644 Père Mersenne affirmed* that, out of the 56 primes $q \not> 257$, only 12 values of q gave M_q *prime*, viz.,

$$q = 1, 2, 3, 5, 7, 13, 17, 19, 31, 67, 127, 257;$$

and that the remaining 44 values of q (< 257) gave M_q *composite.* The grounds for this assertion are not known, and it has not been found possible (even yet) to completely test it.

Up to the present time—

12 numbers (M_q) have been *proved prime,* viz. those given by

$$q = 1, 2, 3, 5, 7, 13, 17, 19, 31, 61, 89, 127;$$

29 numbers (M_q) have been *proved composite,* viz. those given by

$q = 11, 23, 29, 37, 41, 43, 47, 53, 59, 67, 71;$ [11 completely factorised]

$q = 73, 79, 83, 97, 113, 131, 151, 163, 173, 179, 181, 191, 197, 211, 223, 233, 239, 251;$ [18 with one or more factors found].

Thus, out of the 12 numbers *affirmed prime* by Mersenne, 10 have been *proved prime,* one (M_{67}) has been *proved composite,* and one (M_{257}) remains *unverified.* Also, out of the 44 numbers *affirmed composite* by him, two (M_{61}, M_{89}) have been *proved prime,* 28 have been *proved composite,* and only 14 remain unverified (as to prime or composite character), viz. those given by

$$q = 101, 103, 107, 109, 137, 139, 149, 157, 167, 193, 199, 227, 229, 241.$$

All possible divisors under one million have been tried† for these 14 numbers: no divisors (< 1 million) were found.

* *Cogitata Physico-mathematica,* Paris, 1644, *Præf. gen.* Art. 19.

† By the present writer, with assistance; all the work has been done twice. For the "trial divisors" >200,000 two lists were prepared, one by M. A. Gérardin (of Nancy, France), one by the present writer, and then collated: the lists of "trial divisors" have been published in the Journal *Sphinx-Œdipe,* see references in the bibliography below.

Thus, up to the present, three mistakes have been found in Mersenne's classification, viz. M_{67} proved* composite, and M_{61}, M_{89} proved prime.

The Table below shows the prime factors† of all these numbers so far as now known. The names of the discoverers of the prime M_q, and of the factors of the composite M_q, are shown by their initials (according to the scheme below), and the dates of publication (or discovery) are annexed. A bibliography is also given below of authorities for each prime M_q and for each prime factor of a M_q.

REFERENCES.

B.	Bickmore, C. E., 1896.	G.	Gérardin, A., 1912.	Pw.	Powers, R. E., 1911.
Co.	Cole, F. N., 1903.	Fq.	Fauquembergue, E., 1912.	Ra.	Ramesam, 1912.
C.	Cunningham, Allan, Lt.-Col., R.E., 1895–1912.	La.	Landry, F., 1867.	R.	Reuschle, C. G., 1856.
		Ll.	Le Lasseur, 1878.	S.	Seelhoff, P., 1886.
E.	Euler, L., 1732–1750.	Lu.	Lucas, Ed., 1876–1891.	T.	Tarry, H., 1911.
F.	Fermat, 1640.	P.	Plana, Baron, 1859.	W.	Woodall, H. J., 1911.

Factors of $M_q = (2^q - 1)$, [*q prime*].

q	$M_q = (2^q - 1)$		q	$M_q = (2^q - 1)$	
1	1; $=p$		107		
2	3; $=p$		109		
3	7; $=p$		113	3391.23279.65993.	C, 1908–9; Ll, 1878
5	31; $=p$		127	prime	Lu, 1877
7	127; $=p$		131	263.	E, 1732
11	23.89;	F, 1640	137		
13	8191; $=p$	E, 1750	139		
17	131071; $=p$	E, 1750	149		
19	524287; $=p$	E, 1750	151	18121.55871.	C, 1909; Ll, 1883
23	47.178481;	F, 1640	157		
29	233.1103.2089;	E, 1732–50	163	150287.	C, 1908
31	prime	E, 1772	167		
37	223.616318177;	F, 1640	173	730753.	C, G, 1912
41	13367.164511353;	P, 1859	179	359.1433.	R, 1856; E, 1732
43	431.9719.2099863;	La, 1869; E, 1732	181	43441	W, 1911
47	2351.4513.13264529;	R, 1856	191	383	E, 1732
53	6361.69431.20394401;	La, 1869	193		
59	179951.3203431780337;	Ll, 1879	197	7487.	C, 1895
61	prime	Co, 1903; S, 1886	199		
67	193707721.761838257287;	Co, 1903; Lu, 1876	211	15193.	Ll, 1883
71	228479.48544121.212885833;	Ra, 1912; C, 1909	223	18287.	Ll, 1883
73	439.	E, 1732	227		
79	2687.	Ll, 1878	229		
83	167.	E, 1732	233	1399.	Ll, 1882
89	prime	Pw, T, 1911; Fq, 1912; Lu, 1891	239	479.1913.5737.	B, 1896; R, 1856; E, 1732
97	11447.	Ll, 1883	241		
101			251	503.54217.	C, 1909; E, 1732
103			257		

Addendum. Mr R. E. Powers has (quite recently) determined that M_{257} also has no divisor <1 million.

* M_{67} was the first mistake discovered (by Ed. Lucas in 1876; the factors were found by F. N. Cole in 1903).

† The semi-colon on right of the factors indicates *complete* factorisation.

BIBLIOGRAPHY.

FERMAT, *Opera Mathemat.*, Toulouse, 1640, p. 164, or Brassine's *Précis*, Paris, 1853, p. 144.

EULER, L., *Hist. de l'Académ. des Sciences*, Berlin, 1774, p. 36.

—— *Comment. Arithm.*, Petrop., 1849, Vol. I, pp. 2, 104.

REUSCHLE, C. G., *Neue Zahlentheoretische Tabellen*, Stuttgart, 1856, pp. 21, 22, 42—53.

PLANA, BARON, *Memoria della Reale Accadem. delle Scienze*, Turin, Ser. 2, Vol. XX, 1863, p. 130.

LANDRY, F., *Décomposition des Nombres* $(2^n \pm 1)$ *en leurs facteurs premiers, etc.*, Paris, 1869, pp. 6, 7.

LUCAS, ED., *Sur la théorie des nombres premiers*, Turin, 1876, p. 11.

—— *Recherches sur plusieurs ouvrages de Léonard de Pise*, Rome, 1877.

—— *Théorie des Nombres*, Vol. I, Paris, 1891, p. 376.

LE LASSEUR, quoted in following Works of Ed. Lucas :—

—— *American Journal of Math.*, Vol. I, 1878, p. 236.

—— *Sur la série récurrente de Fermat*, Rome, 1879, pp. 7, 8, 9.

—— *Recréations mathématiques*, Vol. I, Paris, 1882, p. 241 ; Vol. II, Paris, 1883, p. 230.

SEELHOFF, P., *Zeitschrift für Mathem. und Phys.*, Leipzig, 1886, p. 178.

BICKMORE, C. E., *On the numerical factors of* $(a^n - 1)$, in *Messenger of Mathematics*, Vol. XXV, 1896, p. 19.

COLE, F. N., *Amer. Math. Soc. Bulletin*, Vol. X, 1903, p. 137.

CUNNINGHAM, ALLAN, LT.-COL., R.E., *Proc. Lond. Math. Soc.*, V, 26, 1895, p. 261 ; Ser. 2, Vol. 6, 1908, p. xxii, etc.

—— *Nature*, Vol. 78, 1908, p. 23.

—— *Sphinx-Œdipe*, Nancy, t. iii, 1908, p. 161 ; t. vii, 1912, pp. 15, 38, 58.

GÉRARDIN, A., *Sphinx-Œdipe*, Nancy, t. iii, 1908, pp. 118—120, 161—165, 177—182 ; t. iv, 1909, pp. 1—5.

WOODALL, H. J., *Proc. Lond. Math. Soc.*, Ser. 2, Vol. 9, 1911, p. xvi.

POWERS, R. E., *Amer. Math. Monthly*, Vol. 18, 1911, p. 195.

TARRY, H., *Sphinx-Œdipe*, Nancy, 1911, p. 192.

RAMESAM, *Nature*, Vol. 89, 1912, p. 87.

FAUQUEMBERGUE, E., *Sphinx-Œdipe*, Nancy, 1912, p. 22.

SOME GENERAL TYPES OF FUNCTIONAL EQUATIONS

By Griffith C. Evans.

I. The Functional Equation in one unknown.

In an endeavour to establish existence theorems for certain integro-differential equations I found it necessary to consider the subject of implicit functional equations, and was able to obtain some theorems relative to it. I discovered, however, that the principal method—an extension of one due to Goursat in treating ordinary implicit functions—and the principal theorem had been already developed by Professor Volterra this last spring at Paris in a course of lectures, the relevant portion of which he has very kindly put at my disposal* ! In this article I shall therefore consider a more special type of implicit equation, where the results admit of complete specification and immediate application.

Professor Volterra developed the fundamental theory of what he called functions depending on other functions, and functions of curves, in several papers published in the *Rendiconti* of the Royal Academy of the Lincei in the year 1887†. The relationships of functions depending on other functions are called "fonctionelles" by Hadamard in his Lessons on the Calculus of Variations‡.

1. If we have two functions $u(t)$ and $\phi(t)$ so related that the value of $u(t)$ for a particular value of t depends on all the values of $\phi(t)$ throughout a certain interval ab, we say that u is a "functional" of ϕ in ab and write, with Volterra,

$$u(t) = F \left| \left[\overset{b}{\underset{a}{\phi(\tau)}} \right] \right|,$$

or, more simply,

$$u = F[\overset{b}{\underset{a}{\phi(\tau)}}].$$

One example of such a functional is

$$u(t) = \int_a^b K(t, \tau)\, \phi(\tau)\, d\tau.$$

Another, of exceedingly special character, is $\overset{b}{\underset{a}{\max}} |\phi(t)|$, an expression which we are going to use to denote the upper limit of $|\phi(t)|$ in the interval ab. An entity of still more complicated character is expressed by $\overset{b}{\underset{a}{\max}} |F[\overset{d}{\underset{c}{\phi}}]|$. Under some conditions, as

* For an early example of such an implicit equation see V. Volterra: Sur les fonctions qui dépendent d'autres fonctions [*C.R.* 142, 691—695 (1906)]. See also the extension of the theory of integral equations to non-linear equations by E. Schmidt, T. Lalesco and others. [For literature, see e.g. *Théorie des équations intégrales*, T. Lalesco, Paris, 1912.]

† V. Volterra: Sopra le funzioni che dipendono da altre funzioni [*Rend. della R. Acc. dei Lincei*, Vol. III. fasc. 4, 6, 7, agosto, 1887], Sopra le funzioni dipendenti da linee [*Ibid.* Vol. III. fasc. 9, 10, novembre, 1887].

‡ J. Hadamard: *Leçons sur le calcul des variations*, Bk II. ch. VII. Paris, 1910.

Volterra has shown, the functional can be developed in terms of its argument by means of an infinite series of multiple definite integrals; and thus is defined by an extension of Taylor's theorem a class of functionals analogous to the analytic functions. In this article we shall consider a function to be a special case of a functional.

The difference between the behaviours of linear integral equations with constant and with variable limits is well known. Although for certain questions one may be regarded as a special case of the other, the results in the special case are generally so much simpler that it is conveniently treated by itself. In respect to the linear integral equation the difference arises from the fact that when we have a variable upper limit the interval under consideration may be kept as small as we please, while with constant limits it is always the fixed length ab. Thus for the equation

$$u(t) = \phi(t) + \int_a^t K(t, \tau)\, u(\tau)\, d\tau \quad \text{......................(1)},$$

having obtained the solution $u(t)$ for the values $a \leqq t \leqq t_1$, we may rewrite it in the form

$$u(t) = \phi_1(t) + \int_{t_1}^t K(t, \tau)\, u(\tau)\, d\tau,$$

in which the $\phi_1(t)$ is a new known function, and proceed as before.

This same advantage of specialization appears in the more general case of functional equations. It is obviously this same property which is characteristic of the equation in functionals

$$u(t) = F[u\overset{t}{\underset{t_0}{(}}\tau)] \quad \text{.................................(2)},$$

which has the variable upper limit t. For having obtained the solution as far as $t = t_1$ the equation becomes

$$u(t) = G[u\overset{t}{\underset{t_1}{(}}\tau)],$$

and we may always keep the interval under consideration as small as we please. It remains then merely to define a hypothesis that shall ensure the existence of a solution of the equation for an arbitrarily small region.

This condition consists of two parts, one relating to the region of definition of the F and one relating, so to speak, to the stability of the equilibrium, somewhat similar to the ordinary Cauchy-Lipschitz condition. For the purpose of the applications we shall consider an argument depending on $n+1$ variables, for instance $u(x, y, z, t)$ and a functional F over a certain space of n dimensions, for instance $S: x, y, z$, and over a variable interval $t_0 t$.

2. CONDITION I. *Related to the functional* $F[\overset{t}{\underset{St_0}{u}}]$ *there is a field for* u, *which we shall denote by* Σ_{xyztu}, *finitely or infinitely extended in regard to its components, such that the following property holds:*

The interval $t_0 t_1$ *can be divided into a finite number of sub-intervals* $t_0\tau_1, \tau_1\tau_2, \ldots, \tau_n t_1$, *in such a way that if*

$$u_1(t) = u_2(t) = F \quad \textit{for} \quad t_0 \leqslant t \leqq \tau_i$$

$$\textit{and} \quad u_1(t),\, u_2(t) \quad \textit{lie in} \quad \Sigma_{xyztu} \quad \textit{for} \quad \tau_0 < t \leqslant \tau_{i+1},$$

it follows that

$$\left.\begin{aligned} &\text{(i)}\quad \underset{St_0}{\overset{t}{F[u_1]}},\quad \underset{St_0}{\overset{t}{F[u_2]}} \ \textit{lie in}\ \Sigma_{xyztF} \\ &\text{(ii)}\quad \underset{t_0}{\overset{t}{\max}} \left| \underset{St_0}{\overset{t}{F[u_1]}} - \underset{St_0}{\overset{t}{F[u_2]}} \right| \leqq A \underset{\tau_i}{\overset{t}{\max}} \left| u_1(t) - u_2(t) \right| \end{aligned}\right\} \ \textit{for}\ \tau_i < t \leqslant \tau_{i+1},$$

in which A is a positive constant less than 1.

3. Other conditions of similar character for subsequent use are the following:

Condition I′. This condition is the obvious generalization of I to n functions $u, v, \ldots, n$ functionals $F, G, \ldots$, and their corresponding field

$$\Sigma_{xyztuv\ldots}.$$

We shall replace (ii) by the statement

$$\underset{t_0}{\overset{t}{\max}} \left| F[u_1, v_1, \ldots] \underset{t_0}{\overset{t}{-}} F[u_2, v_2, \ldots] \right| \leqq A \left(\underset{\tau_i}{\overset{t}{\max}} \left| u_1 - u_2 \right| + \underset{\tau_i}{\overset{t}{\max}} \left| v_1 - v_2 \right| + \ldots\right),$$

$$\underset{t_0}{\overset{t}{\max}} \left| G[u_1, v_1, \ldots] \underset{t_0}{\overset{t}{-}} G[u_2, v_2, \ldots] \right| \leqq A \left(\underset{\tau_i}{\overset{t}{\max}} \left| u_1 - u_2 \right| + \underset{\tau_i}{\overset{t}{\max}} \left| v_1 - v_2 \right| + \ldots\right),$$

....

in which for n functions $u, v, \ldots, A$ shall be a constant less than $\frac{1}{n}$.

4. We may regard the condition I as made up of two separate ones, I (i) and I (ii). Let us define II (i) as the condition that we get from I (i) by replacing the Σ_{xyztF} in (i) by another field H_{xyztF}. And let us define II (ii) as the condition that we get from I (ii) by replacing the A in (ii) by another positive constant B, otherwise unrestricted.

Similarly we can form conditions II′ (i) and II′ (ii). The conditions II′ apply obviously to the case where the number of functionals in the system is not the same as the number of arguments.

We may also consider the conditions I, II, I′, II′ as applicable to functionals with constant limits ab, where, since there is no possibility of subdivision, the whole interval ab must be considered at one time.

5. We have at once, by the usual method of successive approximations*, the following theorem:

Theorem 1. *If $\underset{St_0}{\overset{t}{F[u]}}$ satisfies the condition* I *in the interval $t_0 t_1$, there exists one and only one solution of the equation*

$$u(x, y, z, t) = F[u(\xi, \underset{S}{\eta}, \zeta, \underset{t_0}{\overset{t}{\tau}})] \quad \ldots\ldots\ldots\ldots\ldots\ldots\ldots\ldots(3),$$

that lies in the field Σ_{xyztu}.

If, in addition, the functional F is continuous in t except for a certain number of fixed values $t', t'', \ldots$, when its argument $u(x, y, z, t)$ lies in Σ and is continuous in t except for a finite number of discontinuities, then the unique solution of (3) in Σ will be continuous in t except at the points $t', t'', \ldots$.

* See Picard: *Traité d'analyse*, Vol. II. note 3 (Appendix).

6. In the case of the linear integral equation (1), with continuous kernel, Σ_{tu} is defined by the relation

$$|u(t)| \leqq \max_{t_0}^{t_i} |u(\tau)| + \epsilon, \qquad (t_i < t \leqq t_{i+1})$$

where ϵ is a positive constant arbitrarily small or large. The sub-intervals can then be chosen in order to make A as small as we please. And this we may regard in general as the characteristic of a non-singular functional with variable limit: that the A may be made as small as we please by taking small enough sub-intervals. It is true, for instance, if the functional possesses a derivative according to the definition of Volterra.

An interesting special case of theorem 1 is the following:

The equation $$P(u(x)) + \int_a^x \sum_{i=1}^k K_i(x, \xi)(u(\xi))^i \, d\xi = 0,$$

where P is a polynomial in u and $P(u(a)) = 0$ has all its p roots distinct, has p and only p continuous solutions $u(x)$ in the neighbourhood of $x = a$, provided that $K(x, \xi)$ is continuous for $a \leqq \xi \leqq x \leqq a'$, $a' > a$.

It is sufficient if $K(x, \xi)$, instead of being continuous, have discontinuities not necessarily finite, distributed regularly*, provided that $|K(x, \xi)| \leqq r(\xi)$, where $\int_a^{a'} r(\xi)\, d\xi$ is convergent.

Similarly we can discuss the equation where appear integrals of higher order.

7. For the sake of completeness in discussing the equation in one unknown let us consider the case of constant limits, and Professor Volterra's theorem on implicit functionals. We see that if for u sufficiently small, $|u| < M$, we have

$$|F[\overset{b}{\underset{a}{u}}]| < M,$$

and $$|F[\overset{b}{\underset{a}{u_1}}] - F[\overset{b}{\underset{a}{u_2}}]| \leqq A \max_a^b |u_1 - u_2|,$$

where A is some constant, we shall have, for λ sufficiently small, one and only one solution of the equation

$$u(x) = \lambda F[\overset{b}{\underset{a}{u}}].$$

This same result can be obtained by introducing the parameter into the functional itself as a new function ϕ on which it explicitly depends. This $F[\overset{b}{\underset{a}{u, \phi}}]$ shall be continuous in u and ϕ†, and shall vanish for $u = 0$, $\phi = 0$. Moreover for ϕ constant and equal to zero, δF shall vanish when $u = 0$. The equation

$$u = F[\overset{b}{\underset{a}{u, \phi}}] \qquad (4)$$

has then one and only one solution in the neighbourhood of $u = 0$, $\phi = 0$ that changes continuously to $u \equiv 0$ when ϕ becomes $\phi \equiv 0$.

* See M. Bôcher, *Introduction to Integral Equations*, Cambridge University Press, 1909, p. 3, for a definition of regularly distributed.

† See V. Volterra, *loc. cit.* agosto, 1887, Nota I.

This lemma is sufficient to establish the desired theorem. Let us suppose that

$$G[u, \overset{b}{\underset{a}{\phi}}] = 0 \quad \ldots\ldots\ldots\ldots\ldots\ldots\ldots\ldots\ldots\ldots\ldots\ldots (5)$$

is satisfied identically for $\phi = 0$, $u = 0$, and that the variation can be written in the form

$$\delta G = \delta u(\xi) + \int_a^b G'[\overset{b}{\underset{a}{u}}, \overset{b}{\underset{a}{\phi}}, \xi, \xi_1]\, \delta u(\xi_1)\, d\xi_1$$
$$+ \int_a^b \Phi'[\overset{b}{\underset{a}{u}}, \overset{b}{\underset{a}{\phi}}, \xi, \xi_1]\, \delta\phi(\xi_1)\, d\xi_1 + A(\xi)\, \delta\phi(\xi),$$

which becomes for $\phi = 0$, $u = 0$,

$$\delta G = \delta u(\xi) + \int_a^b F(\xi, \xi_1)\, \delta u(\xi_1)\, d\xi_1 + \int_a^b \Gamma(\xi_1 \xi_2)\, \delta\phi(\xi_1)\, d\xi_1 + A(\xi)\, \delta\phi(\xi).$$

Professor Volterra shows that if the Fredholm determinant of this integral expression in δu does not vanish, the equation (5) can be rewritten in the form (4), and hence has one and only one solution in the neighbourhood of $u = 0$, $\phi = 0$. In fact, into the equation (5), there can be introduced in this way a large part of the well-known analysis of the Fredholm equation.

The generality of this theorem can be inferred from the fact that if the functional F of equation (3) depends only on u and ϕ and possesses functional derivatives in regard to them both, theorem 1 can be shown to be a special case of this implicit functional theorem. The property of possessing a functional derivative is however a more essential restriction in the case of functionals than is the corresponding property in the case of functions, and is not assumed in §§ 1—6.

II. Systems of Functional Equations.

8. We have the following theorem:

Theorem 2. *If the n functionals* $F[u, \overset{t}{\underset{St_0}{v}}, \ldots]$, $G[u, \overset{t}{\underset{St_0}{v}}, \ldots], \ldots$, *in n arguments $u, v, \ldots$, satisfy condition* I′ *in the interval $t_0 t_1$, there exists one and only one system of solutions of the system of equations*

$$\left.\begin{aligned} u(x, y, z, t) &= F[u(\xi, \underset{S}{\eta}, \zeta, \overset{t}{\underset{t_0}{\tau}}), \quad v(\xi, \underset{S}{\eta}, \zeta, \overset{t}{\underset{t_0}{\tau}}), \ldots] \\ v(x, y, z, t) &= G[u(\xi, \underset{S}{\eta}, \zeta, \overset{t}{\underset{t_0}{\tau}}), \quad v(\xi, \underset{S}{\eta}, \zeta, \overset{t}{\underset{t_0}{\tau}}), \ldots] \\ &\vdots \end{aligned}\right\} \ldots\ldots\ldots\ldots\ldots (6),$$

such that the system $u, v, \ldots$ remains in $\Sigma_{xyztuv\ldots}$.

If, in addition, the functionals $F, G, \ldots$ are continuous in t, except for a certain number of fixed values $t', t'', \ldots$, when their arguments lie in $\Sigma_{xyztuv\ldots}$ and are continuous in t except for a finite number of discontinuities, then the unique system in Σ satisfying (6) will have all its members continuous except at the points $t', t'', \ldots$.

9. **Theorem 3.** *If, in addition to satisfying condition* I′, *the functionals*

$$\bar{u} = F[\overset{t}{\underset{St_0}{\phi}}, \theta, \ldots], \quad \bar{v} = G[\overset{t}{\underset{St_0}{\phi}}, \theta, \ldots], \ldots,$$

are so related that when

$$\theta = L(\phi), \ldots,$$

($L(\phi), \ldots$ being certain differential relations in ϕ) it follows that $\bar{v} = L(\bar{u}), \ldots$, then there is one and only one solution of the equation

$$u(x, y, z, t) = F\overset{t}{\underset{St_0}{[}} u, L(u), \ldots] \quad \ldots\ldots\ldots\ldots\ldots\ldots\ldots\ldots(7),$$

such that the system $u, L(u), \ldots$ remains in $\Sigma_{xyztu,\ L(u),\ \ldots}$.

The clause relating to continuity may also be added to this last theorem provided that there exists at least one system $u, L(u), \ldots$ in Σ such that $u, L(u), \ldots$ are each continuous in t except for a finite number of discontinuities. It is obvious that by assigning different properties of continuity to the functionals $F, G, \ldots$, different restrictions need be imposed on $u, L(u), \ldots$, or in connection with theorem 2, on the functions $u, v, \ldots$.

In regard to this theorem it may seem that the choice of the region Σ is no longer an obvious process. It may be remarked however that for the equations that come up in practice the Σ is generally determined by relations of the form

$$|u - \phi| < M_1,$$
$$|v - \psi| < M_2,$$
$$\vdots$$

where ϕ, ψ are certain functions put in evidence by the character of the special equation.

10. Within this theorem are contained the existence theorems of many types of integro-differential equations. We see for instance that the equation

$$u(t) = \phi(t) + \int_{t_0}^{t} K(t, \tau)\, Q\left(u(\tau), \frac{du(\tau)}{d\tau}\right) d\tau \quad \ldots\ldots\ldots\ldots\ldots\ldots(8),$$

where Q is a polynomial in u and $\frac{du}{dt}$, where $K(t, \tau)$ and $\frac{\partial K(t, \tau)}{\partial t}$ are continuous except for discontinuities regularly distributed, for $t_0 \leqq \tau \leqq t \leqq t_1$, and are in absolute value less than $r(\xi)$, where $\int_{t_0}^{t_1} r(\xi)\, d\xi$ is convergent, and where $K(t, t) = 0$, has one and only one solution finite, with its derivative, and continuous with its derivative except for a finite number of discontinuities. This integral equation has application to the differential equation of Riccati.

Another special case of some interest is the following. Let us consider the integro-differential equation:

$$\alpha(u(t)) + \int_{t_0}^{t} \alpha_1(u(\tau))\, d\tau + \int_{t_0}^{t} d\tau \int_{t_0}^{\tau} \alpha_2(u(\tau), u(\tau'))\, d\tau' + \ldots$$
$$+ \int_{t_0}^{t} d\tau \int_{t_0}^{\tau} d\tau' \int_{t_0}^{\tau^{(k-1)}} \alpha_{k+1}(u(\tau), u(\tau'), \ldots u(\tau^{(k)}))\, d\tau^k = 0 \quad \ldots(9),$$

in which α represents a differential expression of the nth order in $u(t)$, α_1 represents a differential expression of the nth order in $u(\tau)$ with coefficients continuous functions of t and τ, α_2 represents one of the nth order in $u(\tau)$ and $u(\tau')$ with coefficients continuous functions of t, τ and τ', and so on. Let us suppose that the equation

$$\alpha(u(t)) = p(t),$$

in which p is an arbitrary function, can be written in the canonical form

$$\frac{d^n u}{dt^n} = R\left(p, u, \frac{du}{dt}, \ldots, \frac{d^{n-1}u}{dt^{n-1}}\right) \quad \text{.........................(9')},$$

in which R is a rational integral function of its arguments. Then there is one and only one solution, continuous with its first $n-1$ derivatives, that takes on, with its derivatives, at $t = t_0$, the values $u_0, u_0', \ldots u_0^{(n-1)}$ respectively. For the equation (9) may be rewritten in the form of the system of equations following, the first of which is obtained by substituting for $p(t)$ in (9') its functional value determined from (9), and the rest of which are determined by this and the initial conditions:

$$\left.\begin{aligned} \frac{d^n u}{dt} &= F\left[u, \overset{t}{\underset{t_0}{\ldots}}, \frac{d^n u}{dt^n}\right] \\ \frac{d^{n-1}u}{dt} &= u_0^{(n-1)} + \int_{t_0}^{t} F\left[u, \overset{\tau}{\underset{t_0}{\ldots}}, \frac{d^n u}{d\tau^n}\right] d\tau \\ \vdots \quad & \quad \vdots \qquad\qquad \vdots \\ u &= p_0(t) + \int_{t_0}^{t} dt \int_{t_0}^{\tau} d\tau \ldots \int_{t_0}^{\tau^{(n-3)}} F\left[u, \ldots, \frac{d^n u}{d\tau^{(n-2)\,n}}\right] d\tau^{(n-2)} \end{aligned}\right\} \quad \text{.........(10)}.$$

Defining Σ by the relations

$$\left|\frac{d^n u}{dt^n}\right| < M, \quad \left|\frac{d^{n-1}u}{dt^{n-1}}\right| < |u_0^{(n-1)}| + \epsilon, \ \ldots \ |u(t)| < |u_0| + \epsilon,$$

M being chosen sufficiently large (its value depending on ϵ), and ϵ being chosen arbitrarily small or large, we see that the equation for u satisfies the conditions of theorem 3, and has therefore one and only one solution in Σ.

11. Theorem 3 may be still further specialized to advantage. Representing by $\alpha(u)$, $L(u)$, $N(u)$, $L'(u)$ differential expressions in u, we have the following boundary value theorem:

THEOREM 4. *The equation*

$$\alpha(u(x, y, z, t)) = P[u, \overset{t}{\underset{t_0}{L}}(u), \ldots] \quad \text{.......................(11)},$$

under the surface conditions

$$f_\sigma^{(i)}[u, N(u), \overset{t}{\underset{t_0}{\ldots}}, \phi_1, \phi_2, \ldots \phi_k] = 0, \quad i = 1, 2, \ldots, m \quad \text{..............(12)},$$

has one and only one regular solution provided that the equation*

$$\alpha(\bar{u}) = p(x, y, z, t)$$

under the same conditions has one and only one regular solution, that can be written in the form

$$\bar{u} = M[p, \overset{t}{\underset{St_0}{L'}}(p), \ldots],$$

$$L(\bar{u}) = J[p, \overset{t}{\underset{St_0}{L'}}(p), \ldots],$$

$$\vdots \qquad \vdots \qquad \vdots$$

where the functionals M, J, ... are such that $M[P, L'(\overset{t}{\underset{St_0}{P}}), \ldots]$, $J[P, L'(\overset{t}{\underset{St_0}{P}}), \ldots]$, ... *are functionals of u, $L(u)$, ..., that satisfy condition* I'.

* "Regular" means such that u, $L(u)$, ... exist and lie in Σ.

In fact the equations (11), (12) combined are equivalent to the equations

$$u = M[P, L'(\overset{t}{\underset{St_0}{P}}), \ldots],$$

$$L(u) = J[P, L'(\overset{t}{\underset{St_0}{P}}), \ldots],$$

$$\vdots \qquad \vdots \qquad \vdots$$

which by theorem 2 have one and only one solution in Σ.

12. It is easily shown that the integro-differential equation of parabolic type offers an example of the application of this theorem. The equation may be written in the form

$$\frac{\partial^2 u(x, t)}{\partial x^2} = \frac{\partial u(x, t)}{\partial t} - \int_{t_0}^{t} B(t, \tau) \frac{\partial u(x, \tau)}{\partial \tau} d\tau,$$

and be brought immediately under the theorem by the use of the properties of the Green's function belonging to the parabolic differential equation. This is substantially the analysis carried through already in its treatment*.

Perhaps in regard to this parabolic equation it may be opportune to remark that a "closed" form can be found for its solution; i.e., it can be expressed by means of a finite number of quadratures in terms of the solution of a *linear* integral equation and the solution of the partial differential equation of parabolic type. The development of this subject would be extraneous, however, to the scope of the present paper. The remark is introduced to illustrate the fact that although there is no assumption of linearity in the general theorems 1, 2, 3, 4, as soon as we seek for results convenient for calculation or close discussion we must introduce such an hypothesis,—at least in all the methods so far devised†.

13. In order to tell when a given equation comes under the case of theorem 4, it is useful to have theorems about the transitivity of the relations expressed in conditions I and I′. We have for instance the theorem: If $F[\overset{t}{\underset{t_0}{u}}]$ and $G[\overset{t}{\underset{t_0}{u}}]$ satisfy condition I, then

$$F[\overset{t}{\underset{t_0}{G}}[\overset{\tau}{\underset{t_0}{u}}]]$$

satisfies condition I.

In fact we can set up a series of sub-intervals $t_0, \tau_1', \tau_1'\tau_2', \ldots \tau_k' t_1$, where the points $\tau_1', \tau_2', \ldots \tau_k'$ comprise all the points of division belonging to either F or G, taken in order, such that condition I is satisfied by both F and G with this set of intervals. Assuming then that the condition is satisfied by $F[G[u]]$ in the region $t_0 \tau_i'$ we have for the interval $\tau_i \tau_{i+1}$ that if u lies in Σ then $F[G[u]]$ lies in Σ and

$$\max_{t_0}^{t} | F[\overset{t}{\underset{t_0}{G}}[\overset{\tau}{\underset{t_0}{u_1}}]] - F[\overset{t}{\underset{t_0}{G}}[\overset{\tau}{\underset{t_0}{u_2}}]] | \leq A_F \max_{t_0}^{t} | G[\overset{t}{\underset{t_0}{u_1}}] - G[\overset{t}{\underset{t_0}{u_2}}] |$$

$$\leq A_F A_G \max_{\tau_i'}^{t} | u_1 - u_2 |,$$

* Evans: Sull' equazione integro-differenziale di tipo parabolico [*Rendiconti della R. Acc. dei Lincei*, Vol. XXI. fasc. 1°, 1912]. For another treatment see L. Amoroso: Sopra un' equazione integro-differenziale del tipo parabolico [*ibid.* fasc. 2°, *et seq.* 1912].

† Cf. for instance the method of Green's theorem and of partial solutions.

so that the condition is satisfied for the interval $\tau_i' \tau_{i+1}'$. Similarly it is satisfied for the first interval, and is therefore established by mathematical induction throughout. It is obviously inessential that both functionals should commence at the same value $t = t_0$.

The generalization of this theorem to systems of n functionals in n arguments is immediate. We have, still more generally,

THEOREM 5. *If the system of p functionals in n functions*

$$
\begin{aligned}
(a) \quad & F[u, v, \ldots], \\
& G[u, v, \ldots], \\
& \vdots \qquad \vdots
\end{aligned}
$$

satisfies the condition II′ (i), *changing the space* Σ_{xyztuv} *into the space* $H_{xyztFG\ldots}$, *and the system of n functionals in p functions*

$$
\begin{aligned}
(b) \quad & \Phi[\phi, \psi, \ldots], \\
& \Psi[\phi, \psi, \ldots], \\
& \vdots \qquad \vdots
\end{aligned}
$$

satisfies II′ (i), *changing the space* $H_{xyzt\phi\psi\ldots}$ *back again to the space* $\Sigma_{xyzt\phi\psi\ldots}$, *and if in their respective spaces the systems of functionals satisfy condition* II′ (ii), *thus determining two constants* B_1, B_2, *then the system of n functionals in n functions*

$$
\begin{aligned}
(1) \quad & \Phi\big[F[u, v, \ldots], G[u, v, \ldots], \ldots\big], \\
& \Psi\big[F[u, v, \ldots], G[u, v, \ldots], \ldots\big], \\
& \vdots \qquad \vdots
\end{aligned}
$$

and the system of p functionals in p functions

$$
\begin{aligned}
(2) \quad & F\big[\Phi[\phi, \psi, \ldots], \Psi[\phi, \psi, \ldots], \ldots\big], \\
& G\big[\Phi[\phi, \psi, \ldots], \Psi[\phi, \psi, \ldots], \ldots\big], \\
& \vdots \qquad \vdots
\end{aligned}
$$

satisfy the condition I′, *provided that*

$$B_1 B_2 < 1.$$

14. This theorem finds its immediate application in theorem 4, a type of the boundary-value problem. It also finds its application in the following theorem 6, which we call a type of the "canonical" problem. We give a very special theorem, of which the generalizations are obvious.

THEOREM 6. *If when* $\psi(t)$ *lies in a certain* $\sigma_{t,\psi}$, *the equation*

$$\psi(t) + F[u\overset{t}{\underset{t_0}{(\tau)}}] = 0$$

has the n parameter solution

$$u(t) = h[\psi\overset{t}{\underset{t_0}{(t)}}, c_1, c_2, \ldots, c_n]$$

where $c_1, c_2, \ldots, c_n$ *are the n independent parameters, then the equation*

$$G[u\overset{t}{\underset{t_0}{(\tau)}}] + F[u\overset{t}{\underset{t_0}{(\tau)}}] = 0$$

has an n parameter solution, provided that the functional

$$h\,[\overset{t}{\underset{t_0}{G}}\,[\overset{\tau}{\underset{t_0}{u}}],\; c_1,\, c_2,\, \ldots,\, c_n]$$

satisfies condition I *for fixed values of* $c_1, c_2, \ldots, c_n$, *and the functional* $G\,[\overset{t}{\underset{t_0}{u}}]$ *satisfies the condition* II (i) *changing the* $\Sigma_{t,\,u}$ *into the* $\sigma_{t,\,G}$.

Moreover, all solutions of the equation that lie in $\Sigma_{t,\,u}$ *are contained in this n parameter family.*

15. An interesting special case of this theorem is afforded by the equation of the third kind with variable limits

$$f(t)\,u(t) = \phi(t) + \int_{t_0}^{t} K(t, \tau)\,u(\tau)\,d\tau.$$

With the special hypothesis that $K(t_0, t_0) \neq 0$*, the equation can be treated by putting

$$F\,[\overset{t}{\underset{t_0}{u}}(\tau)] = -f(t)\,u(t) + \int_{t_0}^{t} K(t_0, t_0)\,u(\tau)\,d\tau.$$

But a discussion of this special equation would be outside the scope of the present paper.

* See Evans: Volterra's integral equation of the second kind, with discontinuous kernel [*Transactions of the American Math. Soc.* Vol. XII. Oct. 1911, p. 429]. The results there obtained can however be much simplified.

EINE NEUE RANDWERTAUFGABE FÜR DAS LOGARITHMISCHE POTENTIAL

VON H. A. V. BECKH-WIDMANSTETTER.

$z(x, y)$ soll innerhalb eines Kreises mit dem Radius 1 um den Ursprung der Gleichung genügen:

$$\frac{\partial^2 z}{\partial x^2} + \frac{\partial^2 z}{\partial y^2} = 0.$$

Prof. Wirtinger ist gelegentlich auf eine Randbedingung gestossen:

$$x^2 \frac{\partial^2 z}{\partial x^2} + 2xy \frac{\partial^2 z}{\partial x \partial y} + y^2 \frac{\partial^2 z}{\partial y^2} = 0.$$

Das gibt Anlass sich zu vergegenwertigen, dass bisher nur solche Randwertaufgaben behandelt wurden, wo die Funktion selbst oder eine Ableitung nach einer bestimmten Richtung am Rand vorgegeben ist. Der Vortragende glaubt, dass es mathematisches Interesse hat andere Randbedingungen zu behandeln, wie diese:

$$Ax^2 \frac{\partial^2 z}{\partial x^2} + 2Bxy \frac{\partial^2 z}{\partial x \partial y} + Cy^2 \frac{\partial^2 z}{\partial y^2} = F(x, y).$$

A, B, C sind Konstante, die gewissen Bedingungen nicht genügen dürfen, um sich nicht den alten Aufgaben zu nähern.

Es werden Fourier'sche Reihen angesetzt:

$$z = b_0 + \sum_{k=1}^{\infty} r^k (a_k \sin k\phi + b_k \cos k\phi);$$

$$F(x, y) = f(\phi) = d_0 + \sum_{k=1}^{\infty} (c_k \sin k\phi + d_k \cos k\phi).$$

Die Lösung bleibt unbestimmt, bis auf eine Lösung U der Aufgabe:

$$Ax^2 \frac{\partial^2 U}{\partial x^2} + 2Bxy \frac{\partial^2 U}{\partial x \partial y} + Cy^2 \frac{\partial^2 U}{\partial y^2} = 0.$$

Zu deren Bestimmung ergeben sich Gleichungen von der Form

$$p\alpha_k + q\alpha_{k+2} + \alpha_{k+4} = 0,$$

wobei sich p und q einfach aus A, B und C zusammensetzt und

$$\alpha_k = k(k-1)\, a_k \text{ ist.}$$

So kann man U mit Hilfe elementarer Transzendenten linear in 8 willkürlichen Konstanten darstellen.

Für z erhält man Gleichungen wie

$$p\alpha_k + q\alpha_{k+2} + \alpha_{k+4} = c_k.$$

Der Vortragende hat die Methode der Variation der Konstanten verschmäht und eine unmittelbare Behandlung der Differenzengleichung mit konstanten Koeffizienten gegeben, die ihm bequemere Ausdrücke liefert.

Es gelingt die Fourier'sche Reihe für z darzustellen und nach Einführung von

$$c_k = \frac{1}{\pi}\int_{-\pi}^{+\pi} f(\psi) \sin k\psi \, d\psi, \text{ u.s.w.},$$

sogar zu summieren. Es ist so die Lösung bis zur Darstellung als bestimmtes Integral, dem Analogon des Poisson'schen, durchgeführt. Betreffs dieser ungemein komplizierten Ausdrücke verweist der Vortragende auf seine Arbeit in den *Monatsheften f. Math. u. Phys.* XXIII. Jahrg. 1912. Er wünscht das Interesse der Mathematiker auf dieses neue Gebiet zu lenken. Durch die vollständige Durchrechnung einer solchen Aufgabe hofft der Verfasser eine genaue Diskussion des Resultates eventuell eine weniger formale Gewinnung desselben anzuregen und glaubt die formale Rechnung selbst für ähnliche Aufgaben erleichtert zu haben.

A MECHANISM FOR THE SOLUTION OF AN EQUATION OF THE n^{TH} DEGREE

By W. Peddie.

In the well-known system of pulleys illustrated in the diagram below, the free end P of the last cord moves down through a distance $2^n - 1$ if the bar ad be moved up through unit distance. Here n is the total number of pulleys including the fixed one. Suppose now that ad be fixed, while a, b, c, d, are drums on which the

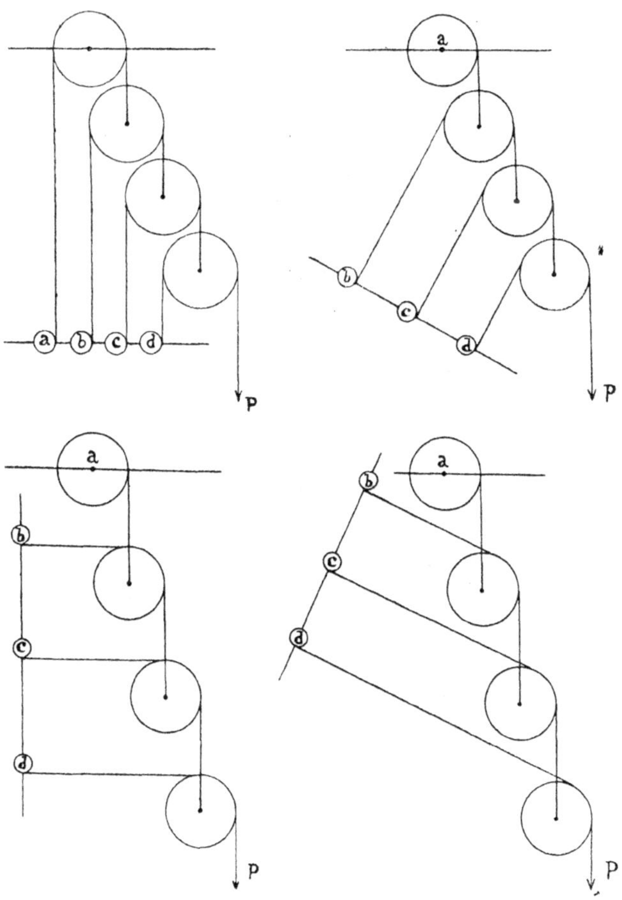

Fig. 1.

respective cords are wound. If a length a be let off the drum a, the free end of the last cord descends by the amount $2^n a$, the number of moving pulleys being n. If, in addition, a length b be unwound from the second pulley, P descends farther by the amount $2^{n-1}b$. If finally, after the various lengths have been let off, or wound on, the first n drums, the $(n+1)$th drum be adjusted so that P retakes its initial position, the equation

$$a \,.\, 2^n + b \,.\, 2^{n-1} + \ldots = 0$$

is satisfied. Thus the arrangement satisfies the conditions imposed by the relation

$$ax^n + bx^{n-1} + \ldots = 0$$

in the particular case in which $x = 2$. The fixed pulley may act the part of the drum a.

If bcd be vertical, the drums b, c, d, must be capable of sliding; and, if they are so slid that the parts of the cords adjacent to them are kept horizontal, the equation is satisfied in the particular case in which $x = 1$.

If the line bcd be fixed at an angle θ to the vertical, the drums being slid into positions in which the parts of the cords adjacent to them are perpendicular to bcd, the corresponding root of the equation is $1 \pm \sin\theta$. The plus sign occurs if bcd slopes downwards to the right; the minus sign occurs if it slopes downwards to the left. Either arrangement may be made the basis of construction of an instrument for finding the roots of an equation. It is convenient to make the axes of the pulleys slide in parallel slots on a metal arm to which the arm bcd is hinged, and the action of gravity on the pulleys is conveniently replaced by the control of cords wound on spring drums. To secure inextensibility thin steel wires may be used instead of cords.

If bcd be a fixed rigid arm, mechanical necessities prevent the axes of the pulleys coinciding with the axis of rotation of the arm carrying the pulleys. Hence, if the original displacements, measured parallel to θ, of the centres of the pulleys from the axis of rotation be α, β, etc., the lengths of wire which have to be let off the corresponding drums are $a - \alpha$, $b - \beta$, etc., instead of a, b, etc.

If desired this may be avoided by the following construction. Let $pqrstuv$ represent the rotating arm; qr, st, and uv, representing the lines along which the axes of the first, second, and third movable pulleys slide. The distances pq, rs, and tu, are of course equal to the radii of the sliding pulleys; more strictly, they are equal to the radius of a pulley plus the semi-diameter of the cord (or fine wire) which is wound upon it. The fixed drum a has its axis coincident with the axis of the hinge p by which the moving arm is attached to the fixed arm pw along which the drum b slides. Arms rxy and tz are hinged to the moving arm at the points r and t; and are compelled to remain parallel to pw by links wx and yz respectively.

The cord which passes over the first movable pulley and is wound on the drum b is guided by a pin carried by the free end of an arm m which is rigidly attached to the axle of the pulley. The pin m is at a distance qp from the line of qr. Thus the free part of the cord between m and b is, when the drum b is slid into position, equal to $pm \sin\theta$; and a similar arrangement is made with each of the movable pulleys.

The mechanism must be so arranged that, when the whole link-work is closed up, θ being zero, the pin m is on the prolongation of the axis p; the pin attached to the

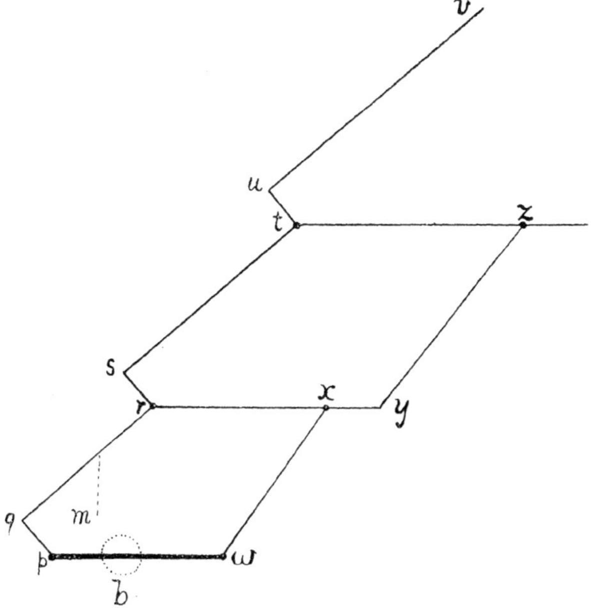

Fig. 2.

second sliding pulley coincides with the prolongation of the axis r; and so on. A length a is then unwound off the drum centred at p, and the first pulley slides out by the distance $pm = a$. So also does the second pulley; but, when the rotating arm is turned through the angle θ, the second pulley is pulled back by the amount $a \sin \theta$. And it moves out through the additional distance b, if that length be unwound off the drum b. Thus the necessary condition is satisfied for the second pulley; and similarly for the others.

The free end of the last cord being wound on a spring drum fixed at the outer end of the movable arm, the angular position of that drum is noted when, θ being zero, all the adjustments of the pulleys have been made, i.e. when m coincides with p, etc. The lengths a, b, etc., having been let off the respective drums, which have been slid if necessary to their appropriate positions, if the drum at the end of the movable arm has retaken its initial angular position, one root of the equation at least is unity. If the end drum is not in its initial position, the angle θ is increased, the drums being slid correspondingly, until the initial position is retaken. The value of $1 - \sin \theta$, which can be indicated directly on a scale by a pointer attached to the movable arm, is then a root of the equation. By appropriate graduation of the end drum, the sum of the terms of the function, for any value of the variable from 0 to 1, can be directly indicated.

It is presumed that the radius, ρ, of any pin m, is so small that $r\theta$ is negligible so far as scale readings are concerned.

If no root of the equation lies within the limits 0 and 1, or at either limit, the equation must be altered so as to realise that condition.

Lastly, the instrument can be also used to solve an equation one degree higher than the number of movable pulleys. Thus one constructed to solve a cubic can

Fig. 3.

solve a quartic. Each term being divided by x, the last term will be, say, $1/x$. If squared paper be fastened on the baseboard of the instrument, and the scale indicating $x = 1 - \sin\theta$ be marked on it, as also the curve $xy = 1$, any value of x, at which the reading of the end drum exceeds its initial reading by the value of y given by the hyperbola, is a root of the quartic.

It may be noted that the instrument, if set to solve a given equation with the left hand side equated to zero, will also give the solution of the equation with the right hand side equal to any constant different from zero, by altering θ until the end drum has a reading differing from its initial reading by the value of that constant. Thus it can be used to trace the value of the function.

SOPRA EQUAZIONI DI TIPO INTEGRALE

Di Vito Volterra.

Lo studio delle *funzioni di linee*, o, come è anche chiamato, dei *funzionali*, che ho cominciato in maniera sistematica dal 1887, mi ha condotto a quello delle equazioni integrali lineari. In virtù dei principii dai quali sono partito, sono stato condotto, per primo, a considerare queste equazioni come il caso limite di equazioni algebriche allorchè il loro numero e quello delle incognite crescono indefinitamente. Tale passaggio al limite è analogo a quello fondamentale del calcolo integrale. Ho poi considerato delle equazioni non lineari in una nota pubblicata nel 1906 nei *Comptes Rendus de l'Académie des Sciences* e dei casi ancora più generali nelle mie lezioni fatte lo scorso inverno alla Sorbona.

Oltre alle equazioni integrali ho studiato le equazioni integro-differenziali dandone la teoria in vari casi, nei quali ho sempre fatto uso del principio da cui ero partito precedentemente. Le ho cioè riguardate come casi limiti di un numero infinitamente crescente di equazioni con un numero pure infinitamente crescente di incognite. Ma io quì desidero di ricordare alcuni teoremi che ho dati recentemente, i quali fanno rientrare tutte le precedenti trattazioni di equazioni integrali e integro-differenziali come casi particolari.

Ho perciò introdotto una speciale operazione che ho chiamato *composizione* che può considerarsi di due tipi diversi, cioè a limiti variabili e a limiti costanti.

Date due funzioni $F_1(x, y)$, $F_2(x, y)$ finite e continue, la composizione a limiti variabili o *composizione di prima specie* consiste nell' operazione

$$\int_x^y F_1(x, \xi)\, F_2(\xi, y)\, d\xi \quad \ldots\ldots\ldots\ldots\ldots\ldots\ldots\ldots(1),$$

mentre quella a limiti costanti o *composizione di seconda specie* consiste nella operazione

$$\int_p^q F_1(x, \xi)\, F_2(\xi, y)\, d\xi \quad \ldots\ldots\ldots\ldots\ldots\ldots\ldots\ldots(2),$$

ove p e q sono quantità costanti.

Ora, se scambiando le due funzioni F_1 e F_2 nella prima formula (1) il resultato non cambia, ho detto che F_1 e F_2 sono *permutabili di prima specie*, mentre se il medesimo scambio non altera il resultato della seconda operazione (2) ho chiamato F_1 e F_2 *permutabili di seconda specie*. Ciò premesso ho dimostrato il teorema che *combinando per somma, per sottrazione o, in generale, combinando lineamente con coefficienti costanti delle funzioni permutabili e combinando mediante composizione delle funzioni permutabili si trovano sempre funzioni permutabili fra loro e colle funzioni primitive.*

Ma le due proprietà più importanti sono le seguenti:

1°. Se

$$a_1x + a_2y + a_3z + \ldots + a_{11}x^2 + a_{22}y^2 + a_{12}xy + \ldots + a_{111}x^3 + a_{123}xyz + \ldots \quad \ldots\ldots(3)$$

è un elemento di una funzione analitica di un numero qualsiasi di variabili e si sostituiscono ad $x, y, z\ldots$ *le espressioni* $xF_1, yF_2, zF_3\ldots$ *ove* $F_1, F_2, F_3\ldots$ *sono funzioni permutabili di prima specie e si interpretano i prodotti e le potenze delle* $F_1, F_2, F_3\ldots$ *(invece che come operazioni algebriche) come operazioni di composizione di prima specie, la serie che si trova è una funzione intera di* $x, y, z\ldots$.

2°. *Se si sostituiscono nella* (3) *a* $x, y, z\ldots$ *le espressioni* $xF_1, yF_2, zF_3\ldots$ *essendo* $F_1, F_2, F_3\ldots$ *funzioni permutabili di seconda specie e si interpretano i prodotti e le potenze delle* $F_1, F_2, F_3\ldots$ *come operazioni di composizione di seconda specie, e, se la serie* (3) *è il rapporto di due funzioni intere, anche la serie che si trova dopo la sostituzione è il rapporto di due funzioni intere di* $x, y, z\ldots$.

L'origine di questi teoremi va ricercata sempre nello stesso principio che corrisponde al solito passaggio al limite di cui abbiamo parlato. Infatti le operazioni di composizione (1), (2) possono riguardarsi come *operazioni limiti di somme.* Si considerino infatti le quantità

$$m_{i_1h},\ n_{i_1h}\ (i_1h = 1, 2, 3 \ldots g).$$

Si può dapprima considerare la somma

$$\sum_{i+1}^{h-1}{}_s\, m_{is}n_{sh} \quad \ldots\ldots(4),$$

e la permutabilità di prima specie sarà data da

$$\sum_{i+1}^{s-1}{}_s\, m_{is}n_{sh} = \sum_{i+1}^{s-1} n_{is}m_{sh} \quad \ldots\ldots(4').$$

Se passiamo al limite, coll' analogo procedimento del calcolo integrale, la operazione (4), da luogo alla composizione di prima specie e la condizione (4′) alla permutabilità di prima specie.

In modo simile la composizione di seconda specie può considerarsi come il limite della operazione

$$\sum_{1}^{g}{}_s\, m_{is}n_{sh},$$

e la permutabilità di seconda specie come la condizione limite di

$$\sum_{1}^{g}{}_s\, m_{is}n_{sh} = \sum_{1}^{g}{}_s\, n_{is}m_{sh}.$$

Ora si può cominciare dallo stabilire i due teoremi precedenti per il caso finito, il che non offre difficoltà, e procedere quindi alla loro estensione al caso infinito.

Una volta stabiliti questi teoremi supponiamo che la serie (3) sia soluzione di un problema algebrico o differenziale. Se noi sostituiamo nelle equazioni algebriche o differenziali, ridotte a forma intera, alle lettere $x, y, z\ldots$ le $xF_1, yF_2, zF_3\ldots$ e interpretiamo i prodotti e le potenze delle $F_1, F_2, F_3\ldots$ come composizioni otteniamo equazioni integrali o equazioni integro-differenziali di cui le soluzioni sono immediatamente date per mezzo di funzioni intere o di rapporti di funzioni intere.

Possono perciò enunciarsi i due principii generali:

Ad ogni problema algebrico o differenziale la cui soluzione conduce a funzioni esprimibili mediante funzioni analitiche corrisponde un problema correlativo integrale o integro-differenziale (a limiti variabili, o di prima specie) la cui soluzione è data da funzioni intere.

Ad ogni problema algebrico o differenziale la cui soluzione conduce a funzioni esprimibili come rapporti di funzioni intere di un certo numero di variabili corrisponde un problema integrale o integro-differenziale di seconda specie (a limiti costanti) *la cui soluzione è pure esprimibile mediante rapporti di funzioni intere delle stesse variabili.*

E' facile riconoscere che il problema della risoluzione delle equazioni integrali lineari non è che un caso particolarissimo fra i problemi generali che sono abbracciati dai due principii precedenti.

La teoria delle funzioni permutabili da luogo a varie questioni che io stesso ho studiato. Essa conduce poi ad un' algebra che il Prof. G. C. Evans ha approfondito in modo molto elegante e che lo ha condotto e resultati molto interessanti. I Prof[ri]. Lauricella, Vessiot, Sinigaglia, Giorgi, Lalesco ed altri si sono pure occupati di questioni relative ad essa.

Mi propongo ora di estendere ulteriormente queste considerazioni. Consideriamo un gruppo continuo di funzioni permutabili, per esempio prendiamo

$$f(u \mid x, y)$$

tale che, u_1 e u_2 essendo due valori qualunque di u, si abbia

$$\int_x^y f(u_1 \mid x, \xi) f(u_2 \mid \xi, y)\, d\xi = \int_x^y f(u_2 \mid x, \xi)\, f(u_1 \mid \xi, y)\, d\xi = f(u_1, u_2 \mid x, y).$$

$f(u_1, u_2 \mid x, y)$ sarà permutabile con $f(u \mid x, y)$, cioè

$$\int_x^y f(u_3 \mid x, \xi)\, f(u_1, u_2 \mid \xi, y)\, d\xi = \int_x^y f(u_1, u_2 \mid x, \xi)\, f(u_3 \mid \xi, y)\, d\xi = f(u_1, u_2, u_3 \mid x, y),$$

e così di seguito.

Ciò premesso, estendiamo, col solito procedimento del passaggio dal finito all' infinito, un teorema dato precedentemente. A tal fine consideriamo la serie analoga a quelle di Taylor che ho dato fino dai miei primi lavori, cioè

$$A + \int_a^b F(u_1) f(u_1)\, du_1 + \int_a^b \int_a^b F'(u_1, u_2)\, f(u_1)\, f(u_2)\, du_1 du_2$$
$$+ \int_a^b \int_a^b \int_a^b F''(u_1, u_2, u_3)\, f(u_1) f(u_2) f(u_3)\, du_1 du_2 du_3 + \ldots,$$

ove le funzioni F sono simmetriche. Supponiamo che essa sia convergente allorchè

$$|f(u)| < M.$$

Sostituiamo a questa serie l' altra

$$A + \int_a^b F(u_1) f(u_1 \mid x, y)\, du_1 + \int_a^b \int_a^b F'(u_1, u_2) f(u_1, u_2 \mid x, y)\, du_1 du_2$$
$$+ \int_a^b \int_a^b \int_a^b F''(u_1, u_2, u_3)\, f(u_1, u_2, u_3 \mid x, y)\, du_1 du_2 du_3 + \ldots.$$

Questa serie sarà convergente comunque grande sia il modulo di $f(u|x, y)$ purchè sia finito.

E' evidente che questo teorema è un' estensione del teorema 1°. E' facile vedere delle applicazioni di questo teorema. Consideriamo l' equazione del tipo trascendente

$$\phi(u|x, y) = f(u|x, y) + \int_a^b F'(u|u_1) f(u_1|x, y)\, du_1$$
$$+ \int_a^b \int_a^b F''(u|u_1, u_2) f(u_1, u_2|x, y)\, du_1 du_2$$
$$+ \int_a^b \int_a^b \int_a^b F'''(u|u_1, u_2, u_3) f(u_1, u_2, u_3|x, y)\, du_1 du_2 du_3 + \ldots \ldots \text{(I}a\text{)}$$

ove $f(u|x, y)$ è la incognita.

Supponiamo che l' insieme delle funzioni date $\phi(u|x, y)$ formi un gruppo continuo di funzioni permutabili, cioè

$$\int_x^y \phi(u_1|x, \xi)\, \phi(u_2|\xi, y)\, d\xi = \int_x^y \phi(u_2|x, \xi)\, \phi(u_1|\xi, y)\, d\xi = \phi(u_1, u_2|x, y).$$

Consideriamo d' altra parte l' equazione

$$\phi(u) = f(u) + \int_a^b F'(u|u_1) f(u_1)\, du_1 + \int_a^b \int_a^b F''(u|u_1, u_2) f(u_1) f(u_2)\, du_1 du_2 + \ldots.$$

Se il determinante dell' equazione integrale

$$f(u) + \int_a^b F'(u|u_1) f(u_1)\, du = \psi(u)$$

è diverso da zero, ho dimostrato che si può dare una soluzione dell' equazione precedente sotto la forma

$$f(u) = \phi(u) + \int_a^b \Phi'(u|u_1)\, \phi(u_1)\, du_1 + \int_a^b \int_a^b \Phi''(u|u_1, u_2)\, \phi(u_1)\, \phi(u_2)\, du_1 du_2 + \ldots,$$

valido finchè il modulo di $\phi(u)$ è inferiore ad un certo limite. Ne viene che la soluzione della (Ia) sarà

$$f(u|x, y) = \phi(u|x, y) + \int_a^b \Phi'(u|u_1)\, \phi(u_1|x, y)\, du_1$$
$$+ \int_a^b \int_a^b \Phi''(u|u_1, u_2)\, \phi(u_1, u_2|x, y)\, du_1 du_2$$
$$+ \int_a^b \int_a^b \int_a^b \Phi'''(u|u_1, u_2, u_3)\, \phi(u_1, u_2, u_3|x, y)\, du_1 du_2 du_3 + \ldots \ldots \text{(I}b\text{)}$$

e non vi sarà più bisogno di alcuna limitazione circa la grandezza del modulo di $\phi(u|x, y)$ purchè finito.

E' facile riconoscere quali sono le estensioni del teorema 2° e degli altri al caso che abbiamo adesso indicato, e le conseguenze ed applicazioni che possono trarsene.

ELLIPTIC AND ALLIED FUNCTIONS; SUGGESTIONS FOR REFORM IN NOTATION AND DIDACTICAL METHOD

By M. M. U. Wilkinson.

"Students of applied mathematics generally acquire their mathematical equipment as they want it for the solution of some definite actual problem." (So wrote Prof. Greenhill, 1892.) An adequate mathematical equipment for the "discussion of definite physical questions" is as necessary now as it was twenty years ago. But there is more for the student of modern science to study now than there was twenty years ago. And he is likely to find that the time he can devote to study is less than the time students could devote to study twenty years ago. Any reform, whether in notation, or in didactical method, which places, or tends to place, more time at the disposal of a student is surely worthy of consideration. The fate of Reforms, generally, is to be reformed away in a few years. But meanwhile they may have been extremely useful.

The object of this paper is to suggest certain reforms, in Notation and Didactical Method, in the elementary treatment of Elliptic and Allied Functions. The Reforms I suggest may have been suggested before. But, if they have been, I do not think attention has been at all generally given to them. I am very anxious that "methods and matter of teaching" should be "progressively adapted" to meet the needs of those who "lack any exceptional capacity." I make three suggestions:

The first is that the notation sn, cn, dn should be replaced by a notation based on the Weierstrassian notation of σ, σ_1, σ_2, σ_3.

The second is that the youthful student should be taught about the sigma function in a somewhat different way from that hitherto adopted.

The third is that more attention should be paid to the case, where

$$k^2 = -\frac{1}{2} + \frac{\sqrt{-3}}{2},$$

than has, I believe, been customary hitherto.

Of course, at first, the four sigmas are only considered in their ratios one to the other. So I propose to represent, in this paper, those ratios thus,

$$\left.\begin{aligned} &\operatorname{bs} u \,.\, \sigma u = \sigma_1 u ; \quad \operatorname{cs} u \,.\, \sigma u = \sigma_2 u ; \quad \operatorname{ds} u \,.\, \sigma u = \sigma_3 u ; \\ &\text{defining bs, cs, ds by the differential equations} \\ &\frac{d \operatorname{bs} u}{du} = -\operatorname{cs} u \operatorname{ds} u ; \quad \frac{d \operatorname{cs} u}{du} = -\operatorname{ds} u \operatorname{bs} u ; \quad \frac{d \operatorname{ds} u}{du} = -\operatorname{bs} u \operatorname{cs} u ; \end{aligned}\right\}\ldots(1);$$

and defining the Weierstrassian function, $\wp u$, thus,

$$\left.\begin{aligned}\frac{d\wp u}{du} &= -2(\wp u - e_1)^{\frac{1}{2}}(\wp u - e_2)^{\frac{1}{2}}(\wp u - e_3)^{\frac{1}{2}}\\ \text{and}\quad \left(\frac{d\wp u}{du}\right)^2 &= 4(\wp^3 u - g_1\wp^2 u - g_2\wp u - g_3)\end{aligned}\right\}\text{..............(2);}$$

$$\left.\begin{aligned}\text{so that}\quad \wp u &= \mathrm{bs}^2 u + e_1 = \mathrm{cs}^2 u + e_2 = \mathrm{ds}^2 u + e_3\\ \text{and}\quad \frac{d\wp}{du} &= -2\,\mathrm{bs}\,u\,\mathrm{cs}\,u\,\mathrm{ds}\,u\end{aligned}\right\}\text{..................(3).}$$

The formulae so obtained are superior, both in elegance and simplicity, to the formulae hitherto used.

Thus, in the addition theory, the old formulae will be replaced by

$$(\wp v - \wp u)\,\mathrm{bs}\,(u+v) = \mathrm{bs}\,u\,\mathrm{cs}\,v\,\mathrm{ds}\,v - \mathrm{bs}\,v\,\mathrm{cs}\,u\,\mathrm{ds}\,u,$$

or

$$\mathrm{bs}\,(u+v)(\mathrm{bs}\,u\,\mathrm{cs}\,u\,\mathrm{ds}\,v + \mathrm{bs}\,v\,\mathrm{cs}\,v\,\mathrm{ds}\,u) = \mathrm{ds}\,u\,\mathrm{bs}\,u\,\mathrm{ds}\,v\,\mathrm{bs}\,v - (e_1 - e_3)\,\mathrm{cs}\,u\,\mathrm{cs}\,v;$$

or

$$\mathrm{bs}\,(u+v)(\mathrm{bs}\,u\,\mathrm{cs}\,v\,\mathrm{ds}\,v + \mathrm{bs}\,v\,\mathrm{cs}\,u\,\mathrm{ds}\,u) = \mathrm{bs}^2 u\,\mathrm{bs}^2 v - (e_1 - e_2)(e_1 - e_3);\quad\text{...(4).}$$

also $\mathrm{bs}\,(u+v)\,\mathrm{bs}\,(u-v)(\wp v - \wp u) = \mathrm{bs}^2 u\,\mathrm{bs}^2 v - (e_1 - e_2)(e_1 - e_3)$,

etc., permuting (b, e_1), (c, e_2), (d, e_3).

These formulae are superior, both in elegance and simplicity, and in other ways, as is obvious if we consider the formulae which result. Thus, Sii', Sij', etc., having the meanings they have in Elementary Treatises on Quaternions, we can take

$$\left.\begin{array}{lll} PSii' = \mathrm{bs}\,v; & PSij' = -\mathrm{cs}\,(u+v); & PSik' = \mathrm{ds}\,u;\\ PSji' = \mathrm{cs}\,u; & PSjj' = \mathrm{ds}\,v; & PSjk' = -\mathrm{bs}\,(u+v);\\ PSki' = -\mathrm{ds}\,(u+v); & PSkj' = \mathrm{bs}\,u; & PSkk' = \mathrm{cs}\,v;\\ \text{where}\quad P^2 = \wp u + \wp v + \wp(u+v) - e_1 - e_2 - e_3 & & \end{array}\right\}\text{...(5);}$$

and if we take

$$\left.\begin{array}{c}\begin{vmatrix} Sii', & Sij', & Sik'\\ Sji', & Sjj', & Sjk'\\ Ski', & Skj', & Skk'\end{vmatrix} = +1,\\ P\,\mathrm{bs}\,v = \mathrm{ds}\,v\,\mathrm{cs}\,v + \mathrm{bs}\,u\,\mathrm{bs}\,(u+v),\\ \text{and require that (for sign)}\\ P(\mathrm{bs}^2 u - \mathrm{bs}^2 v) = \mathrm{bs}\,u\,\mathrm{cs}\,u\,\mathrm{ds}\,u - \mathrm{bs}\,v\,\mathrm{cs}\,v\,\mathrm{ds}\,v\end{array}\right\}\text{.....................(6);}$$

and the well-known equation (see Greenhill, p. 47, or Halphen, I. 25)

$$\frac{\wp' v - \wp' w}{\wp v - \wp w} = \frac{\wp' w - \wp' u}{\wp w - \wp u} = \frac{\wp' u - \wp' v}{\wp u - \wp v}$$

is thus (replacing $-w$ by $u+v$) simply expressed.

Of all the ways of introducing Elliptic Functions to youthful students I think the spherical triangle is the best. Thus, suppose PQR to be a very small spherical triangle, the angles P and R being very small.

R
Q
P

Take $\sin PQ = \dfrac{(e_1 - e_3)^{\frac{1}{2}}}{\mathrm{ds}\,u}$; $\sin QR = \dfrac{(e_1 - e_3)^{\frac{1}{2}}}{\mathrm{ds}\,v}$; $\sin PR = \dfrac{(e_1 - e_3)^{\frac{1}{2}}}{\mathrm{ds}\,(u+v)}$;

$$\sin PRQ = \frac{(e_2-e_3)^{\frac{1}{2}}}{\text{ds}\,u}; \quad \sin QPR = \frac{(e_2-e_3)^{\frac{1}{2}}}{\text{ds}\,v}; \quad \sin PQR = \frac{(e_2-e_3)^{\frac{1}{2}}}{\text{ds}(u+v)};$$

$$\cos PQ = \frac{\text{bs}\,u}{\text{ds}\,u}; \qquad \cos QR = \frac{\text{bs}\,v}{\text{ds}\,v}; \qquad \cos PR = \frac{\text{bs}\,(u+v)}{\text{ds}\,(u+v)};$$

$$\cos PRQ = \frac{\text{cs}\,u}{\text{ds}\,u}; \qquad \cos QPR = \frac{\text{cs}\,v}{\text{ds}\,v}; \qquad \cos PQR = -\frac{\text{cs}\,(u+v)}{\text{ds}\,(u+v)}.$$

But there are other ways, and I think some would prefer one way, and some another.

As an instance of the simplicity obtained by the use of a notation such as I suggest, I would instance

$$\text{bs}\,(\gamma-\alpha)\,\text{cs}\,(\alpha-\beta)+\text{cs}\,(\beta-\gamma)\,\text{ds}\,(\gamma-\alpha)+\text{ds}\,(\alpha-\beta)\,\text{bs}\,(\beta-\gamma)=0 \quad \ldots(7);$$

expressing in one formula six.

And, replacing Halphen's ω, ω', by ω, $i\omega'$ (so that ω, ω', be both positive), we have

$$\left.\begin{aligned} &\text{cs}^2\,\omega = e_1 - e_2; \quad \text{ds}^2\,\omega = e_1 - e_3; \\ \text{and} \quad &\text{bs}\,v\,\text{bs}\,(\omega - v) = \text{cs}\,\omega\,\text{ds}\,\omega; \\ &\text{bs}\,v\,\text{cs}\,(\omega \pm v) = \text{cs}\,\omega\,\text{ds}\,v; \\ &\text{bs}\,v\,\text{ds}\,(\omega \pm v) = \text{ds}\,\omega\,\text{cs}\,v; \\ &\text{etc., etc.} \end{aligned}\right\} \quad \ldots\ldots\ldots\ldots\ldots\ldots\ldots(8).$$

As Greenhill and Halphen both show, all the formulae in the old notation can be readily transformed into the new, since, if

$$\left.\begin{aligned} &\text{sn}\,(e_1-e_3)^{\frac{1}{2}}\,x\,\text{ds}\,u = (e_1-e_3)^{\frac{1}{2}}; \\ \text{we have} \quad &\text{cn}\,(e_1-e_3)^{\frac{1}{2}}\,x\,\text{ds}\,u = \text{bs}\,u; \\ &\text{dn}\,(e_1-e_3)^{\frac{1}{2}}\,x\,\text{ds}\,u = \text{cs}\,u; \end{aligned}\right\} \quad \ldots\ldots\ldots\ldots\ldots\ldots\ldots(9).$$

Another reform I would suggest is, to change the way in which the sigma function is introduced. I would suggest as follows, from (1), we find

$$\left.\begin{aligned} &\frac{1}{\sigma}\frac{d^2\sigma}{du^2} - \left(\frac{1}{\sigma}\frac{d\sigma}{du}\right)^2 + \frac{1}{\text{bs}\,u}\frac{d^2\,\text{bs}\,u}{du^2} - \left(\frac{1}{\text{bs}\,u}\frac{d\,\text{bs}\,u}{du}\right)^2 = \frac{1}{\sigma_1}\frac{d^2\sigma_1}{du^2} - \left(\frac{1}{\sigma_1}\frac{d\sigma_1}{du}\right)^2; \\ \text{now} \quad &\frac{1}{\text{bs}\,u}\frac{d^2\,\text{bs}\,u}{du^2} - \left(\frac{1}{\text{bs}\,u}\frac{d\,\text{bs}\,u}{du}\right)^2 = \text{ds}^2\,u + \text{cs}^2\,u - \frac{\text{cs}^2\,u\,\text{ds}^2\,u}{\text{bs}^2\,u} \\ &= 2\,\text{bs}^2\,u + 2e_1 - e_2 - e_3 - \frac{(\text{bs}^2\,u + e_1 - e_2)(\text{bs}^2\,u + e_1 - e_3)}{\text{bs}^2\,u} \\ &= \text{bs}^2\,u - \frac{(e_1-e_2)(e_1-e_3)}{\text{bs}^2\,u} = \text{bs}^2\,u - \text{bs}^2\,(\omega - u) \\ &= \wp u - \wp\,(\omega - u) \end{aligned}\right\} \quad (10).$$

Hitherto in this paper the sigmas are only connected by their ratios. So we may assume another relation connecting σ and σ_1, provided it is consistent with (1). So assume

$$\left.\begin{aligned} &\frac{1}{\sigma}\frac{d^2\sigma}{du^2} - \left(\frac{1}{\sigma}\frac{d\sigma}{du}\right)^2 = -\wp u; \\ \text{then shall} \quad &\frac{1}{\sigma_1}\frac{d^2\sigma_1}{du^2} - \left(\frac{1}{\sigma_1}\frac{d\sigma_1}{du}\right)^2 = -\wp\,(\omega - u) \end{aligned}\right\} \quad \ldots\ldots\ldots\ldots\ldots\ldots\ldots(11);$$

whence
$$\left.\begin{aligned}\frac{1}{\sigma}\frac{d^3\sigma}{du^3}-\frac{3}{\sigma^2}\frac{d\sigma}{du}\frac{d^2\sigma}{du^2}+2\left(\frac{1}{\sigma}\frac{d\sigma}{du}\right)^3&=-\frac{d\wp}{du}\\ \frac{1}{\sigma}\frac{d^4\sigma}{du^4}-\frac{4}{\sigma^2}\frac{d\sigma}{du}\frac{d^3\sigma}{du^3}+\frac{3}{\sigma^2}\left(\frac{d^2\sigma}{du^2}\right)^2-6\wp^2u&=-\frac{d^2\wp}{du^2}\\ &=-6\wp^2+g_1\wp+\tfrac{1}{2}g_2\end{aligned}\right\}\quad\ldots\ldots\ldots\ldots(12);$$
and (second line above)

so that
$$\sigma\frac{d^4\sigma}{du^4}-4\frac{d\sigma}{du}\frac{d^3\sigma}{du^3}+3\left(\frac{d^2\sigma}{du^2}\right)^2+g_1\left\{\sigma\frac{d^2\sigma}{du^2}-\left(\frac{d\sigma}{du}\right)^2\right\}-\tfrac{1}{2}g_2\sigma^2=0\ \ldots(13).$$

In like manner we can show that, in (13), σ may be replaced by either σ_1, σ_2, or σ_3.

(13) is one of those differential equations which solve themselves. For, assuming initial values for $\frac{d\sigma u}{du}$, $\frac{d^2\sigma u}{du^2}$, $\frac{d^3\sigma u}{du^3}$, and an initial value, not zero, for σu, every subsequent differential coefficient can be found. But we do not get much further that way. But if we assume

$$S=A\exp\left(\tfrac{1}{2}eu^2+bu\right)\sigma(u+\alpha)\quad\ldots\ldots\ldots\ldots\ldots\ldots(14);$$
$$\frac{1}{S}\frac{d^2S}{du^2}-\left(\frac{1}{S}\frac{dS}{du}\right)^2=e+\frac{1}{\sigma(u+\alpha)}\frac{d^2\sigma(u+\alpha)}{du^2}-\left(\frac{1}{\sigma(u+\alpha)}\frac{d\sigma(u+\alpha)}{du}\right)^2.$$

For convenience I represent, thus,

$$\left.\begin{aligned}\tfrac{1}{2}(d-d')^2(SS)&=S\frac{d^2S}{du^2}-\left(\frac{dS}{du}\right)^2=W(2)(SS);\\ \tfrac{1}{2}(d-d')^4(SS)&=S\frac{d^4S}{du^4}-4\frac{dS}{du}\cdot\frac{d^3S}{du^3}+3\left(\frac{dS}{du}\right)^2=W(4)(SS);\\ &\ldots\ldots\ldots\ldots\ldots\ldots\ldots\ldots\\ \tfrac{1}{2}(d-d')^{2n}(SS)&=S\frac{d^{2n}S}{du^{2n}}-2n\frac{dS}{du}\frac{d^{2n-1}S}{du^{2n-1}}+\ldots=W(2n)(SS);\\ &\ldots\ldots\ldots\ldots\ldots\ldots=\ldots\ldots\ldots\end{aligned}\right\}\ldots(15).$$

Then,

$$\left.\begin{aligned}&\frac{W(4)(SS)}{S^2}-6\left(\frac{W(2)(SS)}{S^2}\right)^2=\frac{W(4)(\sigma\sigma)}{\sigma^2}-6\left\{\frac{W(2)(\sigma\sigma)}{\sigma^2}\right\}^2,\\ &\text{so that we have}\\ &\frac{W(4)(SS)}{S^2}=\frac{W(4)(\sigma\sigma)}{\sigma^2}+12e\frac{W(2)(\sigma\sigma)}{\sigma^2}+6e^2;\\ &\text{and}\\ &\frac{W(4)(SS)}{S^2}+4(e_1+e_2+e_3-3e)\frac{W(2)(SS)}{S^2}\\ &=\frac{W(4)(\sigma\sigma)}{\sigma^2}+4(e_1+e_2+e_3-3e)\left[\frac{W(2)(\sigma\sigma)}{\sigma^2}+e\right]+12e\frac{W(2)(\sigma\sigma)}{\sigma^2}+6e^2\\ &=\frac{W(4)(\sigma\sigma)}{\sigma^2}+4(e_1+e_2+e_3)\frac{W(2)(\sigma\sigma)}{\sigma^2}-6e^2+4e(e_1+e_2+e_3)\\ &=-2(e_2e_3+e_3e_1+e_1e_2)-6e^2+4e(e_1+e_2+e_3)\end{aligned}\right\}\ldots(16);$$

so that the effect of the term, in the exponential, $\tfrac{1}{2}eu^2$, is simply to decrease each of e_1, e_2, e_3, by e.

I note here that, if σu has to be $= -\sigma(-u)$, so that its initial value is 0, assuming the initial values of $\frac{d\sigma}{du}, \frac{d^3\sigma}{du^3}, \frac{d^5\sigma}{du^5}$, to be A, B, C, we have

$$-4AB - g_1 A^2 = 0;$$

$$-2AC + 2B^2 - g_2 A^2 = 0;$$

and

$$2g_1 AC - 2g_1 B^2 - 4g_2 AB = 0;$$

whence

$$\frac{g_1}{2AB} = \frac{g_2}{AC - B^2} = -\frac{2}{A^2}.$$

Now the equation

$$\sigma\frac{d^8\sigma}{du^8} - 4\frac{d^2\sigma}{du^2}\frac{d^6\sigma}{du^6} + 3\left(\frac{d^4\sigma}{du^4}\right)^2 + g_1\left[\sigma\frac{d^6\sigma}{du^6} + 2\frac{d\sigma}{du}\frac{d^5\sigma}{du^5} - \frac{d^2\sigma}{du^2}\frac{d^4\sigma}{du^4} - 2\left(\frac{d^3\sigma}{du^3}\right)^2\right]$$
$$- g_2\left[\sigma\frac{d^4\sigma}{du^4} + 4\frac{d\sigma}{du}\cdot\frac{d^3\sigma}{du^3} + 3\left(\frac{d^2\sigma}{du^2}\right)^2\right] = 0$$

fails to give us the initial value of $\frac{d^7\sigma}{du^7}$. But if we assume its initial value to be D, by successive differentiation we can find initial values of $\frac{d^9\sigma}{du^9}, \frac{d^{11}\sigma}{du^{11}}$, etc.

I observe here that by further differentiation we obtain a succession of equations. Thus, writing $W(2n)$ for $W(2n)\sigma\sigma$,

$$W(2) + \sigma^2 . \wp = 0;$$

$$W(4) + g_1 W(2) - \tfrac{1}{2} g_2 \sigma^2 = 0;$$

$$W(6) + g_1 W(4) + 3g_2 W(2) - 12 g_3 \sigma^2 = 0;$$

$$W(8) + g_1 W(6) + 7g_2 W(4) + 48 g_3 W(2) + 12(4g_1 g_3 - g_2^2)\sigma^2 = 0;$$

....

We can readily establish the formulae

$$\sigma\frac{d^2\sigma}{du^2} - \left(\frac{d\sigma}{du}\right)^2 + e_1\sigma^2 + \sigma_1^2 = 0,$$

$$\sigma_1\frac{d^2\sigma_1}{du^2} - \left(\frac{d\sigma_1}{du}\right)^2 + e_1\sigma_1^2 + (e_1 - e_2)(e_1 - e_3)\sigma^2 = 0;$$

with similar formulae for σ_2, σ_3, and can thus check the results obtained by successive differentiation of

$$W(4) + g_1 W(2) - \tfrac{1}{2} g_2 \sigma^2 = 0;$$

which I will call the principal sigma equation.

But, for beginners, I think that, just as the consideration of sines and cosines precedes the consideration of Elliptic Functions, so the consideration of the equation

$$W(4) = 0$$

should precede the consideration of the principal sigma equation.

To prevent confusion, in considering this case, I will put

$$e_1 = e;\quad e_2 = \kappa^4 e;\quad e_3 = \kappa^8 e,$$

where

$$\kappa^4 = \cos\frac{2\pi}{3} + \iota\sin\frac{2\pi}{3},$$

$$\kappa^8 = \cos\frac{4\pi}{3} + \iota\sin\frac{4\pi}{3};$$

so that $$\frac{\kappa^4-\kappa^8}{1-\kappa^8}=\frac{\kappa^4}{1+\kappa^4}=\kappa^2;$$

and replace σ, bs, cs, ds, by τ, bt, ct, dt.

The expansion of τu I find to be

$$\tau u = u - T\frac{u^7}{7\,!} - 6T^2\frac{u^{13}}{13\,!} - 23\,.\,24\,T^3\frac{u^{19}}{19\,!} + 24\,.\,25\,.\,31\,T^4\frac{u^{25}}{25\,!} + \ldots,$$

where, in order that $\frac{d\wp}{du} = -2\sqrt{\wp^3 - e^3}$, we must have

$$T = 24e^3;$$

and it will be readily seen that

$$\text{bt}\, u = \kappa^4 \,\text{ct}\, \kappa^4 u = \kappa^8\, \text{dt}\, \kappa^8 u;$$
$$\text{ct}\, u = \kappa^4 \,\text{dt}\, \kappa^4 u = \kappa^8\, \text{bt}\, \kappa^8 u;$$
$$\text{dt}\, u = \kappa^4 \,\text{bt}\, \kappa^4 u = \kappa^8\, \text{ct}\, \kappa^8 u.$$

And our addition formulae can readily be obtained, thus,

$$\tau(u+v)\,\tau(u-v) = \tau^2 u\,\tau_1^2 v - \tau_1^2 u\,\tau^2 v = \tau^2 u\,\tau_2^2 v - \tau_2^2 u\,\tau^2 v = \tau^2 u\,\tau_3^2 v - \tau_3^2 u\,\tau^2 v;$$
$$\tau_1(u+v)\,\tau_1(u-v) = \tau_1^2 u\,\tau_1^2 v - 3\tau^2 u\,\tau^2 v;$$
$$\tau_2(u+v)\,\tau_2(u-v) = \tau_2^2 u\,\tau_2^2 v - 3\kappa^8\tau^2 u\,\tau^2 v;$$
$$\tau_3(u+v)\,\tau_3(u-v) = \tau_3^2 u\,\tau_3^2 v - 3\kappa^4\tau^2 u\,\tau^2 v;$$

and we may here note the formulae

$$\begin{aligned}&\tau(z+x-2y)\,\tau_1(z-x)\,\tau_2(z-x)\,\tau_3(z-x)\\ &\quad= \tau(x-y)\,\tau_1(x-y)\,\tau_2(y-z)\,\tau_2(z-x)\,\tau_3(y-z)\,\tau_3(z-x)\\ &\quad\quad - \tau(y-z)\,\tau_1(y-z)\,\tau_2(z-x)\,\tau_2(x-y)\,\tau_3(z-x)\,\tau_3(x-y), \text{ etc.};\end{aligned}$$

$$\begin{aligned}&\tau(y+z-2x)\,\tau_1(y-z)\,\tau_2(y-z)\,\tau_3(y-z)\\ &\quad + \tau(z+x-2y)\,\tau_1(z-x)\,\tau_2(z-x)\,\tau_3(z-x)\\ &\quad + \tau(x+y-2z)\,\tau_1(x-y)\,\tau_2(x-y)\,\tau_3(x-y) = 0.\end{aligned}$$

One great interest in these τ functions is that we can split into factors the formulae that present themselves in "Triplication." Thus

$$\begin{aligned}\tau 3u &= 3\tau u\,[\tau_1^8 + 4\tau^2\tau_1^6 + 6\tau^4\tau_1^4 - 3\tau^8]\\ &= 3\tau u\,(\tau_1^2+\tau^2)\,(\tau_1^6 + 3\tau^2\tau_1^4 + 3\tau^4\tau_1^2 - 3\tau^6)\\ &= 3\tau\,(\tau_1^2+\tau^2)\,[(\tau_1^2+\tau^2)^3 - 4\tau^6];\end{aligned}$$

$$\begin{aligned}\tau_1 3u &= \tau_1\,(\tau_1^8 - 18\tau_1^4\tau^4 - 36\tau_1^2\tau^6 - 27\tau^8)\\ &= \tau_1\,(\tau_1^2+3\tau^2)\,(\tau_1^6 - 3\tau_1^4\tau^2 - 9\tau_1^2\tau^4 - 9\tau^6)\\ &= \tau_1\,(\tau_1^2+3\tau^2)\,[\tfrac{4}{3}\tau_1^6 - \tfrac{1}{3}(\tau_1^2+3\tau^2)^3];\end{aligned}$$

$$\tau_2 3u = \tau_2\,(\tau_2^2 + 3\kappa^4\tau^2)\,[\tfrac{4}{3}\tau_2^6 - \tfrac{1}{3}(\tau_2^2+3\kappa^4\tau^2)^3];$$

$$\tau_3 3u = \tau_3\,(\tau_3^2 + 3\kappa^8\tau^2)\,[\tfrac{4}{3}\tau_3^6 - \tfrac{1}{3}(\tau_3^2+3\kappa^8\tau^2)^3].$$

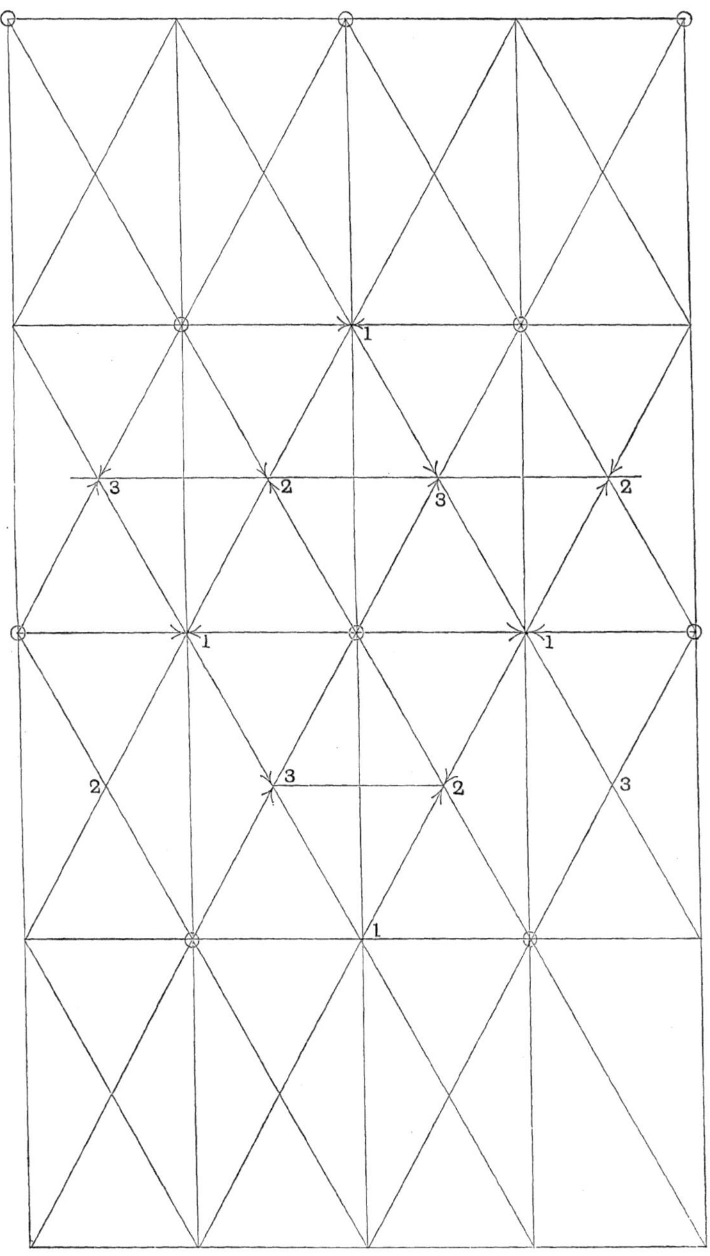
1
3
2
3
2
1
1
2
3
2
3
1

And there are an enormous number of extremely interesting formulae: for instance,

$$\mathrm{bt}\,(\kappa^4+\kappa^2)\,u=\mathrm{bt}\,\kappa^3\sqrt{3}u=\frac{\mathrm{bt}\,\kappa^4u\,\mathrm{ct}\,\kappa^2u\,\mathrm{dt}\,\kappa^2u-\mathrm{bt}\,\kappa^2u\,\mathrm{ct}\,\kappa^4u\,\mathrm{dt}\,\kappa^4u}{\mathrm{bt}^2\,\kappa^2u-\mathrm{bt}^2\,\kappa^4u}$$

$$=\frac{\kappa^4\,\mathrm{bt}\,u\,\mathrm{dt}^2\,u+\mathrm{bt}\,u\,\mathrm{ct}^2\,u}{\kappa^8\,\mathrm{ct}^2\,u-\kappa^4\,\mathrm{dt}^2\,u}$$

$$=\frac{\mathrm{bt}\,u\,(\kappa^4\wp u-e+\wp u-\kappa^4e)}{\kappa^8\,\mathrm{ct}^2u-\kappa^4\,\mathrm{dt}^2\,u}$$

$$=\frac{\mathrm{bt}\,u\,(\wp u-e)}{\sqrt{3}\,.\,\kappa^7\wp u},\ \text{etc.}$$

Since

$$-\wp u=\frac{1}{\tau}\frac{d^2\tau}{du^2}-\left(\frac{1}{\tau}\frac{d\tau}{du}\right)^2=\frac{1}{\tau}\left\{-T\frac{u^5}{5!}-6T^2\frac{u^{11}}{11!}+\dots\right\}-\frac{1}{\tau^2}\left(1-2T\frac{u^6}{6!}-\dots\right)$$

$$=\frac{1}{\tau^2}\left(-1-4T\frac{u^6}{6!}+\dots\right),$$

we have $$\tau^2u\wp u=\tau^2\kappa^4u\wp\kappa^4u=\tau^2\kappa^8u\wp\kappa^8u,$$

and, as $$\tau\kappa^4u=\kappa^4\tau u;\quad \tau\kappa^8u=\kappa^8\tau u,$$

we have $$\wp u=\kappa^8\wp\kappa^4u=\kappa^4\wp\kappa^8u.$$

In the diagram the 0's signify where $\wp u$ is equal to the reciprocal of the differential element, the ones where $\wp u=e$, the twos where $\wp u=\kappa^4e$, and the threes where $\wp u=\kappa^8e$.

SUR LES ÉQUATIONS AUX DÉRIVÉES PARTIELLES DU PREMIER ORDRE À TROIS VARIABLES INDÉPENDANTES

PAR P. ZERVOS.

1. Il est bien connu qu'à l'intégration de toute équation aux dérivées partielles du premier ordre à deux variables indépendantes on peut rattacher l'intégration d'une équation de Monge

$$f(x, y, z, y', z') = 0 \quad \ldots\ldots\ldots\ldots\ldots\ldots\ldots\ldots (1),$$

et inversement, étant donnée une équation de la forme (1) on sait depuis Monge que la solution la plus générale du problème consiste à prendre les équations

$$V = 0, \quad \frac{\Delta V}{\Delta \alpha} = 0, \quad \frac{\Delta^2 V}{\Delta \alpha^2} = 0 \quad \ldots\ldots\ldots\ldots\ldots\ldots\ldots (1'),$$

où $V = 0$ donne l'intégrale complète de l'équation adjointe

$$F(x, y, z, p, q) = 0.$$

2. Pour le cas où le nombre des variables est plus grand que trois ou l'ordre est plus grand qu'un nous avons des résultats très essentiels pour des cas assez généraux de MM. Darboux*, Goursat†, Hadamard‡, Cartan§ et d'autres géomètres.

Sans aborder ici le problème dans toute sa généralité je me propose de faire quelques remarques relatives à l'intégration d'une équation de Monge du deuxième ordre et à quatre variables.

3. M. Goursat dans un très beau Mémoire inséré dans le journal de l'Ecole Polytechnique 1897 a fait remarquer qu'on peut rattacher à l'intégration d'un système linéaire en involution

$$r + \lambda s + \mu = 0$$
$$s + \lambda t + \nu = 0$$

l'intégration d'une équation de Monge du deuxième ordre

$$f(x, y, z, y', z', y'', z'') = 0,$$

* Darboux, "Solutions singulières des équations aux dérivées partielles" (*Mémoires des savants étrangers de l'Académie,*" 1883). Darboux, "Sur la résolution de l'équation $dx^2 + dy^2 + dz^2 = ds^2$ et de quelques équations analogues" (*Journal de Liouville*, 1887).

† Goursat, "Sur le problème de Monge," *Bulletin de la Société Mathématique de France*, 1905.

‡ Hadamard, *Annales de l'Ecole Normale*, 1901.

§ Cartan, *Bulletin de la Société Mathématique*, 1901.

et il a fait montrer que l'intégration de cette équation se ramène alors à l'intégration d'une équation du premier ordre et encore que, étant donné le système en involution, on peut obtenir directement l'équation de Monge sans reconnaître l'intégrale complète.

4. Ici il s'agit de faire remarquer qu'on peut aussi faire correspondre à une équation aux dérivées partielles du premier ordre à trois variables indépendantes une équation de Monge du deuxième ordre en partant d'une intégrale complète de cette équation, et dans ce cas la solution de l'équation de Monge se donne par de formules très simples qui peuvent être considérées comme une extension des formules de Monge (1′).

5. Soit, en effet, une équation aux dérivées partielles du premier ordre

$$F(x_1, x_2, x_3, x_4, p_1, p_2, p_3) = 0 \quad\ldots\ldots\ldots\ldots(1),$$

et $V(x_1, x_2, x_3, x_4, a_1, a_2, a_3) = 0$ une intégrale complète de cette équation.

Outre l'équation $V = 0$ (2) formons les relations suivantes:

$$\sum_{i=1}^{4} \frac{\partial V}{\partial x_i} dx_i = 0 \ldots\ldots(3)$$

$$\sum_{i=1}^{4}\sum_{k=1}^{4} \frac{\partial^2 V}{\partial x_i \partial x_k} dx_i dx_k + \sum_{i=1}^{4} \frac{\partial V}{\partial x_i} d^2x_i = 0 \quad (4)$$

$$\sum_{i=1}^{3} \frac{\partial V}{\partial a_i} da_i = 0 \ldots\ldots(5)$$

$$\sum_{i=1}^{3}\sum_{k=1}^{3} \frac{\partial^2 V}{\partial a_i \partial a_k} da_i da_k = 0 \quad \ldots(6)$$

$$\sum_{i=1}^{3}\sum_{k=1}^{4} \frac{\partial^2 V}{\partial a_i \partial x_k} da_i dx_k = 0 \ldots\ldots(7)$$

$$\sum_{i=1}^{3} \frac{\partial V}{\partial a_i} d^2a_i = 0 \quad \ldots\ldots(8).$$

Les relations (5), (6) et (7) sont homogènes par rapport aux da. On peut donc en général éliminer entre elles les rapports $\frac{da_2}{da_1}$ et $\frac{da_3}{da_1}$; nous trouvons ainsi une équation contenant seulement des x, dx, d^2x et a. Entre cette équation et les équations (2), (3) et (4) éliminons les a; on trouve ainsi, en général, une équation de Monge de la forme

$$f(x_1, x_2, x_3, x_4, dx_1, dx_2, dx_3, dx_4, d^2x_1, d^2x_2, d^2x_3, d^2x_4) = 0 \quad \ldots\ldots(9).$$

Donc à *l'intégrale complète* (2) *de l'équation* (1) *correspond l'équation de Monge* (9).

6. Mais on voit encore que l'équation (8) est une conséquence des équations (5), (6) et (7) et que les quatre équations entre les équations (2), (3), (4), (5), (6), (7) et (8) peuvent être considérées comme indépendantes, car étant données les relations (2), (5), (6) et (8) on aura, comme leurs conséquences, les relations (3), (7) et (4) et comme l'équation (9) est supposée comme résultant de l'élimination des a et da entre les relations (2), (3), (4), (5), (6), (7) nous concluons le théorème suivant:

Une équation de Monge du deuxième ordre correspondante à une équation aux dérivées partielles du premier ordre à trois variables indépendantes, de la manière citée plus haut, aura comme solution celle qui se donne par les équations

$$V = 0, \quad \Sigma \frac{\partial V}{\partial a_i} da_i = 0, \quad \Sigma \frac{\partial^2 V}{\partial a_i \partial a_k} da_i da_k = 0, \quad \Sigma \frac{\partial V}{\partial a_i} d^2a_i = 0, \quad i, k = 1, 2, 3.$$

7. Considérons, d'autre part, les relations suivantes:

$$V=0, \quad \sum_{i=1}^{4} \frac{\partial V}{\partial x_i} dx_i = 0, \quad \sum_{i=1}^{4} \frac{\partial}{\partial x_i}\left(\frac{\frac{\partial V}{\partial a_k}}{\frac{\partial V}{\partial a_3}}\right) dx_i = 0, \quad i = 1, 2, 3, 4, \quad k = 1, 2 \ldots (10).$$

L'élimination des a entre ces relations (10) donnera* une équation de Monge du premier ordre, soit, qu'on a trouvé ainsi, l'équation

$$\phi(x_1, x_2, x_3, x_4, dx_1, dx_2, dx_3, dx_4) = 0 \quad \ldots\ldots\ldots\ldots\ldots\ldots (11).$$

8. Si on considère les x_1, x_2, x_3, x_4 comme de fonctions d'une variable indépendante t on peut énoncer le théorème suivant:

Les fonctions qui satisfont à l'équation (9) et les fonctions qui satisfont à l'équation (11) satisferont aux équations†

$$V=0, \quad \frac{\Delta V}{\Delta \alpha} = 0, \quad \frac{\Delta^2 V}{\Delta \alpha^2} = 0.$$

9. Il est facile de généraliser les résultats précédents et de leur faire des applications intéressantes.

* Voir une Note de M. Bottasso présentée à l'Académie des Sciences de Paris le 13 Juin 1905 et deux Notes de moi présentées à l'Académie le 10 Avril et le 11 Septembre 1905.

† M. Hilbert, dans un travail important ["über den Begriff der Klasse von Differentialgleichungen" (*Festschrift Heinrich Weber*, 5 Marz, 1912)] a donné des résultats très intéressants pour les équations de Monge du deuxième ordre à trois variables. Nous esperons d'y revenir prochainement.

EINDEUTIGKEIT DER ZERLEGUNG IN PRIMZAHLFAKTOREN IN QUADRATISCHEN ZAHLKÖRPERN

VON G. RABINOVITCH.

Ich betrachte im Folgenden quadratische Zahlkörper mit der Diskriminante der Form

$$D = 1 - 4m.$$

Um Missverständnisse zu vermeiden will ich gleich ausdrücklich bemerken, dass ich keine Ideale eingeführt denke und also nur mit Zahlen operiere. Es gilt in diesem Falle der Grundsatz der Zahlentheorie nicht in jedem der betrachteten Zahlkörper, und wenn ein solcher Körper gegeben ist, so kann man sich also fragen, ob jede Zahl dieses Körpers nur auf eine Weise in Primzahlen zerlegt werden kann. Mit dieser Frage beschäftige ich mich im Folgenden. Diese Frage kann auch so formuliert werden: ist die Klassenzahl eines gegebenen Körpers gleich 1 oder grösser?

Dabei wende ich folgende Bezeichnungen an. Lateinische Buchstaben bezeichnen ganze rationale Zahlen, griechische—ganze (im allgemeinen) irrationale. $\bar{\alpha}$ bezeichnet die zu α konjugierte Zahl. Jede Zahl ξ des Körpers kann in der Form

$$\xi = x + y\vartheta$$

dargestellt werden, wo

$$\vartheta = \frac{1 + \sqrt{D}}{2}$$

ist, und umgekehrt ist jede Zahl dieser Form eine ganze Zahl des Körpers. Die Norm der Zahl ξ ist

$$N\xi = x^2 + xy + my^2.$$

In einer Strassburger Dissertation zeigt Herr Jacob Schatunowsky, dass der Euklidsche Algorithmus in solchen Körpern, wenn $m > 3$ ist, im allgemeinen nicht anwendbar ist. Ich versuchte nun diesen Algorithmus zu verallgemeinern. Ein Schritt dieses Algorithmus besteht bekanntlich in der Bildung der Zahl $\alpha - \beta\eta$, welche kleiner als die gegebenen Zahlen α und β ist. Ich habe statt dessen den Ausdruck

$$\alpha\xi - \beta\eta$$

betrachtet. Dabei wurde ich zum folgenden Satze geführt.

Ist es möglich, sobald zwei Zahlen α und β eines Zahlkörpers gegeben sind, von denen keine durch die andere teilbar ist, zwei Zahlen ξ und η desselben Körpers zu finden, welche der Ungleichung

$$0 < N(\alpha\xi - \beta\eta) < N\beta \quad \text{.....}(1)$$

genügen, so ist die Klassenzahl des betreffenden Zahlkörpers gleich 1.

Um das zu beweisen, machen wir die Voraussetzung dass in einem Körper die Ungleichung (1) immer lösbar ist. Wäre die Zerlegung einer Zahl in Primzahlen nicht eindeutig, so müsste es, wie es leicht zu zeigen ist, im betreffenden Körper Zahlen geben, die durch eine Primzahl π teilbar sind, die aber als ein Produkt von zwei Zahlen dargestellt werden können, von denen keine durch π teilbar ist. Betrachten wir die Normen aller Zahlen, die in bezug auf eine Primzahl π diese Eigenschaft haben, so muss es unter diesen Normen eine geben, welche nicht grösser ist, als jede der andern. Es sei dies die Norm der Zahl $\lambda\Lambda$. Nach unserer Voraussetzung müssen zwei Zahlen ξ und η existieren von der Eigenschaft, dass die Zahl

$$\mu = \xi\pi - \eta\lambda \quad \text{.....}(2)$$

den Ungleichungen

$$0 < N\mu < N\pi \quad \text{.....}(3)$$

$$0 < N\mu < N\lambda \quad \text{.....}(4)$$

genügt. Multiplizieren wir (2) mit Λ, so bekommen wir

$$\mu\Lambda = \xi\pi\Lambda - \eta\lambda\Lambda.$$

Der rechte Teil ist durch π teilbar, also muss es auch $\mu\Lambda$ sein. Die Zahl μ ist wegen (3) durch π nicht teilbar, Λ auch nicht, also gehört die Zahl $\mu\Lambda$ zu den betrachteten. Es ist aber wegen (4)

$$N(\mu\Lambda) < N(\lambda\Lambda);$$

das widerspricht der Annahme, dass die Norm der Zahl $\lambda\Lambda$ nicht grösser ist, als die Normen aller anderen solchen Zahlen, also ist die Zerlegung in Primzahlfaktoren eindeutig, die Klassenzahl des betreffenden Körpers ist 1.

Die Umkehrung dieses Satzes ist auch richtig, ich werde sie aber nicht benutzen.

Den eben bewiesenen Satz können wir auch in der folgenden Form aussagen: Ist die Klassenzahl eines Körpers grösser als 1, so gibt es in diesem Körper zwei solche Zahlen α und β, von denen keine durch die andere teilbar ist, dass die Ungleichung

$$0 < N(\alpha\xi - \beta\eta) < N\beta \quad \text{.....}(1)$$

keine Lösungen besitzt. Ich teile nun beide Teile durch $N\beta$ (wie es Schatunowsky in einem ähnlichen Falle tut) und bekomme

$$0 < N\left(\frac{\alpha}{\beta}\xi - \eta\right) < 1 \quad \text{.....}(5).$$

Ich kann jetzt auch so sagen. In einem Zahlkörper, wo die Klassenzahl > 1 ist, muss es *störende* Brüche geben; dabei nenne ich einen Bruch $\frac{\alpha}{\beta}$ störend, wenn es unmöglich ist, die Ungleichung (5) zu befriedigen.

Es leuchtet unmittelbar ein, dass wenn ein Bruch $\frac{\alpha}{\beta}$ störend ist, auch die Brüche

$$\frac{\alpha}{\beta}+\xi, \qquad \frac{\alpha}{\beta}\cdot\xi,$$

wo ξ eine beliebige Zahl bedeutet, störend sein müssen. Indem ich diese Bemerkung anwende, zeige ich leicht, dass wenn in einem Zahlkörper störende Brüche vorhanden sind, notwendig störende Brüche von der Form

$$\frac{p-\vartheta}{q} \quad \ldots\ldots\ldots\ldots\ldots\ldots\ldots\ldots\ldots\ldots(6),$$

wo $p \leqslant q \leqslant m$ ist, existieren müssen.

Unser bisheriges Ergebniss können wir auch so aussprechen. Die Klassenzahl eines Körpers ist grösser oder gleich 1 je nachdem es unter den Brüchen (6) störende gibt oder nicht. Betrachten wir jetzt die Normen der Zähler dieser Brüche, also die $m-1$ Zahl

$$p^2-p+m \qquad (p=1, \ldots, m-1) \quad \ldots\ldots\ldots\ldots\ldots\ldots(7),$$

und nehmen zunächst an, dass alle diese Zahlen Primzahlen sind. Ist dann q eine von 1 verschiedene Zahl $< m$, so sind die Zahlen p^2-p+m und q relative Primzahlen und es gibt zwei Zahlen x und y, die der Gleichung

$$(p^2-p+m)\,x-qy=1$$

genügen. Diese Gleichung können wir aber in der Form

$$\frac{p-\vartheta}{q}(p-\bar{\vartheta})\,x-y=\frac{1}{q}$$

schreiben, und in dieser Form zeigt sie, dass der Bruch $\frac{p-\vartheta}{q}$ nicht störend ist $\left[\text{da } N\left(\frac{1}{q}\right)=\frac{1}{q^2}<1 \text{ ist}\right]$. Es gibt also keine störenden Brüche von der Form (6), also ist die Klassenzahl des Körpers gleich 1.

Ist eine von den Zahlen

$$p^2-p+m \qquad (p=1, \ldots, m-1) \quad \ldots\ldots\ldots\ldots\ldots\ldots(7)$$

(rational) zusammengesetzt, so kann sie einerseits als Produkt von rationalen Zahlen dargestellt werden, anderseits als Produkt der Zahlen $p-\vartheta$ und $p-\bar{\vartheta}$. Es ist leicht zu zeigen, dass $p-\vartheta$ und $p-\bar{\vartheta}$ Primzahlen sind. Wir haben also in der Zahl p^2-p+m ein Beispiel einer Zahl, die in zwei verschiedene Reihen von Primfaktoren zerlegt werden kann, die Klassenzahl ist also grösser als 1.

Wir haben somit den Satz gewonnen:

Die Klassenzahl eines quadratischen Körpers ist grösser oder gleich 1, *je nachdem in der Reihe der Zahlen*

$$p^2-p+m \qquad (p=1, \ldots, m-1)$$

zusammengesetzte Zahlen vorkommen oder nicht.

An dieses Resultat schliesse ich folgende Bemerkungen an.

Es ist längst eine Anzahl von Zahlkörpern mit der Klassenzahl 1 bekannt. Es sind dies die Zahlkörper mit den Diskriminanten

$$-7, \ -11, \ -19, \ -43, \ -67, \ -163.$$

Es ist aber unbekannt, ob es weitere solche Körper gibt und ob die Zahl solcher Körper endlich ist. Der mitgeteilte Satz führt diese Frage auf die folgende Frage über die Verteilung der Primzahlen zurück.

Es gibt einige Primzahlen m, nämlich

$$2, \ 3, \ 5, \ 11, \ 17, \ 41$$

von der Eigenschaft, dass auch die Zahlen

$$p^2 - p + m \qquad (p = 1, \ldots, m-1)$$

Primzahlen sind. Gibt es noch solche Primzahlen, und gibt es deren eine endliche Anzahl oder eine unendliche? Diese Frage ist, soweit ich weiss, ungelöst, und wir haben also für die andere Frage keine Antwort bekommen.

In den speziellen Fällen lässt sich aber die rationalzahlentheoretische Frage leicht beantworten. Die Zahlen der Reihe (7) können auch in folgender Weise bekommen werden. Die erste ist m, die zweite $m+2$, um die dritte zu bekommen addiere ich zur zweiten 4, die vierte bekommt man aus der dritten indem man 6 addiert, dann kommt 8 u.s.f. Als Beispiel wähle ich die Zahl $m = 11$.

$$\begin{array}{l} 11 + \ 2 \\ 13 + \ 4 \\ 17 + \ 6 \\ 23 + \ 8 \\ 31 + 10 \\ 41 + 12 \\ 53 + 14 \\ 67 + 16 \\ 83 + 18 \\ 101 \end{array}$$

Diese kleine Rechnung beweist (da die Zahlen 11, 13 u.s.f. Primzahlen sind), dass der Zahlkörper mit der Diskriminante $D = 1 - 4 . 11 = -43$ die Klassenzahl 1 hat.

Für $m = 41$ bekommen wir die bekannte Funktion

$$p^2 - p + 41,$$

welche für die ersten 40 Werte von p Primzahlen ergiebt.

Zum Schluss erlaube ich mir eine Vermutung auszusprechen. Die oberflächliche Betrachtung der Reihe (7) erlaubt in einem Falle die Klassenzahl des betreffenden Körpers anzugeben, nämlich wenn diese Zahl gleich 1 ist. Vielleicht wird eine genauere Untersuchung dieser Reihe erlauben die Klassenzahl in allen Fällen zu erkennen.

ON AN ELEMENTARY METHOD OF DEDUCING THE CHARACTERISTICS OF THE PARTIAL DIFFERENTIAL EQUATION OF THE SECOND ORDER

$$T_1t_1 + T_2t_2 + T_3t_3 + S_1s_1 + S_2s_2 + S_3s_3 + U_1(t_2t_3 - s_1^2) + U_2(t_1t_3 - s_2^2)$$
$$+ U_3(t_1t_2 - s_3^2) + U_4(t_1s_1 - s_2s_3) + U_5(t_2s_2 - s_1s_3) + U_6(t_3s_3 - s_1s_2) + V = 0 \quad (1).$$

BY J. H. PEEK.

In the above-mentioned equation x, y, z are the independent variables, h is the dependent one, whereas

$$p = \frac{\partial h}{\partial x}, \quad q = \frac{\partial h}{\partial y}, \quad r = \frac{\partial h}{\partial z}, \quad t_1 = \frac{\partial^2 h}{\partial x^2}, \quad t_2 = \frac{\partial^2 h}{\partial y^2}, \quad t_3 = \frac{\partial^2 h}{\partial z^2},$$
$$s_1 = \frac{\partial^2 h}{\partial y \partial z}, \quad s_2 = \frac{\partial^2 h}{\partial x \partial z}, \quad s_3 = \frac{\partial^2 h}{\partial x \partial y},$$

and the coefficients are functions of x, y, z, p, q, r.

The method here developed is quite analogous to that exhibited in Dr Forsyth's treatise on differential equations in the case of two independent variables. We shall see in the following pages that the solution is of the form

$$h = \phi(u, v, w),$$

u, v, w being the integral functions of one set of characteristics. Moreover the existence of the solution is bound to several conditions relating to the coefficients, and u_1, v_1, w_1 are subject to the condition $\left(\frac{u_1 v_1 w_1}{pqr}\right) = 0$ as well as u_2, v_2, w_2.

The notation is held analogous to that of the case of two independent variables, so that z is an independent variable.

We have

$$dp = t_1 dx + s_3 dy + s_2 dz, \quad dq = s_3 dx + t_2 dy + s_1 dz, \quad dr = s_2 dx + s_1 dy + t_3 dz \ldots(\alpha),$$

from which we obtain

$$t_1 = \frac{1}{dx}(dp - s_3 dy - s_2 dz), \quad t_2 = \frac{1}{dy}(dq - s_3 dx - s_1 dz), \quad t_3 = \frac{1}{dz}(dr - s_2 dx - s_1 dy),$$

and after multiplication

$$t_2t_3 - s_1^2 = \frac{1}{dy\,dz}(dq\,dr - s_1 dq\,dy - s_1 dr\,dz - s_2 dq\,dx - s_3 dr\,dx$$
$$+ s_2s_3 dx^2 + s_1s_3 dx\,dy + s_1s_2 dx\,dz),$$

$$t_1 t_3 - s_2^2 = \frac{1}{dx\,dz}(dp\,dr - s_1 dp\,dy - s_2 dp\,dx - s_2 dr\,dz - s_3 dr\,dy$$
$$+ s_2 s_3 dx\,dy + s_1 s_3 dy^2 + s_1 s_2 dy\,dz),$$

$$t_1 t_2 - s_3^2 = \frac{1}{dx\,dy}(dp\,dq - s_1 dp\,dz - s_2 dq\,dz - s_3 dp\,dx - s_3 dq\,dy$$
$$+ s_2 s_3 dx\,dz + s_1 s_3 dy\,dz + s_1 s_2 dz^2).$$

Multiplying the value of t_1 by s_1, of t_2 by s_2, of t_3 by s_3 we obtain

$$t_1 s_1 - s_2 s_3 = \frac{1}{dx}(s_1 dp - s_1 s_3 dy - s_1 s_2 dz - s_2 s_3 dx),$$

$$t_2 s_2 - s_1 s_3 = \frac{1}{dy}(s_2 dq - s_2 s_3 dx - s_1 s_2 dz - s_1 s_3 dy),$$

$$t_3 s_3 - s_1 s_2 = \frac{1}{dz}(s_3 dr - s_2 s_3 dx - s_1 s_3 dy - s_1 s_2 dz).$$

By substitution in the equation (1), it transforms into:

$$s_1(-T_2 dx\,dz^2 - T_3 dx^2 dy + S_1 dx\,dy\,dz - U_1 dq\,dx\,dy - U_1 dr\,dx\,dz - U_2 dp\,dy^2$$
$$- U_3 dp\,dz^2 + U_4 dp\,dy\,dz)$$
$$+ s_2(-T_1 dy\,dz^2 - T_3 dx\,dy^2 + S_2 dx\,dy\,dz - U_1 dq\,dx^2 - U_2 dr\,dy\,dz - U_2 dp\,dx\,dy$$
$$- U_3 dq\,dz^2 + U_5 dq\,dx\,dz)$$
$$+ s_3(-T_1 dy^2 dz - T_2 dx^2 dz + S_3 dx\,dy\,dz - U_1 dr\,dx^2 - U_2 dr\,dy^2 - U_3 dp\,dx\,dz$$
$$- U_3 dq\,dy\,dz + U_6 dr\,dx\,dy)$$
$$+ (T_1 dp\,dy\,dz + T_2 dq\,dx\,dz + T_3 dr\,dx\,dy + U_1 dq\,dr\,dx + U_2 dp\,dr\,dy$$
$$+ U_3 dp\,dq\,dz + V dx\,dy\,dz)$$
$$+ (s_2 s_3 dx + s_1 s_3 dy + s_1 s_2 dz)(U_1 dx^2 + U_2 dy^2 + U_3 dz^2 - U_4 dy\,dz - U_5 dx\,dz - U_6 dx\,dy) = 0.$$

Now we put

$$U_1 dx^2 + U_2 dy^2 + U_3 dz^2 - U_4 dy\,dz - U_5 dx\,dz - U_6 dx\,dy = 0 \quad \ldots\ldots(2).$$

Transforming the coefficients of s_1, s_2, s_3 by means of this relation we obtain

$$s_1 dx\,\{-T_2 dz^2 - T_3 dy^2 + S_1 dy\,dz + U_1(dp\,dx - dq\,dy - dr\,dz) - U_5 dp\,dz - U_6 dp\,dy\},$$
$$s_2 dy\,\{-T_1 dz^2 - T_3 dx^2 + S_2 dx\,dz + U_2(dq\,dy - dp\,dx - dr\,dz) - U_4 dq\,dz - U_6 dq\,dx\},$$
$$s_3 dz\,\{-T_1 dy^2 - T_2 dx^2 + S_3 dx\,dy + U_3(dr\,dz - dp\,dx - dq\,dy) - U_4 dr\,dy - U_5 dr\,dx\},$$

so that the equation (1) may be satisfied by putting

$$V dx\,dy\,dz + T_1 dp\,dy\,dz + T_2 dq\,dx\,dz + T_3 dr\,dx\,dy$$
$$+ U_1 dq\,dr\,dx + U_2 dp\,dr\,dy + U_3 dp\,dq\,dz = 0 \ldots(3),$$

$$U_1 dx^2 + U_2 dy^2 + U_3 dz^2 - U_4 dy\,dz - U_5 dx\,dz - U_6 dx\,dy = 0 \quad \ldots\ldots(4),$$

$$-T_2 dz^2 - T_3 dy^2 + S_1 dy\,dz + U_1(dp\,dx - dq\,dy - dr\,dz) - U_5 dp\,dz - U_6 dp\,dy = 0 \quad (5),$$

$$-T_1 dz^2 - T_3 dx^2 + S_2 dx\,dz + U_2(dq\,dy - dp\,dx - dr\,dz) - U_4 dq\,dz - U_6 dq\,dx = 0 \quad (6),$$

$$-T_1 dy^2 - T_2 dx^2 + S_3 dx\,dy + U_3(dr\,dz - dp\,dx - dq\,dy) - U_4 dr\,dy - U_5 dr\,dx = 0 \quad (7).$$

We will show no $\phi(u, v, w)$ to be a solution which satisfies the equations (3)—(7).

Representing by $\frac{\delta u}{\delta x}$ the expression

$$\frac{\partial u}{\partial x} + p\frac{\partial u}{\partial h} + \frac{\partial u}{\partial p}t_1 + \frac{\partial u}{\partial q}s_3 + \frac{\partial u}{\partial r}s_2 \quad \ldots\ldots\ldots\ldots\ldots\ldots\ldots\ldots(8),$$

we have
$$\frac{\partial\phi}{\partial x}=\frac{\partial\phi}{\partial u}\frac{\delta u}{\delta x}+\frac{\partial\phi}{\partial v}\frac{\delta v}{\delta x}+\frac{\partial\phi}{\partial w}\frac{\delta w}{\delta x},$$
$$\frac{\partial\phi}{\partial y}=\frac{\partial\phi}{\partial u}\frac{\delta u}{\delta y}+\frac{\partial\phi}{\partial v}\frac{\delta v}{\delta y}+\frac{\partial\phi}{\partial w}\frac{\delta w}{\delta y},$$
$$\frac{\partial\phi}{\partial z}=\frac{\partial\phi}{\partial u}\frac{\delta u}{\delta z}+\frac{\partial\phi}{\partial v}\frac{\delta v}{\delta z}+\frac{\partial\phi}{\partial w}\frac{\delta w}{\delta z}.$$

Eliminating ϕ from these equations, u, v, w have to satisfy
$$\begin{vmatrix}\frac{\delta u}{\delta x} & \frac{\delta v}{\delta x} & \frac{\delta w}{\delta x}\\ \frac{\delta u}{\delta y} & \frac{\delta v}{\delta y} & \frac{\delta w}{\delta y}\\ \frac{\delta u}{\delta z} & \frac{\delta v}{\delta z} & \frac{\delta w}{\delta z}\end{vmatrix}=0 \quad \ldots\ldots\ldots\ldots(9).$$

By substitution of the value of $\frac{\delta u}{\delta x}$ from (8) and the corresponding values, (9) takes the form
$$T_1't_1+T_2't_2+T_3't_3+(S_1^1+S_1^2)s_1+(S_2^1+S_2^2)s_2+(S_3^1+S_3^2)s_3+U_1(t_2t_3-s_1^2)$$
$$+U_2(t_1t_3-s_2^2)+U_3(t_1t_2-s_3^2)+(U_4^1+U_4^2)(t_1s_1-s_2s_3)+(U_5^1+U_5^2)(t_2s_2-s_1s_3)$$
$$+(U_6^1+U_6^2)(t_3s_3-s_1s_2)+V'+\left(\frac{uvw}{pqr}\right)\begin{vmatrix}t_1 & s_3 & s_2\\ s_3 & t_2 & s_1\\ s_2 & s_1 & t_3\end{vmatrix}=0 \ (10),$$

in which expression
$$T_1'=\begin{vmatrix}\frac{\partial u}{\partial p} & \frac{\partial u}{\partial y}+q\frac{\partial u}{\partial h} & \frac{\partial u}{\partial z}+r\frac{\partial u}{\partial h}\\ \frac{\partial v}{\partial p} & \frac{\partial v}{\partial y}+q\frac{\partial v}{\partial h} & \frac{\partial v}{\partial z}+r\frac{\partial v}{\partial h}\\ \frac{\partial w}{\partial p} & \frac{\partial w}{\partial y}+q\frac{\partial w}{\partial h} & \frac{\partial w}{\partial z}+r\frac{\partial w}{\partial h}\end{vmatrix},\quad T_2'=\begin{vmatrix}\frac{\partial u}{\partial x}+p\frac{\partial u}{\partial h} & \frac{\partial u}{\partial q} & \frac{\partial u}{\partial z}+r\frac{\partial u}{\partial h}\\ \frac{\partial v}{\partial x}+p\frac{\partial v}{\partial h} & \frac{\partial v}{\partial q} & \frac{\partial v}{\partial z}+r\frac{\partial v}{\partial h}\\ \frac{\partial w}{\partial x}+p\frac{\partial w}{\partial h} & \frac{\partial w}{\partial q} & \frac{\partial w}{\partial z}+r\frac{\partial w}{\partial h}\end{vmatrix},$$
$$T_3'=\begin{vmatrix}\frac{\partial u}{\partial x}+p\frac{\partial u}{\partial h} & \frac{\partial u}{\partial y}+q\frac{\partial u}{\partial h} & \frac{\partial u}{\partial z}\\ \frac{\partial v}{\partial x}+p\frac{\partial v}{\partial h} & \frac{\partial v}{\partial y}+q\frac{\partial v}{\partial h} & \frac{\partial v}{\partial z}\\ \frac{\partial w}{\partial x}+p\frac{\partial w}{\partial h} & \frac{\partial w}{\partial y}+q\frac{\partial w}{\partial h} & \frac{\partial w}{\partial z}\end{vmatrix},\quad S_1^1=\begin{vmatrix}\frac{\partial u}{\partial x}+p\frac{\partial u}{\partial h} & \frac{\partial u}{\partial r} & \frac{\partial u}{\partial z}+r\frac{\partial u}{\partial h}\\ \frac{\partial v}{\partial x}+p\frac{\partial v}{\partial h} & \frac{\partial v}{\partial r} & \frac{\partial v}{\partial z}+r\frac{\partial v}{\partial h}\\ \frac{\partial w}{\partial x}+p\frac{\partial w}{\partial h} & \frac{\partial w}{\partial r} & \frac{\partial w}{\partial z}+r\frac{\partial w}{\partial h}\end{vmatrix},$$
$$S_1^2=\begin{vmatrix}\frac{\partial u}{\partial x}+p\frac{\partial u}{\partial h} & \frac{\partial u}{\partial y}+q\frac{\partial u}{\partial h} & \frac{\partial u}{\partial q}\\ \frac{\partial v}{\partial x}+p\frac{\partial v}{\partial h} & \frac{\partial v}{\partial y}+q\frac{\partial v}{\partial h} & \frac{\partial v}{\partial q}\\ \frac{\partial w}{\partial x}+p\frac{\partial w}{\partial h} & \frac{\partial w}{\partial y}+q\frac{\partial w}{\partial h} & \frac{\partial w}{\partial q}\end{vmatrix},\quad S_2^1=\begin{vmatrix}\frac{\partial u}{\partial r} & \frac{\partial u}{\partial y}+q\frac{\partial u}{\partial h} & \frac{\partial u}{\partial z}+r\frac{\partial u}{\partial h}\\ \frac{\partial v}{\partial r} & \frac{\partial v}{\partial y}+q\frac{\partial v}{\partial h} & \frac{\partial v}{\partial z}+r\frac{\partial v}{\partial h}\\ \frac{\partial w}{\partial r} & \frac{\partial w}{\partial y}+q\frac{\partial w}{\partial h} & \frac{\partial w}{\partial z}+r\frac{\partial w}{\partial h}\end{vmatrix},$$

$$S_2^2 = \begin{vmatrix} \frac{\partial u}{\partial x} + p\frac{\partial u}{\partial h} & \frac{\partial u}{\partial y} + q\frac{\partial u}{\partial h} & \frac{\partial u}{\partial p} \\ \frac{\partial v}{\partial x} + p\frac{\partial v}{\partial h} & \frac{\partial v}{\partial y} + q\frac{\partial v}{\partial h} & \frac{\partial v}{\partial p} \\ \frac{\partial w}{\partial x} + p\frac{\partial w}{\partial h} & \frac{\partial w}{\partial y} + q\frac{\partial w}{\partial h} & \frac{\partial w}{\partial p} \end{vmatrix}, \quad S_3^1 = \begin{vmatrix} \frac{\partial u}{\partial q} & \frac{\partial u}{\partial y} + q\frac{\partial u}{\partial h} & \frac{\partial u}{\partial z} + r\frac{\partial u}{\partial h} \\ \frac{\partial v}{\partial q} & \frac{\partial v}{\partial y} + q\frac{\partial v}{\partial h} & \frac{\partial v}{\partial z} + r\frac{\partial v}{\partial h} \\ \frac{\partial w}{\partial q} & \frac{\partial w}{\partial y} + q\frac{\partial w}{\partial h} & \frac{\partial w}{\partial z} + r\frac{\partial w}{\partial h} \end{vmatrix},$$

$$S_3^2 = \begin{vmatrix} \frac{\partial u}{\partial x} + p\frac{\partial u}{\partial h} & \frac{\partial u}{\partial p} & \frac{\partial u}{\partial z} + r\frac{\partial u}{\partial h} \\ \frac{\partial v}{\partial x} + p\frac{\partial v}{\partial h} & \frac{\partial v}{\partial p} & \frac{\partial v}{\partial z} + r\frac{\partial v}{\partial h} \\ \frac{\partial w}{\partial x} + p\frac{\partial w}{\partial h} & \frac{\partial w}{\partial p} & \frac{\partial w}{\partial z} + r\frac{\partial w}{\partial h} \end{vmatrix}, \quad U_1' = \begin{vmatrix} \frac{\partial u}{\partial x} + p\frac{\partial u}{\partial h} & \frac{\partial u}{\partial q} & \frac{\partial u}{\partial r} \\ \frac{\partial v}{\partial x} + p\frac{\partial v}{\partial h} & \frac{\partial v}{\partial q} & \frac{\partial v}{\partial r} \\ \frac{\partial w}{\partial x} + p\frac{\partial w}{\partial h} & \frac{\partial w}{\partial q} & \frac{\partial w}{\partial r} \end{vmatrix},$$

$$U_2' = \begin{vmatrix} \frac{\partial u}{\partial y} + q\frac{\partial u}{\partial h} & \frac{\partial u}{\partial r} & \frac{\partial u}{\partial p} \\ \frac{\partial v}{\partial y} + q\frac{\partial v}{\partial h} & \frac{\partial v}{\partial r} & \frac{\partial v}{\partial p} \\ \frac{\partial w}{\partial y} + q\frac{\partial w}{\partial h} & \frac{\partial w}{\partial r} & \frac{\partial w}{\partial p} \end{vmatrix}, \quad U_3' = \begin{vmatrix} \frac{\partial u}{\partial z} + r\frac{\partial u}{\partial h} & \frac{\partial u}{\partial p} & \frac{\partial u}{\partial q} \\ \frac{\partial v}{\partial z} + r\frac{\partial v}{\partial h} & \frac{\partial v}{\partial p} & \frac{\partial v}{\partial q} \\ \frac{\partial w}{\partial z} + r\frac{\partial w}{\partial h} & \frac{\partial w}{\partial p} & \frac{\partial w}{\partial q} \end{vmatrix},$$

$$U_4^1 = \begin{vmatrix} \frac{\partial u}{\partial z} + r\frac{\partial u}{\partial h} & \frac{\partial u}{\partial p} & \frac{\partial u}{\partial r} \\ \frac{\partial v}{\partial z} + r\frac{\partial v}{\partial h} & \frac{\partial v}{\partial p} & \frac{\partial v}{\partial r} \\ \frac{\partial w}{\partial z} + r\frac{\partial w}{\partial h} & \frac{\partial w}{\partial p} & \frac{\partial w}{\partial r} \end{vmatrix}, \quad U_4^2 = \begin{vmatrix} \frac{\partial u}{\partial y} + q\frac{\partial u}{\partial h} & \frac{\partial u}{\partial q} & \frac{\partial u}{\partial p} \\ \frac{\partial v}{\partial y} + q\frac{\partial v}{\partial h} & \frac{\partial v}{\partial q} & \frac{\partial v}{\partial p} \\ \frac{\partial w}{\partial y} + q\frac{\partial w}{\partial h} & \frac{\partial w}{\partial q} & \frac{\partial w}{\partial p} \end{vmatrix},$$

$$U_5^1 = \begin{vmatrix} \frac{\partial u}{\partial x} + p\frac{\partial u}{\partial h} & \frac{\partial u}{\partial q} & \frac{\partial u}{\partial p} \\ \frac{\partial v}{\partial x} + p\frac{\partial v}{\partial h} & \frac{\partial v}{\partial q} & \frac{\partial v}{\partial p} \\ \frac{\partial w}{\partial x} + p\frac{\partial w}{\partial h} & \frac{\partial w}{\partial q} & \frac{\partial w}{\partial p} \end{vmatrix}, \quad U_5^2 = \begin{vmatrix} \frac{\partial u}{\partial z} + r\frac{\partial u}{\partial h} & \frac{\partial u}{\partial r} & \frac{\partial u}{\partial q} \\ \frac{\partial v}{\partial z} + r\frac{\partial v}{\partial h} & \frac{\partial v}{\partial r} & \frac{\partial v}{\partial q} \\ \frac{\partial w}{\partial z} + r\frac{\partial w}{\partial h} & \frac{\partial w}{\partial r} & \frac{\partial w}{\partial q} \end{vmatrix},$$

$$U_6^1 = \begin{vmatrix} \frac{\partial u}{\partial y} + q\frac{\partial u}{\partial h} & \frac{\partial u}{\partial r} & \frac{\partial u}{\partial q} \\ \frac{\partial v}{\partial y} + q\frac{\partial v}{\partial h} & \frac{\partial v}{\partial r} & \frac{\partial v}{\partial q} \\ \frac{\partial w}{\partial y} + q\frac{\partial w}{\partial h} & \frac{\partial w}{\partial r} & \frac{\partial w}{\partial q} \end{vmatrix}, \quad U_6^2 = \begin{vmatrix} \frac{\partial u}{\partial x} + p\frac{\partial u}{\partial h} & \frac{\partial u}{\partial p} & \frac{\partial u}{\partial r} \\ \frac{\partial v}{\partial x} + p\frac{\partial v}{\partial h} & \frac{\partial v}{\partial p} & \frac{\partial v}{\partial r} \\ \frac{\partial w}{\partial x} + p\frac{\partial w}{\partial h} & \frac{\partial w}{\partial p} & \frac{\partial w}{\partial r} \end{vmatrix},$$

$$V' = \begin{vmatrix} \frac{\partial u}{\partial x} + p\frac{\partial u}{\partial h} & \frac{\partial u}{\partial y} + q\frac{\partial u}{\partial h} & \frac{\partial u}{\partial z} + r\frac{\partial u}{\partial h} \\ \frac{\partial v}{\partial x} + p\frac{\partial v}{\partial h} & \frac{\partial v}{\partial y} + q\frac{\partial v}{\partial h} & \frac{\partial v}{\partial z} + r\frac{\partial v}{\partial h} \\ \frac{\partial w}{\partial x} + p\frac{\partial w}{\partial h} & \frac{\partial w}{\partial y} + q\frac{\partial w}{\partial h} & \frac{\partial w}{\partial z} + r\frac{\partial w}{\partial h} \end{vmatrix}.$$

In order that $\phi(uvw)$ may be a solution of the equation (1) in which $\begin{vmatrix} t_1 & s_3 & s_2 \\ s_3 & t_2 & s_1 \\ s_2 & s_1 & t_3 \end{vmatrix}$ fails, we must therefore have $\left(\frac{uvw}{pqr}\right) = 0$, this being its coefficient in the equation (10).

The functions u, v, w have to satisfy the equations

$$\left(\frac{\partial u}{\partial x} + p\frac{\partial u}{\partial h}\right) dx + \left(\frac{\partial u}{\partial y} + q\frac{\partial u}{\partial h}\right) dy + \left(\frac{\partial u}{\partial z} + r\frac{\partial u}{\partial h}\right) dz + \frac{\partial u}{\partial p} dp + \frac{\partial u}{\partial q} dq + \frac{\partial u}{\partial r} dr = 0 \quad (11),$$

$$\left(\frac{\partial v}{\partial x} + p\frac{\partial v}{\partial h}\right) dx + \left(\frac{\partial v}{\partial y} + q\frac{\partial v}{\partial h}\right) dy + \left(\frac{\partial v}{\partial z} + r\frac{\partial v}{\partial h}\right) dz + \frac{\partial v}{\partial p} dp + \frac{\partial v}{\partial q} dq + \frac{\partial v}{\partial r} dr = 0 \quad (12),$$

$$\left(\frac{\partial w}{\partial x} + p\frac{\partial w}{\partial h}\right) dx + \left(\frac{\partial w}{\partial y} + q\frac{\partial w}{\partial h}\right) dy + \left(\frac{\partial w}{\partial z} + r\frac{\partial w}{\partial h}\right) dz + \frac{\partial w}{\partial p} dp + \frac{\partial w}{\partial q} dq + \frac{\partial w}{\partial r} dr = 0 \quad (13).$$

The solution of these equations with respect to dp, dq, dr is

$$\left(\frac{uvw}{pqr}\right) dp = - U_1' dx + U_6^1 dy + U_5^2 dz,$$

$$\left(\frac{uvw}{pqr}\right) dq = U_6^2 dx - U_2' dy + U_4^1 dz,$$

$$\left(\frac{uvw}{pqr}\right) dr = U_5^1 dx + U_4^2 dy - U_3' dz.$$

The solution with respect to dx, dy, dz gives three equations more, as well as those with respect to dp, dq, dz and to dp, dr, dy. Putting now $\left(\frac{uvw}{pqr}\right) = 0$, we obtain subsequently the following 12 equations:

$$U_1' dx - U_6^1 dy - U_5^2 dz = 0 \quad \text{..................(14)},$$

$$- U_6^2 dx + U_2' dy - U_4^1 dz = 0 \quad \text{..................(15)},$$

$$- U_5^1 dx - U_4^2 dy + U_3' dz = 0 \quad \text{..................(16)},$$

$$V' dx + T_1' dp + S_3^1 dq + S_2^1 dr = 0 \quad \text{..................(17)},$$

$$V' dy + S_3^2 dp + T_2' dq + S_1^1 dr = 0 \quad \text{..................(18)},$$

$$V' dz + S_2^2 dp + S_1^2 dq + T_3' dr = 0 \quad \text{..................(19)},$$

$$U_3' dp + U_5^2 dr + T_2' dx - S_3^1 dy = 0 \quad \text{..................(20)},$$

$$U_3' dq + U_4^1 dr - S_3^2 dx + T_1' dy = 0 \quad \text{..................(21)},$$

$$U_2' dp + U_6^1 dq + T_3' dx - S_2^1 dz = 0 \quad \text{..................(22)},$$

$$U_4^2 dq + U_2' dr - S_2^2 dx + T_1' dz = 0 \quad \text{..................(23)},$$

$$U_6^2 dp + U_1' dq + T_3' dy - S_1^1 dz = 0 \quad \text{..................(24)},$$

$$U_5^1 dp + U_1' dr - S_1^2 dy + T_2' dz = 0 \quad \text{..................(25)}.$$

From the equations (14)—(25) we derive other ones, from which the indices 1 and 2 in the right upper corner of the coefficients have disappeared, according to the relations

$$S_1^1 + S_1^2 = S_1', \quad S_2^1 + S_2^2 = S_2', \quad S_3^1 + S_3^2 = S_3',$$

$$U_4^1 + U_4^2 = U_4', \quad U_5^1 + U_5^2 = U_5', \quad U_6^1 + U_6^2 = U_6'.$$

Calculating $(14)\,dx + (15)\,dy + (16)\,dz$,

we obtain $U_1'dx^2 + U_2'dy^2 + U_3'dz^2 - U_4'dydz - U_6'dxdy = 0$ ....(26).

From $(16)\,dr + (20)\,dx + (21)\,dy$,

we obtain

$$T_2'dx^2 + T_1'dy^2 - S_3'dxdy + U_3'(dpdx + dqdy - drdz) + U_4'dydr + U_5'dxdr = 0 \quad ....(27),$$

from $(15)\,dq + (22)\,dx + (23)\,dz$,

$$T_3'dx^2 + T_1'dz^2 - S_2'dxdz + U_2'(dpdx + drdz - dqdy) + U_4'dqdz + U_6'dxdq = 0 \quad ....(28),$$

from $(14)\,dp + (24)\,dy + (25)\,dz$,

$$T_3'dy^2 + T_2'dz^2 - S_1'dydz + U_1'(dqdy + drdz - dpdx) + U_5'dpdz + U_6'dpdy = 0 \quad ....(29).$$

Multiplying the equations (14) to (25) respectively by $dqdr$, $dpdr$, $dpdq$, $dydz$, $dxdz$, $dxdy$, $dqdz$, $dpdz$, $drdy$, $dpdy$, $drdx$, $dqdx$, and adding we obtain

$$V'dxdydz + T_1'dpdydz + T_2'dqdxdz + T_3'drdxdy + U_1'dxdqdr + U_2'dydpdr + U_3'dzdpdq = 0 \quad ......(30).$$

Comparing the equations (26) to (30) with (3) to (7) we see, that besides the condition $\left(\frac{uvw}{pqr}\right) = 0$, the conditions

$$\frac{T_1'}{T_1} = \frac{T_2'}{T_2} = \frac{T_3'}{T_3} = \frac{S_1'}{S_1} = \frac{S_2'}{S_2} = \frac{S_3'}{S_3} = \frac{U_1'}{U_1} = \frac{U_2'}{U_2} = \frac{U_3'}{U_3} = \frac{U_4'}{U_4} = \frac{U_5'}{U_5} = \frac{U_6'}{U_6} = \frac{V'}{V} \quad ...(31)$$

must be satisfied in order that $\phi(uvw)$ may be a solution of the equation (1), viz. 12 equations corresponding as to the number to the equations (14) to (25). In the case that those 12 conditions should be satisfied the solution of (1) is identical to that of the equation

$$T_1't_1 + T_2't_2 + T_3't_3 + S_1's_1 + S_2's_2 + S_3's_3 + U_1'(t_2t_3 - s_1^2) + U_2'(t_1t_3 - s_2^2) + U_3'(t_1t_2 - s_3^2) + U_4'(t_1s_1 - s_2s_3) + U_5'(t_2s_2 - s_1s_3) + U_6'(t_3s_3 - s_1s_2) + V' = 0 \quad ......(32).$$

This is the same as equation (10) in the case $\left(\frac{uvw}{pqr}\right) = 0$, and its solution is $\phi(uvw)$ under condition that $\left(\frac{uvw}{pqr}\right) = 0$, which therefore is a solution of (1) under the same condition.

In the case that the coefficients U_1, U_2, U_4, U_5, U_6 do not contain p, q, r, the condition $\left(\frac{uvw}{pqr}\right)$ is satisfied by choosing an integral of the equation (2) as u, v or w.

In general it will not be possible to integrate the equations (3) to (7). The first condition that this may be the case is that these equations can be transformed into other ones of the first degree as to the differentials dx, dy, dz, dp, dq, dr. It is not easy to discern from the equations in the form hitherto obtained if this may be so. I have sought therefore to put them in a form more apt to that purpose, which is arrived at in the following manner. We calculate

$$U_2\{(3)\,dx + (6)\,drdy + (7)\,dqdz\} + (6)\,T_1dydz.$$

After transforming the coefficient of $dq\,dr$ by means of (4) and after division by $dy\,dz$ we derive

$$(U_2V - T_1T_3)dx^2 - T_1^2dz^2 + T_1S_2dx\,dz - T_1U_4dq\,dz - 2T_1U_2dr\,dz + (U_2S_3 - U_6T_1)\,dq\,dx$$
$$+ U_2S_2dr\,dx - U_2^2dr^2 - U_2U_3dq^2 - U_2U_4dq\,dr = 0 \ldots\ldots (33).$$

In the same manner

$$U_3\{(3)\,dy + (7)\,dp\,dz + (5)\,dr\,dx\} + (7)\,T_2dx\,dz$$

gives

$$(U_3V - T_1T_2)\,dy^2 - T_2^2dx^2 + S_3T_2dx\,dy + (U_3S_1 - U_4T_2)dr\,dy - U_5T_2dr\,dx - 2U_3T_2dp\,dx$$
$$+ U_3S_3dp\,dy - U_3^2dp^2 - U_1U_3dr^2 - U_3U_5dp\,dr = 0 \ldots\ldots (34),$$

and $$U_1\{(3)\,dz + (5)\,dq\,dx + (6)\,dp\,dy\} + (5)\,T_3dx\,dy,$$

$$(U_1V - T_2T_3)\,dz^2 - T_3^2dy^2 + S_1T_3dy\,dz + (U_1S_2 - U_5T_3)\,dp\,dz - U_6T_3dp\,dy - 2U_1T_3dq\,dy$$
$$+ U_1S_1dq\,dz - U_1^2dq^2 - U_1U_2dp^2 - U_1U_6dp\,dq = 0 \ldots\ldots (35).$$

The equations (33) to (35) may be written

$$(U_2V - T_1T_3 + \tfrac{1}{4}S_2^2)\,dx^2 + (\tfrac{1}{4}U_4^2 - U_2U_3)\,dq^2 + (U_2S_3 - U_6T_1 - \tfrac{1}{2}U_4S_2)\,dx\,dq$$
$$-(T_1dz - \tfrac{1}{2}S_2dx + \tfrac{1}{2}U_4dq + U_2dr)^2 = 0 \ldots\ldots (36),$$

$$(U_3V - T_1T_2 + \tfrac{1}{4}S_3^2)\,dy^2 + (\tfrac{1}{4}U_5^2 - U_1U_3)\,dr^2 + (U_3S_1 - U_4T_2 - \tfrac{1}{2}U_5S_3)\,dy\,dr$$
$$-(T_2dx - \tfrac{1}{2}S_3dy + \tfrac{1}{2}U_5dr + U_3dp)^2 = 0 \ldots\ldots (37),$$

$$(U_1V - T_2T_3 + \tfrac{1}{4}S_1^2)\,dz^2 + (\tfrac{1}{4}U_6^2 - U_1U_2)\,dp^2 + (U_1S_2 - U_5T_3 - \tfrac{1}{2}U_6S_1)\,dz\,dp$$
$$-(T_3dy - \tfrac{1}{2}S_1dz + \tfrac{1}{2}U_6dp + U_1dq)^2 = 0 \ldots\ldots (38).$$

In the same manner the combinations

$$U_3\{(3)\,dx + (6)\,dr\,dy + (7)\,dq\,dz\} + (7)\,T_1dy\,dz,$$
$$U_1\{(3)\,dy + (7)\,dp\,dz + (5)\,dr\,dx\} + (5)\,T_2dx\,dz,$$
$$U_2\{(3)\,dz + (5)\,dq\,dx + (6)\,dp\,dy\} + (6)\,T_3dx\,dy,$$

give respectively

$$(U_3V - T_1T_2 + \tfrac{1}{4}S_3^2)\,dx^2 + (\tfrac{1}{4}U_4^2 - U_2U_3)\,dr^2 + (U_3S_2 - U_5T_1 - \tfrac{1}{2}U_4S_3)\,dx\,dr$$
$$-(T_1dy - \tfrac{1}{2}S_3dx + \tfrac{1}{2}U_4dr + U_3dq)^2 = 0 \ldots\ldots (39),$$

$$(U_1V - T_2T_3 + \tfrac{1}{4}S_1^2)\,dy^2 + (\tfrac{1}{4}U_5^2 - U_1U_3)\,dp^2 + (U_1S_3 - U_6T_2 - \tfrac{1}{2}U_5S_1)\,dy\,dp$$
$$-(T_2dz - \tfrac{1}{2}S_1dy + \tfrac{1}{2}U_5dp + U_1dr)^2 = 0 \ldots\ldots (40),$$

$$(U_2V - T_1T_3 + \tfrac{1}{4}S_2^2)\,dz^2 + (\tfrac{1}{4}U_6^2 - U_1U_2)\,dq^2 + (U_2S_1 - U_4T_3 - \tfrac{1}{2}U_6S_2)\,dz\,dq$$
$$-(T_3dx - \tfrac{1}{2}S_2dz + \tfrac{1}{2}U_6dq + U_2dp)^2 = 0 \ldots\ldots (41).$$

The six equations (36) to (41) are symmetrical with respect to the variables. We see from them that the condition for being dissolvable into factors of the first degree is that the first three terms in each of them be a quadric, viz. that the product of the first two coefficients be equal to four times the quadric of the third. In that case the equations are the sum or the difference of two quadrics, dissolvable in real or complex factors of the first degree.

I need scarcely mention that three of these factors from three different equations must be an exact differential in order that $\phi\,(u, v, w)$ may be a solution.

As every combination u, v, w must satisfy the above six equations only three of them can be independent. Confining therefore the consideration to three of them, we must conclude in general eight different combinations of three linear factors to exist. Should each of these combinations furnish an integrable system, then there would be as many different intermediate integrals of the equation (1). How far this is possible I cannot decide. From the general theory we may conclude, if one of the systems furnishes four integrable combinations, all the remaining systems to be identical with it (*Encyclopädie der Mathematischen Wissenschaften*, German edition, II. i. 3. page 393).

If in each of three equations the first three coefficients should vanish, the equation (1) is satisfied only by one combination u, v, w.

The case of two independent variables x and y is derived from the more general case of three variables by equalizing to zero in (1) T_3, S_1, S_2, U_2, U_4, U_5 and U_6; and by putting in equations (36) to (41)

$$dz = 0, \quad dr = 0,$$

so that these equations reduce to the following two:

$$(U_3V - T_1T_2 + \tfrac{1}{4}S_3^2)\,dy^2 - (T_2dx - \tfrac{1}{2}S_3dy + U_3dp)^2 = 0,$$

$$(U_3V - T_1T_2 + \tfrac{1}{4}S_3^2)\,dx^2 - (T_1dy - \tfrac{1}{2}S_3dx + U_3dp)^2 = 0.$$

These equations are identical with those which in the theory of the equation with two variables result after introduction of the values of λ, obtained from the solution of an algebraical equation of the second degree. In our theory we avoid the introduction of λ, which introduction from an elementary point of view may seem to be not quite justifiable.

The equation $$a^2t_1 + b^2t_2 + c^2t_3 = 0 \quad \text{.............................(42)}$$

has no intermediate integral derivable from the above method. We may conclude this from the equations (36) to (41) as well as (5) to (7), which both reduce to

$$b^2dx^2 + a^2dy^2 = 0, \quad c^2dx^2 + a^2dz^2 = 0, \quad c^2dy^2 + b^2dz^2 = 0.$$

Multiplying the first by c^2, the second by b^2 and taking the difference, we obtain

$$c^2dy^2 - b^2dz^2 = 0,$$

which in combination with the third would furnish

$$dy = 0, \quad dz = 0 \quad \text{...............................(43).}$$

These in combination with the equation (3), which reduces to

$$a^2dp\,dy\,dz + b^2dq\,dx\,dz + c^2dr\,dx\,dy = 0,$$

and is therefore identically satisfied by (43), give no system of three integrable combinations.

The equation (42) is of special interest for the case that $a^2 = b^2 = c^2$, which equation, known under the form

$$\nabla_2 V = 0,$$

is integrable in a direct manner after a method given in my brief treatise, *Applications importantes de la théorie du quaternion exponentiel*, edited by H. Eisendrath, Amsterdam. It contains a general method of constructing cases of motion of unlimited fluids*. It was in searching after a method for the further case of free or fixed boundaries that I wished to solve the equation (1). However the solution of the latter problem, as it seems to me, is to be sought for in another direction, the equation concerned not giving rise to integrable combinations of the equations (36) to (41). The solution given above seems however not to be found anywhere, so that the communication may be of some interest for other cases.

Without difficulty we may extend the above method to the more general case that the equation (1) contains the further term

$$W\begin{vmatrix} t_1 & s_3 & s_2 \\ s_3 & t_2 & s_1 \\ s_2 & s_1 & t_3 \end{vmatrix}.$$

Resolving the three equations (α) with respect to dx, dy, dz, we find

$$\begin{vmatrix} t_1 & s_3 & s_2 \\ s_3 & t_2 & s_1 \\ s_2 & s_1 & t_3 \end{vmatrix} = \frac{1}{3dx}\begin{vmatrix} dp & s_3 & s_2 \\ dq & t_2 & s_1 \\ dr & s_1 & t_3 \end{vmatrix} + \frac{1}{3dy}\begin{vmatrix} t_1 & dp & s_2 \\ s_3 & dq & s_1 \\ s_2 & dr & t_3 \end{vmatrix} + \frac{1}{3dz}\begin{vmatrix} t_1 & s_3 & dp \\ s_3 & t_2 & dq \\ s_2 & s_1 & dr \end{vmatrix},$$

and by development,

$$\begin{vmatrix} t_1 & s_3 & s_2 \\ s_3 & t_2 & s_1 \\ s_2 & s_1 & t_3 \end{vmatrix} = \frac{dp}{3dx}(t_2t_3 - s_1^2) + \frac{dq}{3dy}(t_1t_3 - s_2^2) + \frac{dr}{3dz}(t_1t_2 - s_3^2)$$

$$- \frac{dq\,dy + dr\,dz}{3dy\,dz}(t_1s_1 - s_2s_3) - \frac{dp\,dx + dr\,dz}{3dx\,dz}(t_2s_2 - s_1s_3) - \frac{dp\,dx + dq\,dy}{3dx\,dy}(t_3s_3 - s_1s_2).$$

Substituting the values of the minors from page 422 we obtain

$$\begin{vmatrix} t_1 & s_3 & s_2 \\ s_3 & t_2 & s_1 \\ s_2 & s_1 & t_3 \end{vmatrix} = \frac{1}{dx\,dy\,dz}\{dp\,dq\,dr - s_1dp\,(dq\,dy + dr\,dz) - s_2dq\,(dp\,dx + dr\,dz)$$

$$- s_3dr\,(dp\,dx + dq\,dy) + (s_2s_3dx + s_1s_3dy + s_1s_2dz)\,(dp\,dx + dq\,dy + dr\,dz)\}.$$

By substitution in equation (1) we find that equation (3) is to be amplified by the term $W\,dp\,dq\,dr$, (4) by $W(dp\,dx + dq\,dy + dr\,dz)$, whereas the equations (5), (6), (7) are to be amplified respectively by $-W\,dp^2$, $-W\,dq^2$, $-W\,dr^2$. The equations (14), (15) and (16) obtain the additional terms $W'dp$, $W'dq$, $W'dr$, consequently the equations (26) to (30) are to be modified in a manner quite analogous to that of the equations (3) to (7), whereas in the equations (36) to (41) the coefficient in the second term is to be modified, viz. $\frac{1}{4}U_4^2 - U_2U_3$ being amplified by WT_1, $\frac{1}{4}U_5^2 - U_1U_3$ by WT_2, $\frac{1}{4}U_6^2 - U_1U_2$ by WT_3.

* In this brief treatise I proved every function of a certain variable, under condition that $a^2 - b^2 - c^2 = 0$, to satisfy the equation $\nabla_2 V = 0$. For constructing cases of motion it is however necessary, that the function should be developable in series according to positive or negative powers of the variable, which powers prove to be spherical harmonics. In spite of the fact that my developments have given rise to an objection in the *Jahrbuch für die Fortschritte der Mathematik*, Bd. 40, pp. 146—148, they are quite reliable, the result of the integration over surfaces excluding the lines along which a certain denominator is zero, being 0, as I proved in a letter to the critic concerned. The *Jahrbuch* not inserting anticritiques I propose to revert to the subject in a subsequent publication.

ON POWERS OF NUMBERS WHOSE SUM IS THE SAME POWER OF SOME NUMBER

BY ARTEMAS MARTIN.

"The powers that be."—*Romans*, xiii, 1.

Square numbers whose sum is a square, and cube numbers (more than two) whose sum is a cube, have been known for centuries.

So far as the present writer is aware, the late Dr David S. Hart of Stonington, Connecticut, was the first to find biquadrate (fourth-power) numbers whose sum is a biquadrate number. Dr Hart communicated the sets of such numbers that he found to the writer who made them known in a paper read before the Mathematical Section of the Philosophical Society of Washington, Nov. 30, 1887, entitled 'On nth-Power Numbers whose Sum is an nth Power,' an abstract of which was published in the *Bulletin* of the Society, Vol. X, pp. 107—110 of the Proceedings of the Mathematical Section. In that paper the writer found fifth-power numbers whose sum is a fifth power. See also *Educational Times Reprint*, Vol. L (London, 1889), pp. 74—75.

The methods (which are tentative) employed by Dr Hart are as follows:

First Method. Put $S =$ the sum to x terms of the nth powers of the natural series 1^n, 2^n, 3^n, 4^n, 5^n, etc. to x^n; assume an auxiliary formula $(s+m)^n - s^n = a$; then the difference d of the two formulas $S - a, = S - (s+m)^n + s^n$, must, if possible, be separated, by trial or otherwise, into nth-power numbers, all different. Then

$$S - [S - (s+m)^n + s^n] + s^n = (s+m)^n \quad \ldots\ldots\ldots\ldots\ldots\ldots \text{(I)},$$

where x may be any number, and s, m any numbers that will make

$$[(s+m)^n - s^n] < S, \; s \text{ being not less than } x.$$

Second Method. Put the auxiliary quantity $(s+m)^n - s^n = d'$, and divide d', if possible, into nth-power numbers as before directed for d;

$$\text{then these } n\text{th powers} + s^n = (s+m)^n \quad \ldots\ldots\ldots\ldots\ldots\ldots \text{(II)}.$$

As an example of the first method, for biquadrates, take $n = 4$, $s = 14$, $m = 1$, $x = 9$; then $S = 15333$, $a = 15^4 - 14^4 = 12209$, $S - a = 3124 = 7^4 + 5^4 + 3^4 + 2^4 + 1^4$; $S - (1^4 + 2^4 + 3^4 + 5^4 + 7^4) + 14^4 = 15^4$, or $4^4 + 6^4 + 8^4 + 9^4 + 14^4 = 15^4$, five fourth-power numbers whose sum is a fourth power, which are the smallest known all different.

For an example of the second method take $s=44$, $m=1$, $n=4$; then

$$d' = 45^4 - 44^4 = 352529 = 24^4 + 12^4 + 2^4 + 1^4;$$

therefore
$$1^4 + 2^4 + 12^4 + 24^4 + 44^4 = 45^4,$$

another set of five biquadrate numbers whose sum is a biquadrate.

In the paper read before the Philosophical Society of Washington I used the notation

$$S_{x,n} = 1^n + 2^n + 3^n + 4^n + 5^n + \dots + x^n,$$

but have long since discarded it and use

$$S(x^n) = 1^n + 2^n + 3^n + 4^n + 5^n + \dots + x^n,$$

which is simpler and much easier to put in type.

Dr Hart contributed a paper, 'Square Numbers whose Sum is a Square,' to the *Mathematical Magazine*, edited and published by the writer of the present paper, Vol. I, No. 1 (January, 1882, Erie, Pa.), pp. 8—9; and a paper on 'Cube Numbers whose Sum is a Cube,' to Vol. I, No. 11 (July, 1884, Erie, Pa.), pp. 173—176. In these papers Dr Hart used the methods described above.

I published a paper, 'About Biquadrate Numbers whose Sum is a Biquadrate,' in the *Mathematical Magazine*, Vol. II, No. 10 (Jan., 1896, Washington, D.C.), pp. 173—184, in which I employed the general tentative formula

$$S(x^n) - r = b^n \quad \dots\dots\dots\dots\text{(III)},$$

devised by me in 1887, where

$$S(x^n) = 1^n + 2^n + 3^n + 4^n + 5^n + 6^n + \dots + x^n;$$

b must be greater than x, and r must be separated, if possible, into nth-power numbers all different and none greater than x^n; take these nth-powers from $S(x^n)$ and the sum of the remaining nth powers will be $= b^n$.

When $n=4$, we have

$$S(x^4) = 1^4 + 2^4 + 3^4 + 4^4 + 5^4 + \dots + x^4 = \tfrac{1}{30}x(x+1)(2x+1)(3x^2+3x-1).$$

Examples. 1. Let $n=4$, $x=14$; then $S(14^4) = 127687$.

Take $b=15$, then

$$r = 77062 = 13^4 + 12^4 + 11^4 + 10^4 + 7^4 + 5^4 + 3^4 + 2^4 + 1^4;$$

therefore
$$1^4 + 2^4 + 3^4 + 4^4 + \dots + 14^4 - (1^4 + 2^4 + 3^4 + 5^4 + 7^4 + 10^4 + 11^4 + 12^4 + 13^4)$$
$$= 4^4 + 6^4 + 8^4 + 9^4 + 14^4 = 15^4,$$

five biquadrate numbers whose sum is a biquadrate, the same as found by formula (I).

2. Let $n=4$, $x=28$; then $S(x^4) = 3756718$. Take $b=35$, then

$$r = 2256093 = 27^4 + 25^4 + 24^4 + 23^4 + 20^4 + 19^4 + \dots + 5^4 + 3^4 + 2^4 + 1^4;$$

therefore
$$1^4 + 2^4 + 3^4 + 4^4 + 5^4 + \dots + 28^4 - (1^4 + 2^4 + 3^4 + 5^4 + \dots + 19^4 + 20^4 + 23^4$$
$$+ 24^4 + 25^4 + 27^4) = 4^4 + 21^4 + 22^4 + 26^4 + 28^4 = 35^4,$$

another set of five fourth powers whose sum is a fourth power.

This set of numbers could have been more easily found by formula (II), which may be simplified as follows:

Take any two numbers p and q, and put $p^n - q^n = d$; then, transposing, we have

$$q^n + d = p^n,$$

where d must, if possible, be separated into nth powers, all different and q not among them, and we shall have

$$q^n + (\text{those } n\text{th powers}) = p^n.$$

3. Find eight biquadrate numbers whose sum is a biquadrate.

Let $p = 31$, $q = 26$, $n = 4$; then will

$$d = 31^4 - 26^4 = 466545 = 20^4 + 18^4 + 17^4 + 16^4 + 14^4 + 10^4 + 8^4,$$

and $$8^4 + 10^4 + 14^4 + 16^4 + 17^4 + 18^4 + 20^4 + 26^4 = 31^4.$$

4. Find ninety biquadrate numbers whose sum is a biquadrate number.

Using formula (III), let $x = 100$; then $S(100^4) = 2050333330$. Take $b = 212$ and we have $r = 30370194 = 72^4 + 42^4 + 24^4 + 14^4 + 10^4 + 8^4 + 4^4 + 3^4 + 2^4 + 1^4$; taking these biquadrates from

$$S(100^4) = 1^4 + 2^4 + 3^4 + 4^4 + 5^4 + 6^4 + \ldots + 100^4$$

we have left

$$5^4 + 6^4 + 7^4 + 9^4 + 11^4 + 12^4 + 13^4 + 15^4 + 16^4 + 17^4 + \ldots + 23^4 + 25^4 + 26^4$$
$$+ 27^4 + \ldots + 41^4 + 43^4 + 44^4 + \ldots + 71^4 + 73^4 + 74^4 + \ldots + 100^4 = 212^4.$$

The sets of biquadrates whose sum is a biquadrate given above in Examples 1 and 3, were first found by Dr Hart and communicated to the writer previous to 1887. He also found and communicated the following sets:

$$1^4 + 2^4 + 12^4 + 24^4 + 44^4 = 45^4,$$
$$4^4 + 8^4 + 13^4 + 28^4 + 54^4 = 55^4,$$
$$1^4 + 8^4 + 12^4 + 32^4 + 64^4 = 65^4;$$
$$2^4 + 8^4 + 15^4 + 16^4 + 24^4 + 32^4 = 35^4,$$
$$2^4 + 10^4 + 12^4 + 13^4 + 18^4 + 34^4 = 35^4;$$
$$2^4 + 6^4 + 8^4 + 10^4 + 12^4 + 20^4 + 21^4 = 25^4,$$
$$6^4 + 7^4 + 10^4 + 12^4 + 14^4 + 22^4 + 42^4 = 43^4;$$
$$4^4 + 5^4 + 8^4 + 10^4 + 12^4 + 20^4 + 24^4 + 30^4 + 36^4 + 44^4 = 51^4;$$
$$2^4 + 8^4 + 10^4 + 12^4 + 14^4 + 15^4 + 16^4 + 20^4 + 22^4 + 24^4 + 26^4 + 32^4 + 36^4 = 45^4;$$
$$1^4 + 3^4 + 5^4 + 6^4 + 7^4 + 10^4 + 12^4 + 14^4 + 15^4 + 16^4 + 18^4 + 19^4$$
$$+ 20^4 + \ldots + 28^4 + 30^4 + 31^4 + 32^4 + \ldots + 41^4 + 42^4 = 72^4.$$

There was published in the *Mathematical Magazine*, Vol. II, No. 12 (Sept. 1904), pp. 285—296, an able and valuable paper, 'On Biquadrate Numbers,' by Mr Cyrus B. Haldeman, of Ross, Butler Co., Ohio, in which he finds, by rigorous methods of solution, many sets of 5, 6, 7, 8, 9, etc., biquadrates whose sum is a biquadrate, from general formulas each of which will give an infinite number of sets. He was the first, so far as the writer knows, to find such numbers by a general formula.

I will give here one of Mr Haldeman's general formulas for finding five biquadrates whose sum is a biquadrate taken from page 290, Vol. II, of the *Mathematical Magazine*, which is as follows:

$$(8s^2+40st-24t^2)^4+(6s^2-44st-18t^2)^4+(14s^2-4st-42t^2)^4$$
$$+(9s^2+27t^2)^4+(4s^2+12t^2)^4=(15s^2+45t^2)^4,$$

where s and t may have any values taken at pleasure.

Examples. 1. If $s=1$ and $t=0$, we have

$$4^4+6^4+8^4+9^4+14^4=15^4,$$

the same as previously found by other methods.

2. Taking $s=2$, $t=-1$, we find

$$22^4+28^4+63^4+72^4+94^4=105^4.$$

The writer contributed to the International Congress of Mathematicians held at Paris, France, August 9—12, 1900, a paper entitled 'A rigorous Method of Finding Biquadrate Numbers whose Sum is a Biquadrate,' which was published in the *Proceedings of the Congress*, pp. 239—248, and later, with alterations, additions, and corrections, republished in Vol. II of the *Mathematical Magazine*, pp. 325—352, under the title 'About Biquadrate Numbers whose Sum is a Biquadrate.—II,' where a great number of sets of biquadrates whose sum is a biquadrate are found by rigorous formulas.

See also, *Educational Times Reprint*, Vol. XX (London, 1874), p. 55.

Below, one of the many methods of solution is given.

To find five biquadrate numbers whose sum is a biquadrate number.

Solution. Let v, w, x, y, z be the roots of the required biquadrates, and s the root of their sum; then

$$v^4+w^4+x^4+y^4+z^4=s^4 \quad \text{.............................(1).}$$

Assume

$$v=2a, \quad w=a-b, \quad x=a+b, \quad z=\frac{y^2-e^2}{2e}, \quad s=\frac{y^2+e^2}{2e};$$

then, by substitution we have the condition

$$(2a)^4+(a-b)^4+(a+b)^4+y^4+\left(\frac{y^2-e^2}{2e}\right)^4=\left(\frac{y^2+e^2}{2e}\right)^4 \quad \text{...........(2).}$$

Expanding, transposing and uniting terms,

$$2(3a^2+b^2)^2+y^4=\frac{y^2(y^4+e^4)}{2e^2} \quad \text{..........................(3).}$$

Transposing y^4 and multiplying by $2e^2$,

$$4e^2(3a^2+b^2)^2=y^2(y^4+e^4)-2e^4y^4=y^2(y^4-2e^2y^2+e^4)$$
$$=y^2(y^2-e^2)^2.$$

Extracting square root,

$$2e(3a^2+b^2)=y(y^2-e^2);$$

whence

$$b^2=\frac{y(y^2-e^2)}{2e}-3a^2 \quad \text{.................................(4).}$$

In (4), take $y = 2e$ and we get

$$b^2 = 3e^2 - 3a^2 = 3(e+a)(e-a) = \frac{m^2(e-a)^2}{n^2}, \text{ say;}$$

from which $$a = \frac{(m^2 - 3n^2)e}{m^2 + 3n^2}.$$

Taking $e = 2(m^2 + 3n^2)$ we have

$$a = 2(m^2 - 3n^2), \quad b = \frac{m(e-a)}{n} = 12mn;$$

$$\therefore \quad v = 2a = 4(m^2 - 3n^2), \quad w = a - b = 2m^2 - 12mn - 6n^2,$$

$$x = a + b = 2m^2 + 12mn - 6n^2, \quad y = 2e = 4(m^2 + 3n^2),$$

$$z = \frac{y^2 - e^2}{2e} = \frac{3}{2}e = 3(m^2 + 3n^2), \quad s = \frac{y^2 + e^2}{2e} = \frac{5}{2}e = 5(m^2 + 3n^2).$$

Substituting in (1) we have the following identity:

$$(4m^2 - 12n^2)^4 + (2m^2 - 12mn - 6n^2)^4 + (4m^2 + 12n^2)^4$$
$$+ (2m^2 + 12mn - 6n^2)^4 + (3m^2 + 9n^2)^4 = (5m^2 + 15n^2)^4 \ldots\ldots(5),$$

where m and n may have any values whatever chosen at pleasure.

Examples. 1. Taking $m = 2$, $n = 1$, we have

$$4^4 + 21^4 + 22^4 + 26^4 + 28^4 = 35^4.$$

2. Taking $m = 1$, $n = 2$, we get

$$2^4 + 39^4 + 44^4 + 46^4 + 52^4 = 65^4.$$

3. Taking $m = 4$, $n = 1$, we find

$$22^4 + 52^4 + 57^4 + 74^4 + 76^4 = 95^4.$$

To the International Mathematical Congress held in connection with the World's Columbian Exposition, at Chicago, U.S., August, 1893, the writer contributed a paper, 'On Fifth-Power Numbers whose Sum is a Fifth Power,' which was published in the *Congress Mathematical Papers*, Vol. I, pp. 168—174, and later republished, with additions and corrections, in the *Mathematical Magazine*, Vol. II, No. 11 (December, 1898), pp. 201—208.

Examples. 1. Using the formula

$$q^5 + d = p^5,$$

take $p = 12$, $q = 11$, and we have

$$d = 12^5 - 11^5 = 87781 = 9^5 + 7^5 + 6^5 + 5^5 + 4^5;$$

therefore $$4^5 + 5^5 + 6^5 + 7^5 + 9^5 + 11^5 = 12^5,$$

a set of six fifth-power numbers whose sum is a fifth power, and the least known.

2. Take $p = 30$ and $q = 29$; then

$$d = 30^5 - 29^5 = 3788851 = 19^5 + 16^5 + 11^5 + 10^5 + 5^5;$$

therefore $$5^5 + 10^5 + 11^5 + 16^5 + 19^5 + 29^5 = 30^5,$$

another set of six fifth powers whose sum is a fifth power. These are the only sets of six now known, but others probably exist.

3. Take $p = 32$, $q = 31$; then

$$d = 4925281 = 18^5 + 16^5 + 15^5 + 14^5 + 13^5 + 11^5 + 10^5 + 8^5 + 7^5 + 6^5 + 3^5;$$

$$\therefore \quad 3^5 + 6^5 + 7^5 + 8^5 + 10^5 + 11^5 + 13^5 + 14^5 + 15^5 + 16^5 + 18^5 + 31^5 = 32^5,$$

twelve fifth powers whose sum is a fifth power.

4. Using formula (III), take $x = 35$, $n = 5$; then

$$S(35^5) = 333263700.$$

Take $b = 50$, then

$$r = 20763700 = 26^5 + 24^5 + 14^5 + 11^5 + 10^5 + 9^5 + \ldots + 2^5 + 1^5,$$

and we have

$$12^5 + 13^5 + 15^5 + 16^5 + 17^5 + \ldots + 23^5 + 25^5 + 27^5 + 28^5 + 29^5 + \ldots + 34^5 + 35^5 = 50^5,$$

21 fifth-power numbers whose sum is a fifth power.

A paper, 'About Sixth-Power Numbers whose Sum is a Sixth Power,' was contributed by the writer to the Fourth Summer Meeting of the American Mathematical Society, held at Toronto, Canada, August, 1897, and published in the *Mathematical Magazine*, Vol. II (January, 1904), pp. 265—271.

Examples. 1. Using formula (III), take $x = 23$, $n = 6$; then

$$S(23^6) = 563637724.$$

Take $b = 28$, then we have

$$r = 81747420 = 19^6 + 17^6 + 14^6 + 11^6 + 10^6 + 8^6 + 3^6;$$

therefore

$$1^6 + 2^6 + 4^6 + 5^6 + 6^6 + 7^6 + 9^6 + 12^6 + 13^6 + 15^6 + 16^6 + 18^6 + 20^6 + 21^6 + 22^6 + 23^6 = 28^6,$$

sixteen sixth-power numbers whose sum is a sixth power. From this set many others may be obtained.

2. Multiplying the set found above by 3^6 and also by 14^6, and substituting for the value of 84^6 in the second product, we have

$$\begin{aligned} 3^6 + 6^6 + 12^6 + 14^6 + 15^6 + 18^6 + 21^6 + 27^6 + 28^6 + 36^6 + 39^6 + 45^6 \\ + 48^6 + 54^6 + 56^6 + 60^6 + 63^6 + 66^6 + 69^6 + 70^6 + 98^6 + 120^6 + 168^6 \\ + 182^6 + 210^6 + 224^6 + 252^6 + 280^6 + 294^6 + 308^6 + 322^6 = 392^6, \end{aligned}$$

thirty-one sixth powers whose sum is a sixth power.

These two sets of sixth powers whose sum is a sixth power were communicated to the New York (now American) Mathematical Society, October 3, 1891, in a paper, 'On Powers of Numbers whose Sum is the Same Power of Some Number,' which paper was published in the *Quarterly Journal of Mathematics*, Vol. XXVI. (London, 1893), pp. 225—227. See also, *Bulletin of the New York Mathematical Society*, Vol. I, No. 2, p. 55.

Many other sets of fifth powers whose sum is a fifth power, and of sixth powers whose sum is a sixth power, are contained in the papers by me noted above; but space considerations preclude their reproduction here.

The writer is not aware that any person besides himself has ever discovered sets of sixth-power numbers whose sum is a sixth power.

In 1910, Dr Edouard Barbette, Professor of Mathematics, Institut Francken, Liége, Belgium, published a monograph of 154 quarto pages, entitled 'Les sommes de $p^{\text{ièmes}}$ puissances distinctes égales à une $p^{\text{ième}}$ puissance.'

On page 141 he finds

$$4^4 + 6^4 + 8^4 + 9^4 + 14^4 = 15^4,$$

and on page 146,

$$4^5 + 5^5 + 6^5 + 7^5 + 9^5 + 11^5 = 12^5,$$

by practically the same tentative methods, including notation, used by me in my paper published in 1887, in which the above numbers were found 23 years before the appearance of Dr Barbette's book.

I have deposited in the Library of the British Museum copies of all my papers and work on the subject of the present paper (except the paper published in the *Quarterly Journal of Mathematics* of which I had no reprints), where they can be seen by members of the Congress and others who may wish to consult them.

SUR L'*INTÉGRATION LOGIQUE* DES ÉQUATIONS DIFFÉRENTIELLES ORDINAIRES

PAR JULES DRACH.

Je me propose de donner ici un aperçu rapide des points essentiels de la théorie d'intégration que j'ai développée depuis 1893 pour les équations différentielles ordinaires (ou encore les équations linéaires aux dérivées partielles). Eu égard au caractère particulier de cette théorie, je lui ai donné le nom d'*intégration logique* par opposition aux termes *intégration géométrique* ou *intégration par séries* qui me paraissent devoir caractériser les anciennes méthodes de Lagrange, Cauchy, Jacobi, etc....où l'on ne fait pas intervenir la nature des coefficients ou celle des solutions.

L'*intégration logique* de l'ensemble des équations différentielles ordinaires aboutit à partager cet ensemble en *types irréductibles* et *irréductibles les uns aux autres* et à *caractériser chacun de ces types.* Pour une équation déterminée

$$\frac{d^n y}{dx^n} = f\left(x, y, \frac{dy}{dx}, \ldots, \frac{d^{n-1}y}{dx^{n-1}}\right)$$

où la fonction f fait partie d'un domaine de rationalité bien défini, on pourra indiquer une méthode régulière (malheureusement théorique dans les cas généraux) pour reconnaître à quel type elle appartient—c'est-à-dire, au fonds, quelles sont les transcendantes *les plus simples* qui permettent d'exprimer rationnellement les éléments de la solution générale.

D'une manière générale, si l'on considère une équation aux dérivées partielles

$$X(z) = \frac{\partial z}{\partial z} + A_1 \frac{\partial z}{\partial x_1} + \ldots + A_n \frac{\partial z}{\partial x_n} = 0 \ldots\ldots\ldots\ldots\ldots\ldots\ldots (a)$$

dont les coefficients sont des fonctions des $(n+1)$ arguments $x, x_1, \ldots, x_n$ appartenant à un certain domaine de rationalité $[\Delta]$, le système fondamental de solutions $z_1, z_2, \ldots, z_n$ de l'équation (a) que l'on peut regarder comme *le plus simple* est défini par des relations :

$$\Omega_i\left(z_1, \ldots, z_n, \frac{\partial z_1}{\partial x_1}, \ldots, \frac{\partial z_n}{\partial x_n}, \ldots\right) = \alpha_i(x, x_1, \ldots, x_n)$$

$$(i = 1, \ldots, k)$$

dans lesquelles les premiers membres sont tous les *invariants différentiels rationnels et rationnellement distincts,* d'un certain groupe de transformations Γ des éléments $z_1, z_2, \ldots, z_n$ regardés comme fonction des variables $x_1, x_2, \ldots, x_n$ non transformées ; les seconds membres sont des fonctions de $x, x_1, \ldots, x_n$ qui appartiennent au domaine de rationalité $[\Delta]$. Le système précédent est *irréductible,* c'est-à-dire que toute relation de même nature compatible avec les précédentes (vérifiée ainsi que ces dernières par un système fondamental au moins) en est une conséquence nécessaire; il est également *primitif*, c'est-à-dire qu'on ne peut abaisser l'ordre des équations du système ou augmenter le nombre de ces équations qui sont d'un ordre donné en passant à un autre système fondamental. Je dis que le groupe Γ est *le groupe de rationalité de l'équation* et que les solutions les plus simples de (a) sont des fonctions de $(n+1)$ arguments $x, x_1, \ldots, x_n$, *attachées* au groupe Γ; ces fonctions sont en général définies *simultanément* et ne peuvent être séparées.

On peut prendre pour groupe Γ un des types de groupes déterminés *a priori* par S. Lie—mais la théorie actuelle s'établit de façon directe et algébriquement; elle redonnerait donc ces types s'il était nécessaire.

Sophus Lie avait appliqué lui-même sa théorie des groupes de transformations à l'étude des équations différentielles, mais ses travaux, *entièrement distincts de ceux dont il s'agit ici*, ne s'appliquaient qu'à des équations *particulières* parmi celles que nous étudions et ne donnaient que des résultats incomplets, qui valables dans un cas *idéalement général* ne subsistaient plus nécessairement pour un exemple particulier. La théorie des groupes avait donc donné des conclusions intéressantes pour l'intégration mais ne paraissait pas *essentielle* pour l'étude des équations différentielles. En la retrouvant dans ses traits fondamentaux *à partir de ces équations* j'espère avoir établi qu'elle est une discipline *inséparable de l'étude des transcendantes du Calcul Intégral.*

On remarquera l'analogie profonde entre ces résultats et ceux que l'on doit à Galois pour la *résolution logique* des équations algébriques. C'est en effet l'étude de la théorie algébrique de Galois et de l'extension remarquable, maintenant classique, de cette théorie *aux équations différentielles linéaires*, faite par M. Emile Picard en 1887, qui m'a conduit à rechercher les raisons cachées de la perfection et du caractère définitif de ces théories et les conditions les plus générales sous lesquelles ces qualités peuvent être conservées.

Équation du Premier Ordre.

I. *Formes-types et Groupes correspondants.*

1. Considérons une équation du premier ordre

$$\frac{dy}{dx} = A(x, y) \quad \ldots\ldots\ldots\ldots\ldots\ldots\ldots\ldots (1)$$

où nous supposons d'abord, *pour fixer les idées*, A rationnel en x et y; nous définirons son intégrale générale par une relation

$$z(x, y) = \text{const.}$$

où z est une solution particulière quelconque de l'équation aux dérivées partielles

$$X(z) = \frac{\partial z}{\partial x} + A(x, y)\frac{\partial z}{\partial y} \quad \ldots\ldots\ldots\ldots\ldots\ldots (a).$$

Cette dernière donne l'expression de toutes les dérivées de z prises une fois au moins par rapport à x, au moyen des dérivées prises seulement par rapport à y et des variables x, y. S'il existe par suite une relation

$$F\left(x, y, z, \frac{\partial z}{\partial x}, \frac{\partial z}{\partial y}, \ldots\right) = 0 \quad \ldots\ldots\ldots\ldots\ldots\ldots (a'),$$

rationnelle par rapport à tous les éléments qu'elle renferme et *compatible avec l'équation* (a) *sans en être une conséquence nécessaire* on pourra supposer qu'il y figure seulement les dérivées $\frac{\partial z}{\partial y}, \frac{\partial^2 z}{\partial y^2}, \ldots$ prises par rapport à y (et aussi l'écrire sous forme entière).

Supposons d'abord que, quel que soit le choix de la solution z (autre qu'une constante), il n'existe pas de relation telle que (a'), je dis alors que l'équation (a) est *générale*. Toutes ses solutions particulières sont des transcendantes de *même nature* ; elles sont définies aux transformations près du groupe ponctuel général à une variable $Z = F(z)$ où F est arbitraire, par l'équation (a) elle-même : je les appelle *fonctions de deux variables* x, y *attachées au groupe* $Z = F(z)$ (F arbitraire) dans le domaine rationnel.

S'il existe des relations telles que (a'), auquel cas je dis que (a) est *spéciale*, considérons celles qui sont d'*ordre minimum* par rapport aux dérivées de z et parmi celles-là, une quelconque de celles où le degré par rapport à la plus haute dérivée de z *est le plus petit possible.* La relation (a') est alors nécessairement d'ordre au plus égal à *trois* et il est possible de choisir z de façon à lui donner l'une des quatre *formes-types* suivantes :

$$(\alpha) \qquad z = R(x, y) = \frac{P(x, y)}{Q(x, y)} \quad \ldots\ldots\ldots\ldots (a'),$$

si elle est d'ordre zéro. Il existe pour l'équation (a) une solution rationnelle ; on peut la prendre de façon que le polynome

$$P(x, y) - zQ(x, y)$$

ne soit pas décomposable quelle que soit la constante z.

La solution générale de (1) est algébrique.

$$(\beta) \qquad \left(\frac{\partial z}{\partial y}\right)^n = K(x, y) \ldots\ldots\ldots\ldots (a'),$$

si elle est d'ordre 1 ; n est un nombre entier positif, supposé le plus petit possible.

La fonction rationnelle $K(x, y)$ doit satisfaire à l'équation *résolvante*

$$X(K) + nK\frac{\partial A}{\partial y} = 0$$

qui ne peut admettre qu'une seule solution rationnelle.

La transcendante z est définie, aux transformations près du groupe linéaire

$$Z = \epsilon z + a \qquad (\epsilon^n = 1) ;$$

on l'obtient par la quadrature

$$z = \int \sqrt[n]{K}\,(dy - A\,dx).$$

En d'autres termes il existe un multiplicateur de l'équation

$$dy - A\,dx = 0,$$

au sens d'Euler, dont la puissance n est rationnelle en x, y.

(γ) Si la relation (a') est du second ordre, on peut prendre pour forme-type

$$\frac{\partial^2 z}{\partial y^2} - J(x, y)\frac{\partial z}{\partial y} = 0 \quad \dots\dots\dots\dots (a'),$$

où $J(x, y)$ est rationnel et doit vérifier la *résolvante*

$$X(J) + J\frac{\partial A}{\partial y} + \frac{\partial^2 A}{\partial y^2} = 0.$$

Cette résolvante ne peut avoir qu'une seule solution rationnelle. La transcendante z est définie, aux transformations près du groupe

$$Z = az + b \qquad (a, b \text{ constantes arbitraires}).$$

On l'obtient par les quadratures superposées

$$z = \int e^{\int Jdy - \left(AJ + \frac{\partial A}{\partial y}\right)dx} (dy - Adx).$$

En d'autres termes, pour l'équation $dy - Adx = 0$, la dérivée logarithmique d'un multiplicateur d'Euler est rationnelle.

On peut encore définir z par le système suivant:

$$\frac{\partial z}{\partial x} = -AK, \qquad \frac{\partial z}{\partial y} = K,$$

$$\frac{\partial K}{\partial x} = -\left(AJ + \frac{\partial A}{\partial y}\right)K, \qquad \frac{\partial K}{\partial y} = JK,$$

où l'on reconnait que K est donné à un facteur constant près multiplicatif; z est ensuite donné à une constante additive près.

(δ) Si la relation (a') est du troisième ordre, on prendra pour forme-type

$$\frac{\partial z}{\partial y}\frac{\partial^3 z}{\partial y^3} - \frac{3}{2}\left(\frac{\partial^2 z}{\partial y^2}\right)^2 - I(x, y)\left(\frac{\partial z}{\partial y}\right)^2 = 0 \quad \dots\dots\dots\dots (a'),$$

où la fonction rationnelle $I(x, y)$ doit vérifier l'équation résolvante

$$X(I) + 2I\frac{\partial A}{\partial y} + \frac{\partial^3 A}{\partial y^3} = 0.$$

Cette équation résolvante ne possède qu'une solution rationnelle. La transcendante z est définie aux transformations près du groupe projectif général $Z = \frac{az + b}{cz + d}$, où a, b, c, d sont des constantes, par les équations (a) et (a').

On peut l'obtenir au moyen des opérations *théoriques* suivantes:—

Détermination de J par le système de deux équations de Riccati:

$$X(J) + J\frac{\partial A}{\partial y} + \frac{\partial^2 A}{\partial y^2} = 0, \qquad \frac{\partial J}{\partial y} = I + \tfrac{1}{2}J^2;$$

Détermination de K au moyen de J, comme plus haut;

Détermination de z au moyen de K, comme plus haut.

Ainsi, à deux quadratures près, la détermination de z se ramène à l'intégration d'un système de deux équations de Riccati dans le cas où ce système ne possède pas de solution rationnelle.

Dans les cas (β), (γ), (δ) je dis que la transcendante z est une fonction de x et y *attachée, dans le domaine rationnel, aux groupes respectifs*

$$Z = \epsilon z + a \ \ (\epsilon^n = 1), \qquad Z = az + b, \qquad Z = \frac{az+b}{cz+d}.$$

II. *Groupes de Rationalité. Solutions Principales.*

2. Les formes-types que nous avons adoptées pour la relation (a') ne se conservent pas quand on remplace la solution z par une autre solution u. Si l'on pose, par exemple,

$$z = \phi(u),$$

on obtient dans le cas (δ)

$$\phi'^2 \{\phi, u\} \left(\frac{\partial u}{\partial y}\right)^4 + \left[\{u, y\} - I\right] \left(\frac{\partial u}{\partial y}\right)^2 \phi'^2 = 0 \quad \ldots\ldots\ldots\ldots\ldots (a'),$$

où $\{\phi, u\}$ représente l'invariant bien connu de Cayley

$$\frac{\phi'''}{\phi'} - \frac{3}{2}\left(\frac{\phi''}{\phi'}\right)^2, \qquad \phi' = \frac{d\phi}{du}, \ldots$$

et cette relation n'est rationnelle que si la fonction ϕ satisfait à une relation

$$\{\phi, u\} = R(u),$$

où R est rationnel.

Dans ce dernier cas (qui se présente toujours si ϕ est rationnel) la forme de la relation (a') est l'une de celles qui satisfont aux deux conditions imposées plus haut : d'être d'ordre minimum par rapport aux dérivées de u relatives à y et d'être de degré le plus petit par rapport à la dérivée $\dfrac{\partial^p u}{\partial y^p}$ d'ordre le plus élevé.

On peut l'écrire sous la forme

$$\frac{\dfrac{\partial^3 u}{\partial y^3}}{\dfrac{\partial u}{\partial y}} - \frac{3}{2}\left(\frac{\dfrac{\partial^2 u}{\partial y^2}}{\dfrac{\partial u}{\partial y}}\right)^2 + R(u)\left(\frac{\partial u}{\partial y}\right)^2 = I(x, y),$$

où le premier membre est *l'invariant caractéristique* du groupe de transformations (u, v) défini par

$$\frac{\dfrac{d^3 u}{dv^3}}{\dfrac{du}{dv}} - \frac{3}{2}\left(\frac{\dfrac{d^2 u}{dv^2}}{\dfrac{du}{dv}}\right)^2 + R(u)\left(\frac{du}{dv}\right)^2 = R(v),$$

quand on étend ce groupe en y regardant u comme fonction de la variable y non transformée.

Toutes les fois où nous choisissons la relation (a') de façon à satisfaire aux deux conditions précitées, elle exprime *que l'invariant caractéristique d'un certain groupe de transformations en* (u, v) a une valeur rationnelle en x, y ; *cette valeur est, suivant les cas*, K, J ou I.

Je dis que le *groupe de rationalité de la solution* u est le groupe en (u, v), *à équation de définition rationnelle,* dont l'invariant caractéristique est connu rationnellement en x, y.

Parmi ces groupes de rationalité, nous avons choisi plus haut les *groupes-types* que l'on peut ici caractériser en disant que leur invariant caractéristique ne dépend pas de u mais seulement des dérivées $\frac{\partial u}{\partial y}, \frac{\partial^2 u}{\partial y^2}, \ldots$ Ces groupes de rationalité *types* correspondent à la solution z que l'on peut regarder comme *la plus simple.* Bien entendu, cette solution la plus simple n'est définie qu'aux transformations près qui conservent les formes-types des relations (a').

3. Il est essentiel de montrer quels sont les groupes de rationalité qui se présentent pour la *solution principale, au point* $x = x_0$, de l'équation

$$X(z) = \frac{\partial z}{\partial x} + A(x, y) \frac{\partial z}{\partial y} = 0,$$

c'est-à-dire pour la solution qui pour $x = x_0$ prend la valeur y.

(α) Dans le cas où z est rationnel en x, y cette solution principale u est définie par $R(x, y) = R(x_0, u)$, c'est-à-dire est algébrique.

(β) Si z est défini aux transformations près $Z = \epsilon z + a$, $(\epsilon^n = 1)$ on a, pour déterminer la solution principale, les équations

$$X(u) = 0, \qquad K(x_0, u) \left(\frac{\partial u}{\partial y}\right)^n = K(x, y);$$

elle est donc définie aux transformations près du groupe (u, v) où

$$K(x_0, u) \left(\frac{du}{dv}\right)^n = K(x_0, v),$$

transformations qui sont en général transcendantes.

(γ) Si z est défini aux transformations près $Z = az + b$, la solution principale u est déterminée par les équations

$$X(u) = 0, \qquad \frac{\frac{\partial^2 u}{\partial y^2}}{\frac{\partial u}{\partial y}} + J(x_0, u) \frac{\partial u}{\partial y} = J(x, y)$$

aux transformations près (u, v), qui satisfont à

$$\frac{\frac{d^2 u}{dv^2}}{\frac{du}{dv}} + J(x_0, u) \frac{du}{dv} = J(x_0, v).$$

(δ) Enfin lorsque z n'est défini qu'aux transformations près du groupe projectif $Z = \frac{az + b}{cz + d}$, la solution principale u de (1) est déterminée par les équations

$$X(u) = 0, \quad \frac{\frac{\partial^3 u}{\partial y^3}}{\frac{\partial u}{\partial y}} - \frac{3}{2} \left(\frac{\frac{\partial^2 u}{\partial y^2}}{\frac{\partial u}{\partial y}}\right)^2 + I(x_0, u) \left(\frac{\partial u}{\partial y}\right)^2 = I(x, y)$$

aux transformations près (u, v) du groupe qui a pour *équation de définition*

$$\frac{\frac{d^3u}{dv^3}}{\frac{du}{dv}} - \frac{3}{2}\left(\frac{\frac{d^2u}{dv^2}}{\frac{du}{dv}}\right)^2 + I(x_0, u)\left(\frac{du}{dv}\right)^2 = I(x_0, v).$$

L'importance de la considération des solutions principales $u(x, y, x_0)$ résulte de ce que la solution de l'équation (1) $\frac{dy}{dz} = A(x, y)$ qui prend la valeur y_0 pour $x = x_0$ est donnée par la relation implicite

$$u(x, y, x_0) = y_0$$

qui, d'après une remarque de Jacobi, s'écrit aussi sous forme résolue

$$y = u(x_0, y_0, x).$$

La solution y de l'équation (1) qui prend en $x = x_0$ la valeur y_0 est donc une fonction des trois arguments x_0, y_0, x dont la nature peut être mise en évidence, dans tous les cas de réduction.

Observons que le groupe de rationalité de u dépend des valeurs des invariants rationnels K, J, I, ce qui complique singulièrement l'étude des propriétés de y alors que celles de z sont si simples.

4. Il convient de remarquer ici que lorsque le groupe de rationalité d'une solution, et par suite *le groupe de rationalité-type* de l'équation

$$\frac{dy}{dx} = A(x, y) \quad \ldots\ldots\ldots\ldots\ldots\ldots\ldots\ldots\ldots\ldots(1)$$

est déterminé, il y a lieu d'étudier les réductions qui peuvent se produire dans la difficulté de la recherche de z à partir des invariants du groupe-type. Ces réductions correspondent à la possibilité d'obtenir z à l'aide de fonctions d'un seul argument attachées au groupe-type ou à l'un de ses sous-groupes. Par exemple, si K ou $\left(\frac{\partial z}{\partial y}\right)^n$ est rationnel, il peut se faire que z puisse s'obtenir par des quadratures de différentielles algébriques portant sur des fonctions d'un argument $\omega(x, y)$, rationnel en x et y.

De même si $J(x, y)$ est rationnel, on voit que, dans tous les cas, l'introduction de la transcendante e^u, fonction d'une variable u attachée au groupe $u' = au$, permet d'obtenir K par une quadrature

$$K = e^{\omega}, \quad \omega = \int J\,dy - \left(AJ + \frac{\partial A}{\partial y}\right) dx,$$

et il y aura lieu de chercher si cette dernière quadrature ne peut pas s'exécuter par l'introduction de logarithmes d'arguments rationnels en x, y.

De même encore, dans le cas (δ), si par exemple I dépend de y seul, on voit que J qui vérifie $\frac{\partial J}{\partial y} = I + \frac{1}{2}J^2$ s'obtiendra en y par l'intégration d'une seule équation de Riccati rationnelle en y et sa détermination complète se ramènera à l'intégration d'une autre équation de Riccati en x seul.

Mais à un premier examen, on peut regarder ces réductions, qui ne modifient pas le groupe de rationalité, comme secondaires.

III. *Extensions du domaine de Rationalité. Adjonction de Transcendantes.*

5. Jusqu'à présent nous avons supposé, dans l'équation

$$\frac{dy}{dx} = A(x, y) \quad \ldots\ldots\ldots\ldots\ldots\ldots\ldots\ldots(1),$$

$A(x, y)$ rationnel. Rien d'essentiel n'est changé à la théorie précédente lorsqu'on *étend* le domaine de rationalité par l'*adjonction* de fonctions algébriques en x, y ou de *transcendantes bien définies.*

Par exemple si la relation algébrique entière

$$f(x, y, \zeta) = 0$$

est irréductible et de degré n en ζ, toute fonction rationnelle en x, y, ζ se ramène *d'une seule manière* à la forme

$$\frac{p_0(x, y) + p_1(x, y)\zeta + \ldots + p_{n-1}(x, y)\zeta^{n-1}}{q(x, y)},$$

où les p et q sont des polynomes en x, y sans diviseur commun. On peut répéter tout ce qui a été dit plus haut, dans le domaine $[\zeta]$, pour une équation

$$\frac{dy}{dx} = A(x, y, \zeta),$$

où A est rationnel en x, y, ζ. Les $n\left(\text{ou } \frac{n}{p}\right)$ équations qu'on obtient en remplaçant ζ par ses n valeurs dans $A(x, y, \zeta)$, ont le même groupe de rationalité.

La théorie s'applique donc à une équation du premier ordre, *de degré quelconque* $F\left(x, y, \frac{dy}{dx}\right) = 0$, que nous écrirons $\frac{dy}{dx} = \zeta$ avec $F(x, y, \zeta) = 0$.

On peut aussi adjoindre au domaine de rationalité des *transcendantes bien définies*; voici ce que nous entendons par là.

Considérons un système différentiel formé d'équations

$$\phi_i\left(x, y, u, \frac{\partial u}{\partial x}, \frac{\partial u}{\partial y}, \frac{\partial^2 u}{\partial x^2}, \ldots\right) = 0,$$
$$(i = 1, \ldots, h),$$

dont les premiers membres sont des polynomes entiers en x, y, u, $\frac{\partial u}{\partial x}$, $\frac{\partial u}{\partial y}$, ..., où u est une fonction de *deux variables* x, y ; on sait par des procédés réguliers qui ne comportent que des opérations rationnelles, reconnaître si ces équations sont *compatibles* c'est-à-dire possèdent au moins une solution $u(x, y)$ dépendant des deux arguments x et y. On sait, dans ce cas, déduire des équations données un système différentiel (Σ) qui définit, dans un domaine algébrique convenable $[\Delta]$, certaines des dérivées de u, que l'on appelle *principales*, au moyen des autres qui sont dites *paramétriques*—et dont sous certaines conditions de convergence, les valeurs pour $x = x_0$, $y = y_0$ peuvent être choisies arbitrairement.

Toute fonction rationnelle des dérivées de u et de x, y, s'exprime *sous une forme unique*, au moyen des dérivées paramétriques et des variables, dans le domaine algébrique $[\Delta]$.

Cela étant rappelé, deux cas peuvent se présenter. Il peut se faire qu'on ne puisse ajouter aux équations de (Σ) aucune équation *nouvelle*, rationnelle par rapport à tous ses éléments [c'est-à-dire aucune relation entière entre les dérivées paramétriques dont les coefficients appartiennent au domaine $[\Delta]$] sans cesser d'avoir un système compatible (le nouveau système n'admettrait plus comme solutions que des constantes, ou des fonctions de x seul ou de y seul). Je dis alors que le système (Σ) envisagé comme définissant des fonctions u de deux variables x, y est *irréductible*. La transcendante u est *bien définie*: le calcul des fonctions rationnelles de x, y, u, $\frac{\partial u}{\partial x}, \frac{\partial u}{\partial y}, \ldots$ (mod. Σ) peut se faire sans ambiguïté.

Si l'on peut ajouter aux équations (Σ) une ou plusieurs équations nouvelles (σ) sans cesser d'avoir un système compatible, les diverses solutions $u(x, y)$ de (Σ) ne se comportent pas de même dans le domaine $[\Delta]$. Je dis que u n'est pas bien défini par (Σ), qui est *réductible*.

Dans ce cas on peut toujours supposer qu'on a ajouté à (Σ) assez d'équations nouvelles (σ) pour former un système $(\Sigma + \sigma)$ irréductible; aux divers systèmes (σ) possibles correspondront autant de types de transcendantes $u(x, y)$ bien définies vérifiant (Σ). Enfin si l'on veut raisonner sur la transcendante $u(x, y)$ qui vérifie les équations (Σ) sans satisfaire à aucune des équations (σ) il faudra compléter les équations (Σ) en ajoutant des *inégalités* qui expriment qu'aucun des premiers membres des équations (σ) n'est nul. Il sera en pratique inutile d'écrire ces inégalités, étant bien entendu que dans le calcul des fonctions rationnelles de u et de ses dérivées $\frac{\partial u}{\partial x}, \frac{\partial u}{\partial y}, \ldots$, les dérivées paramétriques se comporteront comme des indéterminées.

Ce qui vient d'être dit pour les fonctions $u(x, y)$ de deux variables s'applique évidemment aussi aux fonctions d'une seule variable.

L'exemple le plus simple d'une transcendante bien définie est la fonction exponentielle e^x; elle est définie par la relation

$$\frac{\partial u}{\partial x} = u \text{ avec } (u \neq 0)$$

et n'est pas séparable de cu où c est une constante quelconque. Adjoindre par conséquent la transcendante e^x au domaine de rationalité c'est écrire le système irréductible

$$(\Sigma) \quad \frac{\partial u}{\partial y} = 0, \quad \frac{\partial u}{\partial x} = u \quad (u \neq 0)$$

et en tenir compte dans le calcul (addition, multiplication, division, *dérivation*) des fonctions rationnelles de x, y, u, $\frac{\partial u}{\partial x}, \frac{\partial u}{\partial y}, \ldots$.

Si l'on considère une équation

$$\frac{dy}{dx} = A(x, y, u)$$

dont le second membre est rationnel en x, y, u, on peut donc répéter dans le domaine $[u]$ tout ce qui a été dit pour un domaine rationnel *absolu*.

6. A un certain point de vue les fonctions *arbitraires* d'un ou de plusieurs arguments déterminés se comportent comme des transcendantes bien définies. Par exemple pour l'équation de Riccati

$$\frac{dy}{dx} = a_0(x) + a_1(x)\, y + a_2(x)\, y^2 \quad \text{.........................(1)}$$

la résolvante en I possède, lorsque a_0, a_1, a_2 sont arbitraires en x, la seule solution rationnelle $I = 0$. Si l'on cherche à quelle condition la résolvante en J possède une solution rationnelle on trouve que pour cette solution $J = \frac{-2}{y - \xi}$ où $\xi(x)$ est une solution de l'équation de Riccati. Ce n'est donc que dans le domaine $[\xi]$ obtenu par l'adjonction d'une solution particulière de l'équation de Riccati que cette équation se réduit.

De même, si dans une équation (1) $\frac{dy}{dx} = A(x, y)$ dont le groupe de rationalité est l'un des groupes-types (α), (β), (γ), (δ) on remplace x et y par deux fonctions arbitraires $x(u, v)$, $y(u, v)$ des arguments u et v, pour lesquelles $\frac{\partial(x, y)}{\partial(u, v)} \neq 0$, on obtient une nouvelle équation

$$\frac{dv}{du} = m(u, v) = \frac{A(x, y)\frac{\partial x}{\partial u} - \frac{\partial y}{\partial u}}{\frac{\partial y}{\partial v} - A(x, y)\frac{\partial x}{\partial v}} \quad \text{.....................(1')}$$

qui dans le domaine formé par l'adjonction des fonctions arbitraires $x(u, v)$, $y(u, v)$ (ce qui entraîne toujours l'adjonction de leurs dérivées) a le même groupe de rationalité. Le calcul des expressions explicites rationnelles de $\frac{\partial z}{\partial v}$, $\frac{\partial^2 z}{\partial v^2} : \frac{\partial z}{\partial v}$,

$$\frac{\frac{\partial^3 z}{\partial v^3}}{\frac{\partial z}{\partial v}} - \frac{3}{2}\left(\frac{\frac{\partial^2 z}{\partial v^2}}{\frac{\partial z}{\partial v}}\right)^2 = \{z, v\}$$

au moyen des valeurs de

$$K = \frac{\partial z}{\partial y}, \quad J = \frac{\frac{\partial^2 z}{\partial y^2}}{\frac{\partial z}{\partial y}}, \quad I = \{z, y\}$$

se fait sans difficulté.

On peut naturellement supposer que x et y sont des expressions rationnelles en u, v, $\phi_1(u)$, $\phi_2(u)$, ..., $\psi_1(v)$, $\psi_2(v)$, ..., $\chi(u, v)$, ..., où les fonctions ϕ_1, ϕ_2, ..., ψ_1, ψ_2, ..., $\chi(u, v)$ sont arbitraires et former ainsi des équations du premier ordre dépendant de fonctions arbitraires dont le groupe de rationalité est déterminé, mais rien ne serait plus facile que de revenir d'une telle équation à l'équation (1) $\frac{dy}{dx} = A(x, y)$ qui a servi à la construire.

Il est plus intéressant de chercher à exprimer rationnellement, dans un domaine algébrique [Δ], A et K, ou A et J, ou A et I au moyen de x, y, et d'une ou de plusieurs fonctions arbitraires d'une ou de deux variables ainsi que de leurs dérivées jusqu'à un ordre fixé, de manière à satisfaire aux équations résolvantes:

$$X(K) + nK\frac{\partial A}{\partial y} = 0, \quad X(J) + J\frac{\partial A}{\partial y} + \frac{\partial^2 A}{\partial y^2} = 0, \quad X(I) + 2I\frac{\partial A}{\partial y} + \frac{\partial^3 A}{\partial y^3} = 0.$$

On parvient alors à des types généraux d'équations dont le groupe de rationalité est connu et qui ne se ramènent pas à des équations déterminées par un changement explicite des variables.

Toutes les fois où l'on remplace dans l'une de ces équations les arbitraires qui y figurent par des transcendantes *bien définies* on pourra chercher à réduire le groupe de rationalité—et la réduction obtenue sera définitive.

On se trouve ainsi amené à des problèmes "de Diophante" *que l'on peut résoudre méthodiquement,* mais sur lesquels je n'insisterai pas ici. J'ajoute qu'on peut se poser et résoudre des problèmes analogues dans le domaine obtenu par l'adjonction aux variables x, y, d'une ou de plusieurs transcendantes, u, *bien définies* en x, y.

Comme exemple d'une équation renfermant des éléments arbitraires, dont le groupe de rationalité est le groupe projectif général (δ) je donnerai l'équation

$$\begin{vmatrix} y' & 1 & y & y^2 & y^3 \\ \phi' & 1 & \phi & \phi^2 & \phi^3 \\ \psi' & 1 & \psi & \psi^2 & \psi^3 \\ 0 & 0 & 4 & \phi+\psi & \phi^2+\psi^2-4\phi\psi \\ 0 & 0 & 0 & 2 & 3(\phi+\psi) \end{vmatrix} = 0 \quad \ldots\ldots\ldots\ldots\ldots (1)$$

pour laquelle dans le domaine [ϕ, ψ] formé par deux de ses solutions particulières, la résolvante en I possède la solution rationnelle

$$I = -\frac{3}{2} \cdot \frac{1}{(y-\phi)(y-\psi)},$$

unique lorsque ϕ et ψ demeurent arbitraires.

IV. *Détermination du Groupe de Rationalité* (Groupe-type). *Intégration algébrique de l'Équation du Premier Ordre.*

7. Pour déterminer le groupe de rationalité (type) d'une équation

$$\frac{dy}{dx} = A(x, y) \quad \ldots\ldots\ldots\ldots\ldots\ldots\ldots\ldots\ldots\ldots (1),$$

où A appartient à un certain domaine de rationalité [Δ], il est nécessaire de savoir décider, *par un nombre fixé à l'avance de calculs élémentaires,* si l'une des équations résolvantes

$$X(z) = 0, \quad X(K) + nK\frac{\partial A}{\partial y} = 0, \quad X(J) + J\frac{\partial A}{\partial y} + \frac{\partial^2 A}{\partial y^2} = 0, \quad X(I) + 2I\frac{\partial A}{\partial y} + \frac{\partial^3 A}{\partial y^3} = 0$$

possède dans le domaine $[\Delta]$ une solution rationnelle. C'est là un problème qui est loin d'être résolu.

On peut commencer les essais en cherchant si z peut être rationnel, puis si K peut être rationnel, etc.... on est alors certain que l'équation résolvante que l'on étudie (sauf l'équation en z bien entendu) ne peut avoir qu'une seule solution rationnelle. Au contraire si l'on aborde d'emblée l'étude de la résolvante en I et si l'on trouve une solution rationnelle, il peut y en avoir d'autres lorsque l'équation en J possède elle-même une solution rationnelle. On aura donc à étudier, après avoir trouvé I, les deux équations de Riccati qui déterminent J pour décider si elles possèdent ou non une solution J rationnelle; ce n'est que dans le dernier cas que le groupe de rationalité est le groupe projectif général.

Supposons *pour fixer les idées* $A(x, y)$ rationnel, écrivons l'équation (1) $\dfrac{dy}{dx} = \dfrac{\alpha(x, y)}{\beta(x, y)}$ où α, β sont deux polynomes en x, y sans diviseur commun.

Dans l'hypothèse générale où $\dfrac{\partial\beta}{\partial y} \neq 0$ et où β est dépourvu de facteurs multiples on montre aisément que si I est rationnel:

$$I = \frac{\dfrac{\partial^2\beta}{\partial y^2}}{\beta} - \frac{3}{2}\left(\frac{\dfrac{\partial\beta}{\partial y}}{\beta}\right)^2 + \frac{S}{\beta R},$$

et les polynomes S, R sans diviseur commun doivent satisfaire aux relations, où T désigne un polynome auxiliaire:

$$\alpha\frac{\partial R}{\partial y} + \beta\frac{\partial R}{\partial x} = \gamma R, \quad \alpha S + \beta T = R\frac{\partial\beta}{\partial y}\left(\frac{\partial\beta}{\partial x} + \frac{\partial\alpha}{\partial y}\right),$$

$$\alpha\frac{\partial S}{\partial y} + \beta\frac{\partial S}{\partial x} = S\left(\gamma + \frac{\partial\beta}{\partial x} - 2\frac{\partial\alpha}{\partial y}\right) - 3T\frac{\partial\beta}{\partial y}$$
$$+ R\left[\frac{\partial^2\beta}{\partial y^2}\left(\frac{\partial\alpha}{\partial y} + \frac{\partial\beta}{\partial x}\right) + 3\frac{\partial\beta}{\partial y}\frac{\partial}{\partial y}\left(\frac{\partial\alpha}{\partial y} + \frac{\partial\beta}{\partial x}\right) - \beta\frac{\partial^2}{\partial y^2}\left(\frac{\partial\alpha}{\partial y} + \frac{\partial\beta}{\partial x}\right)\right].$$

Si le groupe de rationalité est le groupe projectif général, il n'existe qu'un seul système de polynomes, R, S, T, γ satisfaisant à ces conditions.

On voit que R égalé à zéro définit des solutions particulières algébriques de l'équation (1) et même *toutes les solutions particulières algébriques.* Le polynome R étant connu, la détermination de S et T s'ensuit.

Des conclusions analogues peuvent être énoncées lorsque le groupe de rationalité est le groupe linéaire. L'invariant J, s'il est rationnel s'écrit

$$J = \frac{\dfrac{\partial\beta}{\partial y}}{\beta} - \frac{H}{R},$$

et l'on doit avoir $$\alpha\frac{\partial R}{\partial y} + \beta\frac{\partial R}{\partial x} = \gamma R.$$

Avec
$$\gamma K = \frac{\partial R}{\partial x}\left(\frac{\partial \alpha}{\partial y} + \frac{\partial \beta}{\partial x}\right) + \alpha\left(\frac{\partial K}{\partial y} - \frac{\partial H}{\partial x}\right),$$
$$\gamma H = \frac{\partial R}{\partial y}\left(\frac{\partial \alpha}{\partial y} + \frac{\partial \beta}{\partial x}\right) - \beta\left(\frac{\partial K}{\partial y} - \frac{\partial H}{\partial x}\right),$$
où le polynome auxiliaire K satisfait à la condition
$$\alpha H + \beta K = R\left(\frac{\partial \alpha}{\partial y} + \frac{\partial \beta}{\partial x}\right),$$
qui est d'ailleurs une conséquence des précédentes.

Ces relations expriment simplement l'intégrabilité de l'expression
$$z = \int (\beta dy - \alpha dx)\, e^{-\int \frac{H\,dy + K\,dx}{R}}$$
et l'on reconnait comme plus haut, que R égalé à zéro définit toutes les solutions particulières algébriques de (1).

Enfin si l'équation possède un multiplicateur dont la puissance n soit rationnelle, z est donné par la quadrature
$$z = \int \sqrt[n]{\frac{P}{Q}}\,(\beta dy - \alpha dx),$$
où les polynomes P, Q doivent satisfaire aux relations
$$\alpha \frac{\partial Q}{\partial y} + \beta \frac{\partial Q}{\partial x} = LQ,$$
$$\alpha \frac{\partial P}{\partial y} + \beta \frac{\partial P}{\partial x} = \left[L - n\left(\frac{\partial \beta}{\partial x} + \frac{\partial \alpha}{\partial y}\right)\right] P,$$
qui expriment que $P=0$ et $Q=0$ définissent des solutions particulières algébriques, convenablement associées, de (1); on voit d'ailleurs que toutes les solutions particulières algébriques doivent annuler P ou Q.

Ainsi la difficulté principale est, dans tous les cas, *la détermination de toutes les solutions particulières algébriques de* (1).

La remarque suivante permet quelquefois de les obtenir. Supposons que l'équation $\frac{dy}{dx} = A(x, y)$, à étudier, dépende de certains paramètres λ, μ, ..., et que pour un système de valeurs λ_0, μ_0, ... de ces paramètres on sache déterminer son groupe de rationalité Γ; il est clair que le groupe de rationalité G qui correspond au cas général aura Γ comme sous-groupe. En particulier les solutions algébriques du cas général devront se réduire pour $\lambda = \lambda_0$, $\mu = \mu_0$, ... à des fonctions algébriques connues.

8. On n'a étudié jusqu'à présent que le cas où l'équation
$$X(z) = \beta \frac{\partial z}{\partial x} + \alpha \frac{\partial z}{\partial y} = 0 \qquad\qquad (a)$$
possède une solution z rationnelle, c'est-à-dire où l'équation (1) $\frac{dy}{dx} = \frac{\alpha(x, y)}{\beta(x, y)}$ s'intègre

algébriquement. Dans un Mémoire célèbre (*Bulletin des Sciences Mathématiques*, 1878) Mr Darboux a montré que la connaissance d'un certain nombre de polynomes irréductibles $\phi(x, y)$ satisfaisant à des identités $X(\phi) = M\phi$, c'est-à-dire d'un certain nombre de solutions particulières algébriques de (1) permet de construire la solution générale de (1) ou au moins un multiplicateur. Plus récemment, MM. Poincaré, Painlevé et Autonne, dans des mémoires bien connus, ont cherché à déduire de l'étude des points où $A(x, y)$ est indéterminé (points singuliers) la limitation du degré de l'intégrale algébrique irréductible de (1) en faisant intervenir la forme analytique des intégrales au voisinage des points singuliers.

Je me bornerai à indiquer ici quelques résultats auxquels on parvient par une voie tout élémentaire.

Si l'équation (1) $\dfrac{dy}{dx} = \dfrac{\alpha(x, y)}{\beta(x, y)}$ où α et β sont des polynomes en x, y de degré m s'intègre algébriquement, son intégrale générale peut être définie par une relation

$$P(x, y) + zQ(x, y) = 0,$$

où P, Q sont des polynomes d'un certain degré p et où on peut supposer, d'après Poincaré, que $P + zQ$ n'est pas décomposable pour toute valeur de z. On en déduit aisément les relations

$$Q\frac{\partial P}{\partial x} - P\frac{\partial Q}{\partial x} = L\alpha, \qquad X(P) = MP,$$

$$Q\frac{\partial P}{\partial y} - P\frac{\partial Q}{\partial y} = -L\beta, \qquad X(Q) = MQ,$$

$$\frac{\partial P}{\partial x}\frac{\partial Q}{\partial y} - \frac{\partial P}{\partial y}\frac{\partial Q}{\partial x} = LM, \qquad X(L) = \left(2M - \frac{\partial \alpha}{\partial y} - \frac{\partial \beta}{\partial x}\right)L,$$

d'où l'on peut conclure que L est un polynome de degré $2p - (m + 1)$ divisible par ϕ^{a-1} si l'on a pour une valeur convenable ζ de $z : P + \zeta Q = \phi^a \psi$. Ce polynome L dont l'introduction est due à M. Darboux (*loc. cit.*) qui en a également reconnu la signification profonde, ne renferme d'ailleurs pas d'autres facteurs que ceux qui peuvent être multiples dans $P + zQ$ pour des valeurs convenables de z.

Supposons d'abord qu'il n'existe pour z que *deux valeurs exceptionnelles* pour lesquelles $P + zQ$ possède un diviseur multiple ; on peut admettre qu'elles sont $z = 0$ et $z = \infty$, c'est-à-dire que

$$P = \phi_1^{a_1}\phi_2^{a_2} \dots \phi_h^{a_h},$$

$$Q = \psi_1^{b_1}\psi_2^{b_2} \dots \psi_k^{b_k}$$

où les a et les b sont entiers. On a dans ce cas

$$\Omega = \frac{PQ}{L} = \phi_1\phi_2 \dots \phi_h\psi_1\psi_2 \dots \psi_k$$

et le degré de Ω est égal à $(m + 1)$: le polynome Ω de degré $(m + 1)$ satisfait en outre à l'équation

$$X(\Omega) = \left(\frac{\partial \alpha}{\partial y} + \frac{\partial \beta}{\partial x}\right)\Omega$$

qui exprime que $\dfrac{1}{\Omega}$ est un multiplicateur pour $\beta dy - \alpha dx$.

On sait donc trouver Ω dont le degré est connu et sa décomposition en facteurs indécomposables donne les ϕ et les ψ. Formons les polynomes S_i et T_j de degré $(m-1)$ définis par les identités

$$X(\phi_i) = S_i\phi_i, \qquad X(\psi_j) = T_j\psi_j;$$

on aura immédiatement

$$a_1 S_1 + \ldots + a_h S_h = M,$$
$$b_1 T_1 + \ldots + b_k T_k = M,$$

d'où l'on conclut l'existence d'une relation linéaire à coefficients entiers, positifs ou négatifs, *sans diviseur commun*, entre les S et les T:

$$a_1 S_1 + \ldots + a_h S_h - b_1 T_1 - \ldots - b_k T_k = 0.$$

Cette relation est unique et sa formation effective sépare les T, dont le coefficient est un entier négatif, des S dont le coefficient est positif; elle donne par suite les groupements qui constituent P et Q.

Ainsi l'intégrale $P + zQ = 0$ a pu être formée, mais le *degré des polynomes P et Q n'est connu qu'après la construction effective de ces polynomes.* Le principal objet de Poincaré paraît, au contraire, avoir été de limiter d'avance ce degré.

On aurait pu tout aussi bien partir de l'identité

$$\frac{1}{\Omega}(\beta dy - \alpha dx) = a_1\frac{d\phi_1}{\phi_1} + \ldots + a_h\frac{d\phi_h}{\phi_h} - b_1\frac{d\psi_1}{\psi_1} - \ldots - b_k\frac{d\psi_k}{\psi_k}$$

qui détermine sans ambiguïté les entiers a et b.

9. L'intérêt de la solution précédente, applicable quelle que soit la forme des polynomes α et β et la nature des points communs aux deux courbes $\alpha = 0$, $\beta = 0$, s'augmente si l'on observe que la même méthode peut réussir *alors qu'il existe trois ou quatre valeurs exceptionnelles de z.*

Admettons qu'il existe trois valeurs remarquables de z donnant lieu à des identités

$$P + z_1 Q = A = \phi^a \phi_1^{a_1} \ldots \phi',$$
$$P + z_2 Q = B = \psi^b \psi_1^{b_1} \ldots \psi',$$
$$P + z_3 Q = C = \chi^c \chi_1^{c_1} \ldots \chi',$$

où l'on a mis en évidence les exposants a, b, c, les plus élevés des facteurs indécomposables multiples du premier membre. Peut-on former une expression $\Omega = \dfrac{A^\xi B^\eta C^\zeta}{L^\lambda}$ où ξ, η, ζ, λ sont des entiers, qu'on peut supposer sans diviseur commun, qui se réduise à un polynome *de degré indépendant* de p (degré inconnu, du polynome $P + zQ$) et par conséquent connu?

Il faudra d'abord choisir ξ, η, ζ de façon que

$$\xi + \eta + \zeta = 2\lambda$$

puis il suffira que l'on ait

$$a\xi \geqslant (a-1)\lambda, \qquad b\eta \geqslant (b-1)\lambda, \qquad c\zeta \geqslant (c-1)\lambda.$$

On en déduit aisément

$$\frac{2\xi}{\xi + \eta + \zeta} \geqslant \frac{a-1}{a}, \qquad \frac{2\eta}{\xi + \eta + \zeta} \geqslant \frac{b-1}{b}, \qquad \frac{2\zeta}{\xi + \eta + \zeta} \geqslant \frac{c-1}{c},$$

d'où l'on conclut que a, b, c devront satisfaire à l'inégalité

$$\frac{1}{a}+\frac{1}{b}+\frac{1}{c}\geqslant 1$$

dont les solutions sont respectivement

$$a=2,\quad b=2,\quad c=2,\,3,\,\ldots,\,n,\,\ldots.$$

$$a=2,\quad b=3,\quad c=3,\,4,\,5,\,6.$$

$$a=2,\quad b=4,\quad c=4.$$

$$a=3,\quad b=3,\quad c=3.$$

Dans tous ces cas une *puissance fractionnaire convenable de* Ω *donne un multiplicateur de la différentielle* $\beta dy-\alpha dx$; *bien plus, on connait l'expression en* A/B *de l'intégrale correspondante.* On sait donc former par un nombre limité de calculs, les polynomes A et B.

Voici le résultat le plus intéressant: si l'on suppose $a=2$, $b=2$, $c=n$ on peut, *sans connaître la valeur de* n, former le polynome $\Omega=\frac{ABC^2}{L^2}$, de degré $2(m+1)$; l'expression $\frac{1}{\sqrt{\Omega}}$ est un multiplicateur de la différentielle $\beta dy-\alpha dx$. On a d'ailleurs:

$$\frac{\beta dy-\alpha dx}{\sqrt{\Omega}}=\frac{-d\left(\frac{A}{B}\right)}{\left(\frac{A}{B}+\rho\right)\sqrt{\frac{A}{B}}},\quad \text{si}\ C=A+\rho B,$$

ce qui permettra d'obtenir A et B.

Mais sauf les cas extrêmes, où $\frac{1}{a}+\frac{1}{b}+\frac{1}{c}=1$, on a plusieurs systèmes de valeurs possibles pour ξ, η, ζ, donc au moins deux multiplicateurs, c'est-à-dire l'intégrale algébrique elle-même.

Lorsqu'il existe quatre valeurs exceptionnelles de z pour lesquelles $P+zQ$ renferme un facteur multiple, la méthode précédente ne s'applique plus *que si les facteurs multiples sont doubles.* Si l'on écrit pour les quatre valeurs exceptionnelles

$$P=A=\phi^2\phi',\qquad Q=B=\psi^2\psi',$$

$$P+\lambda Q=C=\chi^2\chi',\qquad P+\mu Q=D=\theta^2\theta',$$

le quotient $\Omega=\frac{ABCD}{L^2}$ est un polynome de degré $2(m+1)$ et $\frac{1}{\sqrt{\Omega}}$ est un multiplicateur de $\beta dy-\alpha dx$. On a d'ailleurs

$$\frac{\beta dy-\alpha dx}{\sqrt{\Omega}}=-\frac{d\left(\frac{A}{B}\right)}{\sqrt{\frac{A}{B}\left(\frac{A}{B}+\lambda\right)\left(\frac{A}{B}+\mu\right)}},$$

ce qui permettra d'obtenir A et B.

V. *Comment on tire parti de la connaissance d'une relation rationnelle vérifiée par une solution de l'équation* $\frac{\partial z}{\partial x} + A\frac{\partial z}{\partial y} = 0$.

10. Jusqu'à présent, nous nous sommes proposé de définir et d'obtenir la solution la plus simple de l'équation $X(z) = \frac{\partial z}{\partial x} + A\frac{\partial z}{\partial y} = 0$, celle qui est attachée *au groupe de rationalité-type.* Mais il y a lieu d'indiquer une méthode régulière qui permette d'utiliser au mieux la connaissance d'une solution quelconque de l'équation précédente, c'est-à-dire en fait, *d'une relation rationnelle quelconque* entre les éléments

$$x,\ y,\ z,\ \frac{\partial z}{\partial y},\ \frac{\partial^2 z}{\partial y^2},\ \dots$$

compatible avec l'équation (*a*).

Soit
$$P\left(x,\ y,\ z,\ \frac{\partial z}{\partial y},\ \frac{\partial^2 z}{\partial y^2},\ \dots\right) = 0,$$

cette relation supposée d'ordre p et mise sous forme entière; la relation

$$X(P) = P_1 = 0$$

est également une relation entière *du même ordre,* vérifiée par la même solution z. Si cette relation n'est pas identique à la précédente, on en peut conclure par élimination de la dérivée d'ordre le plus élevé, une relation entière *d'ordre inférieur* $Q\left(x,\ y,\ z,\ \frac{\partial z}{\partial y},\ \dots\right) = 0$ compatible avec (*a*); on raisonnera sur celle-ci comme sur la première et il est clair qu'on trouvera ainsi une relation entière

$$S = 0$$

telle que l'on ait identiquement $X(S) = MS$ (où M ne dépend que de x et y) et dont toutes les autres sont des conséquences.

Regardons maintenant dans S, z comme fonction d'un argument u que nous ne précisons pas: on pourra mettre S sous la forme

$$S = \sum_{i=1}^{h} l_i\left(z,\ \frac{\partial z}{\partial u},\ \dots\right)\xi_i\left(x,\ y,\ \frac{\partial u}{\partial y},\ \dots\right)$$

où le nombre h est le *plus petit possible,* ce qui exige que les ξ_i ne soient liées par aucune relation linéaire et homogène à coefficients constants et qu'il en soit de même pour les l_i $\left(\text{qui sont des polynomes en } z,\ \frac{\partial z}{\partial u},\ \frac{\partial^2 z}{\partial u^2},\ \dots\right)$.

On conclut alors de l'identité $\qquad X(S) = MS$

les h identités $\qquad X(\xi_i) = M\xi_i \quad (i = 1,\ \dots,\ h).$

Ces identités expriment simplement que les équations

$$\frac{\xi_i\left(x,\ y,\ \frac{\partial u}{\partial y},\ \dots\right)}{\xi_1\left(x,\ y,\ \frac{\partial u}{\partial y},\ \dots\right)} = \alpha_i(u) \quad (i = 1,\ \dots,\ h-1)$$

sont compatibles avec la relation $X(u) = 0$, lorsque les fonctions $\alpha_i(u)$, en général transcendantes, sont choisies de manière que ces équations soient compatibles quand on y regarde x comme un paramètre et y comme la seule variable. Si l'on détermine par conséquent ces fonctions $\alpha_i(u)$ de manière que le système précédent possède la solution u qui pour $x = x_0$ se réduit à la fonction de y définie par $y = \phi(u)$, c'est-à-dire si l'on pose

$$\alpha_i(u) = \frac{\xi_i\left(x_0, \phi(u), \left(\frac{\partial u}{\partial y}\right), \dots\right)}{\xi_1\left(x_0, \phi(u), \left(\frac{\partial u}{\partial y}\right), \dots\right)}$$

en calculant $\left(\frac{\partial u}{\partial y}\right)$, $\left(\frac{\partial^2 u}{\partial y^2}\right)$, ... par les formules

$$1 = \phi'(u)\left(\frac{\partial u}{\partial y}\right), \qquad 0 = \phi''(u)\left(\frac{\partial u}{\partial y}\right)^2 + \phi'(u)\left(\frac{\partial^2 u}{\partial y^2}\right), \dots$$

on a des équations *rationnelles* lorsque $\phi(u)$ est rationnel.

Le système rationnel ainsi obtenu *définit, dans tous les cas, u aux transformations près d'un groupe.* On peut donc en déduire par un procédé régulier, pour une solution convenable z, l'expression rationnelle en x, y, de l'un des invariants que nous avons désigné par K, J, I (si l'on n'en déduit pas z lui-même).

Ainsi, dans tous les cas, la connaissance d'une relation rationnelle entre $x, y, z, \frac{\partial z}{\partial y}, \dots$ compatible avec l'équation (a) entraîne par un calcul régulier, la connaissance de l'invariant rationnel de l'un des groupes-types.

On ne peut pas affirmer que ce groupe-type est le groupe de rationalité de l'équation (a); il y aura donc lieu d'étudier la détermination ultérieure, dans le domaine de rationalité adopté, des invariants J, K ou z.

Toutes les fois où l'on a pu attribuer à une solution particulière z de l'équation (a) $X(z) = \frac{\partial z}{\partial x} + A\frac{\partial z}{\partial y} = 0$, une propriété qui se traduit *en dernière analyse* par une relation rationnelle entre $x, y, z, \frac{\partial z}{\partial y}, \dots$ on est donc certain que l'on se trouve dans l'un des cas de réduction indiqués au début; c'est-à-dire que l'équation

$$\frac{dy}{dx} = A(x, y) \quad \dots\dots\dots\dots\dots\dots (1),$$

est *spéciale* et que la détermination de z se ramène, dans le cas le moins avantageux, à des quadratures et à l'intégration de deux équations de Riccati.

Supposons, par exemple, qu'il existe une solution z de (a) qui soit un polynome entier en y d'ordre n; l'équation $\frac{\partial^{n+1} z}{\partial y^{n+1}} = 0$ doit être compatible avec (a). En appliquant la méthode précédente on trouve (même sans faire intervenir u) d'abord la condition définitive de compatibilité et ensuite l'expression explicite de

$$\frac{\partial^2 z}{\partial y^2} : \frac{\partial z}{\partial y} = J.$$

De même si l'on suppose qu'une certaine solution z s'exprime rationnellement en y, $z = \frac{P(y)}{Q(y)}$, où P et Q sont d'ordre n par exemple, la même méthode conduit, quand n est donné, à l'équation de condition que doit vérifier $A(x, y)$ et ensuite à l'expression rationnelle de $I(x, y)$ dans le cas le plus défavorable.

Bien entendu, si l'on ne fixe pas n, on ne peut *qu'affirmer la réduction* sans donner l'expression de I ou la condition à laquelle A doit satisfaire.

VI. *Exemples de déterminations du Groupe-type de Rationalité.*

11. Je me propose d'indiquer ici quelques exemples, empruntés à la Géométrie, d'équations du premier ordre dont j'ai pu déterminer le groupe de rationalité.

Le premier est relatif à l'équation différentielle qui définit *les lignes de courbure de la surface des ondes de Fresnel.*

La surface étant définie par les équations

$$X^2 + Y^2 + Z^2 = \beta, \qquad \frac{aX^2}{a-\beta} + \frac{bY^2}{b-\beta} + \frac{cZ^2}{c-\beta} = 0,$$

et la variable $\sqrt{\alpha}$ désignant la distance de l'origine au plan tangent en (X, Y, Z), M. Darboux (*Leçons sur la théorie des surfaces*, t. IV, note VIII) pose $\beta = \alpha + \frac{u}{\alpha}$ et trouve pour équation différentielle des lignes de courbure

$$\phi \left(\frac{du}{d\alpha}\right)^2 - \phi' u \left(\frac{du}{d\alpha}\right) + u\phi + \frac{u^2}{2}\phi'' + \frac{u^3}{24}\phi^{(\mathrm{IV})} = 0 \ldots\ldots\ldots\ldots\ldots (1),$$

où les accents indiquent des dérivées par rapport à α et où

$$\phi = \alpha f(\alpha) = \alpha(\alpha - a)(\alpha - b)(\alpha - c).$$

Dans le cas où $f(\alpha)$ se réduit à un polynome du second degré, un artifice ingénieux lui permet de ramener l'intégration de (1) à des quadratures de différentielles algébriques. A notre point de vue, si l'on pose alors

$$\frac{du}{d\alpha} + \varpi = 0 \quad \text{avec} \quad \phi\varpi^2 + u\phi'\varpi + u\phi + \frac{u^2}{2}\phi'' = 0,$$

l'équation $-\frac{\partial z}{\partial \alpha} + \varpi \frac{\partial z}{\partial u} = 0$ est telle que la résolvante en $\left(\frac{\partial z}{\partial u}\right)^3$ possède dans le domaine $[\varpi]$ la solution rationnelle

$$\left(\frac{\partial z}{\partial u}\right)^3 = \frac{1}{u^3(1-\varpi)};$$

le groupe de rationalité est donc $Z = \epsilon z + a$ $(\epsilon^3 = 1)$. La même forme de multiplicateur convient à toutes les équations $\frac{du}{d\alpha} + \varpi = 0$ où ϖ est une fonction de u et de α définie par la relation implicite $u^3(1-\varpi) = \varpi^3\phi\left(\alpha + \frac{u}{\varpi}\right)$, quelle que soit la fonction ϕ.

C'est l'étude du cas précédent, envisagé comme cas limite, qui m'a conduit à la solution du cas général où ϕ est du quatrième degré.

Supposons que ϕ commence par le terme α^4, l'équation à étudier se conserve par la transformation $u = \frac{\phi}{t}$ et si l'on pose

$$\omega^2 = -4\phi u^3 + u^2(\phi'^2 - 2\phi\phi'') - 4\phi^2 u$$

la même transformation remplace l'équation $\frac{du}{d\alpha} = \frac{u\phi' + \omega}{2\phi}$ par la même équation en t, où ω est changé de signe. Pour l'équation

$$X(z) = \frac{\partial z}{\partial \alpha} + \frac{u\phi' + \omega}{2\phi}\frac{\partial z}{\partial u} = 0$$

la résolvante en $\left(\frac{\partial z}{\partial u}\right)^3$ possède dans le domaine $[\omega]$ la solution rationnelle

$$\left(\frac{\partial z}{\partial u}\right)^3 = \frac{\phi}{u(K_0 - \omega K_1)}$$

avec $$K_0 = u[\phi'(u^2 + \phi) + u\phi(2\alpha + f'')], \quad K_1 = u^2 - \phi;$$

le groupe de rationalité est donc, comme tout-à-l'heure, $Z = \epsilon z + a$ $(\epsilon^3 = 1)$. Ainsi, dans le cas général, la transcendante z est donnée par la quadrature

$$dz = \sqrt[3]{\frac{\phi}{u(K_0 - \omega K_1)}}\left(du - \frac{u\phi' + \omega}{2\phi}d\alpha\right)$$

attachée à la surface (ω, u, α) du neuvième degré.

12. Un autre exemple intéressant est celui de *l'équation différentielle des lignes asymptotiques des surfaces générales du troisième degré.*

J'ai pensé que sur les surfaces générales du troisième degré les seules lignes asymptotiques algébriques devaient être les 27 droites.

L'examen d'un cas particulier, celui de la surface

$$z = xy + \frac{x^3 + y^3}{6}$$

dont les lignes asymptotiques sont transcendantes et données par la quadrature

$$df = \frac{y\,dy + (1 + \sqrt{1 - xy})\,dx}{(P + \omega Q)^{\frac{1}{3}}} = 0,$$

où $P = y^3 - 4 + 3xy$, $Q = xy - 4$, $\omega = \sqrt{s^2 - rt}$ (quadrature qui se ramène à celle d'une différentielle binôme non exprimable) a confirmé cette hypothèse et m'a montré comment les 27 droites intervenaient.

Voici le résultat général : Soit $\Omega(x, y, z) = 0$, l'équation, mise sous forme entière, d'une surface du troisième degré ; en différentiant totalement quatre fois cette équation, on obtient un résultat qui, pour $\sigma_2 = r + 2sm + tm^2 = 0$, se réduit à

$$\frac{\partial \Omega}{\partial z}\sigma_4 + 4\sigma_3 X\left(\frac{\partial \Omega}{\partial z}\right) = 0,$$

où l'on a posé $X(f) = \frac{df}{dx} + m\frac{df}{dy} = \frac{\partial f}{\partial x} + m\frac{\partial f}{\partial y} + \frac{\partial f}{\partial m}\left(\frac{\partial m}{\partial x} + m\frac{\partial m}{\partial y}\right)$

et où $\sigma_3 = \frac{d^3z}{dx^3} = \alpha + 3\beta m + 3\gamma m^2 + \delta m^3, \qquad \sigma_4 = \frac{d^4z}{dx^4} = \lambda_0 + 4\lambda_1 m + \ldots.$

L'équation précédente exprime simplement que *pour chaque ligne asymptotique* $\left(\frac{\partial \Omega}{\partial z}\right)^4 \sigma_3$ *est l'inverse du cube d'un multiplicateur.*

En d'autres terms $\left(\frac{\partial\Omega}{\partial z}\right)^{-\frac{4}{3}} \sigma_3^{-\frac{1}{3}}(dy - m\,dx) = df$

est une différentielle exacte (les surfaces réglées sont exclues); le groupe de rationalité dans le domaine $[z, \sqrt{s^2 - rt}]$ est simplement $f' = \epsilon f + a$, $(\epsilon^3 = 1)$.

Un résultat analogue peut s'énoncer pour les lignes qui satisfont à l'équation différentielle de degré $(n-1)$

$$\sigma_{n-1} = \frac{d^{n-1}z}{dx^{n-1}} = 0$$

sur les surfaces algébriques de degré n, dont l'équation peut s'écrire

$$\Omega = a_0 z^2 + \phi_1(x, y)\, z + \phi_n(x, y) = 0.$$

Ces lignes sont données par la quadrature de différentielle algébrique

$$df = \left(\frac{\partial\Omega}{\partial z}\right)^{-\left(\frac{n+1}{n}\right)} \sigma_n^{-\frac{1}{n}}(dy - m\,dx) = 0,$$

où m désigne une racine de l'équation

$$\sigma_{n-1} = \frac{\partial^{n-1}z}{\partial x^{n-1}} + \frac{(n-1)}{1} m \frac{\partial^{n-1}z}{\partial x^{n-2}\partial y} + \ldots = 0.$$

Les surfaces pour lesquelles σ_n s'annule en même temps que σ_{n-1} sont naturellement exclues. Pour en revenir aux lignes asymptotiques des surfaces du troisième degré, j'observerai qu'une réduction du groupe de rationalité de leur équation ne peut se produire que si toutes ces lignes sont algébriques ; j'ai étudié en détail, à ce point de vue, les vingt et un types projectifs de ces surfaces (non réglées).

On déduit aisément des résultats précédents, par l'emploi d'une transformation dualistique (ou d'une transformation de Lie qui change les droites en sphères) d'autres résultats relatifs aux lignes asymptotiques (ou aux lignes de courbures) de surfaces algébriques; mais nous ne pouvons pas y insister ici, pas plus que sur des observations analogues aux précédentes relatives à d'autres formes de l'équation différentielle des lignes asymptotiques ou de courbure.

Je donnerai cependant, à cause de son élégance, la proposition suivante dont la forme seule est nouvelle. Soit $\Omega(x, y, z) = 0$, l'équation générale d'une surface du second degré et (1) $\frac{dy}{dx} = m$ avec $1 + m^2 + (p + qm)^2 = 0$, l'équation différentielle de ses lignes de longueur nulle. On trouve sans difficulté

$$\frac{\partial m}{\partial y} = \frac{1}{2} \frac{\sigma_2' X(m)}{\sigma_2}$$

avec $\sigma_2 = r + 2sm + tm^2$, $\sigma_2' = 2(s + tm)$, $X(f) = \frac{\partial f}{\partial x} + m\frac{\partial f}{\partial y} + \frac{\partial f}{\partial m} X(m)$.

Différentions totalement trois fois l'équation $\Omega = 0$; le résultat où l'on fait $\frac{dy}{dx} = m$, exprime que, pour la forme différentielle $dy - m\,dx$, $\sigma_2^{-\frac{1}{2}} \left(\frac{\partial\Omega}{\partial z}\right)^{-\frac{3}{2}}$ est un multiplicateur

Ainsi les lignes de longueur nulle des surfaces du second degré sont données par la quadrature

$$df = \frac{dy - m\,dx}{\sqrt{\sigma_2 \left(\frac{\partial \Omega}{\partial z}\right)^3}} = 0\,;$$

le groupe de rationalité de l'équation (1) dans le domaine $[z, m]$ est

$$f' = \epsilon f + a, \quad (\epsilon^2 = 1).$$

VII. *Classification des points singuliers. Forme analytique des Intégrales dans le voisinage des points singuliers.*

13. La théorie d'intégration que nous venons d'exposer pour l'équation

$$\frac{dy}{dx} = A\,(x, y) \quad \ldots\ldots\ldots\ldots\ldots\ldots\ldots\ldots (1)$$

a mis en évidence l'intérêt qu'il y a à définir la solution générale de cette équation par une relation implicite

$$z\,(x, y) = \text{const.},$$

où z vérifie
$$X\,(z) = \frac{\partial z}{\partial x} + A \frac{\partial z}{\partial y} = 0.$$

En choisissant la solution z *la plus simple*, on a défini la solution u principale en $x = x_0$ par la relation $z\,(x, y) = z\,(x_0, u)$ et il y aura lieu de dégager de cette relation implicite les propriétés de u ou de y.

Cette méthode n'est pas limitée au cas où $A\,(x, y)$ est une fonction *bien définie*; elle conduit dans l'hypothèse où $A\,(x, y)$ est, dans le voisinage de chaque point, méromorphe ou algébroïde, *à une classification précise des points singuliers de* (1) dont je veux indiquer les traits essentiels.

Si au voisinage d'un point $x = x_0$, $y = y_0$, A ou $\frac{1}{A}$ est holomorphe, on sait que u est holomorphe en x, y, y_0; le point x_0, y_0 est *ordinaire*.

Nous dirons qu'un système de valeurs x_0, y_0 *est singulier*, pour une fonction $A\,(x, y)$ *méromorphe*, lorsqu'au point, $x = x_0$, $y = y_0$, A est indéterminé. Il conviendra d'ajouter, pour envisager les cas où x_0 ou bien y_0 seraient infinis, les systèmes analogues pour la transformée de (1) par $x = \frac{1}{X}$, $y = \frac{1}{Y}$. Considérons, au voisinage d'un point singulier x_0, y_0, l'équation

$$X\,(z) = \frac{\partial z}{\partial x} + A\,(x, y) \frac{\partial z}{\partial y} = 0 \quad \ldots\ldots\ldots\ldots\ldots (a)\,;$$

il peut se faire qu'il existe une solution z de cette équation, qui soit *méromorphe* au voisinage de x_0, y_0. Dans ce cas il en existera une infinité, puisque $\phi\,(z)$ où ϕ est méromorphe en z est méromorphe, en x, y, avec z. Je dis alors que le point x_0, y_0 est un *point singulier apparent*.

On devra donc comme *problème préliminaire* décider si l'équation (a) peut posséder une solution z holomorphe au voisinage de $x = x_0$, $y = y_0$, ou au moins *méromorphe*, c'est-à-dire si en posant $A = \frac{\alpha}{\beta}$ où α et β sont holomorphes, les deux équations

$$\alpha \frac{\partial P}{\partial y} + \beta \frac{\partial P}{\partial x} = MP, \qquad \alpha \frac{\partial Q}{\partial y} + \beta \frac{\partial Q}{\partial x} = MQ$$

peuvent posséder, pour une même détermination de la fonction holomorphe M, deux solutions holomorphes distinctes P et Q (ou une solution holomorphe $P + \lambda Q$ dépendant linéairement d'une constante λ).

Il est facile de former des exemples de points singuliers *apparents*; on n'a qu'à partir des expressions de P et Q et prendre par exemple

$$\beta = P \frac{\partial Q}{\partial y} - Q \frac{\partial P}{\partial y}, \qquad \alpha = -\left(P \frac{\partial Q}{\partial x} - Q \frac{\partial P}{\partial x}\right),$$

lorsque ces expressions n'ont pas de facteur commun, ou si l'on se donne

$$P = \phi_1^{a_1} \phi_2^{a_2} \dots \phi_h^{a_h}, \quad Q = \psi_1^{b_1} \psi_2^{b_2} \dots \psi_k^{b_k},$$

les ϕ et les ψ étant holomorphes pour $x = x_0$, $y = y_0$ avec $P(x_0, y_0) = Q(x_0, y_0) = 0$ prendre pour α et β les quotients des expressions précédentes par

$$L = \phi_1^{a_1 - 1} \phi_2^{a_2 - 1} \dots \psi_1^{b_1 - 1} \psi_2^{b_2 - 1} \dots.$$

Supposons que le point (x_0, y_0) ne soit pas un point singulier apparent: il peut se faire que l'équation résolvante en K

$$X(K) + nK \frac{\partial A}{\partial y} = 0$$

possède une solution K méromorphe au voisinage de $(x = x_0, y = y_0)$ pour une valeur positive de n aussi petite que possible.

Dans ce cas, *il ne peut en exister qu'une seule*: s'il y en avait deux, leur quotient serait une solution méromorphe de l'équation (a); on aurait donc un point singulier apparent. Je dis alors que le point singulier (x_0, y_0) *est du premier ordre.* Pour reconnaître un tel point, on aura d'abord à rechercher si la résolvante en K peut posséder une solution *holomorphe* au voisinage de (x_0, y_0) pour une valeur entière de n *positive* ou *négative* mais différente de 1. Si cela n'a pas lieu, on posera $K = \frac{P}{Q}$, P et Q s'annulant tous deux en (x_0, y_0) et l'on cherchera s'il est possible de trouver P et Q holomorphes de manière à vérifier les deux équations

$$\alpha \frac{\partial Q}{\partial y} + \beta \frac{\partial Q}{\partial x} = \gamma Q,$$

$$\alpha \frac{\partial P}{\partial y} + \beta \frac{\partial P}{\partial x} = \left[\gamma - n\left(\frac{\partial \alpha}{\partial y} + \frac{\partial \beta}{\partial x}\right)\right] P$$

pour une même fonction holomorphe γ.

Il est bien aisé de former des exemples de points singuliers du premier ordre—où l'équation (1) s'intègre par l'adjonction de transcendantes attachées au groupe $Z = \epsilon z + \alpha$, $(\epsilon^n = 1)$ et ne dépendant que d'un argument.

Ainsi, si l'on prend l'équation $X \frac{dY}{dX} + Y(1 - XY) = 0$, où X, Y sont des fonctions holomorphes de x et y s'annulant pour $x=0$, $y=0$, on a pour une telle équation dont la solution générale est définie par

$$z = \frac{1}{XY} + \log X,$$

l'expression méromorphe $\frac{\partial z}{\partial y} = K = -\frac{\left(X\frac{\partial Y}{\partial y} + Y\frac{\partial X}{\partial y}\right)}{X^2Y^2} - \frac{\frac{\partial X}{\partial y}}{X}$.

Un autre exemple, simple et général, d'un point singulier du premier ordre est donné, en $x=0$, $y=0$, par l'équation $\frac{dy}{dx} = \frac{\alpha}{\beta}$, où $\alpha = \lambda x + \dots$, $\beta = \mu y + \dots$, les termes non écrits étant du second ordre au moins, *lorsque* $\frac{\mu}{\lambda}$ *n'est ni une quantité réelle négative, ni un entier positif, ni l'inverse d'un tel entier.* MM. Poincaré et Picard ont établi que l'on peut alors poser

$$z = \frac{P(x, y)^\lambda}{Q(x, y)^\mu},$$

P et Q étant holomorphes au voisinage de $x=0$, $y=0$ et s'annulant en ce point. Dans le cas le plus général où $\mu : \lambda$ est complexe, ou bien réel et irrationnel, on a donc en prenant $Z = \log z$,

$$\frac{\partial Z}{\partial y} = K = \lambda\frac{\frac{\partial P}{\partial y}}{P} - \mu\frac{\frac{\partial Q}{\partial y}}{Q}.$$

Le cas où $\mu : \lambda$ est un entier positif ou l'inverse d'un tel entier et où l'intégrale z peut s'écrire sous la forme

$$z = \frac{S(x, y)}{X(x, y)} + h \log X(x, y)$$

où h est une constante, S et X des fonctions holomorphes donne encore un point singulier du premier ordre. Si $h=0$ avec $\mu = \lambda$ (point *dicritique* de M. Autonne) on a simplement un *point singulier apparent.*

S'il n'existe pas pour la résolvante en K, de solution méromorphe au voisinage de (x_0, y_0), il peut se faire qu'il en existe un pour la résolvante en J,

$$J = \frac{\partial^2 z}{\partial y^2} : \frac{\partial z}{\partial y}, \quad X(J) + J\frac{\partial A}{\partial y} + \frac{\partial^2 A}{\partial y^2} = 0.$$

Il est clair qu'il n'en peut exister qu'une seule, sans quoi on retomberait dans le cas précédent. Je dis alors que le point $(x = x_0,\ y = y_0)$ est *singulier du second ordre.* On donnerait aisément les équations résolvantes à vérifier par des fonctions holomorphes, s'annulant toutes deux lorsque J et $\frac{1}{J}$ ne peuvent ni l'un ni l'autre être holomorphes.

Pour obtenir un exemple simple de point singulier du second ordre, on n'a qu'à former une équation (a) $X(z)=0$, dont le groupe de rationalité soit le groupe linéaire et pour laquelle J soit le quotient de deux polynomes s'annulant en x_0, y_0.

Je donnerai simplement l'équation

$$\frac{dy}{dx}=\frac{2b(2x+y)+ax^2+\frac{a-2c}{3}xy+\frac{a-c}{3}y^2}{2b(x-2y)+\frac{a+c}{3}x^2-\frac{c+2a}{3}xy+cy^2} \quad \text{..............(1)},$$

pour laquelle $J=\frac{\frac{\partial\beta}{\partial y}}{\beta}-\frac{x}{x^2+y^2}$ ou encore $z=\int e^{-\operatorname{arctg}\frac{y}{x}}(\beta dy-\alpha dx)$.

Supposons enfin que l'équation résolvante en J ne possède pas de solution méromorphe au voisinage du point x_0, y_0, il peut se faire que la résolvante en I,

$$I=\{z, y\}=\frac{\frac{\partial^3 z}{\partial y^3}}{\frac{\partial z}{\partial y}}:\frac{3}{2}\left(\frac{\frac{\partial^2 z}{\partial y^2}}{\frac{\partial z}{\partial y}}\right)^2,$$

c'est-à-dire $X(I)+2I\frac{\partial A}{\partial y}+\frac{\partial^3 A}{\partial y^3}=0$, en possède une. Elle ne pourra en avoir qu'une seule. Le point $x=x_0$, $y=y_0$ sera dit alors *singulier du troisième ordre.*

Par exemple, si l'on envisage l'équation de Riccati

$$\frac{dy}{dx}=\frac{a_0(x)+a_1(x)y+a_2(x)y^2}{c(x)} \quad \text{........................(1)},$$

où a_0, a_1, a_2, c sont des fonctions holomorphes de x, dont les extrêmes a_0 et c s'annulent pour $x=0$, on sait que l'équation aux dérivées partielles en I admet la solution $I=0$; le point singulier est en général du troisième ordre et si l'on observe que l'expression de J peut s'écrire $J=\frac{-2}{y-X}$ où X doit vérifier l'équation (1) on peut affirmer que pour que le point $x=0$, $y=0$ soit seulement du second ordre il faut et il suffit que l'équation (1) possède une solution particulière X *méromorphe* au voisinage de $x=0$. Ce n'est évidemment que dans des cas exceptionnels que cette circonstance se présente.

Lorsqu'un point singulier n'est ni du premier ni du second ni du troisième ordre, je dirai qu'il est *général.* On peut également construire sans difficulté des exemples de points singuliers généraux.

(Il serait facile de donner une classification analogue, en supposant A *algébroïde* au voisinage de chacun des points singuliers.)

L'intérêt de cette classification réside évidemment dans ce fait que *s'il existe une relation*

$$F\left(x, y, z, \frac{\partial z}{\partial y}, \frac{\partial^2 z}{\partial y^2}, \ldots\right)=0,$$

compatible avec l'équation

$$X(z)=\frac{\partial z}{\partial x}+A\frac{\partial z}{\partial y}=0,$$

qui soit rationnelle en $\qquad z, \frac{\partial z}{\partial y}, \frac{\partial^2 z}{\partial y^2}, \ldots$

et dont les coefficients sont *méromorphes* en x, y, au *voisinage de* $x = x_0$, $y = y_0$, *cette relation a l'une des formes-types obtenues au début.* Il suffit de remplacer le mot *rationnel* par *rationnel au voisinage de* $x = x_0$, $y = y_0$; les démonstrations subsistent. L'extension du domaine de rationalité peut également se faire de la même manière. A chaque point singulier correspond ainsi un groupe-type dont l'invariant est rationnel au *voisinage* de ce point: on en déduira une représentation analytique de l'intégrale en faisant appel à des transcendantes simples (logarithmes, exponentielles, etc.) qui subissent précisément *au voisinage du point singulier considéré* les mêmes transformations que z.

Le guide naturel dans cette recherche est la théorie de Fuchs, pour les équations différentielles linéaires du second ordre. (Cf. aussi F. Marotte, Annales de la Faculté des Sciences de Toulouse, 1898.)

Il me reste à faire observer que si l'on suppose $A = \frac{\alpha}{\beta}$, au voisinage du point singulier considéré, quotient de deux fonctions holomorphes quelconques, il ne sera pas possible de réaliser effectivement la classification précédente: l'existence d'un point singulier non général, ou d'un point singulier apparent, peut entraîner une infinité de conditions distinctes (théorie des *centres* de Poincaré).

Mais il n'en sera plus de même si l'on admet simplement que α et β sont rationnels ou font partie d'un domaine $[\Delta]$ bien défini. La plus grande difficulté à résoudre (formation des conditions d'existence de développements holomorphes *convergents* au voisinage du point singulier) amène seulement à distinguer pour chaque degré des cas *généraux* (où α et β commencent par des termes *de ce degré déterminé,* dont les coefficients ne satisfont pas à certaines conditions d'égalité et d'inégalité, et où les termes de degré supérieur ont des coefficients arbitraires) et des cas particuliers où ces conditions n'étant plus satisfaites, tout se passe comme si le degré déterminé était augmenté d'une ou de plusieurs unités. Mais il peut se faire qu'il y ait indéfiniment des cas d'exception. Ce n'est donc que si α et β sont *entièrement déterminés* qu'on peut espérer, dans tous les cas, épuiser la question.

L'emploi des fonctions *majorantes* en donne alors le moyen: il s'agit de ranger les dérivées de la fonction inconnue de façon que le rang croisse avec l'ordre de dérivation et que toutes les équations qui déterminent les dérivées d'un certain ordre donnent l'expression de chacune d'elles au moyen des dérivées de rang précédent. Les exemples classiques dus à Poincaré et à M. Picard montrent comment on peut ainsi *majorer* les modules de chacune de ces dérivées et trouver des conditions *suffisantes* pour la convergence des fonctions inconnues.

Équation d'ordre quelconque.

I. *Systèmes irréductibles réguliers. Groupe de rationalité.*

14. Soit une équation différentielle ordinaire

$$y^{(n)} = f(x, y, y', \ldots, y^{(n-1)}) \ldots\ldots\ldots\ldots\ldots\ldots\ldots\ldots (1),$$

où f appartient à un certain domaine de rationalité; je définirai sa solution générale par n relations

$$\phi_i(x, y, y', \ldots, y^{(n-1)}) = c_i \qquad (i = 1, \ldots, n),$$

où les ϕ_i forment un *système fondamental* de solutions de l'équation

$$X(z) = \frac{\partial z}{\partial x} + \frac{\partial z}{\partial y} y' + \ldots + \frac{\partial z}{\partial y^{(n-1)}} f = 0 \quad \ldots\ldots\ldots\ldots\ldots\ldots (2),$$

et je me propose de *fixer les caractères du système fondamental le plus simple de l'équation* (2).

Prenons tout de suite au lieu de (2), l'équation

$$X(z) = \frac{\partial z}{\partial x} + A_1 \frac{\partial z}{\partial x_1} + \ldots + A_n \frac{\partial z}{\partial x_n} = 0 \ldots\ldots\ldots\ldots\ldots\ldots (\mathrm{I}),$$

où les A sont des fonctions de $x, x_1, \ldots, x_n$ appartenant à un certain domaine de rationalité $[\Delta]$ [*pour fixer les idées*, on pourrait supposer simplement les A rationnels]; les éléments $z_1, z_2, \ldots, z_n$ de tout système fondamental vérifient les relations

$$\frac{D}{1} = \frac{D_1}{A_1} = \ldots = \frac{D_n}{A_n} \quad \ldots\ldots\ldots\ldots\ldots\ldots\ldots\ldots (a),$$

avec $D \neq 0$, où les D_i sont des déterminants fonctionnels dont la formation est immédiate; par exemple

$$D = \frac{\partial (z_1, \ldots, z_n)}{\partial (x_1, \ldots, x_n)}.$$

En d'autres termes les quotients $\frac{D_i}{D}$ sont les *invariants différentiels* indépendants du groupe ponctuel général Γ_n

$$Z_i = \phi_i(z_1, \ldots, z_n) \qquad (i = 1, \ldots, n),$$

étendu en regardant les z comme fonction des $(n+1)$ variables $x, x_1, \ldots, x_n$ non transformées. Si le système (a) est *irréductible*, c'est-à-dire si toute relation rationnelle $\left(\text{entre } z_1, z_2, \ldots, z_n \text{ et leurs dérivées } \frac{\partial z_1}{\partial x_1}, \ldots, \frac{\partial z_n}{\partial x_n}, \ldots\right)$ d'ordre quelconque dont les coefficients sont des fonctions des variables $x, x_1, \ldots, x_n$ appartenant au domaine $[\Delta]$ *compatible** avec les équations (a), en est une conséquence nécessaire, je dis que l'équation (I) est *générale*. Tous les systèmes fondamentaux $z_1, \ldots, z_n$ sont des transcendantes de même nature, ce sont des *fonctions des* $(n+1)$ *variables* $x, x_1, \ldots, x_n$ attachées dans *le domaine* $[\Delta]$, *au groupe ponctuel général* Γ_n.

Tous les invariants différentiels rationnels de ce groupe Γ_n sont des fonctions rationnelles des quotients $\frac{D_i}{D}$ et de leurs dérivées et sont par suite connus rationnellement dans $[\Delta]$.

Si le système (a) est *réductible*, c'est-à-dire s'il existe des relations rationnelles, *compatibles* avec les équations (a) sans en être une conséquence nécessaire, on peut distinguer certains systèmes fondamentaux des autres.

En ajoutant aux relations (a) les relations nouvelles dont nous supposons l'existence, on obtiendra un système (S) que l'on peut supposer complété de façon à ce qu'il soit *irréductible*. Aux divers systèmes fondamentaux correspondront divers systèmes (S); il s'agit de choisir celui qu'on devra regarder comme le *plus simple*.

* J'entends par là simplement qu'un *système fondamental* $z_1, \ldots, z_n$, *au moins*, la vérifie.

Pour cela, s'il existe des systèmes (S) renfermant des équations d'ordre zéro (c'est-à-dire rationnelles en $z_1, \ldots, z_n$) on prendra tous ceux (S_0) qui en renferment le plus grand nombre, parmi ceux-là on prendra tous les systèmes (S_1) qui renferment le plus grand nombre d'équations du premier ordre, puis parmi ceux-là on prendra tous ceux (S_2) qui renferment le plus grand nombre d'équations du second ordre, et ainsi de suite.

Le raisonnement ne se continue pas indéfiniment. M. Tresse a établi en effet qu'à partir d'un certain ordre assignable, toutes les équations d'un système compatible quelconque sont des conséquences nécessaires des équations d'ordre inférieur. D'ailleurs tout système (S) renfermera certainement des équations d'ordre 0, 1, 2 ou 3.

On obtient ainsi des systèmes (S_p) irréductibles, qui déterminent tous un même ensemble de *dérivées principales*—(au sens de MM. Méray et Riquier). On peut former pour ces systèmes (S_p) continués jusqu'à un ordre convenable (et même pour chaque ordre de dérivation) une résolvante algébrique qui définit une combinaison linéaire et homogène à coefficients rationnels arbitraires, de ces dérivées principales au moyen des dérivées paramétriques. Si ces résolvantes n'ont pas le même degré, on prendra enfin parmi les systèmes (S_p) ceux (Σ) qui donnent des résolvantes de degré minimum.

Ce sont ces systèmes (Σ) que j'appelle *systèmes irréductibles réguliers* : le nombre des équations de chaque ordre et le degré de la résolvante algébrique sont les mêmes pour tous ces systèmes.

On peut encore dire que ces systèmes (Σ) sont à la fois *irréductibles* et *primitifs* en entendant par là qu'aucune transformation

$$z_i = \phi_i(Z_1, \ldots, Z_n) \qquad (i = 1, \ldots, n),$$

où les ϕ_i sont définis par un système différentiel rationnel quelconque, ne peut jamais augmenter le nombre des équations du système (Σ) qui sont d'un ordre donné, ou abaisser le degré de la résolvante algébrique correspondante, en conservant les nombres analogues attachés aux ordres plus petits.

15. Les systèmes irréductibles *réguliers* sont susceptibles d'une forme remarquable qui justifie leur choix.

On peut écrire les équations nouvelles (Σ) sous la forme *canonique*

$$\Omega_i\left(z_1, \ldots, z_n, \frac{\partial z_1}{\partial x_1}, \ldots\right) = \alpha_i(x, x_1, \ldots, x_n) \qquad (i = 1, \ldots, k) \ldots\ldots(\Sigma),$$

dans laquelle les Ω_i sont des invariants différentiels, rationnellement distincts, d'un groupe Γ de transformations én $z_1, \ldots, z_n$ formant pour ce groupe Γ un *système complet* d'invariants et les α_i des fonctions de $x, x_1, \ldots, x_n$ appartenant au domaine de rationalité adopté $[\Delta]$.

Les transformations (z, Z) du groupe Γ sont entièrement définies par les relations

$$\Omega_i\left(z_1, \ldots, z_n, \frac{\partial z_1}{\partial x_1}, \ldots\right) = \Omega_i\left(Z_1, \ldots, Z_n, \frac{\partial Z_1}{\partial x_1}, \ldots\right) \qquad (i = 1, \ldots, k),$$

où les variables $x_1, \ldots, x_n$ n'interviennent qu'en apparence. Il suffira d'y faire $x_1 = z_1, \ldots, x_n = z_n$ pour obtenir la *forme canonique* des *équations de définition* du groupe Γ, au sens de Lie.

J'ai appelé le groupe Γ *groupe de rationalité* de l'équation *spéciale* (I); il possède les propriétés fondamentales du groupe de rationalité (au sens de Galois) des équations algébriques:

1°. Tout invariant rationnel de Γ est égal à une fonction des variables $x, x_1, \ldots, x_n$ rationnelle dans le domaine $[\Delta]$.

2°. Toute fonction rationnelle de $z_1, \ldots, z_n$ et de leurs dérivées, à coefficients rationnels dans $[\Delta]$, et qui est égale à une fonction rationnelle dans $[\Delta]$, est une fonction rationnelle des invariants de Γ à coefficients rationnels dans $[\Delta]$.

Je dis que les transcendantes $z_1, \ldots, z_n$ sont des *fonctions des* $(n+1)$ *variables* $x, x_1, \ldots, x_n$ *attachées, dans le domaine* $[\Delta]$, *au groupe* Γ.

On doit observer tout de suite que le groupe Γ n'est déterminé qu'à certaines transformations près: toute transformation (z, z') qui change un système irréductible régulier (Σ) en un système de même nature (Σ') change le groupe Γ en un groupe *homologue* (Γ') qui sera le groupe de rationalité pour les solutions de (Σ'). Il faut et il suffit pour cela que les équations de (Γ) soient transformées par (z, z') en équations rationnelles de même ordre.

J'ajoute en passant que si le système (Σ) renferme des équations d'ordre zéro, en nombre p, elles peuvent s'écrire $z_1 = R_1, \ldots, z_p = R_p$, où les R sont rationnels dans $[\Delta]$, les autres équations de (Σ) ne renfermant plus que $z_{p+1}, \ldots, z_n$ et le groupe *intransitif* Γ correspondant est formé de transformations

$$Z_i = z_i \quad (i = 1, \ldots, p), \qquad Z_{p+j} = \phi_j(z_1, \ldots, z_n) \quad (j = 1, \ldots, n-p).$$

II. *Forme normale d'un système irréductible régulier.*

16. Une méthode particulière, dont je vais indiquer l'essentiel, m'a permis d'obtenir pour les systèmes irréductibles réguliers une forme *normale, nouvelle,* qui met en évidence les propriétés précédentes. Soit $P\left(Z, \frac{\partial Z}{\partial x}, x\right)$ un polynome entier par rapport aux éléments $Z_1, \ldots, Z_n$ et à leurs dérivées d'ordre quelconque relatives aux variables $x_1, \ldots, x_n$ dont les coefficients sont des fonctions rationnelles de $x, x_1, \ldots, x_n$ dans le domaine $[\Delta]$. Si l'on regarde les Z comme des fonctions indépendantes des n arguments $z_1, \ldots, z_n$ (qu'on ne précise pas davantage) et si l'on exécute les transformations

$$\frac{\partial Z}{\partial x_i} = \frac{\partial Z}{\partial z_1}\frac{\partial z_1}{\partial x_i} + \ldots + \frac{\partial Z}{\partial z_n}\frac{\partial z_n}{\partial x_i},$$
$$\ldots\ldots\ldots\ldots\ldots\ldots,$$
$$\frac{\partial^2 Z}{\partial x_i \partial x_j} = \frac{\partial^2 Z}{\partial z_1^2}\frac{\partial z_1}{\partial x_i}\frac{\partial z_1}{\partial x_j} + \ldots + \frac{\partial Z}{\partial z_1}\frac{\partial^2 z_1}{\partial x_i \partial x_j} + \ldots,$$
$$\ldots\ldots\ldots\ldots\ldots\ldots,$$

on pourra toujours écrire une identité

$$P\left(Z, \frac{\partial Z}{\partial x}, x\right) = \sum_1^h l_i\left(Z, \frac{\partial Z}{\partial z}\right)\xi_i\left(x, \frac{\partial z}{\partial x}\right),$$

où les l_i ne dépendent que des Z et de leurs dérivées par rapport aux z et les ξ_i que des variables et des dérivées $\frac{\partial z}{\partial x}$; nous supposerons que le *nombre h est le plus petit possible*, ce qui exige que les l_i d'une part, les ξ_i d'autre part ne soient liés par aucune relation linéaire et homogène à coefficients constants.

Les ξ_i, qui ne sont d'ailleurs définis qu'à une transformation linéaire à coefficients constants près, seront dits *coordonnées* du polynome P; leur introduction met en évidence de façon simple la manière dont se comportent les polynomes tels que P, quand on exécute sur les Z une transformation ponctuelle.

Soit $L_i(Z)$ ce que devient $l_i\left(Z, \frac{\partial Z}{\partial z}\right)$ quand on y fait $z_1 = Z_1, \ldots, z_n = Z_n$; on a évidemment

$$P\left(Z, \frac{\partial Z}{\partial x}, x\right) = \sum_1^h L_i(Z)\, \xi_i\left(x, \frac{\partial Z}{\partial x}\right).$$

D'autre part si les z' sont des fonctions quelconques des z, on a encore

$$\sum_1^h l_i\left(Z, \frac{\partial Z}{\partial z}\right) \xi_i\left(x, \frac{\partial z}{\partial x}\right) = \sum_1^h l_i\left(Z, \frac{\partial Z}{\partial z'}\right) \xi_i\left(x, \frac{\partial z'}{\partial x}\right),$$

d'où l'on conclut que la *transformation* (z, z') *fait subir aux* ξ_i *et aux* l_i *deux transformations linéaires adjointes.*

Je ne veux pas insister ici sur l'étude générale des fonctions rationnelles en $Z, \frac{\partial Z}{\partial x}, \ldots$ dans un domaine $[\Delta]$ faite en partant de la considération des *coordonnées*: groupe des transformations qu'admet P, permutation des polynomes P entre eux, formation du système différentiel qui définit P quand on y regarde les z comme donnés et les Z comme arbitraires, ... Il s'agit simplement de montrer l'usage des coordonnées ξ_i pour l'étude des systèmes irréductibles réguliers.

Supposons d'abord que les éléments $Z_1, \ldots, Z_n$ constituent un système fondamental *déterminé*, pour une équation

$$X(Z) = \frac{\partial Z}{\partial x} + A_1 \frac{\partial Z}{\partial x_1} + \ldots + A_n \frac{\partial Z}{\partial x_n} \quad \ldots\ldots\ldots\ldots\ldots\ldots\ldots\text{(I)},$$

dont les coefficients appartiennent au domaine $[\Delta]$, et que de plus le polynome P satisfasse à une identité

$$X(P) = MP,$$

qui est la condition nécessaire et suffisante pour que l'équation *unique*

$$P\left(Z, \frac{\partial Z}{\partial x}, x\right) = 0$$

forme avec le système

$$X(Z_i) = 0 \qquad (i = 1, \ldots, n) \quad \ldots\ldots\ldots\ldots\ldots\ldots\ldots\text{(II)}$$

un système complètement intégrable.

On en conclura que les éléments $z_1, \ldots, z_n$ qui représentent un système fondamental *quelconque* satisfont aux identités

$$\frac{X(\xi_1)}{\xi_1} = \ldots = \frac{X(\xi_h)}{\xi_h} = M(x, x_1, \ldots, x_n),$$

ou encore que les équations

$$\frac{\xi_i\left(x, \frac{\partial z}{\partial x}\right)}{\xi_1\left(x, \frac{\partial z}{\partial x}\right)} = \alpha_i(z_1, \ldots, z_n) \qquad (i = 2, \ldots, h)$$

sont toujours vérifiées, pour un choix convenable des fonctions (en général transcendantes) α_i par un système fondamental particulier quelconque.

Si l'on veut que le système précédent admette la solution qui pour $x = x_0$ satisfait aux conditions

$$z_i(x_0, x_1, \ldots, x_n) = \phi_i(x_1, \ldots, x_n) \qquad (i = 1, \ldots, n),$$

il suffira de poser

$$\alpha_i = \frac{\xi_i\left(x_0, x_1, \ldots, x_n, \frac{\partial \phi_1}{\partial x_1}, \ldots\right)}{\xi_1\left(x_0, x_1, \ldots, x_n, \frac{\partial \phi_1}{\partial x_1}, \ldots\right)},$$

et de calculer les seconds membres au moyen des z, en partant des formules

$$z_i = \phi_i(x_1, \ldots, x_n) \qquad (i = 1, \ldots, n)$$

résolues en $x_1, \ldots, x_n$.

Pour que le système précédent soit rationnel, il suffira de choisir un système fondamental qui, pour $x = x_0$, satisfait aux conditions

$$x_i = R_i(z_1, \ldots, z_n) \qquad (i = 1, \ldots, n),$$

où les R_i sont rationnels. C'est ce qui arrive toujours pour la solution *principale en* $x = x_0$ de l'équation (I) qui est définie par

$$x_i = z_i \text{ pour } x = x_0.$$

17. Considérons maintenant un système irréductible régulier (Σ) vérifié par $Z_1, \ldots, Z_n$ et supposons que jusqu'à un certain ordre N, tel que les équations distinctes d'ordre $(N+p)$ s'obtiennent en dérivant simplement jusqu'à cet ordre les équations d'ordre N, le système (Σ) renferme k relations entières, *rationnellement distinctes,*

$$f_1 = 0, \ldots, f_k = 0 \ldots\ldots\ldots\ldots\ldots\ldots\ldots\ldots(\Sigma)$$

(c'est-à-dire qu'aucun des polynomes f_i ne s'exprime sous forme entière avec les autres).

Remarquons tout de suite que s'il existe pour l'équation (I) p solutions rationnelles distinctes $z_1 = R_1, \ldots, z_p = R_p$ l'adjonction au domaine $[\Delta]$ de p fonctions algébriques ($x_1, \ldots, x_p$, par exemple, des arguments $x_{p+1}, \ldots, x_n, z_1, \ldots, z_p$) ramène l'équation (I) à une équation à $(n-p+1)$ variables seulement, dépendant des paramètres $z_1, \ldots, z_p$ dans le domaine modifié $[\Delta']$. On peut donc supposer en modifiant $[\Delta]$, qu'il n'existe pas de solutions z_i rationnelles dans ce domaine $[\Delta]$.

J'ajoute en passant qu'il y aura lieu de chercher après avoir déterminé $R_1, \ldots, R_p$ s'il n'est pas possible d'exprimer rationnellement ces fonctions au moyen de p autres $r_1, \ldots, r_p$ appartenant au même domaine, sans que la réciproque soit vraie ; ce n'est qu'à cette condition que les solutions rationnelles $z_1, \ldots, z_p$ seront *les plus simples.*

Formons le polynome $$P = u_1 f_1 + \ldots + u_k f_k$$
où les u désignent des polynomes arbitraires en $x, x_1, \ldots, x_n$ et mettons en évidence les *coordonnées* de ce polynome P, en introduisant les éléments $z_1, \ldots, z_n$ d'un système fondamental *quelconque*, on aura

$$P\left(Z, \frac{\partial Z}{\partial x}, x\right) = \sum_1^h l_i\left(Z, \frac{\partial Z}{\partial z}\right) \xi_i\left(x, \frac{\partial z}{\partial x}\right),$$

et le système pourra se remplacer par les équations

$$P = \sum_1^h l_i \xi_i = 0,$$

$$X(P) = P_1 = \sum_1^h l_i X(\xi_i) = 0,$$

$$\ldots\ldots\ldots\ldots\ldots\ldots\ldots\ldots$$

$$X(P_{k-2}) = P_{k-1} = \sum_1^h l_i X_{k-1}(\xi_i) = 0.$$

La relation $P_k = \sum_1^h l_i X_k(\xi_i) = 0$ étant une conséquence des précédentes, on en conclut l'existence d'*identités*

$$X_k(\xi_i) = \beta_0 \xi_i + \beta_1 X(\xi_i) + \ldots + \beta_{k-1} X_{k-1}(\xi_i) \qquad (i = 1, \ldots, h).$$

Il est aisé de montrer que *le nombre h des coordonnées de P est nécessairement égal à $(k+1)$*. Si l'on avait $h \leqslant k$ on déduirait des équations de (Σ) une ou plusieurs relations rationnelles qui seraient vérifiées par les éléments $z_1, \ldots, z_n$ d'un système *fondamental quelconque et qui ne se réduiraient pas à des identités*—ce qui est impossible.

Si l'on avait $h = k + r$ on pourrait former en partant du système précédent un système rationnel comprenant rk équations d'ordre au plus égal à N. En rangeant les ξ dans un ordre convenable on peut montrer que chaque coordonnée dont l'indice dépasse $(k+1)$ donne une équation au moins distincte rationnellement des k premières formées en partant de $\xi_1 \ldots \xi_k \xi_{k+1}$.

Le système (Σ) étant résolu par rapport à $l_1, \ldots, l_k$ pourra donc s'écrire :

$$\frac{l_i\left(Z, \dfrac{\partial Z}{\partial z}\right)}{l_{k+1}\left(Z, \dfrac{\partial Z}{\partial z}\right)} = -\frac{\Delta_i\left(x, \dfrac{\partial z}{\partial x}\right)}{\Delta\left(x, \dfrac{\partial z}{\partial x}\right)} \qquad (i = 1, \ldots, k) \quad \ldots\ldots\ldots\ldots(\Omega).$$

C'est la forme nouvelle que je voulais signaler ; je l'appelle *forme normale* des équations (Σ). On conclut immédiatement des équations (Ω) que l'on peut trouver un système irréductible régulier (Σ') possédant la solution $z_1, \ldots, z_n$ qui satisfait pour $x = x_0$ aux conditions

$$x_i = R_i(z_1, \ldots, z_n),$$

où les R_i sont n fonctions rationnelles indépendantes. Il suffira de poser

$$\Delta_i\left(x, \frac{\partial z}{\partial x}\right) = \alpha_i(z_1, \ldots, z_n)\, \Delta\left(x, \frac{\partial z}{\partial x}\right),$$

en calculant les α_i par les formules

$$\alpha_i = \frac{\Delta_i\left(x_0, x_1, \ldots, x_n, \frac{\partial z_1}{\partial x_1}, \ldots\right)}{\Delta\left(x_0, x_1, \ldots, x_n, \frac{\partial z_1}{\partial x_1}, \ldots\right)},$$

où l'on remplace les x_i et les $\frac{\partial z_i}{\partial x_k}$ par leurs expressions en $z_1, \ldots, z_n$.

En particulier, il existe un système irréductible régulier (Σ_0) qui possède la *solution principale en* $x = x_0$, c'est-à-dire celle qui satisfait pour cette valeur de x aux conditions

$$x_i = z_i \ (i = 1, \ldots, n);$$

on l'obtient en prenant

$$\alpha_i = \frac{\Delta_i(x_0, x_1, \ldots, x_n, 1, \ldots)}{\Delta(x_0, x_1, \ldots, x_n, 1, \ldots)},$$

où le second membre se déduit de $\frac{\Delta_i}{\Delta}$ en faisant

$$x_i = z_i, \quad \frac{\partial z_i}{\partial x_k} = 0, \quad \frac{\partial z_i}{\partial x_i} = 1, \ldots.$$

Le système *initial* (Σ) peut s'écrire, en faisant $z_i = Z_i$,

$$\frac{L_i(Z)}{L_{k+1}(Z)} = -\frac{\Delta_i\left(x, \frac{\partial Z}{\partial x}\right)}{\Delta\left(x, \frac{\partial Z}{\partial x}\right)},$$

et l'on passe de l'une de ses solutions $z_1, \ldots, z_n$ à une autre solution $Z_1, \ldots, Z_n$ en satisfaisant simplement aux relations

$$\frac{l_i\left(Z, \frac{\partial Z}{\partial z}\right)}{l_{k+1}\left(Z, \frac{\partial Z}{\partial z}\right)} = \frac{L_i(z)}{L_{k+1}(z)} \quad \ldots\ldots\ldots\ldots\ldots\ldots\ldots\ldots\ldots(\Omega_1),$$

qui sont les *équations de définition du groupe de rationalité* Γ, correspondant à (Σ). On reconnaît là l'ensemble des transformations (z, Z) qui n'altèrent pas la *forme* de la relation unique

$$P = 0.$$

Enfin, si l'on observe que dans la *forme normale* (Ω) les éléments $z_1, \ldots, z_n$ peuvent être regardés comme arbitraires et si l'on y pose

$$z_i = x_i, \quad \frac{\partial z_i}{\partial x_k} = 0, \quad \frac{\partial z_i}{\partial x_i} = 1, \ldots$$

on obtient la nouvelle forme

$$\frac{l_i\left(Z, \frac{\partial Z}{\partial x}\right)}{l_{k+1}\left(Z, \frac{\partial Z}{\partial x}\right)} = -\frac{\Delta_i(x, x_1, \ldots, x_n, 1, \ldots)}{\Delta(x, x_1, \ldots, x_n, 1, \ldots)} \quad \ldots\ldots\ldots\ldots\ldots\ldots(\Omega'),$$

qui prouve que les *invariants différentiels du groupe de rationalité* Γ (ce sont les premiers membres) sont, dans le domaine de rationalité adopté, des fonctions rationnelles de $x, x_1, \ldots, x_n$. Cette forme (Ω') est la *forme canonique* signalée plus haut.

Il convient de signaler ici que les *invariants relatifs* $l_i\left(Z, \frac{\partial Z}{\partial x}\right)$ du groupe Γ subissent, quand on exécute sur les variables x (non transformées par Γ) une transformation quelconque du groupe ponctuel général Γ_n, *une simple transformation linéaire et homogène.*

Les invariants absolus de Γ, $\frac{l_i}{l_{k+1}}$, subissent donc une transformation projective.

En adjoignant au besoin la racine p-ième d'une fonction du domaine de rationalité $[\Delta]$, on peut toujours s'arranger pour que $l_{k+1}\left(Z, \frac{\partial Z}{\partial x}\right)$ soit une puissance entière du déterminant fonctionnel $\frac{\partial (Z_1, \ldots, Z_n)}{\partial (x_1, \ldots, x_n)}$.

III. *Formation des résolvantes.*

18. Considérons le polynome P qui nous a servi à obtenir la *forme normale* du système irréductible régulier (Σ): si l'on met en évidence les *coordonnées* on a

$$P = \sum_1^h l_i\left(Z, \frac{\partial Z}{\partial z}\right) \xi_i\left(x, \frac{\partial z}{\partial x}\right),$$

et nous avons observé qu'une transformation (z, z') donne lieu à l'identité

$$\sum_1^h l_i\left(Z, \frac{\partial Z}{\partial z}\right) \xi_i\left(x, \frac{\partial z}{\partial x}\right) = \sum_1^h l_i\left(Z, \frac{\partial Z}{\partial z'}\right) \xi_i\left(x, \frac{\partial z'}{\partial x}\right).$$

On en conclut que l'on a des identités:

$$l_i\left(Z, \frac{\partial Z}{\partial z}\right) = \sum_1^h \lambda_{i,j}\left(\frac{\partial z}{\partial z'}\right) l_j\left(Z, \frac{\partial Z}{\partial z'}\right) \qquad (i = 1, \ldots, h),$$

où les z et les z' sont des variables quelconques. Faisons dans ces identités $z'_1 = x_1, \ldots, z'_n = x_n$ et en même temps $z_1 = x'_1, \ldots, z_n = x'_n$: elles s'écriront

$$l_i\left(Z, \frac{\partial Z}{\partial x'}\right) = \sum_1^h \lambda_{i,j}\left(\frac{\partial x}{\partial x'}\right) l_j\left(Z, \frac{\partial Z}{\partial x}\right) \qquad (i = 1, \ldots, h).$$

Si l'on pose maintenant

$$x'_i = x_i + \delta x_i \qquad (i = 1, \ldots, n),$$

en définissant les δx_i par

$$\frac{\delta x}{1} = \frac{\delta x_1}{A_1} = \ldots = \frac{\delta x_n}{A_n},$$

où l'on suppose δx infiniment petit, ce qui revient à prendre

$$x'_i = x_i + X(x_i)\,\delta x$$

avec
$$X(f) = \frac{\partial f}{\partial x} + A_1 \frac{\partial f}{\partial x_1} + \ldots + A_n \frac{\partial f}{\partial x_n},$$

$Z_1, \ldots, Z_n$ seront des invariants pour la transformation infinitésimale précédente et l'on aura, par exemple, en écrivant

$$\delta\left(dZ - \frac{\partial Z}{\partial x_1}dx_1 \ldots - \frac{\partial Z}{\partial x_n}dx_n\right) = 0,$$

$$\delta\left(\frac{\partial Z}{\partial x_i}\right) = \delta x \, . \, X\left(\frac{\partial Z}{\partial x_i}\right) = -\left(\frac{\partial A_1}{\partial x_i}\frac{\partial Z}{\partial x_1} + \ldots + \frac{\partial A_n}{\partial x_i}\frac{\partial Z}{\partial x_n}\right)$$

.... etc. ....

Les formules classiques du changement de variables (x, x') nous permettent aisément d'obtenir des expressions $\mu_{ij}\left(\frac{\partial x'}{\partial x}\right)$ telles que l'on ait identiquement

$$\mu_{i,j}\left(\frac{\partial x'}{\partial x}\right) = \lambda_{i,j}\left(\frac{\partial x}{\partial x'}\right) \qquad (i, j = 1, \ldots, h);$$

on les aurait d'ailleurs immédiatement en résolvant les équations évidentes

$$\sum_1^h \lambda_{i,j}\left(\frac{\partial x}{\partial x'}\right)\lambda_{j,s}\left(\frac{\partial x'}{\partial x}\right) = \epsilon_{js} \qquad (j, s = 1, \ldots, h),$$

avec $\epsilon_{js} = 0$ pour $j \neq s$, $\epsilon_{jj} = 1$.

Les formules $\qquad \dfrac{\partial x'_i}{\partial x_k} = \delta x \, . \, \dfrac{\partial A_i}{\partial x_k}, \quad \dfrac{\partial x'_i}{\partial x_i} = 1 + \delta x \, . \, \dfrac{\partial A_i}{\partial x_i}$

....

donnent alors aisément le coefficient de δx dans le développement de $\mu_{i,j}\left(\frac{\partial x'}{\partial x}\right)$:

$$\mu_{i,j}\left(\frac{\partial x'}{\partial x}\right) = [\mu_{i,j}] + \delta x \, . \, \omega_{ij} + \ldots.$$

Ce coefficient ω_{ij} est un polynome formé avec les dérivées des A en $x_1, \ldots, x_n$. Il nous suffit d'égaler les coefficients de δx dans les deux membres de toutes les équations qui définissent les $l_i\left(Z, \frac{\partial Z}{\partial x'}\right)$ pour obtenir

$$X\left[l_i\left(Z, \frac{\partial Z}{\partial x}\right)\right] = \sum_{j=1}^h \omega_{ij}(x, x_1, \ldots, x_n)\, l_j\left(Z, \frac{\partial Z}{\partial x}\right) \qquad (i = 1, \ldots, h) \ldots (\mathrm{R}_0).$$

Ce système (R_0) est le *système résolvant* des *invariants relatifs* $l_i\left(Z, \frac{\partial Z}{\partial x}\right)$ *du groupe de rationalité* Γ: on voit qu'il est *linéaire et homogène.*

Pour obtenir le système résolvant dont dépendent les *invariants absolus*

$$\frac{l_i\left(Z, \frac{\partial Z}{\partial x}\right)}{l_{k+1}\left(Z, \frac{\partial Z}{\partial x}\right)}$$

nous poserons, en supprimant l'indice $(k+1)$ pour avoir des formules plus simples,

$$\frac{l_i\left(Z, \frac{\partial Z}{\partial x}\right)}{l\left(Z, \frac{\partial Z}{\partial x}\right)} = \Lambda_i \qquad (i = 1, \ldots, k) \ldots\ldots\ldots\ldots (\Sigma),$$

et le système précédent s'écrira

$$lX(\Lambda_i) + \Lambda_i X(l) = l\Sigma_j \omega_{ij}(x)\Lambda_j$$
$$X(l) = l\Sigma_j \omega_j \Lambda_j,$$

d'où encore

$$X(\Lambda_i) + \Lambda_i \Sigma_j \omega_j \Lambda_j = \Sigma_j \omega_{ij} \Lambda_j \qquad (i = 1, \dots, k) \quad \dots\dots\dots\dots(\mathrm{R}).$$

Pour que ce système résolvant (R) soit linéaire il faut et il suffit que l'on puisse choisir l'un des $l_i\left(Z, \frac{\partial Z}{\partial x}\right)$ de façon que tous les ω_{ij} correspondants soient nuls—c'est-à-dire que cet $l_i\left(Z, \frac{\partial Z}{\partial x}\right)$ ne renferme pas de dérivées des Z.

19. Lorsque le groupe de rationalité est Γ *le système résolvant ne peut posséder qu'*UNE SEULE SOLUTION RATIONNELLE. Si l'on avait en effet un autre système

$$\frac{l_i\left(z, \frac{\partial z}{\partial x}\right)}{l\left(z, \frac{\partial z}{\partial x}\right)} = \lambda_i \qquad (i = 1, \dots, k) \quad \dots\dots\dots\dots(\sigma),$$

vérifié par un système fondamental au moins $z_1, \dots, z_n$, il est impossible qu'il existe des solutions communes aux deux systèmes (σ), (Σ).

Nous avons vu plus haut que ces deux systèmes peuvent recevoir la forme

$$\frac{L_i(Z)}{L_{k+1}(Z)} = -\frac{\Delta_i\left(x, \frac{\partial Z}{\partial x}\right)}{\Delta\left(x, \frac{\partial Z}{\partial x}\right)}$$

où $-\frac{\Delta_i}{\Delta}$ se réduit à Λ_i pour $Z_p = x_p \qquad (p = 1, \dots, n)$

et

$$\frac{L_i(z)}{L_{k+1}(z)} = -\frac{\delta_i\left(x, \frac{\partial z}{\partial x}\right)}{\delta\left(x, \frac{\partial z}{\partial x}\right)}$$

où $-\frac{\delta_i}{\delta}$ se réduit à λ_i pour $z_p = x_p\,(p = 1, \dots, n)$. Si tous les λ_i ne sont pas égaux aux Λ_i de même indice, les expressions $\frac{\delta_i}{\delta}$ et $\frac{\Delta_i}{\Delta}$ *ne sont pas identiques,* mais les identités

$$X\left(\frac{\Delta_i}{\Delta}\right) = 0, \quad X\left(\frac{\delta_i}{\delta}\right) = 0,$$

nous permettent de former un système *rationnel* comprenant les équations, en nombre $2k$:

$$\frac{\Delta_i\left(x, \frac{\partial u}{\partial x}\right)}{\Delta\left(x, \frac{\partial u}{\partial x}\right)} = \Phi_i(u), \quad \frac{\delta_i\left(x, \frac{\partial u}{\partial x}\right)}{\delta\left(x, \frac{\partial u}{\partial x}\right)} = \phi_i(u) \qquad (i = 1, \dots, k),$$

et satisfait par un système fondamental. Il suffit toujours de prendre le système fondamental qui satisfait pour $x = x_0$ aux conditions

$$x_p = R_p(u_1, \dots, u_n) \qquad (p = 1, \dots, n)$$

où les R_p sont n fonctions rationnelles distinctes. Le système (Σ) ne serait donc pas *irréductible et régulier.*

IV. *Exemples. Groupes-types de rationalité.*

20. (A) Soit l'équation aux dérivées partielles à trois variables

$$T(f)=\frac{\partial f}{\partial t}+A(u,v,t)\frac{\partial f}{\partial u}+B(u,v,t)\frac{\partial f}{\partial v}=0,$$

dont nous désignons par x, y deux solutions distinctes. Si le groupe de rationalité est (Γ) $Y=\phi(y)$, $X=\dfrac{x}{\phi'(y)}$ où ϕ est arbitraire, les équations de définition

$$\frac{\partial Y}{\partial x}=0, \qquad X\frac{\partial Y}{\partial y}=x,$$

expriment simplement que l'on a $XdY=xdy$, d'où l'on conclut que $x\dfrac{\partial y}{\partial u}$ et $x\dfrac{\partial y}{\partial v}$ sont les deux invariants au moyen desquels tous les autres s'expriment. En écrivant que le système

$$x\frac{\partial y}{\partial u}-\lambda=0, \qquad x\frac{\partial y}{\partial v}-\mu=0, \quad \ldots\ldots\ldots\ldots\ldots\ldots\ldots\ldots(a)$$

reste invariant par l'opération $T(f)$, on a les résolvantes que doivent vérifier λ et μ (rationnels dans le domaine adopté):

$$\left.\begin{aligned}T(\lambda)+\lambda\frac{\partial A}{\partial u}+\mu\frac{\partial B}{\partial u}&=0\\ T(\mu)+\lambda\frac{\partial A}{\partial v}+\mu\frac{\partial B}{\partial v}&=0\end{aligned}\right\}\quad\ldots\ldots\ldots\ldots\ldots\ldots\ldots\ldots(\mathrm{R}).$$

On pourrait aussi déduire de là l'équation unique à laquelle doit satisfaire le quotient

$$\frac{\dfrac{\partial y}{\partial u}}{\dfrac{\partial y}{\partial v}}=\frac{\lambda}{\mu}.$$

On remarquera aisément qu'avec les équations (a) tout système irréductible régulier renferme une autre équation du premier ordre, qui en est une condition d'intégrabilité:

$$\frac{\partial x}{\partial u}\frac{\partial y}{\partial v}-\frac{\partial x}{\partial v}\frac{\partial y}{\partial u}=\frac{\partial \mu}{\partial u}-\frac{\partial \lambda}{\partial v}.$$

La méthode générale conduit alors pour la *forme normale* aux équations

$$\frac{\mu\dfrac{\partial x}{\partial u}-\lambda\dfrac{\partial x}{\partial v}}{X\dfrac{\partial Y}{\partial y}}=\frac{\mu\dfrac{\partial y}{\partial u}-\lambda\dfrac{\partial y}{\partial v}}{-X\dfrac{\partial Y}{\partial x}}=\frac{\dfrac{\partial \mu}{\partial u}-\dfrac{\partial \lambda}{\partial v}}{\dfrac{\partial(X,Y)}{\partial(x,y)}}=\frac{\dfrac{\partial(x,y)}{\partial(u,v)}}{1}\ldots\ldots\ldots\ldots(\Omega),$$

sur lesquelles on raisonnera comme plus haut.

Par exemple, si l'on pose

$$\lambda_0=\lambda(t_0,x,y), \quad \mu_0=\mu(t_0,x,y),$$

le système de solutions (x, y) de $T(f)=0$ qui pour $t=t_0$ prennent respectivement les valeurs u et v (système principal pour $t=t_0$) est défini par les équations

$$\frac{\mu\frac{\partial x}{\partial u}-\lambda\frac{\partial x}{\partial v}}{\mu_0}=\frac{\mu\frac{\partial y}{\partial u}-\lambda\frac{\partial y}{\partial v}}{-\lambda_0}=\frac{\frac{\partial\mu}{\partial u}-\frac{\partial\lambda}{\partial v}}{\frac{\partial\mu_0}{\partial x}-\frac{\partial\lambda_0}{\partial y}}=\frac{\frac{\partial(x, y)}{\partial(u, v)}}{1}.$$

On voit quelle est la complication de ces équations, où interviennent les valeurs particulières de λ et μ; alors que les éléments les plus simples, correspondant au *groupe-type* de rationalité, sont définis par les équations

$$T(Y)=0,\quad \frac{\partial Y}{\partial u}=\frac{\lambda}{\mu}\frac{\partial Y}{\partial v},\qquad X=\frac{\mu}{\frac{\partial Y}{\partial v}},$$

dont les deux premières forment un système complet qui peut être quelconque.

21. (B) Considérons encore le cas où le groupe de rationalité est le groupe fini

$$X=\frac{ax+b}{cx+d},\qquad Y=\frac{ay+b}{cy+d}\quad\ldots\ldots\ldots\ldots(\Gamma),$$

a, b, c, d étant les constantes arbitraires.

Nous définirons ce groupe par l'équation aux différentielles totales

$$\frac{dX\,dY}{(X-Y)^2}=\frac{dx\,dy}{(x-y)^2},$$

qui nous donne immédiatement les trois invariants du premier ordre

$$\frac{\frac{\partial x}{\partial u}\frac{\partial y}{\partial u}}{(x-y)^2}=\lambda,\quad \frac{\frac{\partial x}{\partial u}\frac{\partial y}{\partial v}+\frac{\partial x}{\partial v}\frac{\partial y}{\partial u}}{(x-y)^2}=\mu,\quad \frac{\frac{\partial x}{\partial v}\frac{\partial y}{\partial v}}{(x-y)^2}=\nu\ \ldots\ldots\ldots\ldots(\Sigma).$$

Les fonctions rationnelles λ, μ, ν devront satisfaire au système résolvant :

$$\left.\begin{aligned}&T(\lambda)+2\frac{\partial A}{\partial u}\lambda+\frac{\partial B}{\partial u}\nu=0\\&T(\mu)+2\frac{\partial A}{\partial v}\lambda+\left(\frac{\partial B}{\partial v}+\frac{\partial A}{\partial u}\right)\mu+2\frac{\partial B}{\partial u}\nu=0\\&T(\nu)+\frac{\partial A}{\partial v}\nu+2\frac{\partial B}{\partial v}\mu=0\end{aligned}\right\}\ldots\ldots\ldots\ldots(\mathrm{R}),$$

qui ne possèdera qu'un seul système de solutions rationnelles.

On obtient aisément la *forme normale* du système (Σ), il suffit de résoudre les trois équations

$$\left.\begin{aligned}&\frac{\frac{\partial X}{\partial x}\frac{\partial Y}{\partial x}}{(X-Y)^2}\left(\frac{\partial x}{\partial u}\right)^2+\frac{\frac{\partial X}{\partial x}\frac{\partial Y}{\partial y}+\frac{\partial X}{\partial y}\frac{\partial Y}{\partial x}}{(X-Y)^2}\frac{\partial x}{\partial u}\frac{\partial y}{\partial u}+\frac{\frac{\partial X}{\partial y}\frac{\partial Y}{\partial y}}{(X-Y)^2}\left(\frac{\partial y}{\partial u}\right)^2=\lambda\\&2\frac{\frac{\partial X}{\partial x}\frac{\partial Y}{\partial x}}{(X-Y)^2}\frac{\partial x}{\partial u}\frac{\partial x}{\partial v}+\frac{\frac{\partial X}{\partial x}\frac{\partial Y}{\partial y}+\frac{\partial X}{\partial y}\frac{\partial Y}{\partial x}}{(X-Y)^2}\left(\frac{\partial x}{\partial u}\frac{\partial y}{\partial v}+\frac{\partial x}{\partial v}\frac{\partial y}{\partial u}\right)+2\frac{\frac{\partial X}{\partial y}\frac{\partial Y}{\partial y}}{(X-Y)^2}\frac{\partial y}{\partial u}\frac{\partial y}{\partial v}=\mu\\&\frac{\frac{\partial X}{\partial x}\frac{\partial Y}{\partial x}}{(X-Y)^2}\left(\frac{\partial x}{\partial v}\right)^2+\frac{\frac{\partial X}{\partial x}\frac{\partial Y}{\partial y}+\frac{\partial X}{\partial y}\frac{\partial Y}{\partial x}}{(X-Y)^2}\frac{\partial x}{\partial v}\frac{\partial y}{\partial v}+\frac{\frac{\partial X}{\partial y}\frac{\partial Y}{\partial y}}{(X-Y)^2}\left(\frac{\partial y}{\partial v}\right)^2=\nu\end{aligned}\right\}\ldots(\Omega),$$

par rapport aux trois invariants absolus du groupe. Mais cette résolution n'est pas nécessaire pour l'application de la méthode de transformation des équations (Ω) indiquée plus haut.

Il convient d'observer ici que si l'on définit (Γ) par l'invariance de $\frac{dx\,dy}{(x-y)^2}$ la permutation de x et y est possible, on doit donc la faire figurer explicitement dans (Γ).

Les invariants évidents

$$\frac{\frac{\partial x}{\partial u}}{\frac{\partial x}{\partial v}} = \alpha, \qquad \frac{\frac{\partial y}{\partial u}}{\frac{\partial y}{\partial v}} = \beta,$$

sont donnés par les équations symétriques

$$\alpha + \beta = \frac{\mu}{\nu}, \qquad \alpha\beta = \frac{\lambda}{\nu}.$$

La résolvante, dont dépend α,

$$T(\alpha) + \frac{\partial B}{\partial u} - \alpha \frac{\partial B}{\partial v} = \alpha \left(\alpha \frac{\partial A}{\partial v} - \frac{\partial A}{\partial u} \right),$$

admet deux solutions α, β dont les fonctions symétriques élémentaires sont rationnelles.

Si α et β sont rationnels tous deux, la permutation de x et y n'est pas possible : le système irréductible régulier s'obtiendra en ajoutant aux équations précédentes (Σ) une nouvelle équation rationnelle qui résulte de ce que

$$(\alpha + \beta)^2 - 4\alpha\beta = \frac{\mu^2 - 4\lambda\nu}{\nu^2} = \Delta^2,$$

où Δ est rationnel : cette équation est simplement

$$\frac{\frac{\partial x}{\partial u}\frac{\partial y}{\partial v} - \frac{\partial x}{\partial v}\frac{\partial y}{\partial u}}{(x-y)^2} = \Delta.$$

22. (C) Des circonstances analogues se présentent toutes les fois où le groupe de rationalité (Γ) comprend des transformations finies non engendrées par ses transformations infinitésimales, autrement dit est un *groupe complexe*.

Supposons par exemple que ce soit le groupe :

$$X = f(x), \quad Y = g(y),$$

où f et g sont arbitraires, la permutation de x et y étant possible. Les quotients

$$\frac{\frac{\partial x}{\partial u}}{\frac{\partial x}{\partial v}} = \alpha, \qquad \frac{\frac{\partial y}{\partial u}}{\frac{\partial y}{\partial v}} = \beta,$$

peuvent s'échanger par les transformations du groupe. Ce sont leurs fonctions symétriques qui sont rationnelles, et si l'on pose

$$\alpha\beta = \lambda, \quad \alpha + \beta = \mu$$

le système irréductible régulier est formé des deux relations identiques :

$$\left.\begin{aligned}\left(\frac{\partial x}{\partial u}\right)^2 - \mu\frac{\partial x}{\partial u}\frac{\partial x}{\partial v} + \lambda\left(\frac{\partial x}{\partial v}\right)^2 = 0\\ \left(\frac{\partial y}{\partial u}\right)^2 - \mu\frac{\partial y}{\partial u}\frac{\partial y}{\partial v} + \lambda\left(\frac{\partial y}{\partial v}\right)^2 = 0\end{aligned}\right\}\ldots\ldots\ldots\ldots(\Sigma),$$

avec $\dfrac{\partial(x, y)}{\partial(u, v)} \neq 0$. Le système résolvant dont une seule solution est rationnelle sera

$$\left.\begin{aligned}T(\mu) - \mu\left(\frac{\partial B}{\partial v} - \frac{\partial A}{\partial u}\right) + 2\frac{\partial B}{\partial u} - 2\lambda\frac{\partial A}{\partial v} - \mu^2\frac{\partial A}{\partial v} = 0\\ T(\lambda) + \mu\frac{\partial B}{\partial u} - 2\lambda\left(\frac{\partial B}{\partial v} - \frac{\partial A}{\partial u}\right) - \lambda\mu\frac{\partial A}{\partial v} = 0\end{aligned}\right\}\ldots\ldots\ldots(\mathrm{R}).$$

Lorsque α et β sont rationnels, le système régulier (Σ), au lieu de se composer des *deux* relations linéaires écrites plus haut, comporte *trois* équations quadratiques, l'équation nouvelle étant

$$\frac{\partial(x, y)}{\partial(u, v)} = \alpha - \beta.$$

On verrait sans difficulté que ces équations sont *rationnellement distinctes,* bien qu'elles soient fonctionnellement dépendantes.

Les trois équations du système irréductible régulier (Σ) conduisent aisément à la forme normale (Ω) où interviennent les *quatre* invariants relatifs du groupe

$$\frac{\partial X}{\partial x}\frac{\partial Y}{\partial x}, \quad \frac{\partial X}{\partial x}\frac{\partial Y}{\partial y} + \frac{\partial X}{\partial y}\frac{\partial Y}{\partial x}, \quad \frac{\partial X}{\partial y}\frac{\partial Y}{\partial y} \quad \text{et} \quad \frac{\partial X}{\partial x}\frac{\partial Y}{\partial y} - \frac{\partial X}{\partial y}\frac{\partial Y}{\partial x}.$$

Les remarques que l'on vient de faire suffisent pour rendre compte des diverses circonstances qui peuvent se présenter lorsqu'on veut utiliser pour la théorie de la rationalité les *types* de groupes à n variables (engendrés par leurs transformations infinitésimales) donnés par Lie et M. Cartan, quand ils sont *imprimitifs.*

D'abord un groupe G engendré par des transformations infinitésimales peut être compris dans un *groupe complexe* Γ : on obtient Γ en ajoutant à G un nombre fini de transformations, formant groupe, et appartenant au *plus grand groupe H dans lequel G est invariant.*

On aura ainsi divers types de groupes Γ correspondant aux divers types de groupes finis contenus dans H.

En outre si Γ est imprimitif et se définit comme sous groupe du groupe

$$\begin{aligned}&Z_i = \phi_i(z_1, \ldots z_p) && (i = 1, \ldots, p)\\ &Z_j = \psi_j(z_{p+1}, \ldots, z_{p+q}) && (j = p+1, \ldots, p+q)\\ &\ldots\ldots\ldots\ldots\ldots\ldots\\ &Z_k = \omega_k(z_1, \ldots, z_p, z_{p+1}, \ldots, z_n) && (k = p+q+\ldots+1, \ldots, n),\end{aligned}$$

où les $\phi_i, \psi_i, \ldots, \omega_k \ldots$ sont arbitraires, il peut arriver que certains ensembles de variables z, en même nombre, subissent des transformations appartenant à un même groupe (définies par les mêmes équations différentielles).

Le groupe Γ comprendra alors normalement les permutations qui échangent entre eux les éléments homologues de ces ensembles. Il pourra y avoir une réduction, par adjonction de grandeurs algébriques du domaine de rationalité $[\Delta]$ à un groupe engendré par des transformations infinitésimales, mais la *forme normale* du groupe restera composée d'équations d'ordre élevé en *nombre surabondant*, c'est-à-dire que les premiers membres de ces équations seront rationnellement mais non fonctionellement distincts.

Il est clair que si le groupe Γ est *imprimitif* et compris dans H, on peut obtenir pour le système irréductible régulier correspondant une *forme réduite*, que j'appelle encore *forme-type* et qu'on obtient en faisant jouer à H le rôle joué dans la théorie générale par le groupe ponctuel général Γ_n.

On obtiendra ainsi des *groupes de rationalité-types* identiques à peu de chose près aux types de groupes adoptés par Lie: les invariants rationnellement indépendants de ces groupes Γ sont connus rationnellement dans $[\Delta]$ et réciproquement, mais *le nombre des invariants relatifs de* Γ (invariants qui subissent toujours une transformation linéaire, quand on exécute une transformation de H) peut dépasser le nombre des équations rationnellement distinctes, parce que dans la forme normale réduite les dénominateurs des premiers membres ne sont plus nécessairement identiques.

23. (D) Un exemple simple éclaircira cela :

Supposons pour l'équation

$$T(f) = \frac{\partial f}{\partial t} + A(t, u, v)\frac{\partial f}{\partial u} + B(t, u, v)\frac{\partial f}{\partial v} = 0,$$

dont x, y sont deux solutions distinctes, le groupe de rationalité

$$\left.\begin{aligned} X &= x\phi'(y) + \psi(y) \\ Y &= \phi(y) \end{aligned}\right\} \qquad \ldots\ldots\ldots\ldots\ldots\ldots\ldots\ldots\ldots(\Gamma),$$

où ϕ et ψ sont arbitraires. Sous cette forme on a, pour le groupe, les équations de définition

$$\frac{\partial Y}{\partial x} = 0, \quad \frac{\partial Y}{\partial y} = \frac{\partial X}{\partial x},$$

et l'on obtient immédiatement les invariants fonctionnellement distincts

$$\frac{\dfrac{\partial y}{\partial u}}{\dfrac{\partial y}{\partial v}} = \lambda, \qquad \frac{\dfrac{\partial x}{\partial u}\dfrac{\partial y}{\partial v} - \dfrac{\partial x}{\partial v}\dfrac{\partial y}{\partial u}}{\left(\dfrac{\partial y}{\partial v}\right)^2} = \mu,$$

d'où l'on peut déduire deux relations linéaires

$$\frac{\partial y}{\partial u} - \lambda\frac{\partial y}{\partial v} = 0, \qquad \frac{\partial x}{\partial u} - \lambda\frac{\partial x}{\partial v} - \mu\frac{\partial y}{\partial v} = 0 \quad \ldots\ldots\ldots\ldots\ldots\ldots(\mathrm{A}).$$

En leur appliquant la méthode générale, on trouve

$$\frac{\partial Y}{\partial x}\left(\frac{\partial x}{\partial u} - \lambda\frac{\partial x}{\partial v}\right) + \frac{\partial Y}{\partial y}\left(\frac{\partial y}{\partial u} - \lambda\frac{\partial y}{\partial v}\right) = 0,$$

$$\frac{\partial X}{\partial x}\left(\frac{\partial x}{\partial u} - \lambda\frac{\partial x}{\partial v}\right) + \frac{\partial X}{\partial y}\left(\frac{\partial y}{\partial u} - \lambda\frac{\partial y}{\partial v}\right) - \mu\left(\frac{\partial Y}{\partial x}\frac{\partial x}{\partial v} + \frac{\partial Y}{\partial y}\frac{\partial y}{\partial v}\right) = 0,$$

où l'on voit apparaître quatre expressions linéairement distinctes :

$$\frac{\partial Y}{\partial x}, \quad \frac{\partial Y}{\partial y}, \quad \frac{\partial X}{\partial x}, \quad \frac{\partial X}{\partial y},$$

au lieu de trois que demande la théorie si *le système* (A) *est irréductible régulier.*

Mais on forme aisément le système irréductible régulier qui comporte *trois* relations du premier ordre et s'écrit sous la *forme normale*

$$\frac{\left(\frac{\partial x}{\partial u} - \lambda \frac{\partial x}{\partial v}\right)^2}{\left(\frac{\partial Y}{\partial y}\right)^2} = \frac{\left(\frac{\partial x}{\partial u} - \lambda \frac{\partial x}{\partial v}\right)\left(\frac{\partial y}{\partial u} - \lambda \frac{\partial y}{\partial v}\right)}{-\frac{\partial Y}{\partial y}\frac{\partial X}{\partial x}} = \frac{\left(\frac{\partial y}{\partial u} - \lambda \frac{\partial y}{\partial v}\right)^2}{\left(\frac{\partial Y}{\partial x}\right)^2} = \frac{\mu\left(\frac{\partial x}{\partial u}\frac{\partial y}{\partial v} - \frac{\partial x}{\partial v}\frac{\partial y}{\partial u}\right)}{\left(\frac{\partial Y}{\partial x}\right)^2},$$

où l'échange de x et y est possible.

Ces trois relations sont du type

$$\Delta = h^2 = hk = k^2,$$

mais deux d'entre elles, comprenant Δ, ne permettent jamais de conclure, *sans restriction*, $h = k$.

Si maintenant on considère (Γ) comme sous-groupe de

$$\left.\begin{aligned} Y &= \phi(y) \\ X &= f(x, y) \end{aligned}\right\} \quad \ldots\ldots\ldots\ldots\ldots\ldots\ldots\ldots\ldots\ldots (\mathrm{H}),$$

le système (A) conduit pour la *forme normale réduite* à

$$\frac{\frac{\partial y}{\partial u}}{\lambda \frac{\partial y}{\partial v}} = 1, \quad \frac{\mu \frac{\partial y}{\partial v}}{\frac{\partial x}{\partial u} - \lambda \frac{\partial x}{\partial v}} = \frac{\frac{\partial X}{\partial x}}{\frac{\partial Y}{\partial y}},$$

avec

$$\frac{\partial Y}{\partial y} \neq 0.$$

V. *Réduction du Groupe de Rationalité.*

24. Le problème de *l'intégration logique*, pour une équation

$$X(z) = \frac{\partial z}{\partial x} + A_1 \frac{\partial z}{\partial x_1} + \ldots + A_n \frac{\partial z}{\partial x_n} = 0 \ldots\ldots\ldots\ldots\ldots\ldots (\mathrm{I}),$$

exige la résolution des questions suivantes :

1°. Détermination de tous les *types* Γ de groupes contenus dans le groupe ponctuel général Γ_n, à n variables $z_1, \ldots, z_n$ dont les équations de définition sont rationnelles.

Ce problème peut être regardé comme résolu en principe par les recherches de Lie et de M. Cartan.

2°. Détermination du groupe de rationalité Γ de (I).

Chaque type de groupe (Γ') est caractérisé par un *système complet* d'invariants différentiels, quand on y regarde les z comme des fonctions de n variables $x_1, \ldots, x_n$ *non transformées* : ces invariants différentiels sont liés à ceux de Γ_n qui sont les quotients $\frac{D_i}{D}$ par un *système résolvant* qui possède une solution rationnelle en $x, x_1, \ldots, x_n$ quand Γ' renferme le groupe Γ.

Si Γ' coïncide avec Γ, ce système résolvant ne possède qu'*une seule solution rationnelle*: le groupe de rationalité est donc aussi *le plus petit groupe dont les invariants différentiels sont rationnels dans* $[\Delta]$.

Lorsque Γ est connu la nature des transcendantes $z_1, \ldots, z_n$ *attachés à ce groupe* est fixée: une décomposition du groupe Γ mettra en évidence leurs propriétés essentielles.

Soit Γ_1 un *plus grand sous-groupe invariant* dans Γ dont les équations de définition sont rationnelles; ses invariants différentiels subissent par les transformations de Γ des transformations formant un groupe Γ/Γ_1 qui ne possède pas de sous-groupe invariant maximum (groupe *simple*).

Soit de même Γ_2 un plus grand sous-groupe (à équations rationnelles) invariant dans Γ_1; les invariants de Γ_2 subissent par les transformations de Γ_1 des transformations formant le groupe Γ_1/Γ_2, etc.....

Au bout d'un nombre *limité* d'opérations (la limitation résulte du fait que tout groupe Γ_i possède au moins un invariant rationnel n'appartenant pas aux précédents et de l'application du théorème de M. Tresse) on tombe sur un groupe Γ_p formé de la transformation identique. Le groupe Γ_{p-1} est simple.

Les équations qui constituent un système irréductible régulier (Σ) peuvent donc s'écrire de manière à mettre en évidence:

Un système définissant $z_1, \ldots, z_n$ au moyen des invariants différentiels de Γ_{p-1};

Un système définissant les invariants de Γ_{p-1} au moyen de ceux de Γ_{p-2}, etc.....

Un système définissant les invariants de Γ_1 au moyen de ceux de Γ qui sont rationnels dans le domaine $[\Delta]$.

Je dis que les *transcendantes* $z_1, \ldots, z_n$ *attachées à* Γ *sont amenées à faire partie du domaine de rationalité par des adjonctions successives de transcendantes attachées à des groupes simples.* Mais la réduction précédente, utile pour manifester les propriétés des z, demeure *théorique.* On montre aussi que le nombre des groupes Γ_i et les types des groupes Γ_i/Γ_{i+1} sont, à peu de chose près, déterminés.

Jusqu'à présent nous avons raisonné, dans l'hypothèse d'un groupe de rationalité *intransitif,*

$$Z_1 = z_1, \ldots, Z_p = z_p,$$

$$Z_{p+i} = \phi_{p+i}(z_1, \ldots, z_p, Z_{p+1}, \ldots, Z_n) \qquad (i = 1, \ldots, n-p)$$

sur les transformations des Z_{p+i}, en regardant $z_1, \ldots, z_p$ comme *adjoints* au domaine de rationalité. Mais il serait facile de tenir compte aussi de la *nature des coefficients des transformations précédentes* en $z_1, \ldots, z_p$. Tout se passe comme si l'on étudiait dans l'équation

$$X(Z) = \frac{\partial Z}{\partial x} + A_{p+1}\frac{\partial Z}{\partial x_{p+1}} + \ldots + A_n\frac{\partial Z}{\partial x_n} = 0,$$

dont les coefficients dépendent de $x, x_{p+1}, \ldots, x_n$ et des *paramètres* $z_1, \ldots, z_p$, *les* Z_{p+i} *à la fois comme fonction des* x *et des paramètres* z.

Il est donc nécessaire d'envisager dans la formation du système irréductible régulier (Σ) les éléments $\frac{\partial Z}{\partial z_i}, \ldots, \frac{\partial^2 Z}{\partial z_i \partial z_j}, \ldots$ en même temps que les dérivées $\frac{\partial Z}{\partial x_{p+i}}, \ldots, \frac{\partial^2 Z}{\partial x_{p+i} \partial x_{p+j}}, \ldots$. Cela entraîne quelques modifications qui n'atteignent rien d'essentiel. Des exemples simples suffisent à mettre en évidence ces modifications.

Observons cependant que les transformations des *coefficients* des Z_{p+i} $(i=1, \ldots n-p)$ peuvent être de forme beaucoup plus générale que celles des Z_{p+i} entre eux; comme cela résulte de la recherche des types de groupes intransitifs.

L'exemple évident, signalé par M. Cartan, du groupe

$$X=x, \quad Y=y, \quad Z=z+f(x, y),$$

où f est quelconque, ou bien est la *solution la plus générale d'un système quelconque, linéaire,* d'équations aux dérivées partielles

$$F_i\left(x, y, f, \frac{\partial f}{\partial x}, \frac{\partial f}{\partial y}, \ldots, \frac{\partial^2 f}{\partial x^2}, \ldots\right)=0 \qquad (i=1, \ldots, k),$$

à coefficients dépendant seulement de x *et* y et qui peut se présenter pour l'équation

$$T(z)=\frac{\partial z}{\partial t}+C(x, y, w, t)\frac{\partial z}{\partial w}=0$$

sous des conditions faciles à donner—c'est-à-dire pour une équation de premier ordre

$$\frac{\partial w}{\partial t}=C(x, y, w, t),$$

dépendant de deux paramètres—fait prévoir la complication possible des transformations du groupe Γ de rationalité quand on laisse des paramètres dans l'équation étudiée[1].

VI. *Comment on tire parti de relations connues satisfaites par un système fondamental.*

25. Avant de passer aux applications particulières, je signalerai que la considération des *coordonnées* d'un polynome $P\left(Z, \frac{\partial Z}{\partial x}, x\right)$ permet, par *une méthode régulière de calcul,* de déduire de tout système d'équations rationnelles vérifié par $Z_1, \ldots, Z_n$ et leurs dérivées en $x_1, \ldots, x_n$ un autre système (peut-être *réductible*) ayant la *forme canonique.* Si l'on traite en effet ce système de la même manière que le système

* La théorie précédente *d'intégration logique* des équations linéaires aux dérivées partielles a été résumée dans des *Notes aux Comptes rendus de l'Académie des Sciences* (1893, 1895) et développée dans ma thèse (*Annales de l'École Normale Supérieure*, 1898), la définition des *systèmes irréductibles réguliers* auxquels s'appliquent exclusivement les raisonnements et conclusions de ma thèse a été communiquée en Octobre 1898, à MM. Painlevé et Vessiot, qui avaient amicalement appelé mon attention sur l'ambiguïté de certains énoncés. La théorie des *coordonnées* des polynomes, avec son application à la détermination de la *forme normale* d'un système irréductible régulier, a été exposée dans un mémoire présenté en 1902 à l'Académie des Sciences et qui a commencé à paraître aux *Annales de la Faculté des Sciences de Toulouse* (1908).

régulier (Σ) tout ce que nous avons dit subsiste, fors que le nombre h des coordonnées de P peut être $(k+r)$, où r est quelconque. En résolvant les k relations obtenues par rapport à $l_1, \ldots, l_k$ on a le système

$$l_i = \sum_{j=1}^{r} l_{k+j} \frac{\Delta_{i,j}}{\Delta} \qquad (i=1, \ldots, k)\ldots\ldots\ldots\ldots(\mathrm{A}),$$

d'où l'on déduit les identités en $z_1, \ldots, z_n$

$$X\left(\frac{\Delta_{i,j}}{\Delta}\right)=0,$$

qui permettent de former le système compatible, rationnel:

$$\frac{\Delta_{ij}}{\Delta} = \omega_{ij}(z_1, \ldots, z_n) \qquad (i=1, \ldots, k;\ j=1, \ldots, r)\ldots\ldots\ldots(\mathrm{B}),$$

où les ω_{ij} sont calculés de manière que le système précédent possède la solution qui pour $x=x_0$ satisfait aux conditions

$$x_i = R_i(z_1, \ldots, z_n) \qquad (i=1, \ldots, n),$$

où les R_i sont rationnels et quelconques.

Les équations $l_i = \Sigma l_{k+j}\omega_{ij}$ $(i=1, \ldots, k)$ demeurent inaltérées quand on exécute sur les z une transformation (z, z') qui conduit d'une solution du système (B) à une autre et réciproquement.

Eu égard à la transformation linéaire des l_i quand on passe des z aux z', on trouve que ces transformations (z, z') satisfont aux conditions, en nombre rk:

$$\sum_{p=1}^{k} \lambda_{i,p}\left(\frac{\partial z}{\partial z'}\right)\omega_{pj}(z') + \lambda_{i,k+j}\left(\frac{\partial z}{\partial z'}\right) = \sum_{q=1}^{r}\omega_{iq}(z)\left\{\sum_{m=1}^{k}\lambda_{k+q,m}\left(\frac{\partial z}{\partial z'}\right)\omega_{mj}(z') + \lambda_{k+q,k+j}\left(\frac{\partial z}{\partial z'}\right)\right\}$$
$$(i=1, \ldots, k;\ j=1, \ldots, r)\ldots(\mathrm{C}).$$

On établirait aisément *qu'elles définissent un groupe* Γ *dont on a rationnellement les invariants.* Ce groupe n'est pas nécessairement le groupe de rationalité de $X(z)=0$ puisque (Σ) n'est pas même irréductible.

26. Je voudrais encore faire remarquer qu'il est aisé de *préciser de quel secours peut être l'intégration d'une équation*

$$Y(f) = \frac{\partial f}{\partial x} + B_1\frac{\partial f}{\partial x_1} + \ldots + B_n\frac{\partial f}{\partial x_n} = 0,$$

pour celle d'une autre équation aux mêmes variables

$$Z(g) = \frac{\partial g}{\partial x} + A_1\frac{\partial g}{\partial x_1} + \ldots + A_n\frac{\partial g}{\partial x_n} = 0.$$

Soient respectivement $Y_1, \ldots, Y_n$ et $Z_1, \ldots, Z_n$ les éléments de deux systèmes fondamentaux: s'il existe de tels systèmes pour lesquels les transcendantes Y_i et Z_j ne sont pas *étrangères,* c'est-à-dire pour lesquels on peut ajouter aux équations $Y(Y_i)=0$, $Z(Z_j)=0$ des relations rationnelles en $Y_i, \frac{\partial Y_i}{\partial x_p}, \ldots, Z_j, \frac{\partial Z_j}{\partial x_p}, \ldots$ dont les coefficients appartiennent au domaine de rationalité adopté et renfermant effectivement les deux systèmes de fonctions $Y_1, \ldots, Y_n, Z_1, \ldots, Z_n$; l'application de la méthode générale qui a servi à transformer (Σ) conduit à les mettre sous la forme

$$\Omega_i\left(Z, \frac{\partial z}{\partial x}\right) = \omega_i(x) \qquad (i = 1, \ldots, k),$$

$$\Omega_{k+j}\left(Z, \frac{\partial z}{\partial x}\right) = \omega_{k+j}\left(x, Y, \frac{\partial Y}{\partial x}, \ldots\right) \qquad (j = 1, \ldots, r),$$

où les Ω sont tous les invariants rationnels et rationnellement distincts d'un certain type de groupe Γ de transformations en $Z_1, \ldots, Z_n$ et où les ω sont rationnels dans le domaine $[\Delta']$ formé par l'adjonction à $[\Delta]$ de $Y_1, \ldots, Y_n$. On peut d'ailleurs supposer que les ω_{k+j} sont également des invariants différentiels pour un groupe (γ) de transformations en $Y_1, \ldots, Y_n$.

Enfin le système précédent peut être supposé *irréductible et primitif*; il contient alors normalement des équations

$$\Omega_{k+r+l}\left(Y, \frac{\partial Y}{\partial x}\right) = \omega_{k+r+l}(x) \qquad (l = 1, \ldots, s),$$

qui donnent avec les précédentes l'expression rationnelle de tous les invariants distincts du groupe (γ).

Applications.

I. *Équation du second ordre. Équations différentielles linéaires.*

27. L'étude de l'équation

$$T(f) = \frac{\partial f}{\partial t} + A(t, u, v)\frac{\partial f}{\partial u} + B(t, u, v)\frac{\partial f}{\partial v} = 0,$$

faite d'après les principes précédents, n'entraîne pas d'autre difficulté—et celle-là insurmontable dans le cas général—que de reconnaître si un système résolvant formé, a, ou non, une solution rationnelle. En effet Lie, et après lui M. Cartan, a donné tous les types de groupes finis et infinis à deux variables engendrés par leurs transformations infinitésimales et l'on en déduit sans difficulté les groupes *complexes*. Le nombre de ces types croît assez vite quand on passe d'une variable à deux : il y en a effet une soixantaine et quelques-uns renferment un entier arbitraire. Mais la plupart sont *imprimitifs*, c'est-à-dire que parmi les invariants différentiels figure par exemple :

$$\frac{\partial y}{\partial u} : \frac{\partial y}{\partial v} = \lambda.$$

La solution y de $T(y) = 0$ est donc donnée par un système complet de deux équations —système qui peut être quelconque—et se distingue nettement des autres.

Quand on envisage seulement les groupes primitifs, on n'en trouve que six—dont trois sont finis :

Groupe général.

Groupe dont les transformations multiplient les aires par une constante.

Si l'on pose
$$D = \frac{\partial (x, y)}{\partial (u, v)},$$
$$\frac{1}{D}\frac{\partial D}{\partial u} = \lambda \quad \text{et} \quad \frac{1}{D}\frac{\partial D}{\partial v} = \mu$$
sont rationnels.

Groupe des transformations qui conservent les aires:
$$D = \frac{\partial (x, y)}{\partial (u, v)}$$
est rationnel. C'est le multiplicateur de Jacobi.

Groupe projectif général.

Les invariants s'obtiennent en observant que l'équation aux différentielles totales, $dxd^2y - dyd^2x = 0$, demeure inaltérée par les transformations du groupe.

Groupe linéaire général.

Les invariants s'obtiennent en observant que le système
$$d^2x = 0, \quad d^2y = 0,$$
admet les transformations du groupe.

Groupe linéaire spécial.

Avec les équations précédentes, on a $\frac{\partial (x, y)}{\partial (u, v)} = D$ où D est rationnel, c'est-à-dire que l'élément d'intégrale double $dxdy$ est invariant.

Remarquons en passant que *toutes les fois où le groupe de rationalité est fini* (et cette remarque est générale) les transcendantes x, y peuvent être amenées à faire partie du domaine de rationalité par des *adjonctions de transcendantes attachées à des groupes linéaires.*

Ceci augmente l'importance des recherches faites dans les cas où le groupe de rationalité est linéaire, ce qui arrive toujours lorsque dans l'équation
$$X(z) = \frac{\partial z}{\partial x} + A_1 \frac{\partial z}{\partial x_1} + \dots + A_n \frac{\partial z}{\partial x_n} = 0$$
les A_i sont linéaires en $x_1, \dots, x_n$.

On retrouve ainsi comme cas particulier, soit directement, soit en faisant intervenir *l'équation adjointe* de Lagrange, la théorie de M. Émile Picard pour les équations différentielles linéaires. Mais on n'ajoute rien d'essentiel à cette théorie. Les observations faites par Fuchs et M. Darboux sur les propriétés des intégrales algébriques se présentent simplement de façon nécessaire.

II. *Problème Normal de Lie.*

28. Il est bien clair que la théorie *d'intégration logique* des équations aux dérivées partielles donne le moyen de préciser pour *tous les problèmes de la théorie des groupes de Lie* (le domaine de rationalité étant fixé) la difficulté de la solution, cette difficulté étant toujours caractérisée par un groupe de rationalité. Dans certains cas les avantages qu'elle apporte sont de pure forme, mais il n'en est pas toujours ainsi. Lie

et ses élèves ont fait appel, suivant les méthodes habituelles, à des *changements de variables transcendants* et à des déterminations *successives d'éléments indissolublement liés.*

Par exemple, Lie est revenu à diverses reprises sur l'étude des systèmes complets d'équations linéaires aux dérivées partielles qui admettent des transformations infinitésimales. Il énonce ainsi le *problème normal* de sa théorie.

Une équation

$$A(f) = \Sigma \alpha_\nu (x_1, \ldots, x_{r+1}) \frac{\partial f}{\partial x_\nu} = 0 \quad \ldots\ldots\ldots\ldots\ldots\ldots\ldots\ldots (1)$$

à $(r+1)$ variables admet les r transformations infinitésimales indépendantes

$$X_k(f) = \Sigma \xi_{k\nu}(x_1, \ldots, x_{r+1}) \frac{\partial f}{\partial x_\nu} \qquad (k = 1, \ldots, r),$$

qui satisfont aux conditions

$$(X_i, X_k) = \Sigma c_{iks} X_s(f);$$

on suppose qu'il n'existe aucune identité

$$\phi(x) A(f) + \Sigma \phi_k(x) X_k(f) = 0,$$

où les ϕ sont quelconques et l'on demande *d'utiliser le plus possible la connaissance des $X_k(f)$ pour l'intégration de l'équation.*

En supposant que l'équation $A(f) = 0$ soit *la plus générale possible* parmi celles qui satisfont à ces conditions, Lie a indiqué une méthode de détermination *successive* des solutions de (1), mais le groupe de rationalité n'a pas été mis en évidence. C'est ce groupe, qui est *linéaire*, qui caractérise pour nous la simplification apportée dans l'intégration.

Par exemple, si les X_i forment un groupe *intégrable* dépourvu de transformation distinguée, le système obtenu en ajoutant à (1) les équations

$$X_i(f) + E_i(f) = 0 \qquad (i = 1, \ldots, r)$$

où les $E_i(f)$ définissent le *groupe adjoint*, admet r solutions pour lesquelles on peut prendre

$$Z_1 = e_i - \omega_i(x_1, \ldots, x_{r+1}) \qquad (i = 1, \ldots, r).$$

Les ω sont des fonctions de $x_1, \ldots, x_{r+1}$ *attachées à un groupe linéaire intégrable.*

III. *Groupes infinis simples. Équations les plus générales qui leur correspondent.*

29. Il résulte des recherches de Lie qu'il existe, à n variables, quatre types de groupes infinis *simples*; M. Cartan a démontré qu'il *n'y en a pas d'autres.* Ce sont :

(1°) le groupe ponctuel général Γ_n à n variables;

(2°) le groupe V_n dont les transformations conservent les volumes;

(3°) le groupe W_n des transformations de contact, qui n'altère pas l'équation

$$dz - p_1 dx_1 - \ldots - p_n dx_n = 0$$

où $n = 2k+1$ et où $z, x_1, \ldots, x_k, p_1, \ldots, p_k$ sont les n variables transformées;

(4°) le groupe de transformations à $n=2k$ variables qui conserve l'intégrale double

$$\iint dx_1 dy_1 + \dots + dx_k dy_k.$$

Il est facile d'indiquer des équations *spéciales*, dépendant de fonctions rationnelles arbitraires en nombre aussi grand que possible, dont le groupe de rationalité Γ est l'un de ces groupes simples.

II.—Pour le groupe V_n dont l'équation de définition est

$$\frac{\partial (X_1, \dots, X_n)}{\partial (x_1, \dots, x_n)} = 1,$$

si l'on regarde les x comme des fonctions de $t, u_1, \dots, u_n$ satisfaisant à

$$T(f) = \frac{\partial f}{\partial t} + A_1 \frac{\partial f}{\partial u_1} + \dots + A_n \frac{\partial f}{\partial u_n} = 0,$$

l'invariant différentiel caractéristique est

$$D = \frac{\partial (x_1, \dots, x_n)}{\partial (u_1, \dots, u_n)},$$

et la résolvante correspondante,

$$T(D) + D\left(\frac{\partial A_1}{\partial u_1} + \dots + \frac{\partial A_n}{\partial u_n}\right) = 0,$$

est l'équation au multiplicateur de Jacobi. On obtiendra l'expression la plus générale de $D, A_1, \dots, A_n$ en posant

$$DA_i = \frac{\partial \omega_i}{\partial t} \qquad (i = 1, \dots, n)$$

avec

$$D + \frac{\partial \omega_1}{\partial u_1} + \frac{\partial \omega_2}{\partial u_2} + \dots + \frac{\partial \omega_n}{\partial u_n} = 0.$$

L'équation cherchée est donc

$$DT(f) = \frac{\partial (\omega_1, f)}{\partial (t, u_1)} + \frac{\partial (\omega_2, f)}{\partial (t, u_2)} + \dots + \frac{\partial (\omega_n, f)}{\partial (t, u_n)} = 0,$$

où les ω sont des fonctions arbitraires en $u_1, \dots, u_n$ et t, *dans le domaine de rationalité adopté.*

III.—Si l'on veut que l'équation

$$dz - p_1 dx_1 - \dots - p_k dx_k = 0$$

soit une relation invariante entre $(2k+1) = n$ solutions, formant un système fondamental de l'équation

$$T(f) = \frac{\partial f}{\partial t} + A_1 \frac{\partial f}{\partial u_1} + \dots + A_n \frac{\partial f}{\partial u_n} = 0,$$

il faut et il suffit pour cela que les rapports des expressions

$$\frac{\partial z}{\partial u_i} - p_1 \frac{\partial x_1}{\partial u_i} - \dots - p_k \frac{\partial x_k}{\partial u_i}$$

soient des invariants pour le groupe de rationalité. On pourra donc poser

$$dz - p_1 dx_1 - \ldots - p_k dx_k = \rho\,(\lambda_1 du_1 + \ldots + \lambda_{2k} du_{2k} + du_{2k+1}),$$

où les λ_i seront des fonctions rationnelles. Ces invariants devront satisfaire aux équations résolvantes :

$$T(\rho\lambda_i) + \frac{\partial A_1}{\partial u_i}\rho\lambda_1 + \ldots + \frac{\partial A_{2k}}{\partial u_i}\rho\lambda_{2k} + \frac{\partial A_{2k+1}}{\partial u_i}\rho = 0 \qquad (i = 1, \ldots, 2k),$$

$$T(\rho) + \frac{\partial A_1}{\partial u_{2k+1}}\rho\lambda_1 + \ldots + \frac{\partial A_{2k}}{\partial u_{2k+1}}\rho\lambda_{2k} + \frac{\partial A_{2k+1}}{\partial u_{2k+1}}\rho = 0,$$

d'où l'on conclut simplement :

$$T(\lambda_i) + \frac{\partial A_i}{\partial u_i}\lambda_1 + \ldots + \frac{\partial A_{2k+1}}{\partial u_i} = \lambda_i \left\{\frac{\partial A_1}{\partial u_{2k+1}}\lambda_1 + \ldots + \frac{\partial A_{2k+1}}{\partial u_{2k+1}}\right\} \quad (i = 1, \ldots, 2k) \ldots (\mathrm{R}).$$

Pour résoudre le système (R), je pose

$$A_{2k+1} = -\{\lambda_0 + \lambda_1 A_1 + \ldots + \lambda_{2k} A_{2k}\},$$

où λ_0 est une nouvelle inconnue et l'on reconnait facilement que *toutes les dérivées des A_i disparaissent des équations* (R). Ces équations sont donc simplement $2k$ équations linéaires aux inconnues $A_1, A_2, \ldots, A_{2k}$ et λ_0 demeure arbitraire.

Si l'on pose, pour la symétrie, $t = u_0$ et

$$\omega_{ij} = \frac{\partial\lambda_i}{\partial u_j} - \frac{\partial\lambda_j}{\partial u_i} + \lambda_i \frac{\partial\lambda_j}{\partial u_{2k+1}} - \lambda_j \frac{\partial\lambda_i}{\partial u_{2k+1}} \qquad (i, j = 0, 1, \ldots, 2k),$$

le système qui définit les A_i peut s'écrire

$$\omega_{i0} + \omega_{i1} A_1 + \ldots + \omega_{i,2k} A_{2k} = 0 \qquad (i = 1, \ldots, 2k).$$

On aura donc pour l'équation cherchée

$$\Omega T(f) = \begin{vmatrix} \frac{\partial f}{\partial t} & \frac{\partial f}{\partial u_1} & \cdots & \frac{\partial f}{\partial u_{2k}} & \frac{\partial f}{\partial u_{2k+1}} \\ \omega_{10} & 0 & \cdots & \omega_{1,2k} & 0 \\ \cdots & \cdots & \cdots & \cdots & \cdots \\ \cdots & \cdots & \cdots & \cdots & \cdots \\ \omega_{2k,0} & \omega_{2k,1} & \cdots & 0 & 0 \\ \lambda_0 & \lambda_1 & \cdots & \lambda_{2k} & 1 \end{vmatrix} = 0,$$

avec $\Omega = \| \omega_{ij} \|$.

Quand les λ sont pris arbitrairement dans le domaine de rationalité, les A_i sont également rationnels et le groupe de rationalité est bien W_n. Ce résultat est d'ailleurs lié à la réduction de $\lambda_1 du_1 + \ldots + du_{2k+1}$ à sa forme canonique. Il est nécessaire de supposer les λ choisis de façon que la réduite soit bien à $(2k+1)$ éléments. On peut appliquer ceci à partir de $k = 1$ et former, par exemple, une équation

$$T(f) = \frac{\partial f}{\partial t} + A\frac{\partial f}{\partial u} + B\frac{\partial f}{\partial v} + C\frac{\partial f}{\partial w} = 0,$$

dépendant de trois fonctions rationnelles arbitraires de u, v, w, t et dont le groupe de rationalité est défini par

$$dZ - PdX = \sigma(dz - pdx),$$

c'est-à-dire est le groupe des transformations de contact du plan.

IV.—Supposons enfin qu'il s'agisse d'obtenir une équation

$$T(f) = \frac{\partial f}{\partial t} + A_1 \frac{\partial f}{\partial u_1} + \ldots + A_{2k} \frac{\partial f}{\partial u_{2k}} = 0,$$

dont le groupe de rationalité soit formé des transformations qui conservent *l'élément d'intégrale double*

$$dx_1 dy_1 + \ldots + dx_k dy_k$$

(les x et les y étant $2k$ solutions convenables de $T(f) = 0$). Si u_i et u_j sont deux quelconques des variables $u_1, \ldots, u_{2k}$ l'expression

$$\frac{\partial(x_1, y_1)}{\partial(u_i, u_j)} + \frac{\partial(x_2, y_2)}{\partial(u_i, u_j)} + \ldots + \frac{\partial(x_k, y_k)}{\partial(u_i, u_j)}$$

est un invariant; désignons par μ_{ij} son expression rationnelle en $t, u_1, \ldots, u_{2k}$. On forme aisément le système des résolvantes pour les μ_{ij}:

$$T(\mu_{ij}) + \frac{\partial A_1}{\partial u_i}\mu_{1j} + \ldots + \frac{\partial A_{2k}}{\partial u_i}\mu_{2k,j} = \frac{\partial A_1}{\partial u_j}\mu_{1i} + \ldots + \frac{\partial A_{2k}}{\partial u_j}\mu_{2k,i} \quad (i, j = 1, \ldots, 2k) \ldots (\text{R}),$$

mais ici se présente pour la recherche des A une difficulté qui tient à ce que *les μ_{ij} ne sont pas arbitraires.*

Ils doivent satisfaire à toutes les identités

$$\frac{\partial \mu_{ij}}{\partial u_l} + \frac{\partial \mu_{jl}}{\partial u_i} + \frac{\partial \mu_{li}}{\partial u_j} = 0 \qquad (i \neq j \neq l = 1, \ldots, 2k),$$

et ces conditions nécessaires *sont suffisantes.* Observons qu'en posant

$$x_1 \frac{\partial y_1}{\partial u_j} + \ldots + x_k \frac{\partial y_k}{\partial u_j} = \sigma_j,$$

ces fonctions σ sont données par les équations—évidemment compatibles:

$$\frac{\partial \sigma_j}{\partial u_i} - \frac{\partial \sigma_i}{\partial u_j} = \mu_{ij} \qquad (i, j = 1, \ldots, 2k).$$

On passe de la solution particulière $\sigma_i, \ldots$ de ces relations à la solution *la plus générale* $\Sigma_i, \ldots$ en posant

$$\Sigma_i = \sigma_i + \frac{\partial \Omega}{\partial u_i} \qquad (i = 1, \ldots, 2k),$$

où Ω est arbitraire en $u_1, \ldots, u_{2k}$ et t.

On peut donc dire que *l'expression*

$$\sigma_1 du_1 + \ldots + \sigma_{2k} du_{2k}$$

est définie à une différentielle totale additive près $d\Omega$ par les équations précédentes; il en est donc ainsi de

$$x_1 dy_1 + \ldots + x_k dy_k.$$

En d'autres termes, les éléments X, Y les plus généraux qui satisfont aux conditions imposées sont liés à un système particulier x, y par l'équation aux différentielles totales

$$X_1 dY_1 + \ldots + X_k dY_k = x_1 dy_1 + \ldots + x_k dy_k + d\Omega,$$

où Ω est arbitraire en t, $u_1, \ldots, u_{2k}$ ou bien encore en $x_1, \ldots, x_k, y_1, \ldots, y_k$ et t.

Le groupe de rationalité peut donc aussi se définir comme *l'ensemble des transformations* en $x_1, \ldots, x_k, y_1, \ldots, y_k$ qui *conservé* À UNE DIFFÉRENTIELLE EXACTE près *la forme*

$$x_1 dy_1 + \ldots + x_k dy_k.$$

Soit alors
$$x_1 \frac{\partial y_1}{\partial u_i} + \ldots + x_k \frac{\partial y_k}{\partial u_i} = \sigma_i + \frac{\partial \omega}{\partial u_i} \quad (i = 1, \ldots, 2k),$$

où ω et les x, y sont inconnus; les résolvantes correspondantes sont

$$T\left(\sigma_i + \frac{\partial \omega}{\partial u_i}\right) + \frac{\partial A_1}{\partial u_i}\left(\sigma_1 + \frac{\partial \omega}{\partial u_i}\right) + \ldots + \frac{\partial A_{2k}}{\partial u_i}\left(\sigma_{2k} + \frac{\partial \omega}{\partial u_{2k}}\right) = 0 \quad (i = 1, \ldots, 2k) \ldots (\mathrm{R}).$$

Si l'on pose
$$T(\omega) + A_1 \sigma_1 + \ldots + A_{2k} \sigma_{2k} = \phi,$$

où nous pouvons cette fois regarder ϕ comme connu, les A sont donnés par le système

$$\frac{\partial \sigma_i}{\partial t} + \frac{\partial \phi}{\partial u_i} + A_1 \left(\frac{\partial \sigma_i}{\partial u_1} - \frac{\partial \sigma_1}{\partial u_i}\right) + \ldots + A_{2k}\left(\frac{\partial \sigma_{i'}}{\partial u_{2k}} - \frac{\partial \sigma_{2k}}{\partial u_i}\right) = 0 \quad (i = 1, \ldots, 2k).$$

L'équation cherchée est donc explicitement

$$MT(f) = \begin{vmatrix} \frac{\partial f}{\partial t} & \frac{\partial f}{\partial u_1} & \cdots & \frac{\partial f}{\partial u_{2k}} \\ \frac{\partial \sigma_1}{\partial t} + \frac{\partial \phi}{\partial u_1} & 0 & \cdots & \mu_{1,2k} \\ \cdots & \cdots & \cdots & \cdots \\ \frac{\partial \sigma_{2k}}{\partial t} + \frac{\partial \phi}{\partial u_{2k}} & \mu_{2k,1} & \cdots & 0 \end{vmatrix} = 0,$$

où $M = \|\mu_{ij}\|$; elle dépend des $2k$ fonctions $\sigma_1, \ldots, \sigma_{2k}$ arbitraires *dans le domaine de rationalité.* La fonction ϕ, si elle n'est pas la dérivée en t d'une fonction du domaine, doit être conservée. Lorsque $\phi = \frac{\partial \psi}{\partial t}$ où ψ appartient au domaine, on peut supprimer ϕ de l'équation en modifiant les σ_i.

Tout se passe ici pour les σ comme dans le problème où il s'agit de déterminer σ_1, σ_2, μ_{12} en u_1, u_2 dans un certain domaine de rationalité, de façon que $\frac{\partial \sigma_1}{\partial u_2} - \frac{\partial \sigma_2}{\partial u_1} = \mu_{12}$; on n'a pas toutes les solutions en se donnant arbitrairement σ_1, σ_2 dans le domaine.

Les transformations infiniment petites du groupe de rationalité (bien connu dans la théorie des transformations de contact) sont obtenues en résolvant

$$\delta(x_1 dy_1 + \ldots + x_k dy_k) = d\Omega \, . \, \delta u,$$

ce qui donne
$$x'_i = x_i + \frac{\partial \phi}{\partial y_i} \delta u, \quad y'_i = y_i - \frac{\partial \phi}{\partial x_i} \delta u \quad (i = 1, \ldots, k),$$

où ϕ demeure arbitraire.

IV. *Systèmes complets d'équations linéaires.*

30. Un système complet, mis sous la forme de Jacobi, et composé de p relations à $(m+p)$ variables, résolubles en $\frac{\partial z}{\partial x_1}, \ldots, \frac{\partial z}{\partial x_p}$, définit des transcendantes $Z_1, \ldots, Z_m$ fonctions de $(m+p)$ variables et *attachées au groupe ponctuel général* Γ_m *ou à l'un des sous-groupes types* Γ, dans le domaine de rationalité $[\Delta]$ dont font partie les coefficients.

La théorie de la rationalité pourrait, en effet, se développer comme pour une seule équation, p des variables ($x_1, \ldots, x_p$ par exemple) jouant le rôle de la variable unique x et les autres $x_{p+1}, \ldots, x_{p+m}$ jouant le rôle des *variables principales.*

Il est bon d'observer que les transcendantes $Z_1, \ldots, Z_m$ seront définies par des systèmes auxiliaires Σ *identiques* à ceux qui se présentent pour $(m+1)$ variables : il y aura seulement à envisager pour chacune d'elles les p relations

$$X_1(Z)=0, \ldots, X_p(Z)=0,$$

qui sont les seules où interviennent $\frac{\partial Z}{\partial x_1}, \ldots, \frac{\partial Z}{\partial x_p}$.

Une réduction *théorique* du problème permettra de n'ajouter ces relations que successivement : d'abord toutes celles qui renferment des dérivées en x_1, c'est-à-dire $X_1(z_i)=0$, puis toutes celles qui renferment des dérivées en x_2, etc. C'est à ce même point de vue théorique qu'un système complet intégrable, de deux équations de Riccati à deux variables, se ramène, d'après M. Darboux, à deux équations successives à une seule variable.

Cette observation s'applique encore pour ramener, d'une manière générale, la détermination des fonctions de plusieurs variables *attachées à un groupe fini,* à une succession de déterminations de *fonctions d'une seule variable attachées au même groupe,* dans des domaines convenablement choisis.

On peut aussi envisager l'une des équations du système et fixer son groupe de rationalité, sachant qu'elle possède des solutions qui satisfont aux $(m+p-1)$ autres —c'est-à-dire appliquer à une seule équation la théorie générale.

La méthode de Jacobi peut d'ailleurs être présentée comme une simple application de cette théorie générale. Supposons, par exemple, que pour l'équation

$$X(f)=\frac{\partial f}{\partial x}+A_1\frac{\partial f}{\partial x_1}+\ldots+A_n\frac{\partial f}{\partial x_n}=0$$

on sache qu'une solution au moins, f, vérifie une équation linéaire qu'on écrit

$$B(f)=\frac{\partial f}{\partial x_1}+\ldots+b_n\frac{\partial f}{\partial x_n}=0,$$

on en déduira immédiatement, en appliquant $X(f)$, ce qui donne

$$X\left(\frac{\partial f}{\partial x_i}\right)+\Sigma_j\frac{\partial A_j}{\partial x_i}\frac{\partial f}{\partial x_j}=0 \qquad (i=1, \ldots, n),$$

$$X[B(f)]=B'(f)=b'_1\frac{\partial f}{\partial x_1}+\ldots+b'_n\frac{\partial f}{\partial x_n}=0,$$

où $$b'_i=X(b_i)-B(A_i) \qquad (i=1, \ldots, n).$$

($B'(f)$ est précisément le *crochet* de Jacobi $[X, B]$ obtenu d'un seul coup.)

Si $B'(f)$ est distinct de $B(f)$, on formera de même $X[B'(f)] = B''(f)$ et ainsi de suite. On ne s'arrête que lorsque $B^{(k)}(f)$ est combinaison linéaire de $B, B', \ldots, B^{(k-1)}$.

Le système ainsi obtenu s'écrit aussi, en regardant f comme fonction de $z_1, \ldots, z_n$:

$$\frac{\partial f}{\partial z_1} B(z_1) + \ldots + \frac{\partial f}{\partial z_n} B(z_n) = 0,$$

$$\ldots\ldots\ldots\ldots\ldots\ldots\ldots\ldots\ldots\ldots,$$

$$\frac{\partial f}{\partial z_1} B^{(k-1)}(z_1) + \ldots + \frac{\partial f}{\partial z_n} B^{(k-1)}(z_n) = 0,$$

et puisque l'opération $X(f)$ ne l'altère pas, on peut en le résolvant par rapport à $\frac{\partial f}{\partial z_1}, \ldots, \frac{\partial f}{\partial z_k}$ en tirer un système rationnel (Σ) comprenant $k(n-k)$ relations du premier ordre, dont on pourra disposer de manière qu'il admette une solution $z_1, \ldots, z_n$ satisfaisant pour $x = x_0$ à des conditions initiales données:

$$x_i = R_i(z_1, \ldots, z_n) \qquad (i = 1, \ldots, n).$$

Mais parmi tous ces systèmes de $k(n-k)$ relations figurent également ceux qui résultent des hypothèses $\frac{\partial f}{\partial z_1} = \ldots = \frac{\partial f}{\partial z_k} = 0$, f arbitraire en $z_{k+1}, \ldots, z_{n-1}$, c'est-à-dire les équations linéaires

$$B^{(i)}(z_l) = 0, \qquad (i = 0, 1, \ldots, (k-1)), (l = k+1, \ldots, n),$$

qui expriment que $z_{k+1}, \ldots, z_n$ sont des solutions communes aux $(k+1)$ équations

$$B^{(i)}(f) = 0, \quad X(f) = 0.$$

Ce sont donc toutes les solutions de ce système, et il est nécessaire et suffisant, pour que l'hypothèse du début soit justifiée, que $k < n$.

V. *Équations aux dérivées partielles du premier ordre, non linéaires.*

31. On sait qu'une *intégrale complète* de l'équation

$$Z(z, x_1, \ldots, x_p, p_1, \ldots, p_n) = a \ \ldots\ldots\ldots\ldots\ldots\ldots\ldots\ldots (1),$$

où $p_i = \frac{\partial z}{\partial x_i}$, peut, d'après Lie, être définie par les équations

$$X_i(z, x_1, \ldots, x_n, p_1, \ldots, p_n) = a_i \qquad (i = 1, \ldots, n),$$

où les Z et les X satisfont à une identité

$$dZ - P_1 dX_1 - \ldots - P_n dX_n = \rho(dz - p_1 dx_1 - \ldots - p_n dx_n) \ldots\ldots\ldots (2),$$

dans laquelle les P sont connus rationnellement du moment où les X le sont.

Les Z, X, P vérifient les conditions

$$[Z, X_i] = [X_i, X_k] = [P_i, P_k] = [P_i, X_k] = 0,$$

$$[P_i, X_i] = \rho, \quad [P_i, Z] = \rho P_i,$$

$$[Z, \rho] = \rho \frac{\partial Z}{\partial z} - \rho^2, \quad [X_i, \rho] = \rho \frac{\partial X_i}{\partial z}, \quad [P_i, \rho] = \rho \frac{\partial P_i}{\partial z},$$

qui sont suffisantes pour l'existence d'une identité (2) et où l'on a posé, suivant l'usage :

$$[F, \Phi] = \sum_{i=1}^{n} \left\{ \frac{\partial F}{\partial p_i} \left(\frac{\partial \Phi}{\partial x_i} + p_i \frac{\partial \Phi}{\partial z} \right) - \frac{\partial \Phi}{\partial p_i} \left(\frac{\partial F}{\partial x_i} + p_i \frac{\partial F}{\partial z} \right) \right\}.$$

L'équation $$[Z, f] = 0 \quad \ldots\ldots (1)$$

à $(2n+1)$ variables possède donc les solutions $X_1, \ldots, X_n$, Z et $\frac{P_2}{P_1}, \ldots, \frac{P_n}{P_1}$.

Pour fixer le groupe de rationalité qui leur correspond j'observe qu'un autre système de déterminations,

$$dZ - Q_1 dY_1 - \ldots - Q_n dY_n = \sigma (dz - p_1 dx_1 - \ldots - p_n dx_n),$$

donne $$dZ - P_1 dX_1 - \ldots - P_n dX_n = \frac{\rho}{\sigma} (dZ - Q_1 dY_1 - \ldots - Q_n dY_n).$$

Les X et les P sont donc définis, par l'identité (2), à une transformation de contact près *qui conserve* Z. Il est aisé d'en déduire la transformation la plus générale subie par $X_1, \ldots, X_n, \frac{P_2}{P_1}, \ldots, \frac{P_n}{P_1}$; on n'a qu'à former les équations finies d'une transformation de contact en partant des *équations directrices, parmi lesquelles on fera figurer* $Z = z$.

Les éléments $X_1, \ldots, X_n, \frac{P_2}{P_1}, \ldots, \frac{P_n}{P_1}$ sont donc en général *inséparables* : bien que les $(n-1)$ derniers s'expriment avec les premiers, les conditions auxquelles ils doivent satisfaire *n'isolent pas les* X *et on ne peut songer à les déterminer isolément.*

Cela n'aurait de sens que si le groupe de rationalité était imprimitif, les X étant transformés entre eux.

Cependant comme les P font partie du domaine de rationalité défini par les X on n'augmente pas la difficulté du problème en se proposant de les obtenir tous.

32. Examinons d'un peu plus près l'équation à deux variables

$$Z(x, y, z, p, q) = a,$$

et l'équation linéaire correspondante

$$[Z, F] = 0 \quad \ldots\ldots (\mathrm{I}).$$

Soit $$dZ - P dX - Q dY = \rho (dz - p dx - q dy);$$

si nous posons $P = TQ$, il existera pour (I) un système de 3 solutions X, Y, T qui satisfont aux deux relations

$$[X, Y] = 0, \quad [T, Y + TX] = 0.$$

Si ξ, η, θ désigne un système particulier de valeurs de X, Y, T on a évidemment

$$X = X(\xi, \eta, \theta, Z), \quad Y = Y(\xi, \eta, \theta, Z), \quad T = T(\xi, \eta, \theta, Z),$$

et ces trois fonctions de quatre variables devront satisfaire aux relations

$$\frac{\partial (X, Y)}{\partial (\xi, \theta)} - \theta \frac{\partial (X, Y)}{\partial (\eta, \theta)} = 0,$$

$$\frac{\partial (T, Y)}{\partial (\xi, \theta)} - \theta \frac{\partial (T, Y)}{\partial (\eta, \theta)} + T \left\{ \frac{\partial (T, X)}{\partial (\xi, \theta)} - \theta \frac{\partial (T, X)}{\partial (\eta, \theta)} \right\} = 0,$$

qui sont les *équations de définition* du *groupe de rationalité*. Il est facile d'obtenir ses transformations infinitésimales; soit

$$X = \xi + \epsilon x, \quad Y = \eta + \epsilon y, \quad T = \theta + \epsilon t,$$

l'une de ces transformations où x, y, t sont des fonctions à déterminer de ξ, η, θ et Z; on aura explicitement

$$x = \frac{\partial \phi}{\partial \theta}, \quad y = \phi(\xi, \eta, \theta, Z) - \theta \frac{\partial \phi}{\partial \theta},$$

$$t = \theta \frac{\partial \phi}{\partial \eta} - \frac{\partial \phi}{\partial \xi},$$

où ϕ demeure arbitraire en ξ, η, θ, Z.

Parmi les réductions qui peuvent se présenter pour le groupe Γ, lorsque Z est choisi de manière particulière, je signalerai celle qui correspond à l'existence d'une solution rationnelle ρ pour l'équation

$$[Z, \rho] = \rho \frac{\partial Z}{\partial z} - \rho^2.$$

Les éléments P, Q, X, Y sont donnés par

$$P dX + Q dY = dZ - \rho(dz - p dx - q dy),$$

donc définis à une transformation homogène de contact près à 2 variables et $X, Y, \frac{Q}{P}$ sont donnés à une *transformation de contact du plan près*.

Si l'équation à étudier est

$$F(x, y, z, p, q) = 0,$$

nous pouvons l'écrire *en adjoignant la fonction algébrique* q au domaine de rationalité:

$$q = f(x, y, z, p).$$

La méthode de Lagrange conduit à chercher une fonction $\phi(x, y, p)$ telle que

$$dz = p dx + f dy,$$

où p est donné par $\phi(x, y, z, p) = a$, soit intégrable par une seule relation en x, y, z, dépendant d'une nouvelle constante. La fonction p est solution de l'équation

$$[q - f, \phi] = 0,$$

qui s'écrit développée

$$A(\phi) = \frac{\partial \phi}{\partial y} - \frac{\partial f}{\partial p}\frac{\partial \phi}{\partial x} + \left(f - p\frac{\partial f}{\partial p}\right)\frac{\partial \phi}{\partial z} + \left(\frac{\partial f}{\partial x} + p\frac{\partial f}{\partial z}\right)\frac{\partial \phi}{\partial p}.$$

Elle admet en général trois solutions *inséparables*. Désignons par $\psi(x, y, z, p)$ une seconde solution; pour que les deux équations

$$\phi(x, y, z, p) = a, \quad \psi(x, y, z, p) = b,$$

soient compatibles, il faut

$$\frac{\dfrac{\partial \phi}{\partial x} + p\dfrac{\partial \phi}{\partial z}}{\dfrac{\partial \psi}{\partial x} + p\dfrac{\partial \psi}{\partial z}} = \frac{\dfrac{\partial \phi}{\partial p}}{\dfrac{\partial \psi}{\partial p}},$$

et l'on reconnait immediatement qu'alors

$$dz - p\,dx - f(x, y, z, p)\,dy = \lambda(d\phi - \omega\, d\psi),$$

où

$$\frac{\partial \phi}{\partial p} - \omega \frac{\partial \psi}{\partial p} = 0$$

définit la troisième solution. Ces trois éléments ϕ, ω, ψ sont connus à une transformation de contact du plan près: si f est arbitraire *le groupe de rationalité est donc le groupe de transformations de contact du plan.*

Il serait facile d'indiquer tous les types possibles de réduction. Je ferai simplement observer que si z disparaît de l'équation $q = f(x, y, p)$, auquel cas il suffit de chercher ϕ de façon que $\phi(x, y, p) = a$ rende intégrable $p\,dx + f(x, y, p)\,dy$, l'équation qui définit ϕ est

$$A(\phi) = -\frac{\partial f}{\partial x}\frac{\partial \phi}{\partial p} + \frac{\partial f}{\partial p}\frac{\partial \phi}{\partial x} - \frac{\partial \phi}{\partial y} = 0.$$

Elle admet le multiplicateur 1, et le groupe de rationalité du cas général est formé des transformations

$$\frac{\partial(\Phi, \Psi)}{\partial(\phi, \psi)} = 1.$$

33. Des remarques analogues aux précédentes peuvent se faire pour *l'intégration d'un système en involution*

$$Z = a, \quad X_1 = a_1, \ \ldots, \ X_q = a_q.$$

Tout se passe comme si, dans la résolution de l'identité

$$dZ - P_1 dX_1 - \ldots - P_n dX_n = \rho(dz - p_1 dx_1 - \ldots - p_n dx_n),$$

ou dans l'intégration de $[Z, f] = 0$, on connaissait rationnellement $X_1, \ldots, X_q$.

On sait que la transformation infinitésimale *de contact* la plus générale en $z, x_1, \ldots, x_n, p_1, \ldots, p_n$,

$$W(f) = \zeta \frac{\partial f}{\partial z} + \xi_1 \frac{\partial f}{\partial x_1} + \ldots + \varpi_1 \frac{\partial f}{\partial p_1} + \ldots,$$

est définie par les formules

$$\xi_i = \frac{\partial W}{\partial p_i}, \quad \varpi_i = -\left(\frac{\partial W}{\partial x_i} + p_i \frac{\partial W}{\partial z}\right), \quad \zeta = \Sigma p_i \frac{\partial W}{\partial p_i} - W,$$

où W *fonction caractéristique* est quelconque en $z, x_1, \ldots, x_n, p_1, \ldots, p_n$.

Les transformations qui n'altèrent pas les éléments $z, x_1, \ldots, x_q$ satisfont donc aux conditions

$$\frac{\partial W}{\partial p_1} = \ldots = \frac{\partial W}{\partial p_q} = 0, \quad \Sigma p_{q+i} \frac{\partial W}{\partial p_{q+i}} - W = 0,$$

qui expriment que $\quad W = p_n F\left(z, x_1, \ldots, x_n, \frac{p_{q+1}}{p_n}, \ldots, \frac{p_{n-1}}{p_n}\right).$

On déduira de là les transformations infinitésimales subies par

$$x_{q+1}, \ldots, x_n, \frac{p_{q+1}}{p_n}, \ldots, \frac{p_{n-1}}{p_n},$$

qui sont les $(2n-2q-1)$ solutions du système en involution

$$(z, \phi) = (x_i, \phi) = 0 \qquad (i = 1, \ldots, q),$$

autres que les solutions rationnelles $z, x_1, \ldots, x_q$.

En posant $$\omega_{q+i} = \frac{p_{q+1}}{p_n} \qquad (i = 1, \ldots, (n-1)),$$

on aura

$$\xi_{q+i} = \frac{\partial F}{\partial \omega_{q+i}}, \quad \varpi_{q+i} = -\frac{\partial F}{\partial x_{q+i}} + \omega_{q+i}\frac{\partial F}{\partial x_n}, \quad \xi_n = F - \Sigma \omega_{q+i}\frac{\partial F}{\partial \omega_{q+i}} \qquad (i = 1, \ldots, (n-1)),$$

où F est arbitraire en

$$z, x_1, \ldots, x_q, \quad x_{q+1}, \ldots, x_n, \quad \omega_{q+1}, \ldots, \omega_n.$$

Comme F renferme à la fois les x et les ω, ces éléments sont donc en général *inséparables*: pour que les x se déterminent séparément il faut et il suffit que les $\frac{\partial F}{\partial \omega}$ ne dépendent plus des ω, c'est-à-dire que F soit linéaire en $\omega_{q+1}, \ldots, \omega_{n-1}$. Les x subiront alors une transformation ponctuelle et les ω les transformations projectives qui en résultent lorsqu'on *étend* le groupe ponctuel.

Il est bien évident que la *théorie de la rationalité*, à la fois *logique et nécessaire*, permet de discuter les diverses méthodes proposées par Jacobi, Cauchy, Mayer, Lie, etc., pour intégrer les équations (où systèmes d'équations) non linéaires à une inconnue. Les méthodes où l'on se propose la détermination *successive* ou *partielle* des éléments d'une intégrale complète ne se justifient que dans des cas particuliers; il n'y a guère que l'emploi des *caractéristiques de Cauchy* (ou de leurs équivalents) qui peut être légitimé.

VI. *Problème de Pfaff.*

34. Il s'agit de ramener une forme différentielle donnée

$$\Delta = a_1 du_1 + \ldots + a_n du_n,$$

à l'un ou à l'autre des types

$$dy - x_1 dy_1 - \ldots - x_p dy_p,$$
$$x_1 dy_1 + \ldots + x_p dy_p,$$

où les y et les x sont des fonctions *indépendantes* des n variables $u_1, \ldots, u_n$.

1°. Si l'on a $$\Delta = x_1 dy_1 + \ldots + x_p dy_p,$$

les équations $$a_{i1}\xi_1 + \ldots + a_{in}\xi_n = a_i \xi_0 \qquad (i = 1, \ldots, n) \ldots\ldots\ldots\ldots\ldots\ldots (1),$$

où $a_{ik} = \frac{\partial a_i}{\partial u_k} - \frac{\partial a_k}{\partial u_i}$, possèdent $(n-2p+1)$ systèmes de solutions linéairement distinctes les équations

$$X_s(f) = \xi_{1s}\frac{\partial f}{\partial u_1} + \ldots + \xi_{ns}\frac{\partial f}{\partial u_n} = 0 \qquad (s = 1, \ldots, (n-2p+1)) \ \ldots (A)$$

formées avec ces solutions constituent un *système complet* dont les solutions sont

$$y_1, y_2, \ldots, y_p, \frac{x_2}{x_1}, \ldots, \frac{x_p}{x_1}.$$

Le groupe de rationalité de ce système complet est *dans le cas général* (c'est-à-dire lorsque que les a_i sont pris, *dans un domaine de rationalité fixé,* de la manière la plus générale de façon à satisfaire aux conditions précédentes) le groupe général des transformations de contact à $(2p-1)$ variables résultant de l'identité

$$x_1\left(dy_1+\frac{x_2}{x_1}dy_2+\ldots+\frac{x_p}{x_1}dy_p\right)=X_1\left(dY_1+\frac{X_2}{X_1}dY_1+\ldots+\frac{X_p}{X_1}dX_p\right).$$

Les éléments y_i, $\frac{x_i}{x_1}$ sont donc en général *inséparables*; les transformations du groupe de rationalité les échangent les uns dans les autres. Toute tentative de détermination séparée des y et des x ne peut avoir de sens que s'il y a une réduction du groupe de rationalité et s'il devient un *groupe de transformations ponctuelles en* $y_1, \ldots, y_n$ *étendu.*

Quand le système complet (A) est intégré, la fonction x_1 est connue explicitement.

2°. Si l'on a $$\Delta=dy-x_1dy_1-\ldots-x_pdy_p,$$
les équations (1) ne sont satisfaites que si $\xi_0=0$ et possèdent $(n-2p)$ solutions linéairement distinctes; le système complet

$$X_s(f)=\xi_{1s}\frac{\partial f}{\partial u_1}+\ldots+\xi_{ns}\frac{\partial f}{\partial u_n}=0 \quad \ldots\ldots\ldots\ldots\ldots\ldots\ldots(\text{A}),$$

formé avec ces solutions, admet lui-même les solutions

$$y_1, \ldots, y_p, \quad x_1, \ldots, x_p,$$

son groupe de rationalité est *dans le cas général* défini par l'identité

$$X_1dY_1+\ldots+X_pdy_p=x_1dy_1+\ldots+x_pdy_p+d\Omega,$$

où Ω demeure arbitraire en $x_1, \ldots, x_p, y_1, \ldots, y_p$. C'est donc le groupe *simple* pour lequel *l'élément d'intégrale double* $dx_1dy_1+\ldots+dx_pdy_p$ *demeure invariant.*

Ici encore toute tentative de détermination séparée ou successive des éléments $x_1, \ldots, x_p, y_1, \ldots, y_p$ n'a de sens que si le groupe de rationalité se réduit. Dans le cas où ce groupe est réduit à un groupe ponctuel en $y_1, \ldots, y_n$ étendu, on pourra trouver un système définissant seulement les y; les x seront alors connus sans nouvelle intégration. On observera ici que lorsque le système (A) est intégré, y est donné par une quadrature de différentielle totale.

Les remarques précédentes nous permettraient avec le tableau des types de groupes de contact à $2p+1$ ou $2p$ variables d'indiquer sans difficulté tous les cas de réduction qui peuvent se présenter dans la détermination de la forme canonique de Δ.

Si l'expression Δ est *inconditionnelle*, c'est-à-dire si l'on a $n=2p$ ou $n=2p+1$, le système (A) se réduit à une seule équation, qui se trouve pour des raisons évidentes être celle obtenue dans la formation des équations les plus générales à groupes simples.

35. Pour terminer cet exposé, déjà trop long, je me bornerai à signaler que la *théorie de la rationalité* s'applique encore dans l'étude des *groupes de fonctions* introduits par Lie (détermination des fonctions distinguées d'un groupe, réduction du groupe à la forme canonique, etc....), dans celle des équations aux dérivées partielles du second ordre qui peuvent s'intégrer par les méthodes de Monge,

d'Ampère ou de M. Darboux, enfin dans l'étude générale des systèmes d'équations linéaires aux différentielles totales: la transformation du système en un autre où le nombre des différentielles est réduit à sa plus petite valeur peut se faire par l'intégration d'un certain système complet (E. V. Weber, Cartan) dont il y a lieu de préciser le groupe de rationalité.

J'ai indiqué en détail pour le premier ordre, comment, en remplaçant le mot *rationnel* par *rationnel dans un certain voisinage,* on pouvait déduire de la même théorie une classification précise des domaines singuliers d'une équation

$$X(z) = \frac{\partial z}{\partial x} + A_1 \frac{\partial z}{\partial x_1} + \dots + A_n \frac{\partial z}{\partial x_n} = 0,$$

(ou d'un système complet) basée sur la nature des invariants différentiels *rationnels dans ce voisinage.* Le principe de cette classification subsiste dans le cas général, puisque la réduction d'un système irréductible régulier à la forme normale et l'unicité des solutions du système résolvant relatif au groupe Γ (qui remplace le groupe de rationalité au voisinage du domaine étudié) se conservent. Il y a là un domaine immense ouvert à la curiosité des chercheurs—mais où l'on rencontre des difficultés sérieuses, alors que les applications précédentes sont de nature facile.

Enfin le problème qui consiste, dans un domaine de rationalité bien déterminé $[\Delta]$, à reconnaître, *par un nombre limité à l'avance de calculs élémentaires,* quel est le groupe de rationalité d'une équation donnée

$$X(z) = \frac{\partial z}{\partial x} + A_1 \frac{\partial z}{\partial x_1} + \dots + A_n \frac{\partial z}{\partial x_n} = 0,$$

s'impose également à l'attention. La principale difficulté est toujours la détermination de tous les polynomes P irréductibles dans $[\Delta]$, qui satisfont à une identité

$$X(P) = MP.$$

Ce n'est que dans des cas exceptionnels qu'on peut le résoudre.

INDEX TO VOLUME I

PART I

REPORT OF THE CONGRESS

PART II

LECTURES

COMMUNICATIONS

SECTION I (ARITHMETIC, ALGEBRA, ANALYSIS)

CAMBRIDGE: PRINTED BY JOHN CLAY, M.A. AT THE UNIVERSITY PRESS